- Elmworth -
Case Study of a
Deep Basin Gas Field

AAPG Memoir 38

Elmworth, Case Study of a Deep Basin Gas Field

Edited by
John A. Masters

Published by
The American Association of Petroleum Geologists
Tulsa, Oklahoma 74101, U.S.A.

Copyright © 1984
The American Association of Petroleum Geologists
All Rights Reserved
Printed in the United States of America

Elmworth, case study of a deep basin gas field.

(AAPG memoir ; 38)
Includes bibliographies and index.
1. Elmworth Gas Field (Alta.)--Addresses, essays, lectures. 2. Gas fields--Canada, Western--Addresses, essays, lectures. I. Masters, John A. II. American Association of Petroleum Geologists. III. Title: Deep basin gas field. IV. Series.
TN880.E46 1985   553.2'85'097123   84-45745
ISBN 0-89181-315-2

AAPG grants permission for a single photocopy of any article herein for research or non-commercial educational purposes. Other photocopying not covered by the copyright law as Fair Use is prohibited. For permission to photocopy more than one copy of any article, or parts thereof, contact: Permissions Editor, AAPG, P.O. Box 979, Tulsa, Oklahoma 74101.

Association Editor: Richard Steinmetz
Science Director: Edward A. Beaumont
Project Editors: Douglas A. White and Ronald L. Hart
Typography: Tricia Kinion and Eula Matheny
Design and Production: S. Wally Powell

Contents

John A. Masters	Introduction vii
John A. Masters	Lower Cretaceous Oil and Gas in Western Canada 1
D. H. Welte, R. G. Schaefer, W. Stoessinger, M. Radke	Gas Generation and Migration in the Deep Basin of Western Canada 35
Paul C. Jackson	Paleogeography of the Lower Cretaceous Mannville Group of Western Canada 49
David G. Smith, Carl E. Zorn, Robert M. Sneider	The Paleogeography of the Lower Cretaceous of Western Alberta and Northeastern British Columbia In and Adjacent to the Deep Basin of the Elmworth Area 79
Robert M. Gies	Case History for a Major Alberta Deep Basin Gas Trap: The Cadomin Formation 115
R. A. Rahmani	Facies Control of Gas Trapping, Lower Cretaceous Falher A. Cycle, Elmworth Area, Northwestern Alberta 141
Richard Smith	Gas Reserves and Production Performance of the Elmworth/Wapiti Area of the Deep Basin 153
R. E. Wyman	Gas Resources in Elmworth Coal Seams 173
T. B. Davis	Subsurface Pressure Profiles in Gas Saturated Basins 189
Robert M. Sneider, Howard R. King, R. W. Hietala, E. T. Connolly	Integrated Rock-Log Calibration in the Elmworth Field, Alberta — Canada 205
Don L. Myers	Drilling in the Deep Basin 283
John A. Stayura	Completion Practices in the Alberta Deep Basin 291
Kam K. Chiang	The Giant Hoadley Gas Field, South-Central Alberta 297
	Index 315

- Elmworth -
Case Study of a
Deep Basin Gas Field

The discovery of a major oil or gas field onshore in North America is a significant event to geologists. But, the discovery of a supergiant field in a "subtle trap" is so unique that it fulfills the fondest hopes of all explorationists and confirms their faith that the earth holds hidden treasures that cannot be known by statisticians or economists. It is also of major economic importance because of the quantity of hydrocarbons involved and the fact that it opens many minds to other similar accumulations.

Elmworth contains 17 tcf of proved plus probable gas and 1 billion barrels of natural gas liquids with a huge adjoining area which has been only lightly drilled. The field is already known to be the largest gas accumulation in Canada and ranks as one of the largest in North America. Its size and its puzzling trap characteristics make it a worthy subject for careful geologic study.

Ten years ago it would have seemed scientifically irresponsible to suggest that Canada had yet to see its largest gas field and that it would be found in a stratigraphic trap where gas lies downdip from water with no impermeable barrier between. Rational analysis of the future is always constrained by limited knowledge.

Following the discovery of Elmworth in 1976, and extensive development over an area of 5,000 sq mi (12,950 sq km), the upside down trapping conditions of the "Deep basin" have been absorbed into the exploratory thinking of the industry. Similar conditions have been recognized in the tight gas sands of the Rocky Mountain basins, Arkoma, Appalachian, and various foreign basins.

In 1982, Dietrich Welte observed to me that Elmworth represented the largest field and the newest collection of geologic and reservoir engineering data on a tight sand, deep-basin type gas accumulation. Because of the sophisticated data requirements and collection facilities of the Alberta Energy Resources and Conservation Board, the very recent drilling history of the field involving the most modern logging, coring, and testing, as well as Canadian Hunter's high technology development practices, Welte felt that petroleum science would be advanced by an extensive presentation of the field information. Ted Beaumont of the AAPG gave enthusiastic support to the idea. Thus was born the rather unusual concept of a memoir to be formed of a number of papers largely from a single company. Justice demanded that Dr. Welte contribute his knowledge of geochemistry to the volume which he had suggested. I also asked Kam Chiang of Sundance Oil to write on the geology of the Hoadley gas field which is an Elmworth "look-alike" in the same Deep basin. All the other papers were written by Canadian Hunter specialists in various phases of Elmworth geology and engineering.

Every problem presents itself first in a blur of confusion. It is only after some amount of analysis that it breaks down into relatively simple components. Figure 1 gives a picture of the complex of information which we had to deal with initially and decide how to separate into manageable parts. The Elmworth field is a huge area of gas saturation 65 mi (105 km) long and 30 mi (48 km) wide within which there are now 250 recognized pools of

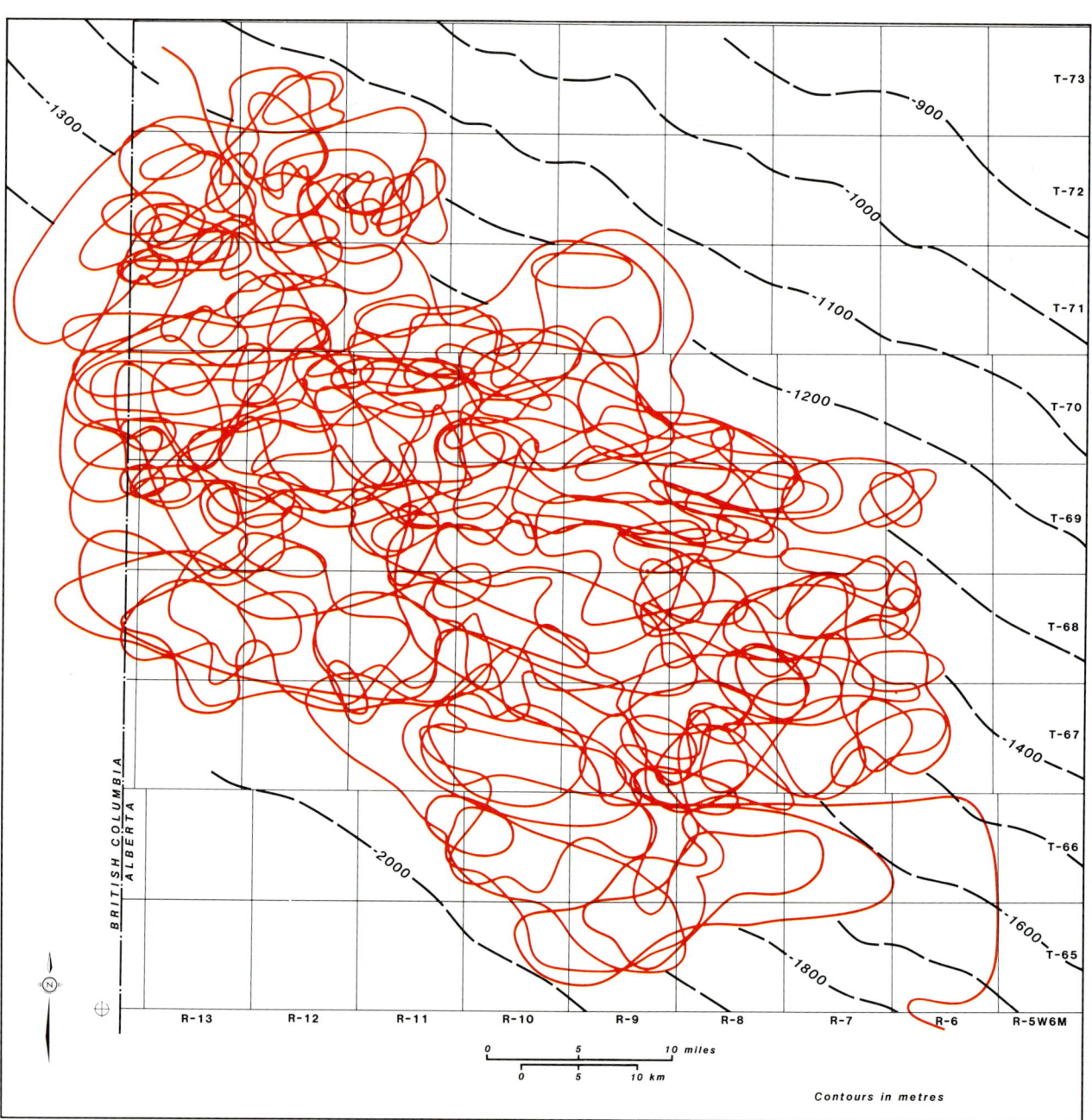

Figure 1. Gas pool outlines, Elmworth area.

commercially recoverable gas in 23 distinct stratigraphic zones. Southwest of a generalized water/gas contact the entire Mesozoic section from 3,000 to 10,000 ft (914 to 3,050 m) contains gas in every stringer of porosity; in fact the entire rock section is saturated with gas. The gas area has a total gas column of 3,000 ft (914 m). Updip, where the sands become more porous and permeable, the rocks are water saturated.

Memoir 38 deals with the generation and distribution of oil and gas in the Lower Cretaceous rocks; geochemistry of the rock section at Elmworth; stratigraphy of the Lower Cretaceous sands; detailed studies of the Cadomin and Falher reservoirs; reserves of gas at Elmworth; measurement of the gas contained in coals; pressure studies; electric log and rock calibrations; drilling and completion practices; and the geology of a similar field at Hoadley.

Finding a giant Deep basin-type gas field is technologically relatively simple, although statistically very rare. We prefer Mesozoic - Tertiary sands, of dominantly low permeability but with well-sorted marine shoreline sands present for moderate reservoir quality, adjacent coal or organic shales, adequate thermal maturity, located structurally deep in the basin, with sub-normal pressures, and anomalous electric log resistivity or, better, small recoveries of gas with no water. The techniques for recognizing the above characteristics are standard operating procedures for most geologists.

Exploiting such a field, however, calls upon some of the most advanced reservoir technology available and requires an unusual amount of coordination between the geological and engineering arms of a company. It is virtually impossible for one man to have all the skills required to analyze, measure, and produce these low-permeability, high-damage reservoirs, so a chain-link team of specialists must be available. Few companies have built, or can hold together, such teams. Perhaps the most significant contribution of this memoir, in fact, is its description of the several areas of expertise that must necessarily bond together in the exploitation of a major Deep basin-type gas field.

John A. Masters
Canadian Hunter Exploration, Ltd.

Lower Cretaceous Oil and Gas in Western Canada

John A. Masters
Canadian Hunter Exploration Ltd.
Calgary, Alberta

"Truth is so large a target that nobody can wholly miss hitting it, but at the same time, nobody can hit all of it...."
Aristotle

The Western Canada sedimentary basin contains tar deposits which exceed by three times the known recoverable oil reserves of the entire world. The tar was originally liquid oil which has been degraded by aerobic bacteria. It was generated principally from Lower Cretaceous but also from Jurassic and Triassic shales. The oil is contained in only 900 ft (275 m) of stratigraphic section above and below the Paleozoic unconformity.

Oil migration paths were northeasterly, directly updip from the oil thermal window. The Athabasca anticline, a drape structure caused by Devonian salt removal, connects southward with the Sweetgrass Arch to form a 600-mi (965-km) long structural barrier on the eastern, updip rim of the basin. Most of the tar deposits are along the anticline or in a giant stratigraphic trap on the Paleozoic unconformity surface on the west flank of the anticline. There is no oil or gas east of the anticline.

In the deepest part of the basin the Mesozoic section generated gas in comparably large volumes. Most of the gas has escaped to the outcrop, a small amount is contained in thousands of conventional stratigraphic pools on the east side, and an enormous volume is contained in tight sands on the west side, or the Deep basin. Most of the reservoirs are Lower Cretaceous sandstones. The tight, gas-saturated sands grade updip into porous water-saturated sands. The trap is not tightly sealed but leaks off at a steady rate. Continuing gas generation keeps the trap pumped full. This bottleneck trap contains 1,750 tcf of gas in place.

Commercial gas accumulations are present in the Deep basin where coarser-grained marine shoreline sands occur. These are most numerous in the Elmworth area but are also present in five specific trends to the south.

The gigantic oil and gas accumulations of the Lower Cretaceous make the Western Canada sedimentary basin the richest hydrocarbon province in the world.

INTRODUCTION

Two important sets of information have recently become available on the petroleum geology of Western Canada. First, petroleum geochemistry has demonstrated that the immense tar deposits on the east flank of the Alberta basin were derived from Lower Cretaceous shales, migrated and accumulated as liquid oil, and were later degraded to tar by water washing and bacterial action. Second, the drilling of some 2,000 new wells in the last five years has confirmed the concept of a vast gas charged area — the Deep basin, on the west flank of the Alberta basin; it too was sourced by Lower Cretaceous shales, and coals.

Thus, in the last several years the Lower Cretaceous has belatedly been recognized as the pre-eminent oil and gas source and reservoir system in Canada, far greater in hydrocarbon volume than the Devonian. In fact, it appears to be the richest in hydrocarbons of any group of rocks in the world.

Studies of the "habitat of oil" have been based fundamentally on the Middle East, Venezuela, Western Siberia, and sometimes the Los Angeles basin because of their presumed dominance in quantities of oil and gas. But our new data tell us that the largest single liquid oil accumulation in the world was the Athabasca anticline and the quantity of oil in eastern Alberta is three times the known ultimate recoverable oil in the world. In addition, the west side of the basin generated comparably vast amounts of gas and contains very large accumulations.

This report attempts for the first time to link conceptually the vast oil accumulations of the east side of the basin to the huge gas fields of the west side. My hope is to stimulate ideas about what hydrocarbons may lie undiscovered in the middle. The report also provides a geologic introduction to subsequent papers on the detailed geology and engineering of Elmworth which promises to become the largest gas field in Canada.

GENERAL GEOLOGY

A cross section (Fig. 1) illustrates the fundamental geology of the Western Canada sedimentary basin. A broad wedge of Paleozoic and Mesozoic sedimentary rocks trends northwesterly, thins eastward to zero on the Canadian shield and thickens westward into the Rocky Mountains. Maximum sedimentary thickness is 19,000 ft (5,800 m). The western margin of the basin is the frontal thrust of the Rocky Mountain overthrust belt. Within the thick sediment wedge a major regional unconformity divides the Paleozoic carbonates from the Mesozoic clastics. Subsidence of the basin ended with the conclusion of the Laramide orogeny which formed the mountains on the west side.

Various permeable strata acted as carrier beds for transporting oil and gas generated in the deep part of the basin updip to the east. Permeability along the Paleozoic unconformity surface was also an important fluid path.

A Precambrian structure map (Fig. 2) shows the general basin shape and principal structural features. The Athabasca anticline which is expressed in Cretaceous beds is superimposed on the map. Oil and gas migrated updip from the deep part of the basin onto a gently westward dipping slope. In the Lower Cretaceous, stratigraphic variations created thousands of small traps on the slope but the bulk of the hydrocarbons moved eastward toward the very long, low-relief Athabasca anticline along the eastern margin of the basin which nearly linked up in a giant crescent with the ancestral Sweetgrass Arch to the south. This 600-mi (965-km) long structural rim was a controlling factor in the accumulation of enormous quantities of Lower Cretaceous oil. Few basins have been provided with such a gigantic collection area on their updip margins that has escaped subsequent erosion.

Figure 3 is a map of the principal oil and gas producing sedimentary basins of North America with the axis of the Athabasca-Sweetgrass anticline shown along the eastern margin of the Western Canada basin. The map demonstrates visually the great length and optimal position of the anticline and the enormous comparative size of the basin.

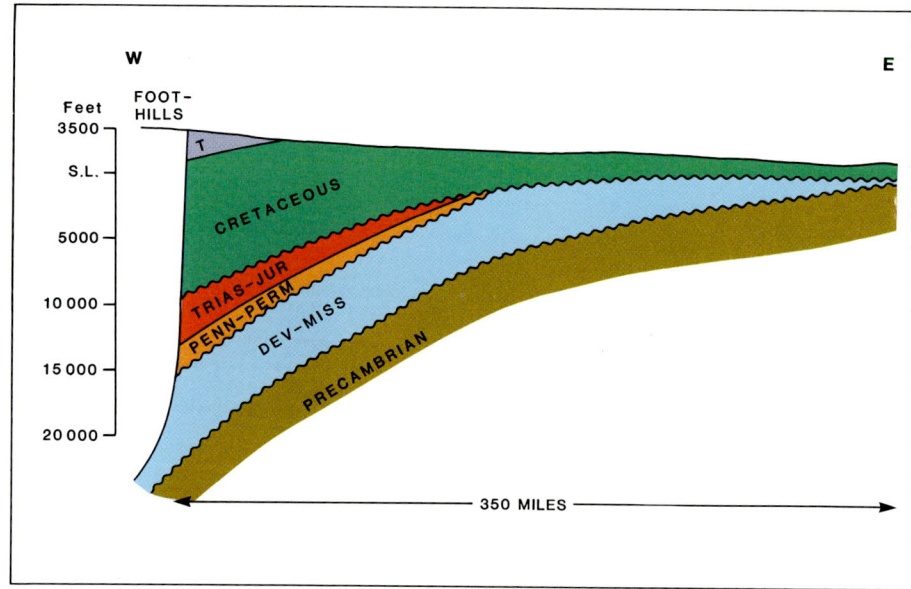

Figure 1. Cross section of Western Canada sedimentary basin across Central Alberta.

STRATIGRAPHIC DISTRIBUTION OF OIL AND GAS

Perhaps the most important new information on the oil and gas habitat of Western Canada comes from petroleum geochemistry. Deroo et al (1977) wrote a pioneering study which demonstrated that the great tar sand deposits of eastern Alberta were once conventional liquid oil derived from the Lower Cretaceous shales and accumulated in porous reservoir sands. The subsequent alteration of the oil to tar was accomplished by "biodegradation and water washing with some possible inorganic oxidation (as the) normal Lower Cretaceous oil (migrated) updip where it encountered fresh water invading the basin from the outcrop area along the shield edge." From a geochemical study of 358 samples, the oils in Alberta were divided into 3 groups. Aromatic fractions provide a recognizable chemical distinction between Upper Cretaceous (Group 1), Lower Cretaceous (Group 2), and Devonian oils (Group 3). The Lower Cretaceous Mannville "was the source of the oil found in the Lower and Upper Mannville reservoirs throughout the region and the Jurassic, Mississippian and Upper Devonian deposits adjacent to the unconformity" (Deroo, et al, 1977). Thus, contrary to previous assertions, the vast quantities of oil in the tar sands did not come from the Paleozoics, although there may have been some contribution. Neither did they originate as tar by some unique process that may be ignored when discussing the critical conditions for creating giant conventional oil accumulations.

Figure 4 shows the stratigraphic distribution of recoverable oil and gas reserves in Alberta (Alberta Conservation Board reports), and the source rocks (Deroo et al, 1977).

Considering only proved recoverable reserves of conventional oil and gas, the Group 2 hydrocarbons derived from Lower Cretaceous shales (and probably Jurassic and Triassic shales) comprise 44% of the total hydrocarbons discovered in Alberta. These figures are representative of Western Canada because the bulk of all the reserves are in Alberta.

The entire Elmworth field area is approximately 5,000 sq mi (12,950 sq km). The gas contract area within which we have done detailed reservoir mapping is 1,500 sq mi (3,885 sq km). Data from 650 wells within this area indicate to the operators established (contractible) reserves of 7.5 trillion cu ft (Canadian Hunter reserve reports). Additional probable reserves within mapped reservoirs are 9.6 tcf, bringing the total proved plus probable to 17.1 tcf. The recoverable natural gas liquids are estimated at 1 billion barrels. A large area to the west and south remains essentially unevaluated. With the addition of Elmworth, the Lower Cretaceous probably contains nearly 50% of the total oil and gas reserves of Western Canada.

Recognizing that the tar deposits were once conventional oil fields makes it necessary to incorporate them too in any comparison of the relative importance of different generating sources. When nearly 3,000 billion barrels of oil (Outtrim and Evans, 1977) are included an entirely different graph scale is required to compare the different groups (Fig 5.) The Lower Cretaceous-Jurassic-Triassic Group 2 hydrocarbons (oil and gas), contained in Lower Cretaceous sands and Paleozoic carbonates at the unconformity, represent over 99% of the total hydrocarbons of Alberta and, in fact, exceed by three times the total produced and remaining oil reserves of the whole world (Nehring, 1978).

Recoverable oil in world reserves averages about one third of oil in place so it is not fair to compare oil in place numbers in

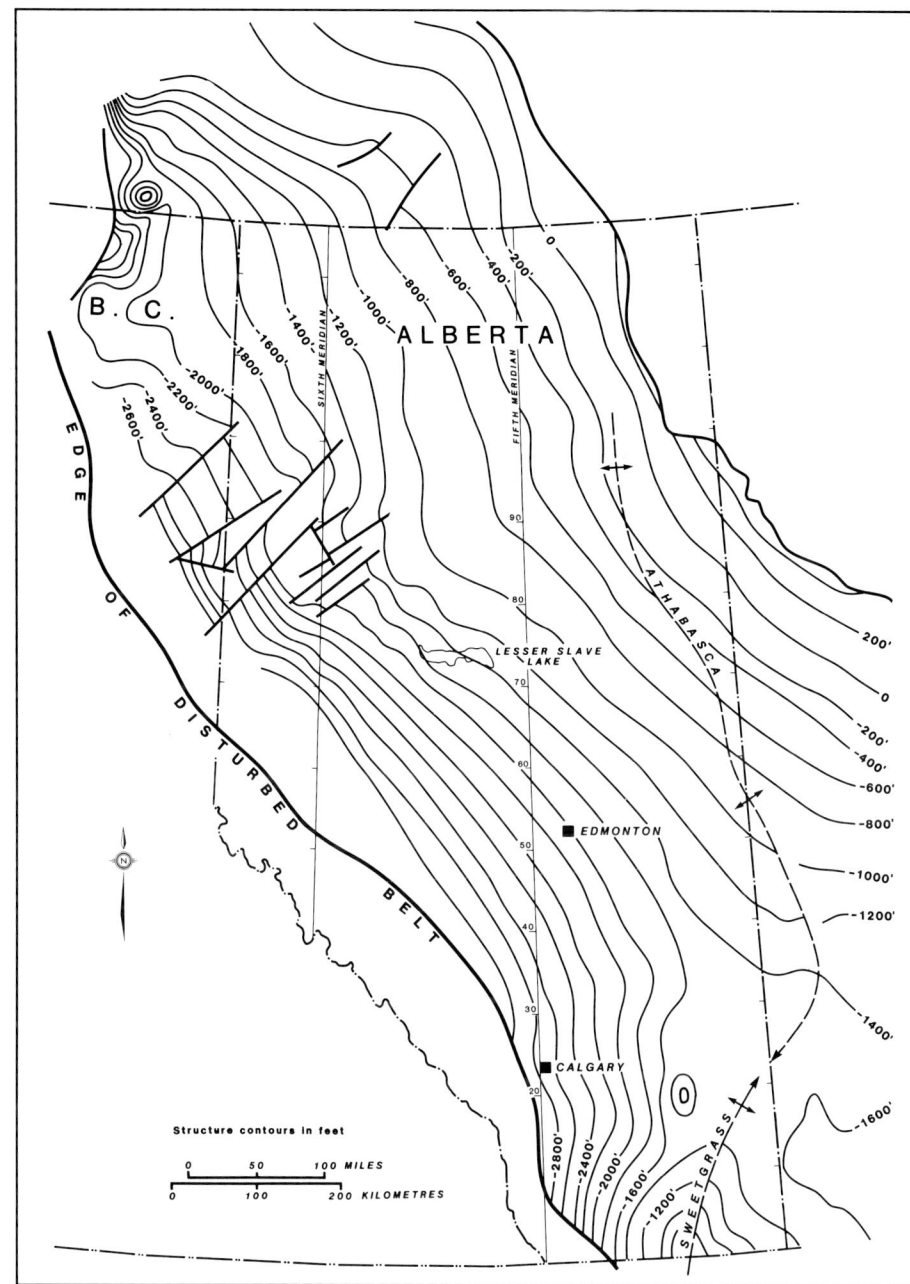

Figure 2. Precambrian structure map (data from Geological Society of Canada map 1251A).

Western Canada; as well it calls for a re-examination of geochemical data from other basins to determine if they too generated hydrocarbons in vastly larger quantities than heretofore suspected. This could have an important influence on our understanding of petroleum migration and trap efficiencies.

PALEOZOIC UNCONFORMITY SURFACE

Figure 6 is a map of the Paleozoic unconformity surface under the Mesozoic beds.

During oil migration the unconformity surface itself must have been an important fluid path because the basal Cretaceous sands are not sufficiently extensive to have transported oil over the vast areas involved. In support of this, the unconformity surface is known to be oil saturated over an enormous area, as shown on the map.

The so-called "Carbonate Triangle" in the Athabasca region contains a very approximate measure of 1,300 billion barrels of oil in place (Outtrim and Evans, 1977). The oil is contained in porosity in the first several hundred feet of rocks below the unconformity over an area of 700 townships (25,000 sq mi; 64,750 sq km) in beds of Permian, Mississippian and Devonian age. This is easily the largest stratigraphic trap or group of traps, both in area and volume, in the world. It bears repeating that the geochemists consider most of this oil to be Lower Cretaceous (and probably Jurassic-Triassic; Deroo et al, 1977).

A number of conventional accumulations occur along the subcrop edges of the Mississippian units, as well as the Devonian Grosmont and Nisku edges. The largest volume traps are in the Mississippian. The total known recoverable reserves in unconformity traps is 9 tcf of gas and 350 million barrels of oil (ERCB, 1982).

A topographic ridge developed on the Paleozoic unconformity trending northwesterly through the middle of the province was important in the framework of all Lower Cretaceous deposition. It is well known that it separated the principal basins of Lower Mannville deposition (Rudkin, 1964), but it also persisted into Upper Mannville as a slight positive element which divided the southward transgressing Clearwater sea.

the Canadian deposits with recoverable oil. On the other hand, the oil in place numbers in Canada refer to a degraded tar which is less than half the original liquid oil. The precise ratio of Alberta tar sand oil to world recoverable oil reserves is unimportant. What is important is to reach some relative understanding of the gigantic size of the tar deposits.

Data to be presented later will demonstrate that in the deep western part of the basin a proportionally vast quantity of gas was also generated. These Mesozoic source rocks, rather obscure by reputation, apparently are the most prolific source rocks for petroleum in the world.

I believe that petroleum geologists are not generally aware of such extraordinary quantities of hydrocarbons anywhere, nor have they ever identified Western Canada with such abundance. This new awareness justifies additional study and emphasis on the geologic history, structure and sedimentary facies of the Lower Cretaceous in

Figure 3. Map of North America sedimentary basins.

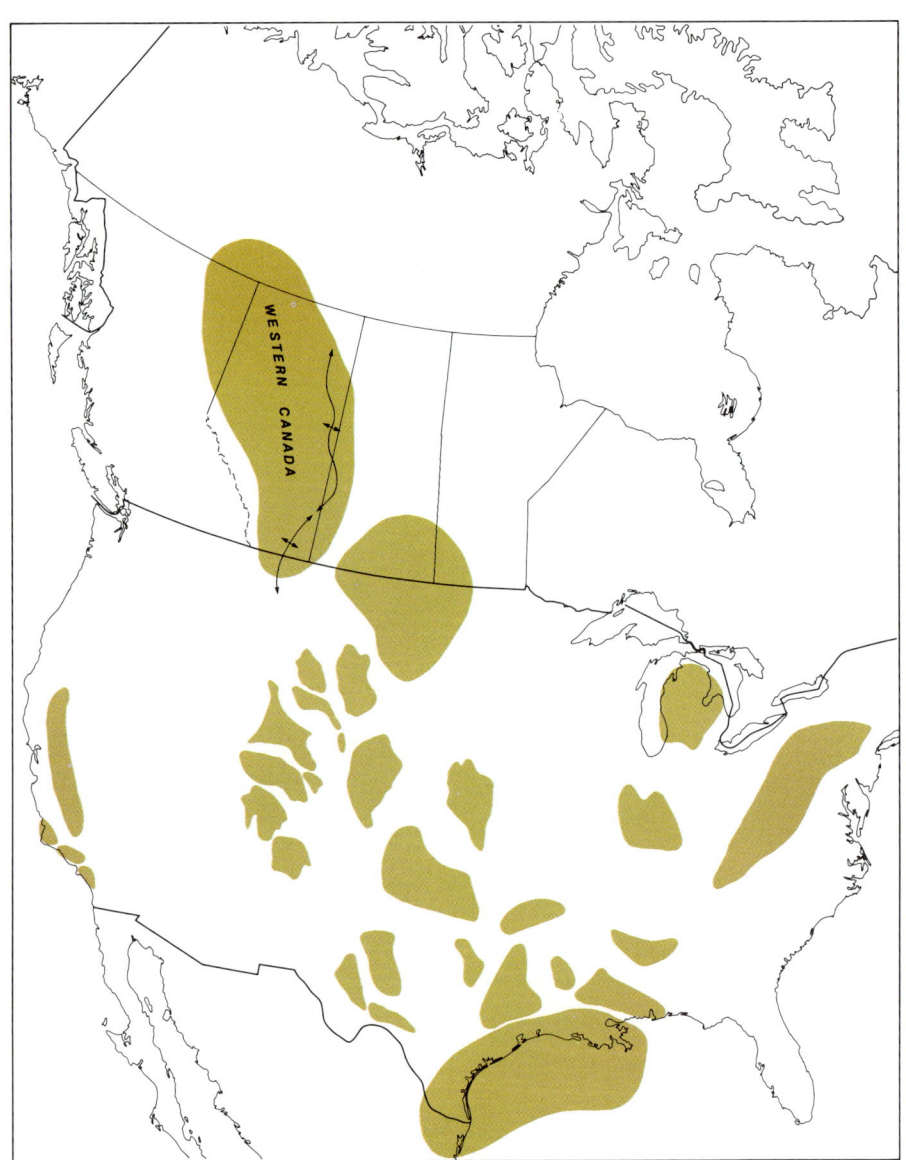

Figure 4. Stratigraphic distribution of recoverable oil and gas reserves in Alberta (from Energy Conservation Board, Alberta, 1981).

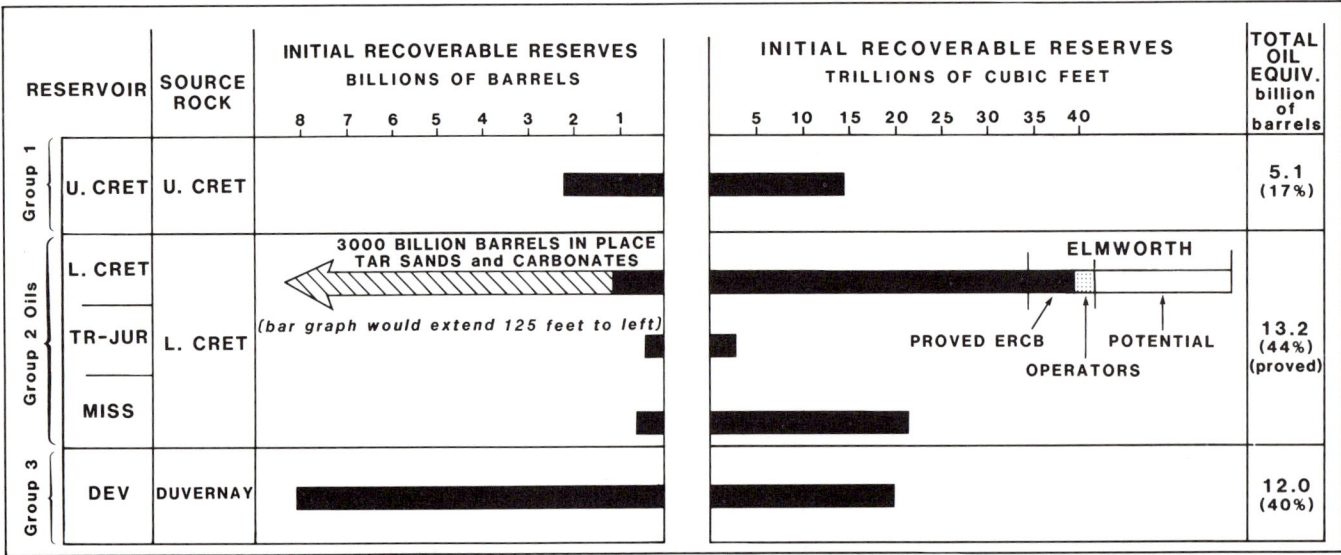

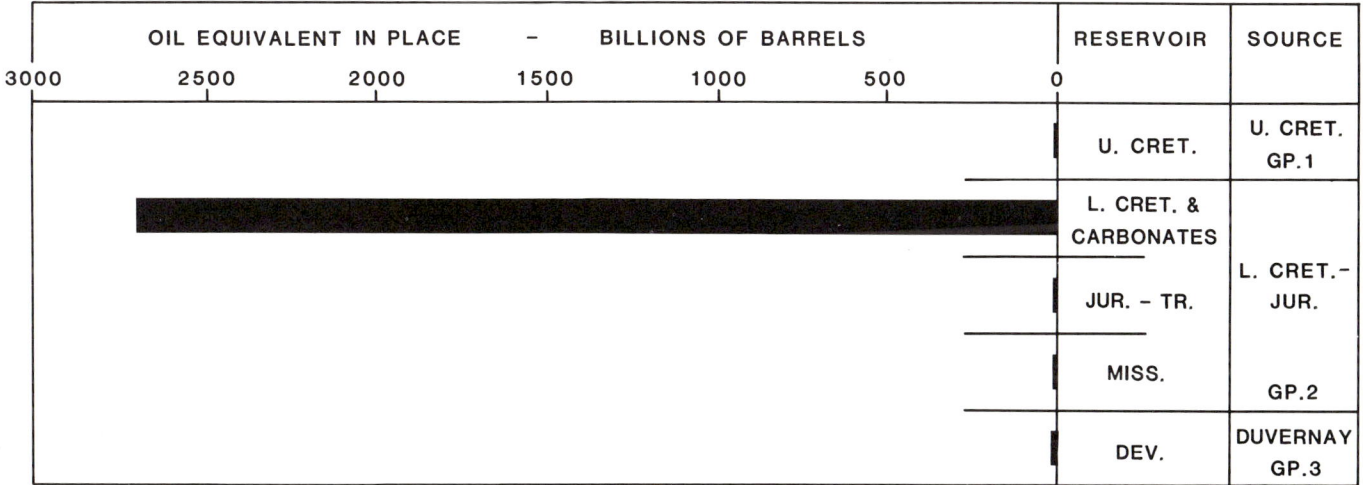

Figure 5. Stratigraphic distribution of all hydrocarbons (gas and oil) in place in Alberta.

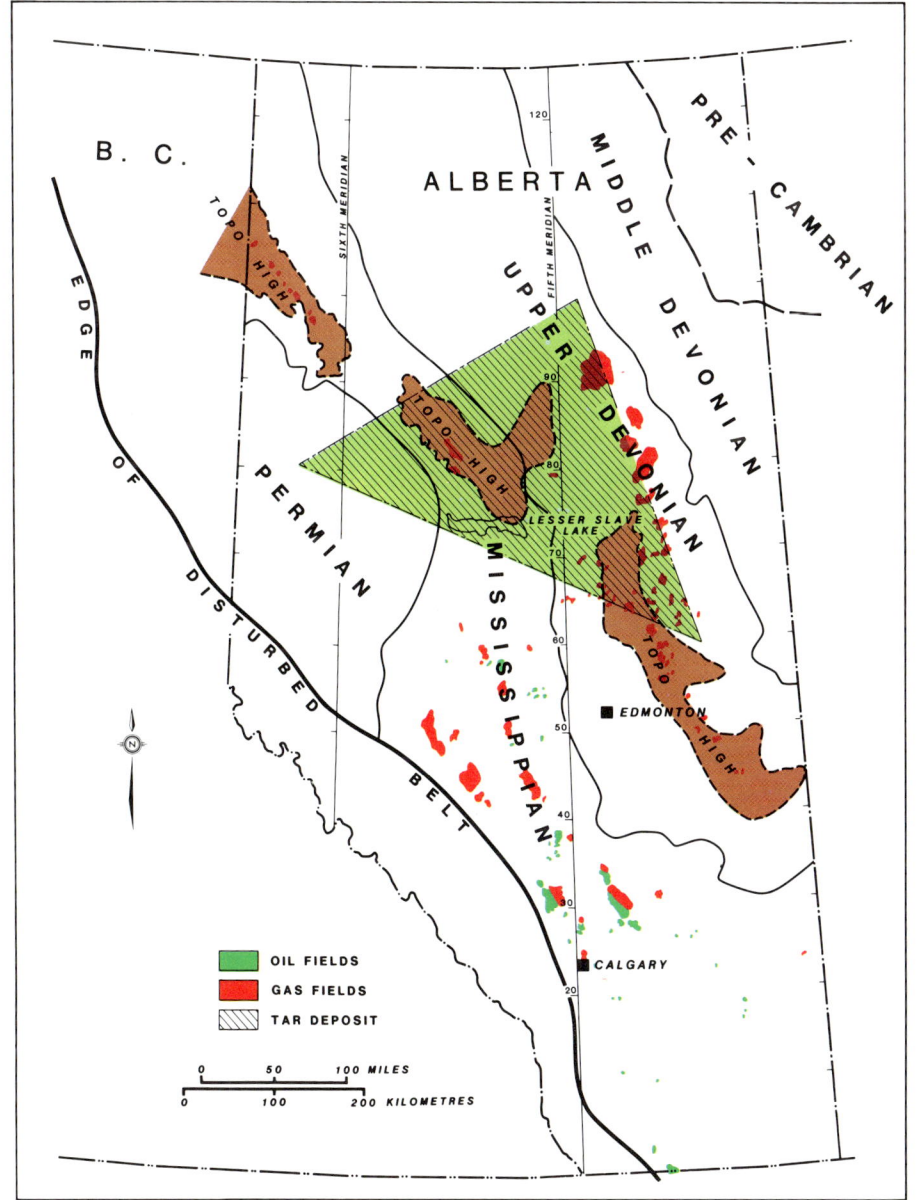

Figure 6. Paleogeologic map of Pre-Mesozoic unconformity surface with principal oil and gas fields related to the subcrop.

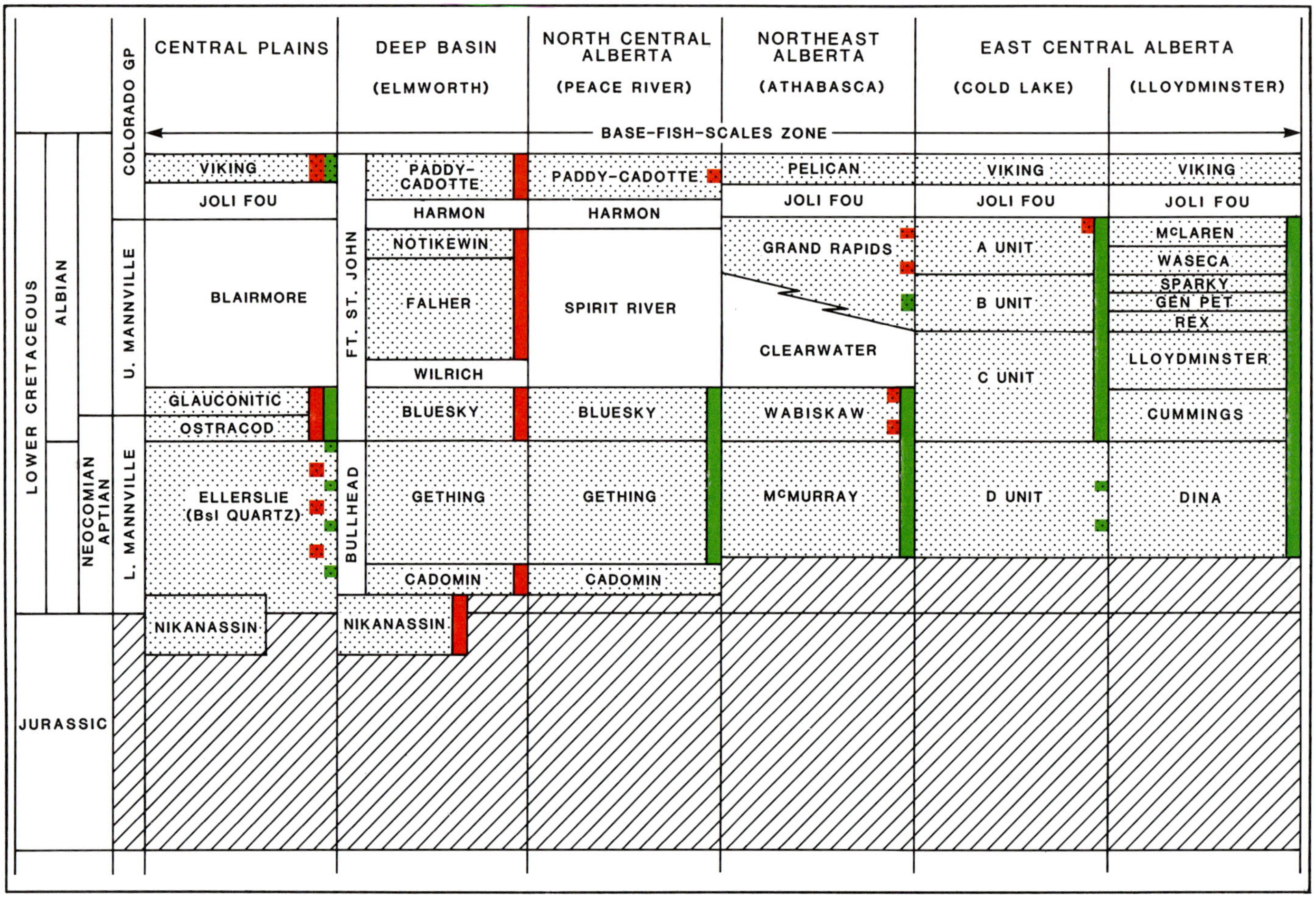

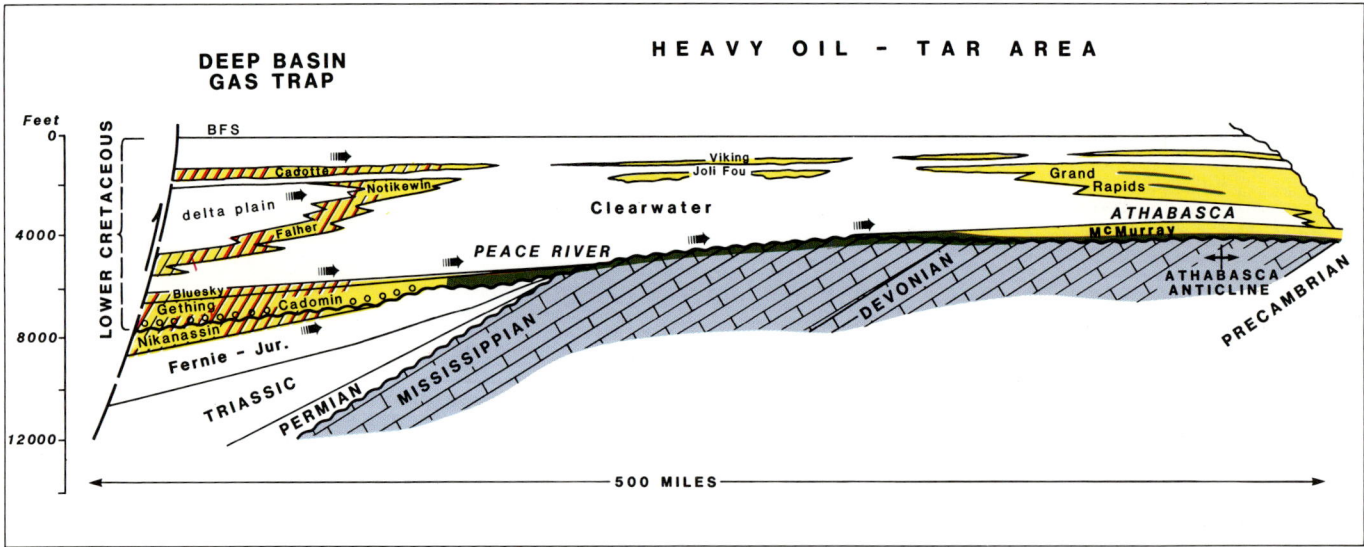

Figure 7. Correlation chart of Lower Cretaceous-Jurassic formations showing principal oil and gas zones. Reservoir beds stippled. Source-seal beds blank.

Figure 8. Diagrammatic cross section across Central Alberta showing Lower Cretaceous sedimentary facies, nomenclature, unconformity surface, direction of hydrocarbon migration, and major oil and gas accumulations.

LOWER CRETACEOUS RESERVOIR ROCKS

A stratigraphic correlation chart of the Lower Cretaceous is shown in Figure 7. The nomenclature is confusing largely because the basin was separated by the northwest-trending ridge on the unconformity surface and sediment was supplied from both the east and west sides of the basin. Reservoir beds and source-seal beds are indicated on the chart by specific patterns. The stratigraphic position of the major oil and gas accumulations are noted.

A diagrammatic cross section (Fig. 8) across central Alberta presents the major sedimentary facies in a simplistic fashion which emphasizes the major source beds, reservoirs, seals, and the major petroleum accumulations. Several of these elements will be described in more detail later, but for now attention is called to the following:

(1) Truncated Jurassic and Triassic source shales with 2% organic carbon lie on the deeper, west side of the basin.
(2) Truncated Paleozoic carbonate beds in the central part of the basin hold a gigantic petroleum accumulation along the unconformity in the Carbonate Triangle.
(3) Bluesky-Gething-Cadomin sandstone beds are gas saturated over a wide area in the Deep basin downdip from water and, to the east, pinch out against a Mississippian ridge where they contain a huge tar deposit at Peace River. East of the ridge lies the McMurray basin where deltaic and fluviatile sands from the Precambrian Shield accumulated and now hold the giant Athabasca tar deposit.
(4) Clearwater marine source shale with 2 to 3% organic carbon transgressed from the north and spread laterally, east and west, interfingering with shoreline sands.
(5) On the west side of the basin the shoreline and continental deposits marginal to the Clearwater sea were Falher sands, conglomerates, shales, and coals, and Notikewan sands. The Falher coals are an important, particularly rich gas source. The west side sands are gas saturated over a wide area in the Deep Basin downdip from water.
(6) On the east side of the basin the shoreline and continental deposits are the Grand Rapids formation which contain the heavy oil of Cold Lake and Lloydminster, and part of the tar deposits of Athabasca.
(7) The Joli Fou marine shale caps the source-reservoir complex. It is probably a source rock too but its principal importance is as a regional seal preventing further upward movement of the Lower Cretaceous hydrocarbons.
(8) The Cadotte-Viking sands are saturated with thermal gas over a wide area on the west side downdip from water and contain extensive conventional fields of gas on water on the east side.
(9) In the downdip area the source shales reached sufficient thermal maturity to generate oil and gas.
(10) The probable major migration path of oil is in the basal sands and along the unconformity.
(11) The total gas in place in the Deep basin area on the west side is approximately 1,750 tcf (calculations presented later). Potential recoverable gas reserves at today's economics may be 45 tcf (adding 60% to Canada's proved remaining gas reserves). On the east side of the basin there are nearly 3,000 billion barrels of degraded oil in place.

Figure 9 is a generalized lithofacies map of the Lower Mannville (Cadomin-Gething). The unconformity surface on which the Lower Mannville was deposited was dominated by a long northwesterly-trending ridge of Paleozoic carbonate with a maximum relief of 500 ft (152 m) across a width of about 30 mi (48 km). The Lower Mannville sediments were dominantly continental and related to large, braided river systems which drained northwesterly toward the early Clearwater sea whose southern shoreline was in northern British Columbia. Sediment entered from both sides of the basin. The west side sands are generally clayey and poorly sorted with only poor to fair porosity and permeability. The east side sands, derived from Precambrian granite or quartzite, are relatively clean and extremely porous and permeable. McMurray sands in the Athabasca tar deposit reach 100 ft (30 m) in thickness and measure up to 35% porosity with darcys of permeability (Jardine, 1974). They are a superb reservoir.

In some places on the west side the entire Lower Mannville sand-shale section of 500 ft (152 m) pinches out against the Paleozoic ridge as at the 90 billion barrel Peace River tar deposit. In other places the ridge was eroded away and sands were continuous across the basin. The entire west flank of the ridge represents a potential area for stratigraphic oil fields in Lower Mannville sands.

Figure 10 is a generalized lithofacies map of the Upper Mannville. It shows the Clearwater sea in the north and its maximum southward transgressive position in central Alberta. The transgression reworked the top of the Gething continental deposits although the Bluesky bar and beach sands were deposited largely in regressive pulses of the overall transgression. Then, as the sea regressed, it built long east- to northeast-trending shoreline sand belts. The southernmost of these are the Glauconite sands of central Alberta. Farther north they become the nine individual Falher sand cycles up to the Notikewin.

Behind the easterly-trending shoreline sands were extensive coal swamps which migrated northward with the receding shoreline. Coals are very prolific gas generators and in western Alberta, where they were buried deeply, they contributed great volumes of gas to the associated Upper Mannville sands.

Traced eastward the shoreline sands are contemporaneous with, although not necessarily connected to, various beach sands of the Grand Rapids and older formations of the Athabasca area.

The chert conglomerates of the Cadomin are well known. Less familiar is the fact that the western terrain continued to supply chert conglomerate to the western side of the basin throughout all of Lower Cretaceous and into Upper Cretaceous at least as late as Cardium time. Cadomin fanglomerates are widespread and conglomerate caps on the upward-coarsening beach sands are present locally in the Gething marine, Bluesky, Falher, Notikewin, and Cadotte; channel conglomerates are seen in the delta plain sediments behind the beaches. The beach conglomerates are the best reservoirs in the Deep basin gas area. In the entire Cretaceous system from Texas to the Arctic the repeated occurrence of chert conglomerates in the stratigraphic column is common only to the Canadian section. It gives the Canadian Cretaceous gas fields

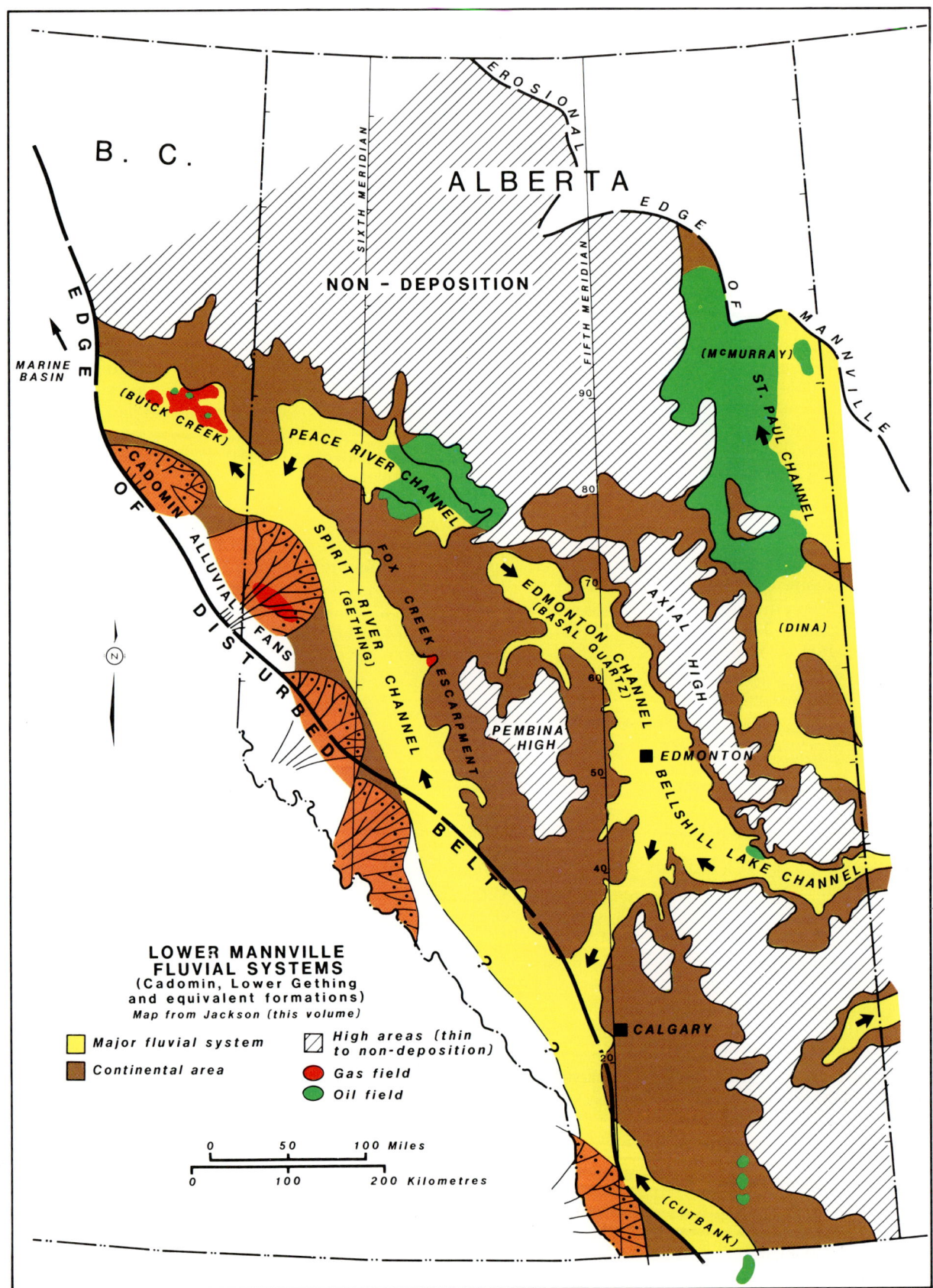

Figure 9. Summary map of Lower Mannville fluvial trends.

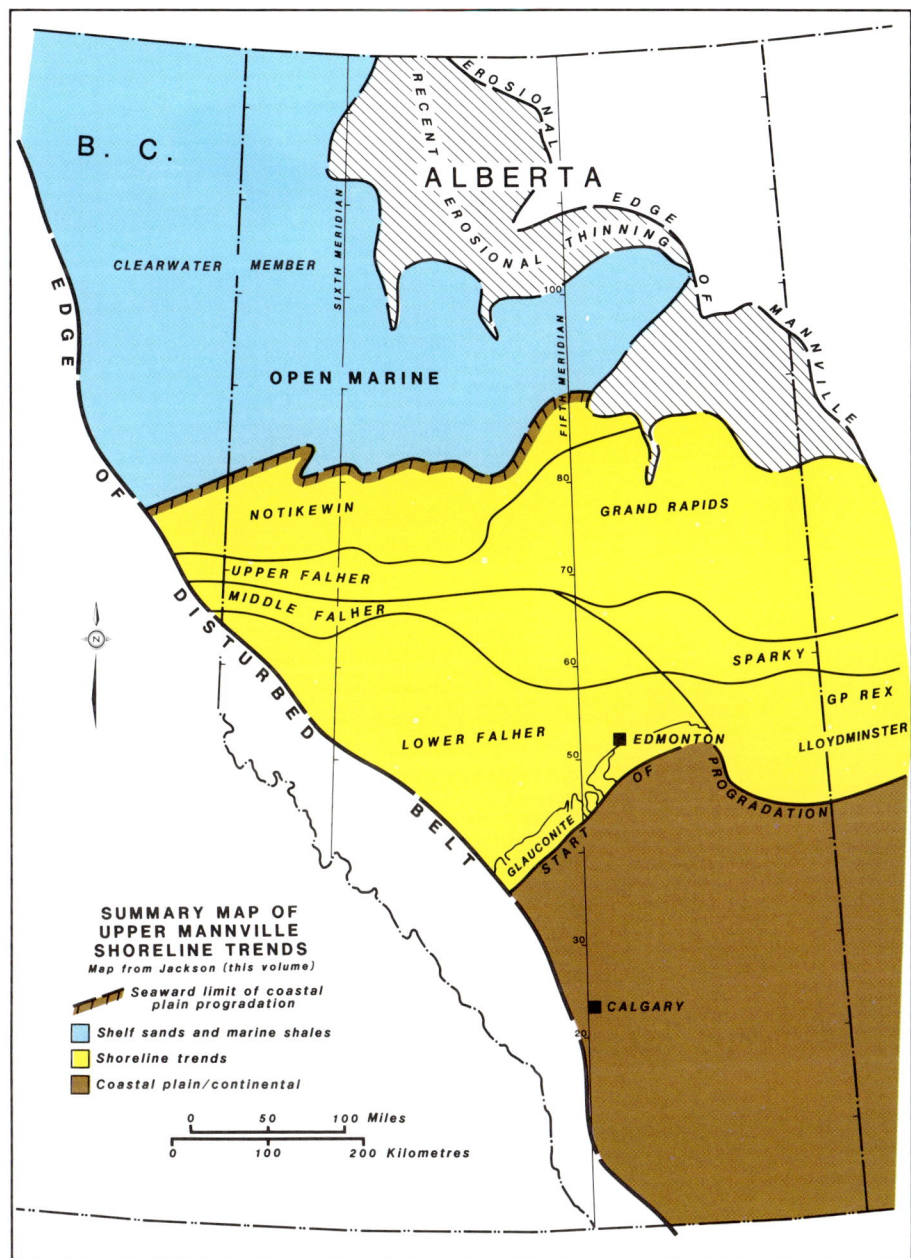

Figure 10. Upper Mannville lithofacies map.

on the west side of the basin an important economic advantage over their American counterparts in the Green River, Uinta, Piceance, etc., basins

The dominant easterly or northeasterly trends of the Lower Cretaceous shoreline sands persisted until the Viking sea transgressed all the way across Alberta to Montana. From then on, Upper Cretaceous sands were deposited dominantly along the northwesterly trends that are so familiar to Canadian geologists.

The accompanying paper by Jackson in this volume provides detailed sand maps of various units with their associated oil and gas accumulations.

SOURCE SHALES AND COALS

Figures 11 and 12 are isopach maps of the Triassic and Jurassic dark shales which contain an average 2% organic carbon (limited measurements by Welte, private work for Canadian Hunter). They probably made a significant hydrocarbon contribution to the Lower Cretaceous sands and the Paleozoic unconformity reservoirs.

Figure 13 is an isopach map of the Clearwater shale. (Refer again to Fig. 8 to emphasize the central importance of this source shale). It thins to zero in central Alberta, hence was not available as a source rock for the Lower Cretaceous fields in southernmost Alberta and northern Montana. For these fields, the only viable source (directly downdip and thermally mature) is the Jurassic Fernie which itself has thinned to only 250 ft (76 m) in the critical area south of Calgary.

The Clearwater ranges in thickness from its zero edge to a maximum of nearly 900 ft (274 m) but its thickness is of importance to this study principally in the area of the oil window, shown on the map, but described in detail later. The Clearwater reaches its maximum thickness in the oil generating belt and averages 500 ft (152 m) over a large area west of Lesser Slave Lake. The organic carbon content of the Clearwater averages at least 2.5% (Deroo et al, 1977), perhaps considerably higher and is most surely Type II although information on kerogen types is not contained in the Deroo report.

The Clearwater marine shale does not play as significant a role in gas generation because its thickness decreases markedly in the gas window to the west. However, it is replaced by a thickening facies equivalent of shoreline and continental sediments which include coals and coaly shales. These provided prolific gas generating material.

Figure 14 is an isopach map of the total coal in the Lower Cretaceous section. Note that the thickest coals are to the north, in the area of the Elmworth gas field. Where the Clearwater sea was at stillstand for a prolonged period a stack of nine Falher shoreline sands were deposited. Behind these beaches back-barrier coalbeds were also stacked. Only the gas window is shown on the coal map because this type source rock produces almost no liquids.

Lower Cretaceous coals exceeding 10 ft (3 m) in thickness and reaching a maximum of 50 ft (15 m) cover an area of 50,000 sq mi (12,950 sq km). The total coal resources are approximately one trillion tons. The United States contains the largest coal reserves in the world with a total coal resource down to 6,000 ft (1,830 m) depth of 4 trillion tons (National Petroleum Council, 1980). These comparative

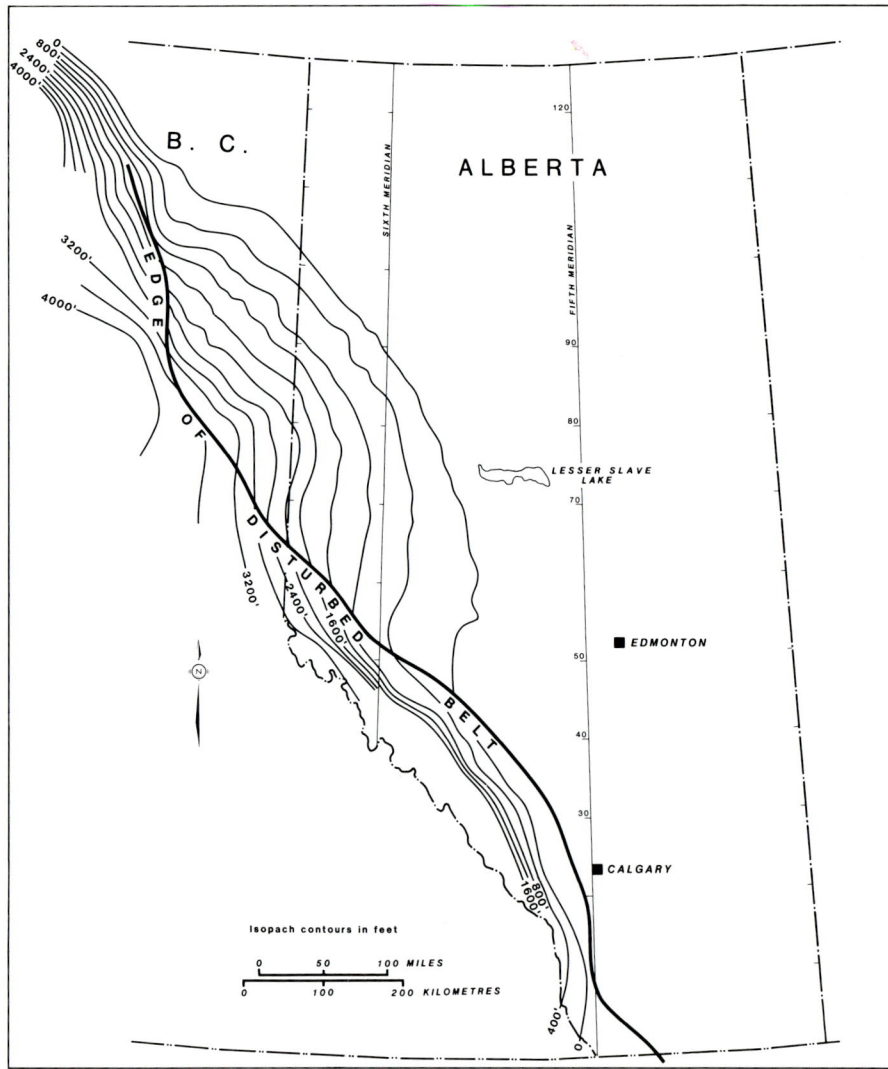

Figure 11. Isopach map of total Triassic (data from Barss, Bert, and Meyers, 1964).

figures indicate that a very large quantity of organic material was deposited as coal in the Lower Cretaceous of Alberta. The western half of the coal deposit, approximately 500 billion tons, is in the gas window. As shown later, the total gas that could have been generated by these vast coal deposits is in the order of 3,000 trillion cu ft. This is a much larger figure than the amount of free and absorbed gas which is contained in the coal today.

SEALS

The marine Joli Fou shale (shown by isopach map in Fig. 15) caps the main complex of source and reservoir beds of the Lower Cretaceous. While the Joli Fou is probably also a source shale it was particularly important as a regional seal for the whole system. This 50-ft (15-m) thick clay shale presents relatively high entry pressure against fluids moving vertically through the Lower Cretaceous.

OIL AND GAS GENERATION

Deroo et al (1977) measured the organic carbon content of 609 rock samples from various formations in the central two-thirds of Alberta. A summary of their data and the area of study are shown on Figure 16.

The entire Cretaceous section from Belly River to the base of Mannville, some 8,000 ft (2,438 m) of predominantly shale, contains over 1% organic carbon in more than 50% of the samples analyzed by Deroo. In the Belly River and Mannville one-third of the samples contain more than 3% organic carbon. In the remaining rock section widespread organic richness was found only in the Duvernay shale of Upper Devonian age. These measurements appear to rule out any appreciable contributions of oil from Mississippian or Lower Devonian rocks. The Duvernay is identified as the principal source for all Devonian fields. There seems little point in arguing about whether some Devonian oil may have contributed to the Lower Cretaceous deposits. It may have, but the Duvernay is only 200 ft (61 m) thick and simply does not have the volume to make a significant contribution to the massive quantities of oil involved in the Lower Cretaceous-Paleozoic unconformity deposits. No other published data is available to suggest any other large volume Paleozoic source.

Figure 17 presents organic carbon measurements from one well in Western Alberta at point A on the location map of Figure 16 (Welte, 1984). This well gives a more quantitative understanding of the distribution of organic material. The Upper Cretaceous averages about 1% organic carbon in the lower 4,000 ft (1,220 m). The Lower Cretaceous-Jurassic averages nearly 10% for 2,500 ft (762 m) through a section of coal and coaly shale. The Triassic contains 2% for another 1,000 ft (305 m). Welte (unpublished letter) says this may be the thickest, moderately rich source section in the world. However, most of the organic material in the area of this well on the west side of the basin is Type III, derived from land plants, which generates mostly gas (Tissot and Welte, 1978). It provided the enormous gas volumes of the Deep basin. Marine conditions prevailed in the sea to the east where the organic material must have been Type II to generate the huge oil volumes of the tar deposits.

The published data on organic carbon measurements is admittedly sparse. I regret that, but believe the averages used in the subsequent calculations are reasonable, though conservative, estimates.

Deroo et al (1977) determined that the beginning of oil generation was at the end of the Cretaceous. Gas generation would have started in the Paleocene. Therefore, an isopach map of total Cretaceous and Paleocene was constructed (Fig. 18) to describe the thickness of overburden on the Lower Cretaceous-Jurassic organic material. The available vitrinite reflectance data at the top of Gething (Deroo et al, 1977; Zhujia-Xiang, 1982) was contoured semi-parallel to the isopach contours. Deviations from the isopach contours are caused by differences in thermal gradient. The vitrinite reflectance contours show thermal maturity at the top of Gething (middle part of Lower Cretaceous). The oil generating window from % Ro (vitrinite reflectance) 0.5 to 1.0 is colored green and the gas window from % Ro 1.0 to 2.0 is red.

The oil and gas "windows" are the temperature-time maturity zones where organic material generates oil first, then gas. Vitrinite reflectance is the optical property of organic material which provides a record of the maximum temperature-time to which the material has been exposed. It is a measure of the degree of "cooking." For a detailed discussion the reader is referred to Tissot and Welte (1978).

The isopach map (Fig. 18) represents a structure map on the base of Cretaceous at the end of Paleocene time. This structure controlled the first hydrocarbon migration. The probable directions of migration are indicated on the map. Note the eastern limit of migration determined by the ancestral Sweetgrass Arch.

However, Laramide deformation at the end of Cretaceous altered the structure somewhat so it is preferable to shift attention to a structure map contoured on top Lower Cretaceous (Fig. 19) to discuss oil and gas generation from that section and subsequent migration and entrapment. The oil and gas windows derived from Figure 18 are superimposed on Figure 19. The entire basin is a simple homocline rising gently eastward toward the Precambrian shield. However, its eastern rim has dip reversal along a 600-mi (965-km) long anticlinal axis. The axis consists of the Sweetgrass Arch in the southeast connecting to the Athabasca anticline which is caused by structural drape over the eastern edge of the Devonian salt beds. The salt has been removed by fresh-water solution over a vast area to the east. The present front creates an extraordinarily long drape structure which is now a short distance west of its position in Tertiary time when it was the eastern barrier to migration and the focus of accumulation of most of the Lower Cretaceous oil.

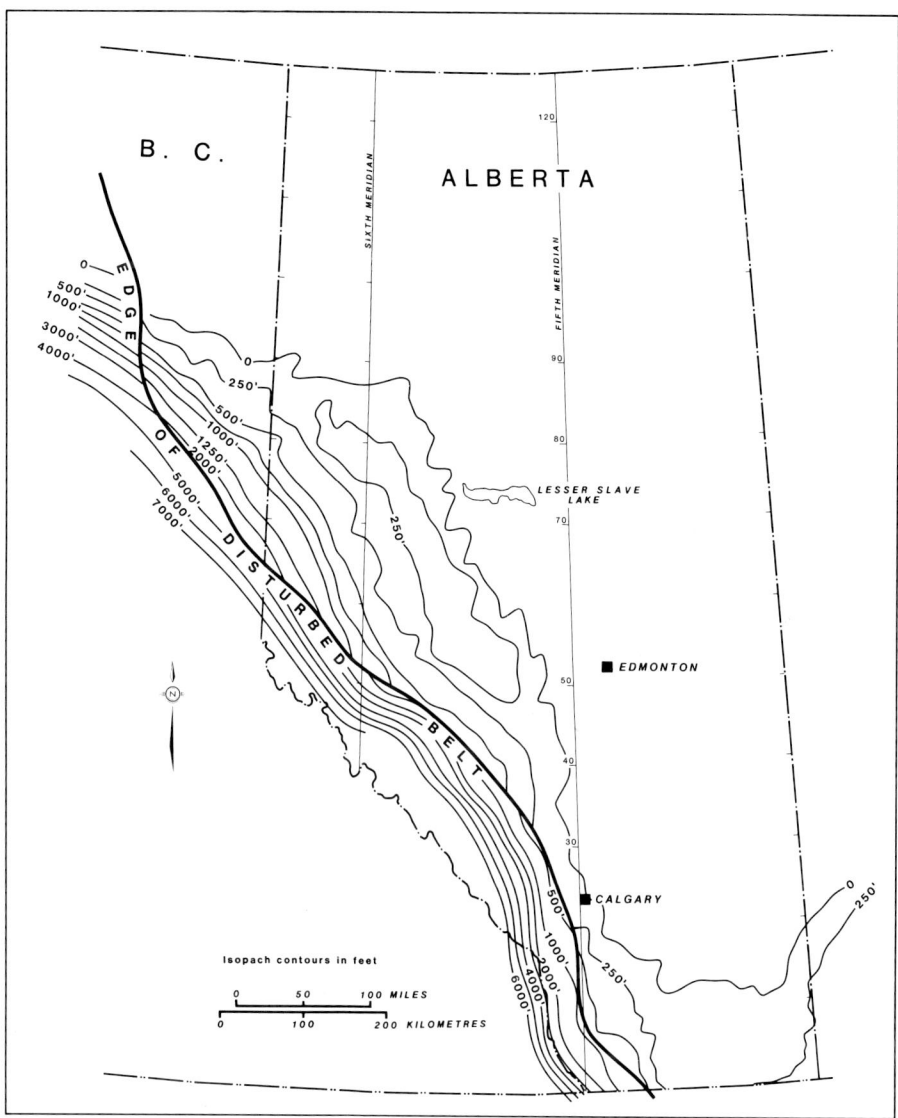

Figure 12. Isopach map of total Jurassic (data from Springer, MacDonald, and Crockford, 1964).

Hydrocarbon migration arrows show the probable routes that oil and gas, migrating under buoyant pressure, would have followed. The hydrocarbons went straight up the dip.

Figure 20 is a cross section across the northern area showing the organic rich section from Upper Cretaceous to Triassic dipping down through the oil and gas generating zones. Note that the organic section is so thick, and the dip so gentle, that the vitrinite reflectance planes intersect the formations at widely spaced locations. Percent Ro 1.0 cuts the top of Gething along the contour line shown on the map, (Fig. 18). But it intersects the base of the Triassic 15 mi (24 km) to the northeast, and the top of the Upper Cretaceous section 40 mi (64 km) to the southwest. Hence, the total gas generating zone, and the oil zone, are smeared out over wider areas of the map.

As demonstrated on Figure 20, three plates of rock (A,B,C) were selected as natural mapping units and by consistency in organic content. Order of magnitude calculations were made of the quantities of oil and gas generated from each plate. The rock intervals are shown in more detail in the type electric log in Figure 21. Only the lower two-thirds of the Upper Cretaceous was measured for gas generation because the upper part of the section does not reach maturity. Welte et al (1984) and Leythaeuser et al (1980) indicated that gas could migrate also by diffusion through the

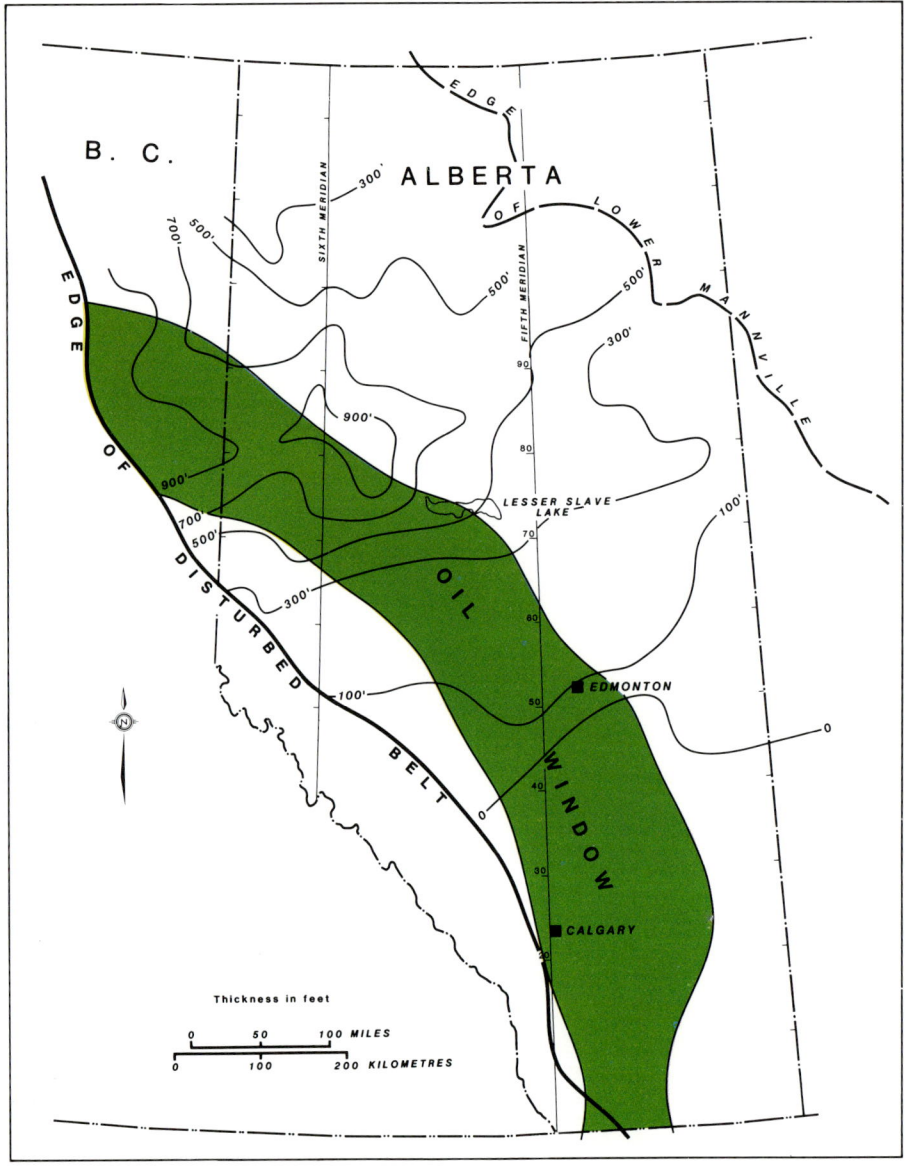

Figure 13. Isopach Clearwater shale, principal Lower Cretaceous oil source.

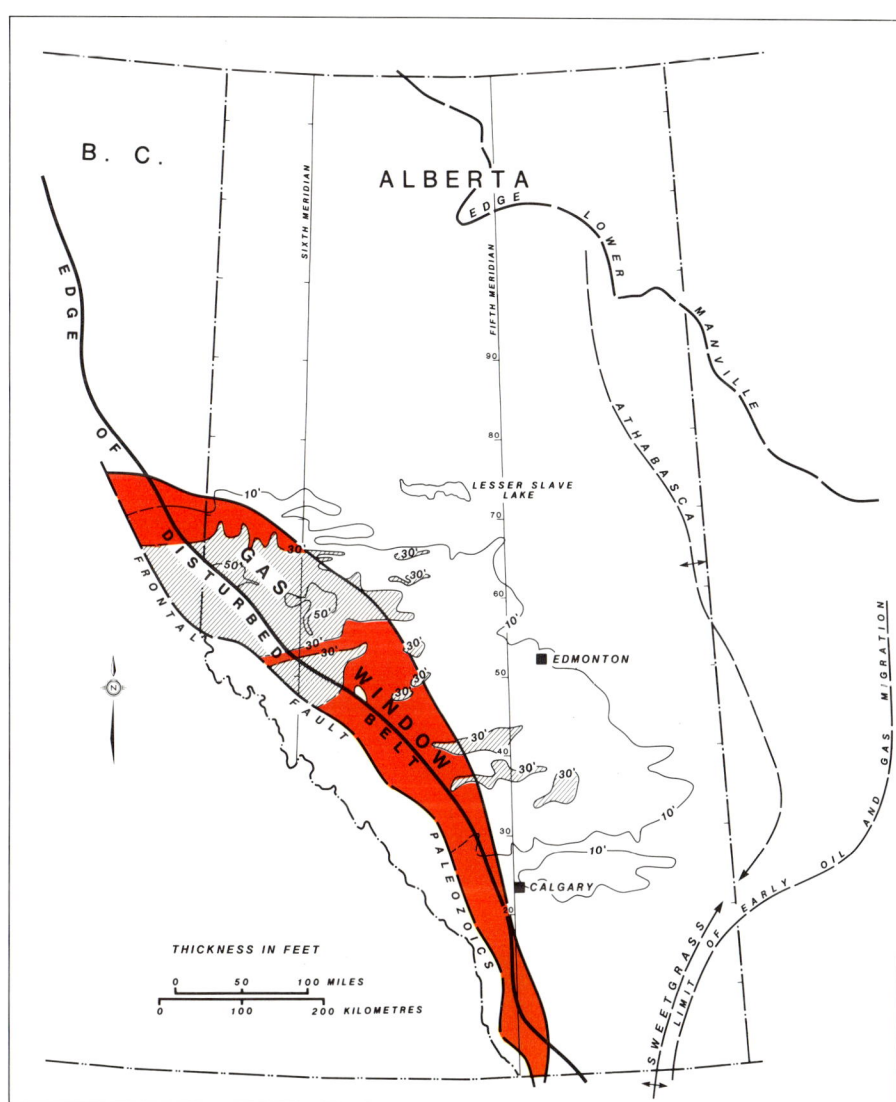

Figure 14. Isopach total Mannville and Jurassic coal.

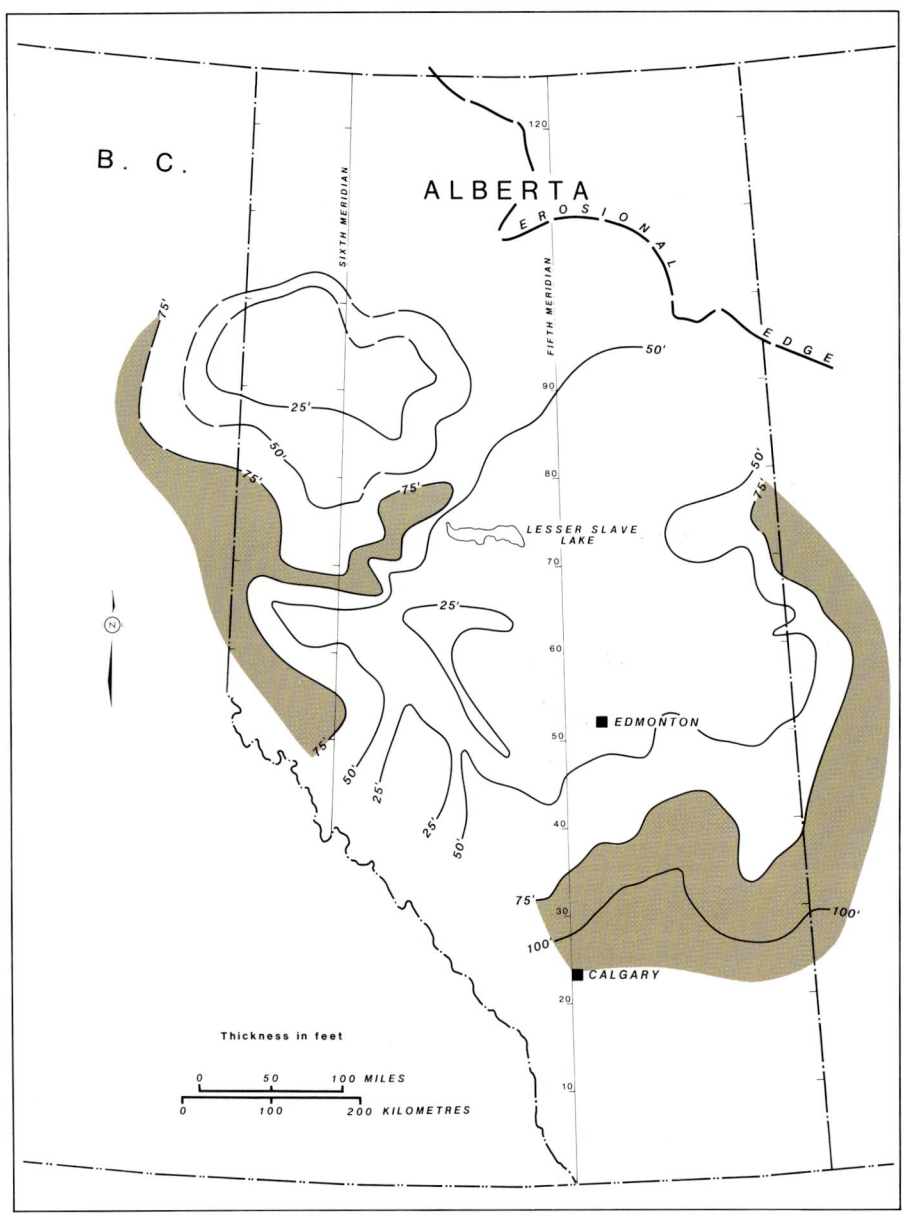

Figure 15. Isopach Joli Fou-Harmon shale, seal for the Lower Cretaceous system.

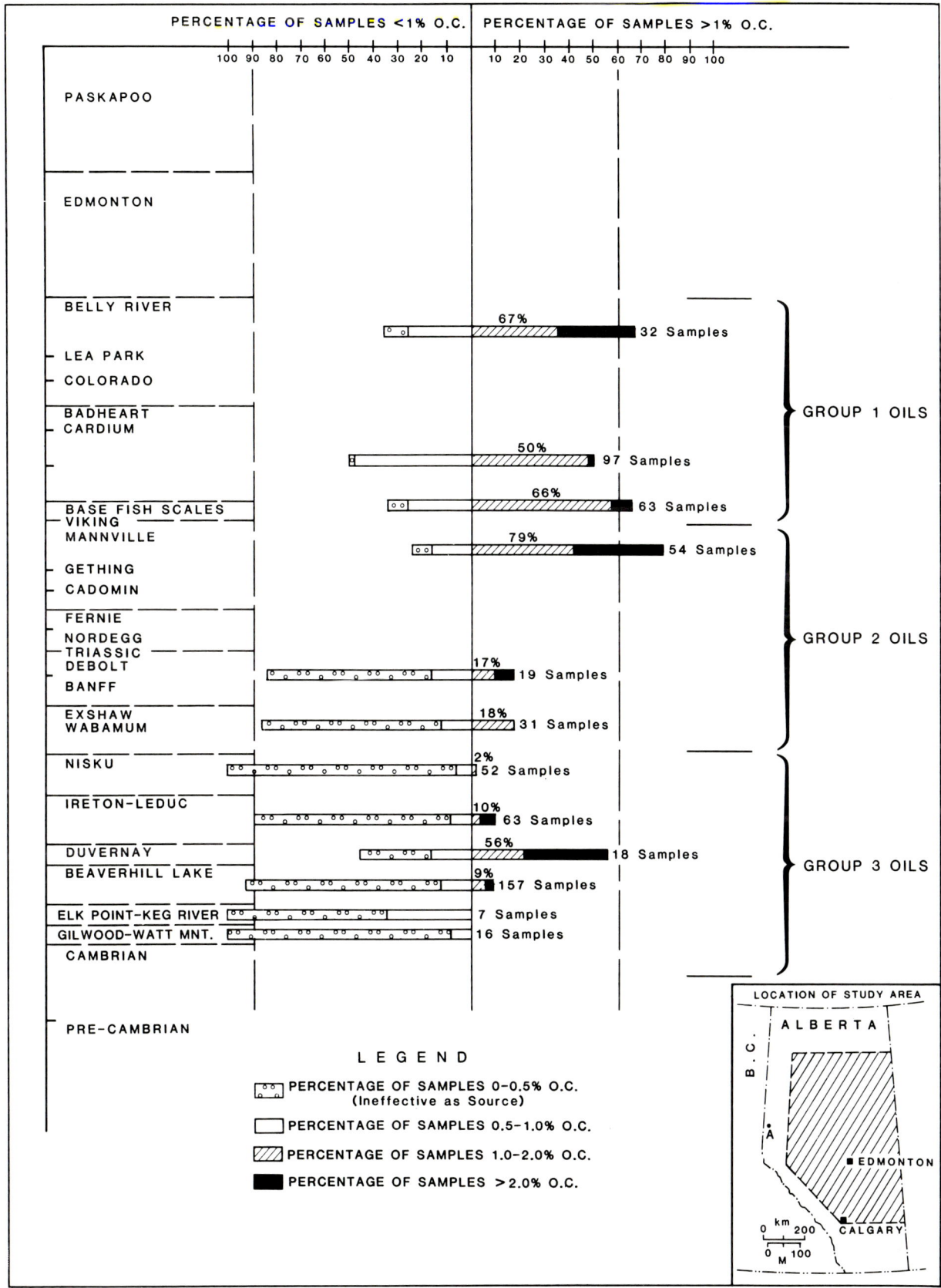

Figure 16. Organic carbon content of 609 samples from Central Alberta.

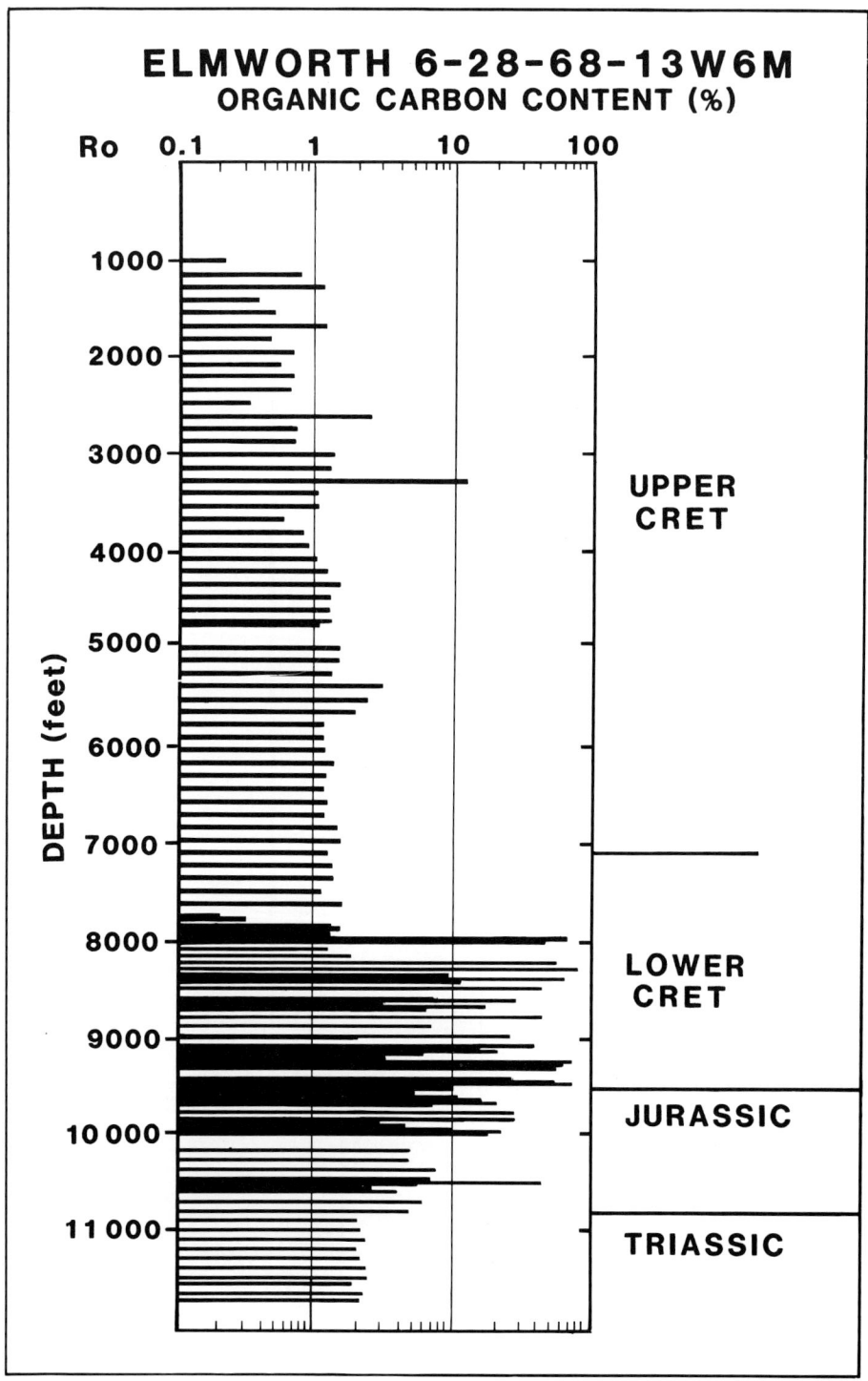

Figure 17. Organic carbon content of a well in the Elmworth field (Welte, 1984).

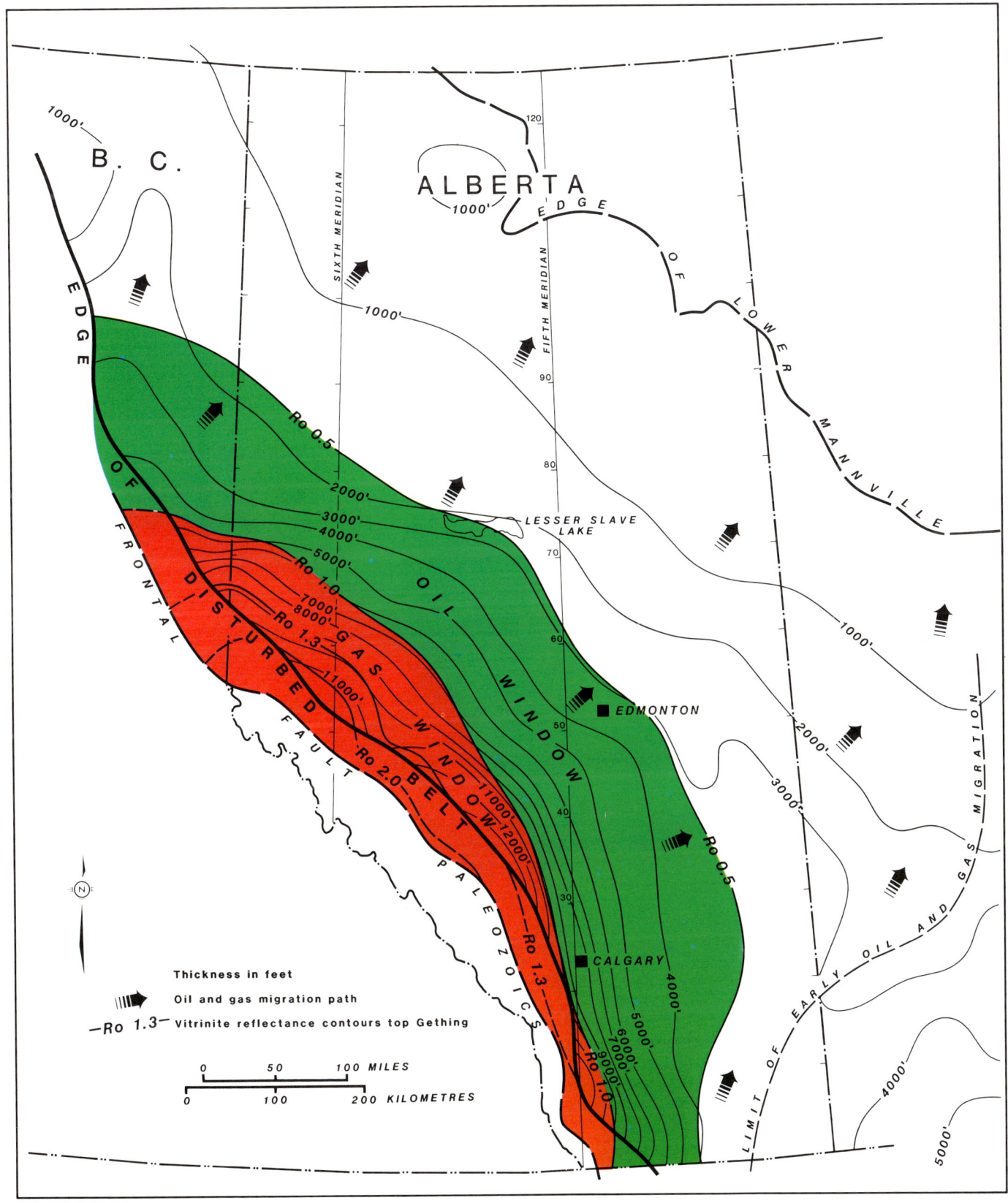

Figure 18. Isopach total Cretaceous and Paleocene with oil and gas generating areas.

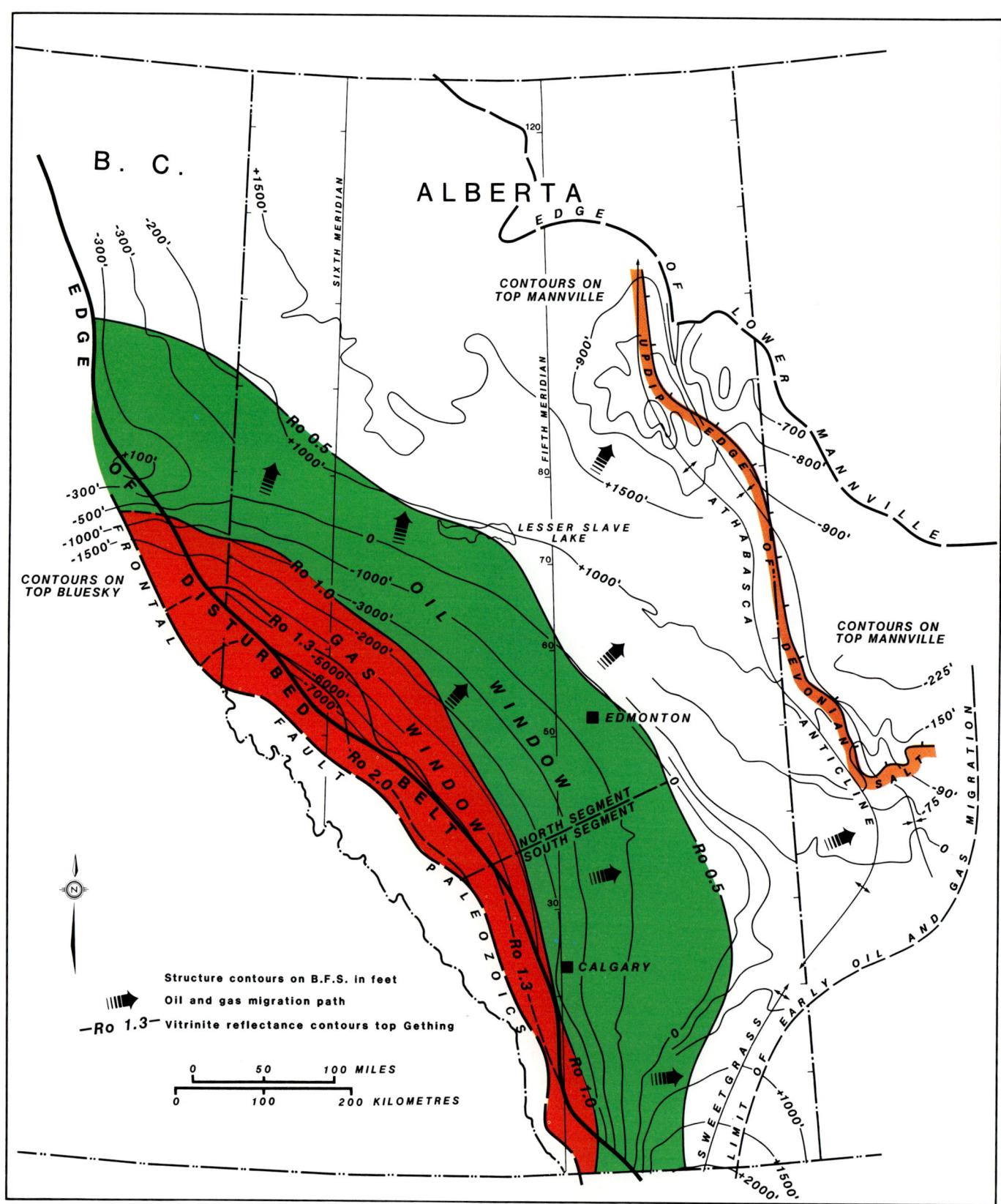

Figure 19. Structure top Lower Cretaceous with oil and gas generating areas.

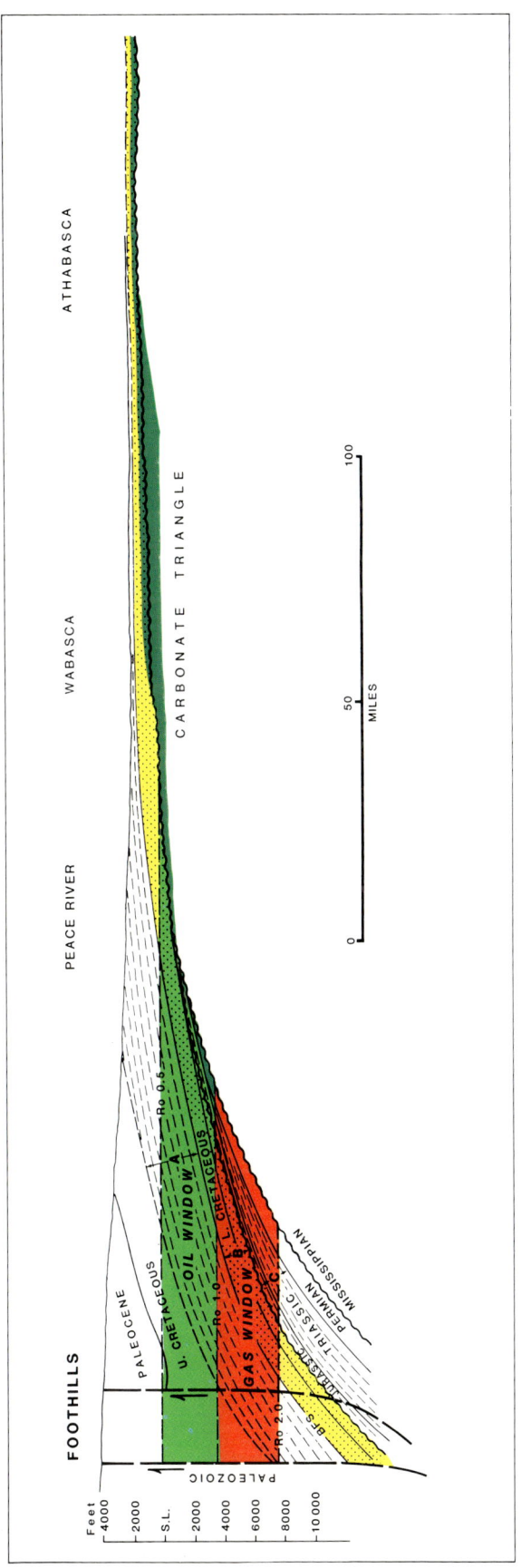

Figure 20. Source rock plates dipping through the oil and gas windows of thermal maturity.

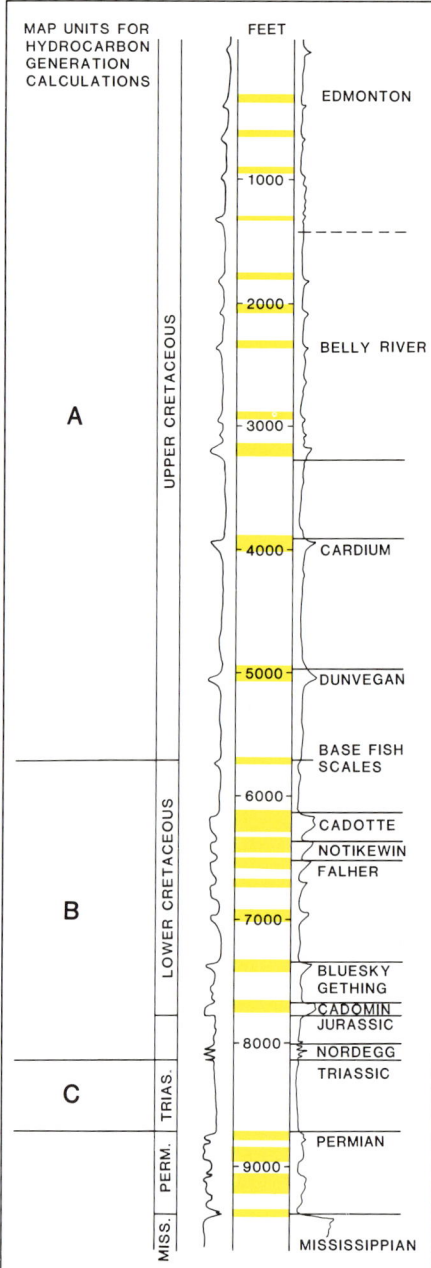

Figure 21. Type electric log, West Central Alberta.

Table 1. Data used in oil and gas calculations.

	Width (Miles)	Organic Carbon (%)	Total Shale (Feet)
GAS WINDOW			
North Segment 300 mi (482 km)			
A. Upper Cretaceous	16 to 37	1.0	4,000
B. Lower Cretaceous Jurassic	16 to 37	7.5	1,500 to 2,000
C. Triassic	16 to 37	2.0	1,000
South Segment 225 mi (362 km)			
A. Upper Cretaceous	10	1.0	1,000
B. Lower Cretaceous-Jurassic	10	2.5	1,000
OIL WINDOW			
North Segment 300 mi (402 km)			
B. Lower Cretaceous-Jurassic	14 to 18	3.0	1,100 to 1,400
C. Triassic	14 to 18	2.0	500 to 2,000
South Segment 225 mi (362 km)			
B. Lower Cretaceous-Jurassic	15 to 30	2.0	1,000

Table 2. Gas generating capability of shale and coal from one sq mi (2.59 sq km) 100 ft (30.48 m) thick (Juntgen and Klein, 1975).

SHALE		COAL	
% Ro Vitrinite Reflectance	Cumulative Bcf gas per 1% Organic Carbon	% Ro Vitrinite Reflectance	Cumulative Bcf gas
1.0 to 1.3	2.7	1.0 to 1.3	270
1.3 to 2.0	3.4	1.3 to 2.0	340

several thousand feet of Upper Cretaceous shale (Unit A) to the Lower Cretaceous reservoir sands. However, oil cannot migrate by diffusion so it was assumed that the oil source section had to be vertically closer to the reservoir sands, especially in a basin with little faulting. Therefore only the Lower Cretaceous-Jurassic and Triassic (Units B and C) were measured for oil generation.

It was necessary to measure the total volumes of rock lying at various % Ro levels in both the oil window and gas window and assign appropriate organic carbon percentages to them. The oil and gas belts were divided into a northern segment 300-mi (482-km) long and a southern segment 225-mi (362-km) long (Fig. 19) and the widths between various % Ro contours measured.

The basic data used in the calculations is recorded in Tables (1 through 5).

While these are only order of magnitude calculations they indicate that the Mesozoic section in western Alberta has a total generation capability in the range of 10,000 trillion cu ft of gas and 7,500 billion barrels of oil. Before considering the implications of these calculations let us discuss the migration and trapping of the oil and gas which resulted in the present distribution of fields.

OIL DISTRIBUTION - MIGRATION AND TRAPPING

Figure 22 is a map of all Lower Cretaceous oil fields and tar deposits (minus the Viking) plus the oil fields and tar deposits of the Paleozoic surface. It is superimposed on the structure and thermal maturity map of Figure 19. Here is nearly 3,000 billion barrels of oil, the largest accumulation of oil in the entire world, largely derived from and entirely contained in the basal 600 ft (183 m) of Cretaceous and the underlying several hundred feet of Paleozoic carbonates.

The Viking fields are omitted from this study because they appear to be in a somewhat different system than the bulk of the Lower Cretaceous. Only a small part of Alberta's Lower Cretaceous oil is in conventional liquid fields. There are about 500 separate pools containing 1.5 billion barrels of recoverable oil. The great bulk of the oil is in degraded tar deposits spread in a vast arc 450 mi (725 km) long from Peace River to Lloydminster. The size of these remnant fields staggers the conventional wisdom of the practicing oil geologist. To emphasize their relative size, Table 6 compares the largest Alberta tar deposits with the largest conventional oil fields in the world.

The three giants of the Middle East are

Table 3. Gas measurements (multiply square miles of area, times hundreds of feet of shale, times a Bcf factor read from Table 2).

TOTAL GAS GENERATED											
North Segment 300 mi						South Segment 225 mi					
	(mi)	(mi)	(00ft)	(Bcf)	(Tcf)		(mi)	(mi)	(00ft)	(Bcf)	(Tcf)
(% Ro 1.0-1.3)						(% Ro 1.0-1.3)					
A.	300 ×	37 ×	40 ×	2.7 =	1,199	A.	225 ×	10 ×	10 ×	2.7 =	61
B.	300 ×	37 ×	15 ×	20.25 =	3,372	B.	225 ×	10 ×	10 ×	6.75 =	152
C.	300 ×	37 ×	10 ×	5.4 =	599						
(% Ro 1.3-2.0)						(% Ro 1.3-2.0)					
A.	300 ×	16 ×	40 ×	3.4 =	652	A.	225 ×	10 ×	10 ×	3.4 =	77
B.	300 ×	16 ×	20 ×	25.5 =	2,448	B.	225 ×	10 ×	10 ×	8.5 =	191
C.	300 ×	16 ×	10 ×	6.8 =	326						
					8,596						481
			GRAND TOTAL		9,077						

Table 4. Oil generating capacity of shale from one sq mi (2.59 sq km) 100 ft (30.48 m) thick (Welte and Yukler, 1981).

% Ro Vitrinite Reflectance	Cumulative MM bbls oil per 1% Organic Carbon
0.5 to 0.7	5.2
0.7 to 1.0	3.14

clustered together as are the Alberta deposits but there is no further comparison. The Alberta deposits dwarf the largest conventional fields in the world. Clearly, the size of the traps is an essential factor in the extraordinarily large accumulations of Alberta. The total area is 25 times larger than the area of the giant conventional fields. Only the Orinoco Heavy Oil Belt in eastern Venezuela is in a size range with the Alberta deposits. Total oil in place there is estimated at 1,000 billion barrels (Holmgren et al, 1975). The basic geology is quite similar to Alberta.

The thermal maturity and structural information in Figure 22 provide a basic explanation of how the extraordinary accumulation in Alberta came about. The

Table 5. Oil measurements (multiply square miles of area, times hundreds of feet of shale, times an oil factor in millions of barrels read from Table 4).

TOTAL OIL GENERATED											
North Segment 300 mi						South Segment 225 mi					
	(mi)	(mi)	(00ft)	(MMBO)	(Billion bbl.)		(mi)	(mi)	(00ft)	(MMBO)	(Billion bbl.)
(% Ro 0.5-0.7)						(% Ro 0.5-0.7)					
B.	300 ×	35 ×	11 ×	15.6 =	1,802	B.	225 ×	50 ×	10 ×	10.4 =	1,170
C.	300 ×	35 ×	7.5 ×	10.4 =	819						
(% Ro 0.7-1.0)						(% Ro 0.7-1.0)					
B.	300 ×	41 ×	14 ×	9.4 =	1,619	B.	225 ×	55 ×	10 ×	6.3 =	778
C.	300 ×	42 ×	17 ×	6.3 =	1,349						
					5,589						1,948
			GRAND TOTAL		7,537						

Table 6. Comparative size of world's largest oil accumulations.

	ALBERTA TAR DEPOSITS (Outtrim and Evans, 1977)			WORLD OIL FIELDS (Nehring, 1978)	
	Oil in Place Billion bbl.	Area in Townships (36 sq mi)		Recoverable Oil Billion bbl.	Area
Carbonate Triangle	1,300	700	Ghawar (Saudi Arabia)	83	36
Athabasca	990	600	Burgan (Kuwait)	72	6
Cold Lake	270	150	Bolivar Coastal (Venezuela)	32	20
Peace River	90	90	Safaniya (Saudi Arabia)	30	4
TOTAL	2,650	1,540	TOTAL	217	66

oil was generated in a belt along the western downdip side of the basin. Note that the downdip edge of the oil window (at % Ro 1.0) coincides with the downdip limit of preserved oil fields. The oil migrated eastward up the gentle homoclinal dip, through stratigraphic complexities which trapped some of the oil in the basal Cretaceous sands and a major amount along the Paleozoic unconformity surface, to the 600-mi (965-km) long anticlinal axis which rimmed the eastern side of the basin and provided the final updip trap. Migration distance from the base of the oil window to Athabasca was a maximum of 250 mi (402 km). The present tar deposits most probably had oil-water contacts which conformed to the structure at the time when the oil was liquid. That structure has changed somewhat to the present day but the oil was "frozen" in place by bacterial degradation, thus recording the original shape of the traps. Note that there is essentially no oil east of the Athabasca anticlinal axis. The axis was an enormously effective trapping agent. There is probably no basin in the world with such an effective trap along the entire updip rim. It could have spilled oil to the east, although there is no trace of it. Oil may also have migrated past the north end of the structure. Undoubtedly, some oil escaped but we clearly have a very unusual situation where a large part of the oil generated and injected by primary migration into the reservoir system has been trapped and preserved, although in a badly degraded state. This provides a nearly unique geochemical model where we know with some accuracy how much oil was injected into the system.

The great tar deposits are in a variety of traps. The simplest is the Athabasca dome, an immense anticlinal closure 150-mi (240-km) long by 70-mi (112-km) wide which now covers 250 townships and apparently covered 600 townships at the time of accumulation in early Tertiary. This is one of the largest closed anticlines in the world in sedimentary rocks and contains the largest single oil accumulation in the world—990 billion barrels in place (Outtrim and Evans, 1977). It was graced with 100 ft (30 m) of superb reservoir sand, a prolific oil source to the west, and a large migration and collection area.

Cold Lake, with 270 billion barrels, is on the southern plunge of the Athabasca axis but its present structural form is so subdued one is willing to guess that at the time of accumulation there was probably some amount of closure. Trapping is influenced significantly by the northward pinchout of marine bar sands along east-west trading shorelines.

The Lloydminster area, reported now to have at least 60 billion barrels of heavy oil in place (Christopher and Knudsen, 1980), is also on the southern plunge of the axis. It is a complex of structural-stratigraphic fields much influenced by shale-filled channels cut through the producing sands.

Peace River, containing 90 billion barrels, is a classic wedge out of the entire Bluesky-Gething-Cadomin sand-shale section, from 500 ft (152 m) to zero, against an old Paleozoic ridge. It reminds one of East Texas although it is a far larger accumulation.

The Carbonate Triangle is an oil-saturated Paleozoic surface with up to 300 ft (91 m) of oil pay covering 700 townships and containing an estimated 1,300 billion barrels of tar. This may be the largest field of all, although it is likely divided into several individual traps. The precise triangle shape on the map is only an approximate outline of the oil-saturated area.

The distribution of tar deposits in the great arc of accumulation from Peace River to Lloydminster leaves little doubt that the "mother lode" of generated oil lay downdip, directly to the southwest, in the area of thickest Clearwater marine shale (see Fig. 13). The migration paths were straight northeast, up the regional dip. This vast flood of oil, on its way to Athabasca, left gigantic remains at Peace River and in the Carbonate Triangle. It is obvious from these two deposits in the migration path that oil moved through both the basal sands and along the unconformity surface. It did not move along restricted channels but more like a huge wave or flood across vast areas.

A substantial amount of oil was generated in the south and moved updip toward the Sweetgrass Arch and toward the southern end of the Athabasca axis. However, the quantities of oil remaining today suggest that the oil charge from the southern end of the basin was not uniquely large. This is due to the very limited thickness of marine shale in the downdip area as described on Figures 11, 12, and 13.

Additional information on Figures 22 portrays water salinities (from Jardine, 1974) and oil gravities. The area of fresh to brackish water (up to 30,000 ppm) extends from the outcrop southwestward and approximately coincides with the tar belt. However, in the south, note the westward projection of brackish water into an area of medium gravity oil. One may conclude that water salinity is not a direct measurement of bacterial activity. We must be careful about rejecting a freshwater area for oil exploration in the Lower Cretaceous lest we overlook a Cutbank (the 200 million barrel field in northern Montana, Fig. 22).

The oil gravity contours on Figure 22 support the story of bacterial degradation developed by the geochemists. There is steadily increasing A.P.I. gravity toward the southwest, away from the fresh water, and toward the generating areas. Only the farthest downdip oil is still in an unaltered state. It is likely that most Lower Cretaceous oil was originally 30 to 40° A.P.I.

The areas in northwest Alberta and northeast British Columbia which are on trend with the western end of the great tar arc, and also directly downdip from the arc, were obviously richly supplied with oil. Future exploration should focus on these areas and on that part of the geologic section adjacent to the unconformity surface.

GAS DISTRIBUTION - MIGRATION AND TRAPPING

Figure 23 is a map of all Lower Cretaceous gas fields (minus the Viking), the few Jurassic fields, and the gas fields of the Paleozoic unconformity surface superimposed on structure and thermal maturity contours. The gas generated in the downdip gas window migrated eastward as far updip as possible, being trapped largely by stratigraphic changes high on the east flank of the basin. The gas does not occupy the huge structural closure on the Athabasca axis, or the great stratigraphic wedge-out trap at Peace River, or the vast unconformity trap in the Carbonate Triangle but, instead, is restricted to much smaller stratigraphic traps largely in thin sands in the upper part of the Lower Cretaceous section downdip from Athabasca. As we have seen before, the bulk of oil migration and trapping was in the thick basal sands of the Lower Cretaceous section and along the Paleozoic unconformity surface, although at Cold Lake and Wabasca considerable oil found its way upward into the Grand Rapids sands.

The updip dry gas area on Figure 23

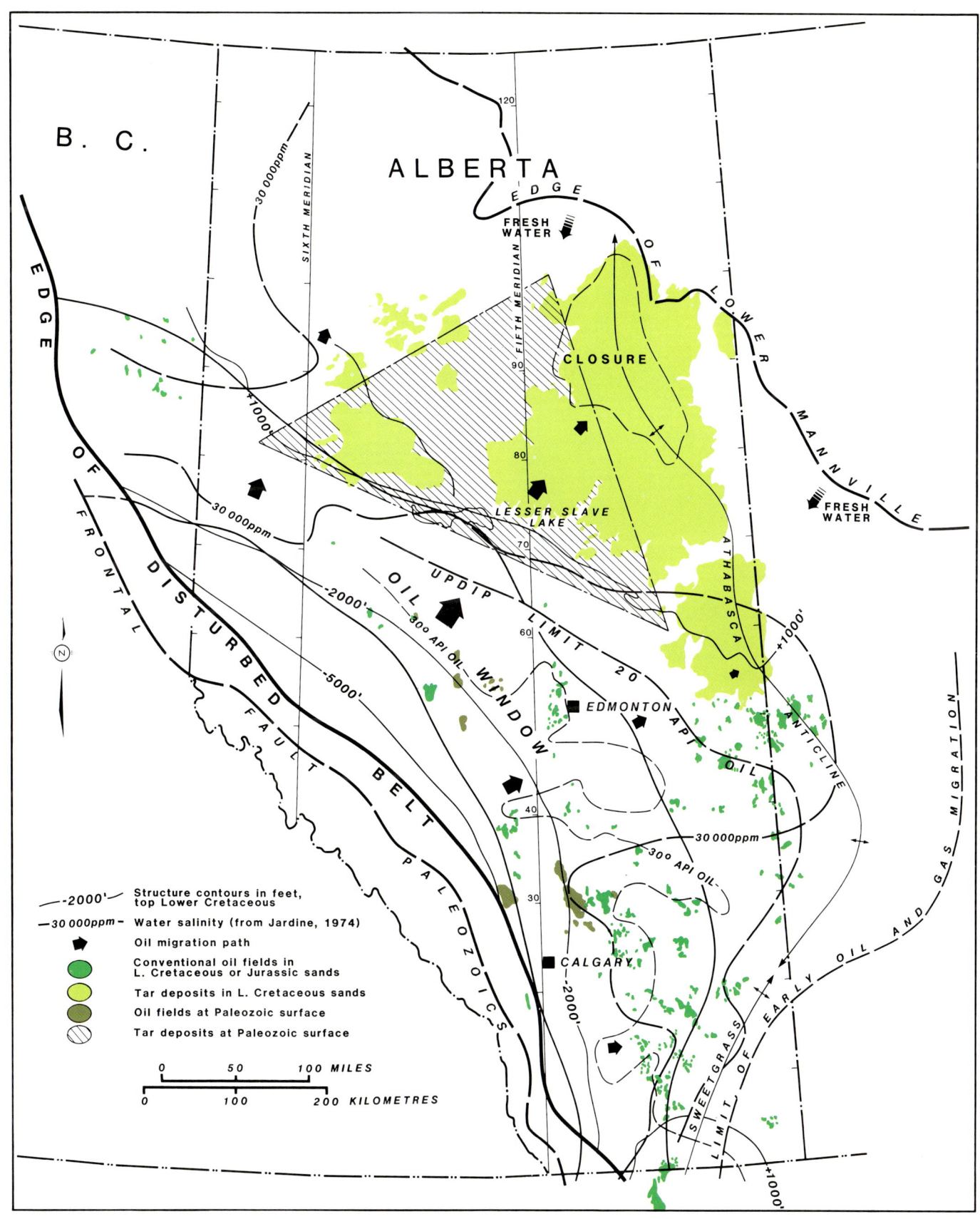

Figure 22. Lower Cretaceous oil.

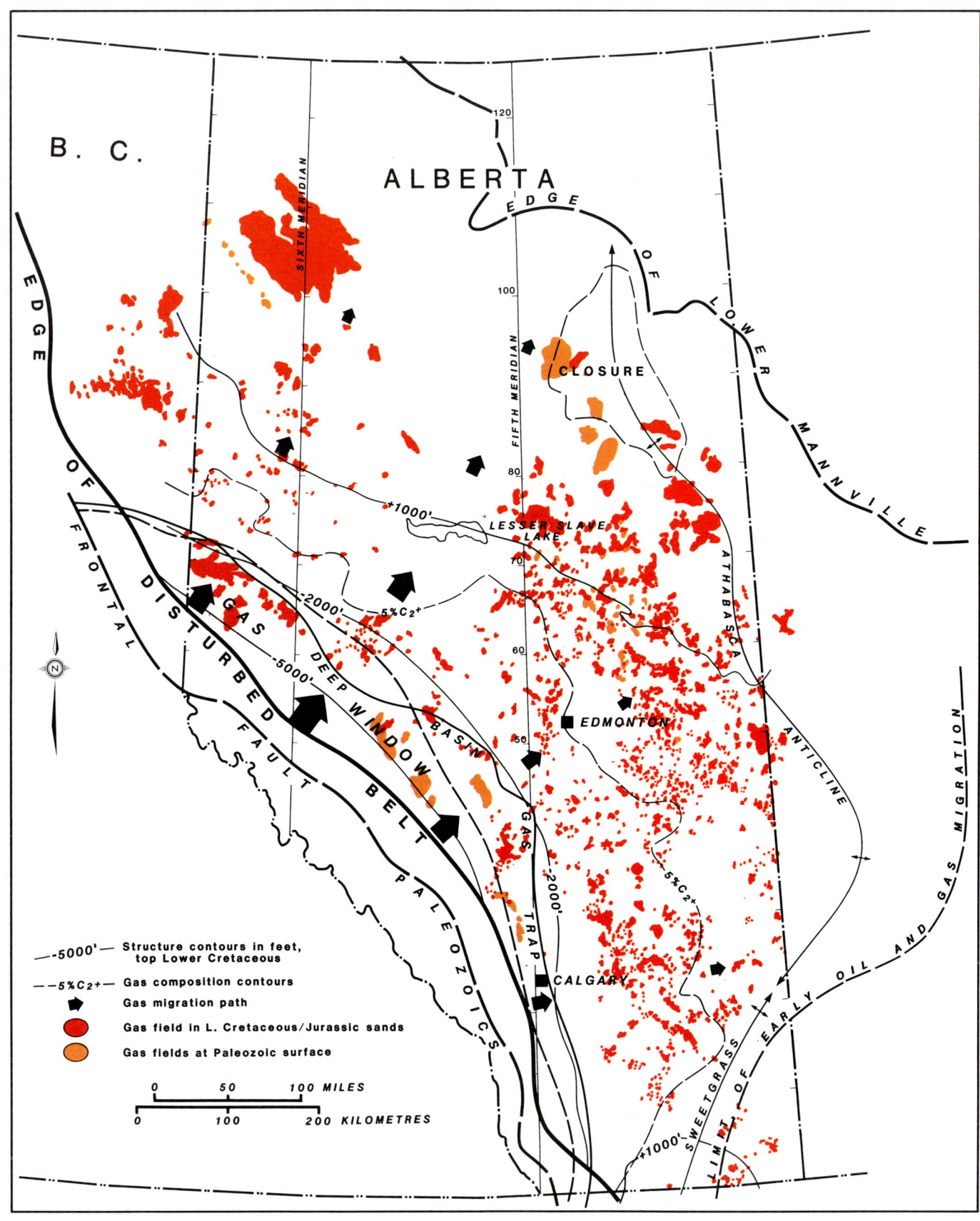

Figure 23. Lower Cretaceous gas.

Lower Cretaceous Oil and Gas in Western Canada 25

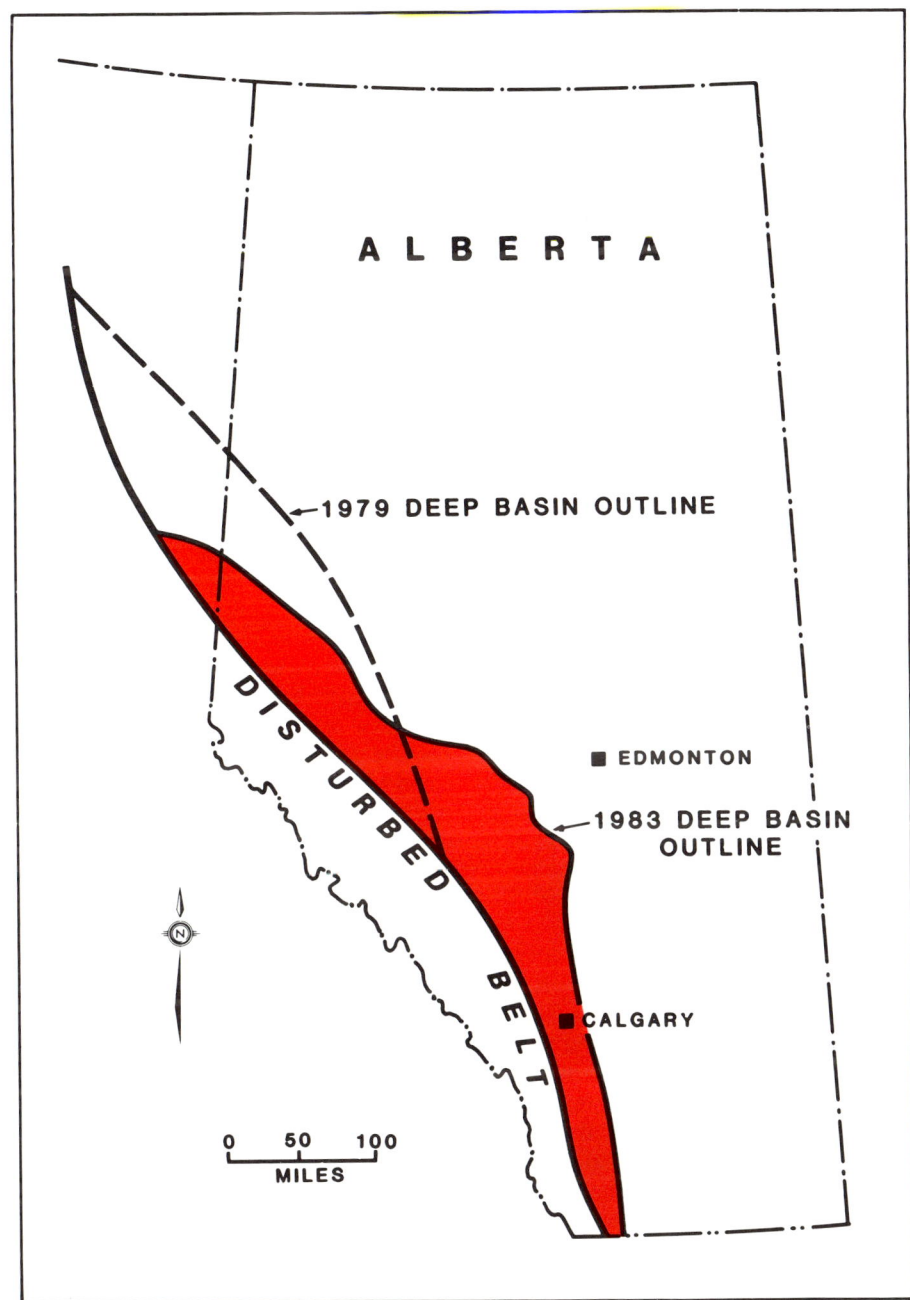

Figure 24. Change in interpretation of the gas-saturated Deep basin.

window. This is recognized on the map by a very large generation arrow in the north segment and a smaller arrow in the south segment. The gas moved straight updip in the same general direction as the oil which preceded it. The principal outlet for gas is at the outcrop in northeastern Alberta where erosion has exposed the Mannville sands in the Athabasca structure, and for 300 mi (482 km) northwestward. To the south, the structural axis blocked gas migration to the east.

Gas migration updip to the east from the gas window is not unimpeded, however. There is an extremely important regional change in porosity-permeability of the Lower Cretaceous sands from the area of the oil window westward into the area of gas generation. According to Masters (1979),

"As the sandstones dip westward into the Deep basin, they become less porous and permeable. This is caused by added clay content in these more rapidly buried sediments, greater compaction, and increased cementation and diagenesis.... (It is in this area of tight sands where) the entire Mesozoic rock section in the Deep basin is saturated with gas below a depth of about 3,500 ft. Within this area it is not possible to drill a dry hole .... every stringer of porosity holds gas."

Welte et al (1984) confirmed these interpretations by actually measuring unusually high volumes of light hydrocarbons from drill cuttings in an Elmworth well through the entire interval from 3,500 ft to 8,000 ft (1,065 to 2,440 m). Water lies updip from the gas in more porous sands. There is no free water below the gas.

The current interpretation of the Deep basin outline, specifically the updip gas-water transition zone, is traced on Figure 23. A relationship to the gas window is suggested by the fact that the gas-water line almost overlies the % Ro 1.0 contour which defines the start of gas generation. The current Deep basin outline, as well as the understanding in 1979, is shown on Figure 24. Nearly 2,000 new wells over the whole area have changed the picture to the 1983 interpretation. The north end of the trap has been cut off by 100 mi (160 km) and the south end extended by 300 mi (480 km).

West of the updip edge of the Deep basin, the entire Mesozoic section generated gas and the deeper part of it continues today to explode gas out of the organic

(northwestern Alberta and east of the 5th Meridian) is extensive, covering 700 by 200 mi (1,125 by 325 km). Within it there are 3,000 individual gas pools in thin sands at shallow depth. There are only two field accumulations which exceed 500 bcf. Still, the total reserve is 17 tcf because of the very large number of small traps.

In the wet gas area southwest of the 5% C_{2+} line there are 34 tcf of gas which includes 9 tcf in Mississippian unconformity stratigraphic traps.

One is tempted to relate the larger accumulation of wet gas to its proximity to the gas window but it is more likely that the accumulation totals are related to trap capacities because an abundance of gas was apparently available for all traps.

The calculations of total gas generated indicate that a gigantic quantity of gas (some 10,000 tcf) was generated, largely from the northern segment of the gas material. This active thermal area is called the "gas furnace."

Our understanding of the trapping conditions which created the vast and thick gas-saturated section downdip from water in the Deep basin has been substantially enlarged. Previously, the updip seal had been tentatively ascribed to "water block" caused by lower relative permeability to gas in the high-water saturation on the updip side (Masters, 1979). Now, Leythaeuser et al (1982), Welte et al (1984), and Gies (1984) have recognized that the trap is "dynamic" in the sense that the tight sand (much of it with the permeability of a silty shale) slows down the passage of gas into the more porous, water-wet sand updip. There is not actually a seal. Gas is continually leaking out updip. But gas is still being generated fast enough that the trap stays filled. A catchy term would be to call it a "bottleneck trap." In Welte's words "the gas saturation of the rock column depends on a dynamic equilibrium between gas generation and gas losses. The low permeabilities and low porosities of the gas saturated part of the rock column are essential for the existence of this unconventional gas deposit. Migration and losses of gas seem to be mainly controlled by diffusion." The coincidence of the Deep basin gas trap and the gas window is explained by this bottleneck concept which requires that the trap be continually fed.

Most of the giant gas accumulations in the world have an especially effective seal of evaporites or permafrost which prevents dissipation of the gas. The Deep basin has an inadequate seal but is continually replenished with new gas.

It needs to be stated clearly again that the Deep Basin accumulation is not all in tight sands. Within the tight sand accumulation, downdip from the bottleneck, there are belts of conventional porosity-permeability rocks. The Falher conglomerate beaches provide superb reservoirs up to 10-mi wide and 30-mi long (16 by 48 km), with permeabilities up to darcys and well tests reaching 40,000 mcf/d at 1,200 psi flowing pressure. Other porous beach reservoirs occur in the Cadotte, Notikewin, and Bluesky sections. These "sweet spots" are analogous to the fracture trends in the San Juan basin tight sands which provide high well deliverabilities and large total recoveries. It is fair to say that most tight sand accumulations would have only marginally economic returns if it were not for the sweet spots. In the Deep basin thin permeable zones such as conglomerates can provide important economic completions because they are in contact with thick sections of overlying or underlying gas-saturated, low-permeability rock. The permeable zone acts as a horizontal fracture into which the adjacent gas need move only tens of feet vertically before it is in a high speed migration path.

It is now a reasonable guess that the Elmworth field in the Deep basin, and its extension westward into British Columbia, will eventually measure at least 20 tcf of economically recoverable gas. Improved economics would increase the number substantially.

Figure 25 is a diagrammatic cross section of the Deep basin gas accumulation. It shows the relatively tight, gas-saturated Mesozoic section lying in the thermal gas window, or furnace. The more porous rocks updip are water saturated. Active gas generation from an extraordinarily thick source section continues to pump gas into the reservoir sands while it leaks slowly away updip. This confluence of tight sands with the gas window occurs in a belt parallel to the Foothills for 500 mi (800 km) from west of Elmworth to the Montana border (see Fig. 23). This is the Deep basin, expanded, and it contains an immense resource of gas in place.

The map in Figure 26 is an isopach of the total gas-saturated sand in the Deep basin. The trap is 500-mi long and 50-mi wide (800 to 80 km). The southern two-thirds averages 150 ft (45 m) of tight gas sand, but the Elmworth area at the north end reaches 800 ft (245 m) of thickness. Several hundred electric log measurements record an average porosity of 6.7% in the sands. The gas-in-place calculation for the entire volume of sand is 1,500 tcf (average total sand 275 ft [84 m], average pressure 3,500 psi, average water saturation 40%). In addition, complete extraction of the free and adsorbed gas from the fractures, pores and bedding planes in the 500 billion tons of coal in the gas window would yield an additional 250 tcf (Wyman, 1984). The total gas resource (gas-in-place with no regard for recovery) in the Deep basin is therefore approximately 1,750 tcf. Over the entire 24,000 sq mi (62,160 sq km) of gas-saturated area, the average gas-in-place is 70 bcf per section. The total does not include significant volumes contained in siltstones and fractured shales. Neither does it include the sandstones in the Foothills belt between the frontal fault and the first Paleozoic fault which would approximately double the total gas volume. The 1,750 tcf projection should replace my previous estimate of 440 tcf as a measure of the "total gas in the Deep Basin" (Masters, 1979).

Commercial gas fields within this vast area are caused by the presence of reservoir-quality sands within the tight sand package where it lies within the gas furnace (Fig. 27). Shorelines were dominantly northeast trending and generally regressed from south to north during most of the Lower Cretaceous. As a result, the favorable sand belts cross the furnace in a northeasterly trend and get progressively younger in a northerly direction. Hence, a series of six composite sand trends through the furnace are now known from Calgary north to Elmworth. Elmworth and Hoadley are the best known but the industry is presently active in several others. These trends usually have one or two beach sandstone pays. Elmworth is uniquely large because it was an area of shoreline stillstand where nine Falher beaches are stacked vertically followed by Notikewin and Cadotte beaches.

Each of the six sand belts are pumped full of gas in trends averaging 50-mi (80-km) long and 10- to 15-mi (16- to 24-km) wide. The total area covered by apparently commercial sand trends is approximately 150 townships (5,400 sq mi; 13,985 sq km) and the total recoverable gas at current economics is estimated to be 45 tcf which includes 20 tcf at Elmworth. This is probably the largest amount of gas presently unaccounted for in official reserve estimates which can be mapped and measured with some precision anywhere in North America. As we have seen, very large quantities of presently noncommercial tight sand gas are also present in the same areas.

The commercial gas area of 5,400 sq mi (13,986 sq km) compares in size with the San Juan basin's 3,600 sq mi (9,325 sq km) (second largest gas field in U.S. at 32 tcf) and Hugoton Panhandle's 8,000 sq mi (20,720 sq km) (largest gas field at 80 tcf) according to Nehring (1981).

Before concluding the gas discussion, it is necessary to introduce a complication into the simple picture of an oil window and a gas window. The area of the gas window depicts Mesozoic sediments which reached depths and temperatures where gas was generated. But, to get there, those

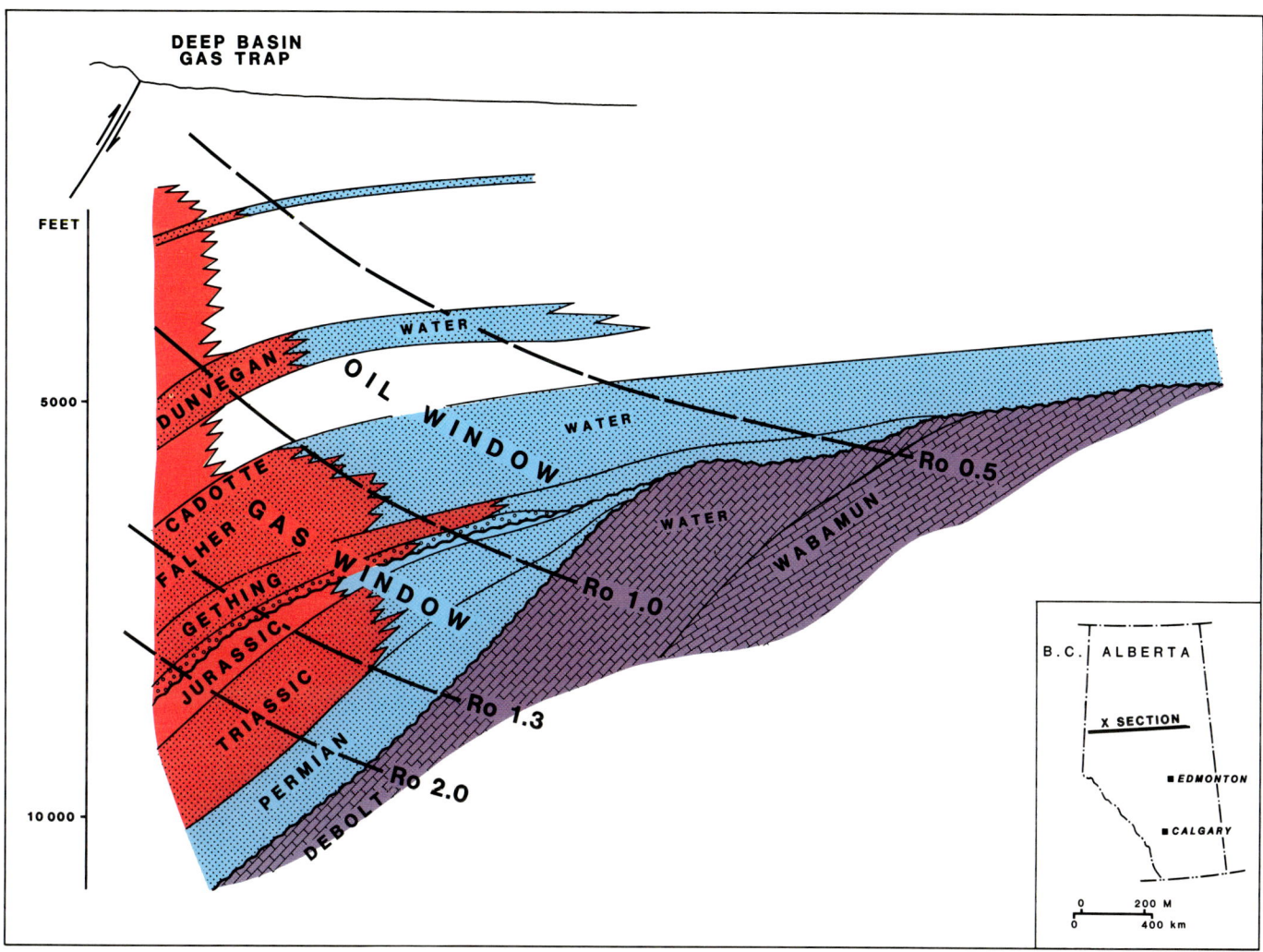

Figure 25. Cross section of Deep basin gas trap with water over gas contact and oil and gas thermal windows.

sediments had to pass through the oil window of temperature. Consequently, this body of sediment was "milked" of its oil before it began to generate gas. The proposition that oil was once generated from the present gas window and perhaps west of it is supported by very large remnant oil accumulations in the Elmworth region. Extensive areas of porosity partially plugged with pyrobitumen are known in the Baldonnel dolomite (Triassic), and in the eastern wedge out of Nikanassin sand (Jurassic). The total area of these remnant fields is over 120 townships and the estimated original liquid oil in place is 55 billion barrels. The oil was either displaced updip by gas with the asphalt remains left behind, or was burned up as the sediments were heated to gas-generating temperatures.

Calculations in Table 5 indicate total generation of 7,500 billion barrels of oil from the oil window. The gas window sediments exceed the oil window in total volume, but the Lower Cretaceous changes facies westward from oil-generating marine to gas-generating continental. However, the thickness of marine Triassic and Jurassic dark shales increases greatly. I will take a long step beyond measured data and suggest that the gas window sediments and those even farther westward generated as much or more oil than the oil window. Hence, I propose total oil generation from the Lower Cretaceous-Jurassic-Triassic of 20,000 billion barrels.

IMPLICATIONS AND QUESTIONS

No claim is made here to a scientific level of accuracy in the calculation of hydrocarbon volumes from the publicly available geochemistry data. Nor is it possible to measure recoverable gas to engineering standards in the currently developing trends in western Alberta. What I have attempted to show is that Alberta contains enormously larger quantities of hydrocarbons than is generally recognized and therefore is a world class geological model for oil and gas generation-migration-accumulation. It also has a unique "answer in the back of the book" in the form of immobile tar deposits which attest to oil volumes in the system which would have been dismissed as unthinkable if the fields had been destroyed in liquid form and no trace left. Finally the basin has a unique anticlinal axis rimming much of the eastern flank which apparently retained most of the oil

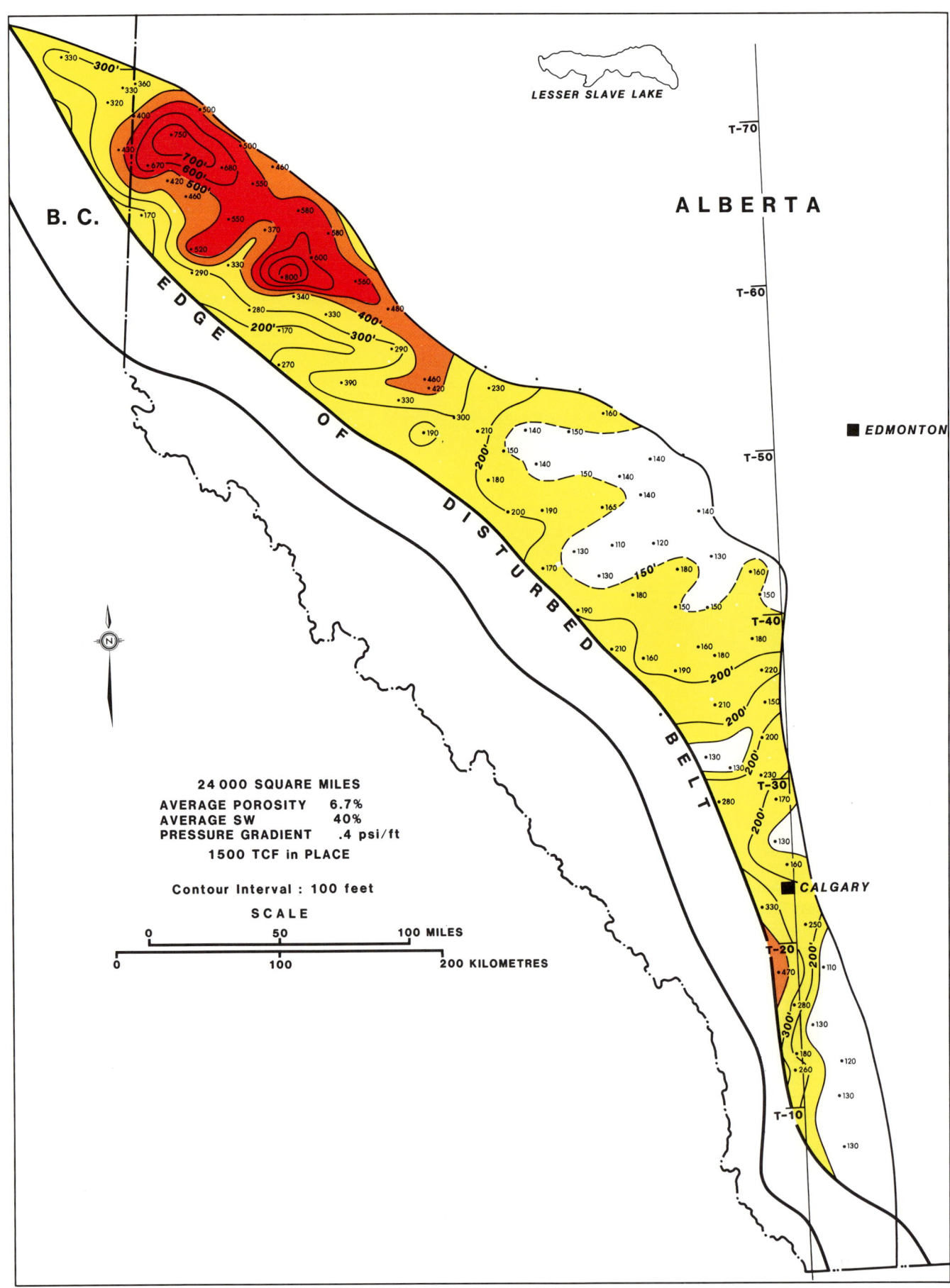

in the basin.

Any geological study of Western Canada must address this problem of stupendous volumes of oil in the system. It is not enough to recognize this or that as a source rock. Additionally, we must know if it is rich, thick, extensive and mature enough to have been a massive oil generator on a scale of thousands of billions of barrels.

My calculations, using published data on organic content, maturation and hydrocarbon yields, indicate total oil generation of 7,500 billion barrels from the oil window and suggest a means of reaching 20,000 billion barrels. (It is a unique feature of this problem that we already know that 20,000 billion is about the answer we have to reach). Assuming an expulsion efficiency from the shales of 35% (Welte, personal communication) that would result in 7,000 billion barrels being injected into the system. If 25% of it was lost that would result in 5,000 billion barrels available to fill the great tar arc. However, we can measure only 2,700 billion barrels of tar in the arc. Actually, the destruction of light ends from a medium gravity oil and extensive degradation would probably have removed about half the total volume (Welte, personal communication) so we can estimate original liquid oil-in-place of about 5,000 billion barrels. The material balance appears to support my oil generation estimates.

I must emphasize that the amount of oil lost in the migration process is unknowable. I suggest 25%, but it may have been substantially more or less.

Moshier and Waples (1982) calculated only 450 billion barrels of oil generated from the Mannville and concluded that it "cannot be the major source of the heavy oils. Dominant contributions probably come from Paleozoic and other Mesozoic rocks." I am sure they can make useful comments on my volumetric measurements. I agree with them about other Mesozoic rocks being important contributors. However, according to the sampling done by Deroo et al (1977), they will have a difficult time finding sufficient volumes of Paleozoic source rocks to make an appreciable contribution to the prodigious quantities of oil needed to satisfy the requirements of the tar deposits. However, the existence of radically different measurements made by sophisticated observers is a caution to everyone that the giant tar fields are not easily explained. But they are a problem of such magnitude and geologic importance that they merit continuing serious study.

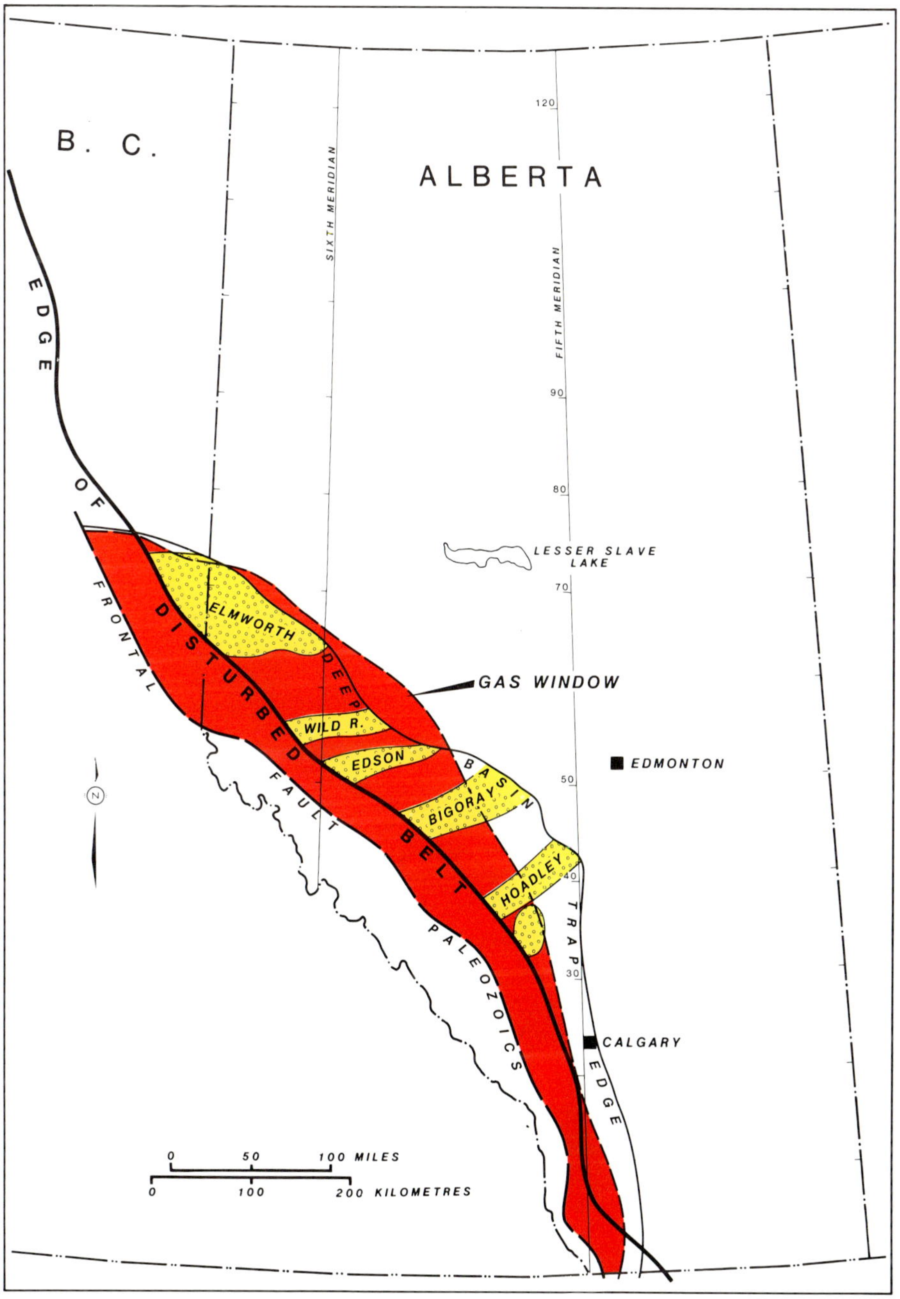

The migration and trapping of the oil appear to be pretty straight forward. We can see clearly where the largest volumes originated and we can trace the principal migration paths. The collecting area along the eastern rim is easily identified. We know that the oil moved into the traps as a liquid at the end of the Cretaceous and was thereafter degraded to tar by aerobic bacteria.

Knowing the location of the main migration paths, both in a geographic sense and also vertically in the stratigraphic section, allows us easily to delineate the rocks and the area that deserves further oil exploration attention in north-

Figure 27. Reservoir sand belts in the gas-saturated Deep basin.

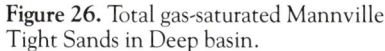

Figure 26. Total gas-saturated Mannville Tight Sands in Deep basin.

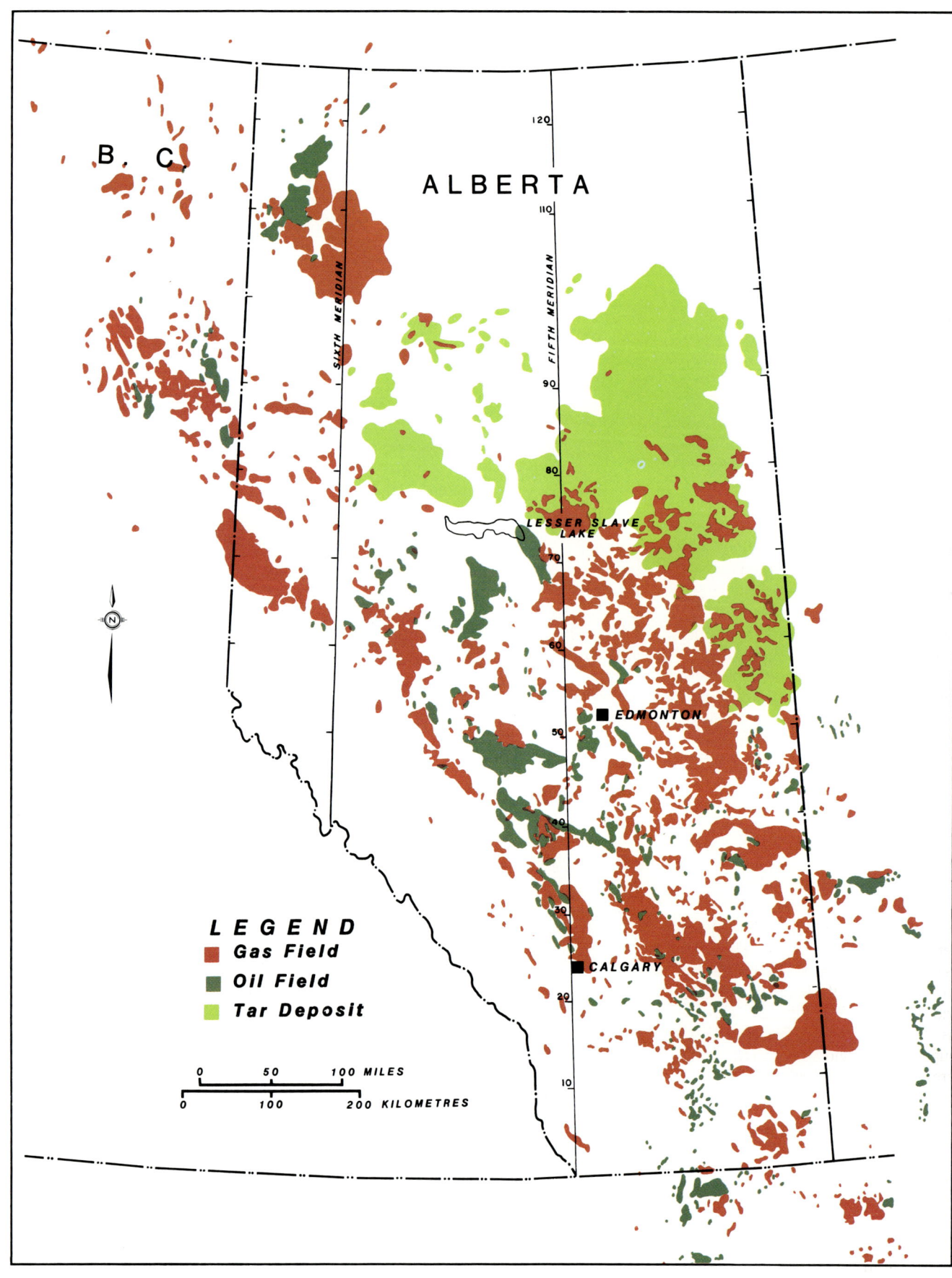

western Alberta and adjacent British Columbia.

Our understanding of the gas system is less precise. At first glance, 10,000 tcf of gas seems to be an overwhelming amount to deal with. But perhaps it is not way out of balance with the oil. Taking all the North American gas and oil fields, the average gas-oil ratio for the total system is 4,500 cu ft per barrel (Nehring, 1982). If we accept that the Lower Cretaceous system contained 5,000 billion barrels of liquid oil that calls proportionally for 22,500 tcf of gas. Of course, the arithmetic is too simple, but we know intuitively that a huge amount of oil would normally be associated with a huge amount of gas. The calculation of 10,000 tcf of gas now has a ring of logic to it. But we still have the problem to dispose of such a huge amount of gas. After all, 10,000 tcf is 100 times the known ultimate recoverable gas reserves of Alberta of 100 tcf (ERCB, 1982). Where did 9,900 tcf go? Even if there are 1,750 tcf stored in the Deep basin (or 3,500 tcf counting the Foothills) that still leaves a huge amount of gas unaccounted for.

Recognizing that a huge volume of gas moved updip, the normal process of differential entrapment (Gussow, 1954) suggests that the great oil accumulations would have been displaced from their traps. It is likely then that by the time (Oligocene) gas was being generated in large volumes, most of the large traps had been pre-empted by oil which was "frozen" in place by degradation. The gas could not displace the viscous oil. Also, at some time in the Tertiary, the Athabasca anticline was breached and the northern outcrop area eroded back to the southwest, behind the probable extension northward of the Athabasca anticline. This opened the basin and allowed a 300-mi (482-km) wide outlet for gas moving updip. With no large traps available, the gas moved to the opening or was lost by diffusion. The model provides us with a lesson in what enormous volumes of hydrocarbons can move through a system and be lost to the atmosphere if traps are not available. We are left with no doubt that had the Alberta basin contained a West Siberia-sized anticline available at the time of gas migration, it would have been filled with a super giant gas field.

◄──────────────

Figure 28. Oil and gas, Western Canada.

The oil and gas geology of Western Canada is usually presented on a map such as Figure 28. The outline of the tar deposits is sketched but in such a way as to draw a clear distinction between them and real oil fields. This is entirely appropriate for most purposes but it does not assist the exploration geologist in making the conceptual link between the two and realizing that they are end members of a continuum, that the tar deposits are not a separate geology entity but are the major element in the oil history of the basin.

Also, the conventional map does not describe our current understanding of the enormous gas resources in the Deep basin, nor does it provide a comparison in size to the United States, which is a familiar standard by which our hydrocarbon volumes may be judged as large or small.

I suggest that the map in Figure 29 provides quite different and more meaningful information to exploration geologists about oil and gas accumulations in Canada and their significance in North America. The tar deposits are, at once, recognized as uniquely large and bespeak the generation and migration of previously unthought of quantities of oil. The area of gas-saturated rock in the Deep basin suggests a gigantic storehouse of gas from which to draw our future commercial reserves. Few geologists can examine this map without reconsidering their views on the Lower Cretaceous as an exploration objective.

To summarize, these studies have provided the following principal conclusions:

(1) The Lower Cretaceous of Alberta (and adjoining stratigraphic sections) is one of the world's richest generators and containers of hydrocarbons.
(2) Some 20,000 billion barrels of oil was probably generated, of which approximately 7,000 billion barrels was expulsed from the shales and entered the reservoir system.
(3) Trapping conditions for the first hydrocarbons in the system were unusually effective because of the Athabasca anticline along the eastern rim. A gigantic stratigraphic trap was also available on the unconformity surface. The resulting accumulation area is 450-mi (725-km) long, up to 200-mi (320-km) wide, and contains nearly 3,000 billion barrels of tar oil. This is the heavy residue of an original accumulation of 5,000 billion barrels of medium gravity oil. An estimated additional 2,000 billion barrels was lost to the outcrop, although this number is subject to great uncertainty.
(4) The main path for the flood of oil and, later, the flood of gas was across the Lesser Slave Lake area straight updip to the northeast. The maximum migration distance was at least 250 mi (400 km).
(5) Proportionally large amounts of gas were generated deeper in the basin from 7,000 ft (2,134 m) of organic shale and coal. Geochemical measurements indicate a total generative capacity of some 10,000 tcf of gas. Most of this hydrocarbon, being a relatively small molecule, was expulsed from the shales and entered the reservoir system mainly by diffusion.
(6) Large volumes of gas, in the range of 1,750 tcf, are contained in the Deep basin in a bottleneck or dynamic trap in gas-saturated tight sands which are slowly leaking off into porous water-saturated sands updip. Continuing gas generation keeps the trap pumped full. The same reservoir sands in the adjoining Foothills belt contain approximately another 1,750 tcf.
(7) Commercially recoverable gas is found primarily in relatively porous beach sands within the Deep basin. Total potential economic reserves (including Elmworth) may be 45 tcf.
(8) About two-thirds of the 10,000 tcf (some part has not yet been generated) passed through the Deep basin and migrated updip toward the eastern shelf. But the large traps were full of tar so the gas sought out the thinner, lenticular sands of the upper part of the section. These provided only about 17 tcf of trap space. Approximately 400 times as much gas escaped at the outcrop to the north, or moved vertically to the surface by diffusion.

The volumes of oil and gas originally generated and put into the reservoir system are so large that they alter some of our basic assumptions about oil and gas provinces. First, I would caution that Alberta alone may not have been so extraordinarily rich in generated hydrocarbons. Perhaps we have not yet measured accurately the amounts of gas and oil originally available in other basins. We may find that those basins with a substantial thickness of

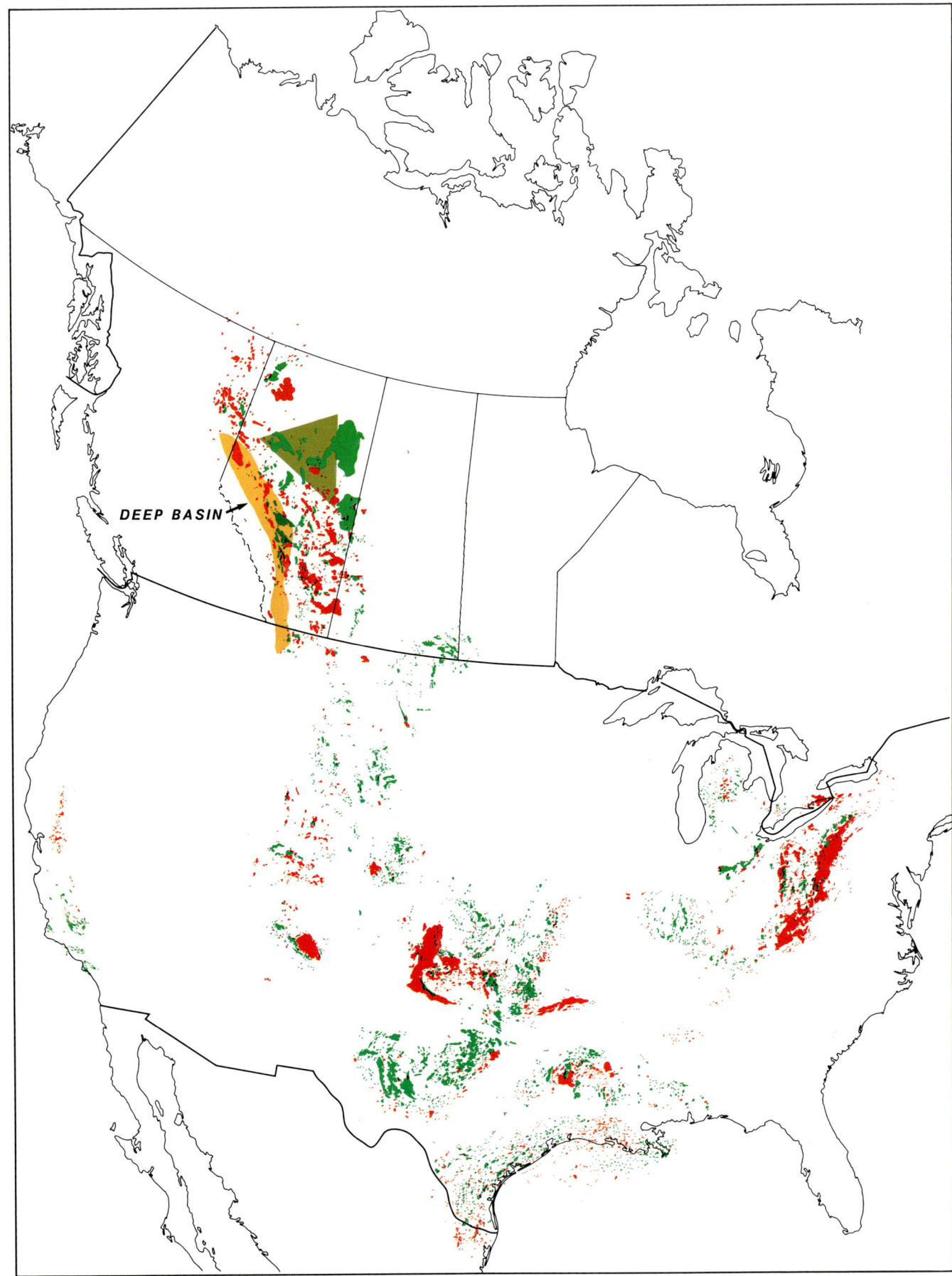

organic-rich rock which reached thermal maturity have generated surprisingly large amounts of hydrocarbons.

Perhaps it is a fruitless exercise to attempt to rank basins for exploration opportunities on their generating potential once a minimum generating capacity has been established. There must be adequate reservoir rocks, of course, but generally the limiting factor is trapping capacity. The whole eastern margin of the Alberta basin was rimmed by a 600-mi (965-km) long anticline. In addition there was a stratigraphic unconformity trap which covered 700 townships. I suggest that the huge oil charge moved updip and was largely captured by these traps. When the great wave of gas followed I conceive that most of the traps were so full of immobilized oil that the bulk of the gas was vented to the atmosphere. Only downdip in a curious inverted trap where water lies over gas was there room for a mammoth gas accumulation.

ACKNOWLEDGEMENTS

Many of the geological staff at Canadian Hunter were involved in the collection of data for this paper. I am particularly indebted to Paul Jackson for his labors in interpreting Lower Cretaceous stratigraphy. The guidance received from Dr. Dietrich Welte of KFA in many discussions about geochemistry was invaluable.

REFERENCES CITED

Barss, D. L., E. W. Bert, and N. Meyers, 1964, Triassic; geological history of Western Canada: Alberta Society of Petroleum Geology, p. 113–136.

Bowman, D. L., 1982, The development of the Deep Basin Natural Gas Project–Western Canada: Eighth Annual Western Region Conference, Engineering Institute of Canada.

Christopher, J. E. and J. E. Knudsen, 1980, Heavy crude oil potential of Saskatchewan, in The future of heavy crude oils and tar sands: New York, McGraw-Hill, Mining Information Service, p. 61–68.

Deroo, G., et al, 1977, The origin and migration of petroleum in the western Canadian sedimentary basin, Alberta–a geochemical and thermal maturation study: Geological Survey of Canada Bulletin, n. 262, 136 p.

ERCB, 1982, Alberta's reserves of crude oil, gas, natural gas liquids, and sulphur at 31 December, 1981: Calgary, Alberta, Energy Resources Conservation Board.

Gies, R. M., 1984, Case history for a major Alberta Deep basin gas trap: the Cadomin formation, in J. A. Masters, ed., Deep basin gas: AAPG Memoir 38, this volume.

Gussow, W. C., 1954, Differential entrapment of oil and gas; a fundamental principle: AAPG Bulletin, v. 38, p. 816–853.

Holmgren, D. A., J. D. Moody, and H. H. Emmerich, 1975, The structural settings for giant oil and gas fields, in Proceedings, Ninth World Petroleum Congress, Volume 2 (exploration and transportation): London, Applied Science Publishers, p. 45–54.

Jardine, D., 1974, Cretaceous oil sands of Western Canada in oil sands–fuel of the future: Canadian Society of Petroleum Geologists, Memoir 3, p. 50–67.

Juntgen, H. and J. Klein, 1975, Genesis of natural gas from carbonaceous sediments: Erdoel Kohle, v. 28, n. 2, p. 65–73.

Leythaeuser, D., R. G. Schaefer, and A. Yukler, 1980, Diffusion of light hydrocarbons through near-surface rocks: Nature, v. 284, p. 522–525.

Masters, J. A., 1979, Deep basin gas trap, Western Canada: AAPG Bulletin, v. 63, p. 152-181.

McCrossan, R. G., and J. W. Porter, 1973, The geology and petroleum potential of the Canadian sedimentary basins–a synthesis, in The future petroleum provinces of Canada–their geology and potential: Canadian Society Petroleum Geologists, Memoir 1, p. 589–720.

Moshier, S. O., and D. W. Waples, 1982, Was the Mannville group the source for Alberta's heavy oils?: AAPG Bulletin (abstract), v. 66, p. 610–611.

National Petroleum Council, 1980, Coal seams, unconventional gas sources: 46 p.

Nehring, R., 1978, Giant oil fields and world resources: Santa Monica, California, Rand Corporation, Report R-2284-CIA, 162 p.

Outtrim, C. P. and R. G. Evans, 1977, Alberta's oil sands reserves and their evaluation, in The oil sands of Canada-Venezuela 1977: Canadian Institute of Mining and Metallurgy, Special Volume 17, p. 36–66.

Rudkin, R. A., 1964, Lower Cretaceous in geological history of Western Canada: Alberta Society of Petroleum Geology, p. 1699.

Springer, G. D., W. D. MacDonald, and M. B. B. Crockford, 1964, Jurassic in geological history of western Canada: Alberta Society Petroleum Geology, p. 137–155.

Tissot, B. P., and D. H. Welte, 1978, Petroleum formation and occurrence: New York, Springer - Verlag, 538 p.

Welte, D. H., and M. A. Yukler, 1981, Petroleum origin and accumulation in basin evolution - a quantitative model: AAPG Bulletin, v. 65, p. 1387–1396.

——, et al, 1984, Gas generation and migration in the Deep basin of western Canada, in J.A. Masters, ed., Deep basin gas: AAPG Memoir 38, this volume.

Wyman, R. E., 1984, Gas resources in Elmworth coal seams, in J.A. Masters, ed., Deep basin gas: AAPG Memoir 38, this volume.

Zhujia-Xiang, 1982, Vitrinite reflectance measurements and geothermal gradients in the Grande Prairie, northwest Alberta Plains, Canada: University of Calgary, Visiting Scholar Report.

Figure 29. Oil and gas, North America.

Gas Generation and Migration in the Deep Basin of Western Canada

D.H. Welte
R.G. Schaefer
W. Stoessinger
M. Radke

*Institute of Petroleum and
Organic Geochemistry
Julich, Federal Republic of Germany*

The Alberta Deep basin, situated along the northeastern front of the Rocky Mountain belt, is the deepest part of the Alberta synclinal sedimentary basin. This trough-shaped deep basin, extending across northwestern Alberta and into northeastern British Columbia, covers an area of 65,000 sq km (25,000 sq mi).

Enormous volumes of natural gas have been found in recent years within the thick, clastic Mesozoic sediments which partly fill the deep basin. These sediments exceed 3,100 m (10,200 ft) in total thickness.

Based on detailed geochemical analyses of more than 300 rock samples (mainly cutting samples) from several wells in the Elmworth gas field, information was obtained on the hydrocarbon source strata and the generation and redistribution of hydrocarbons.

The clastic Mesozoic rock section contains numerous shaly zones which are very rich in organic matter, and also a suite of coal strata. This section, containing mainly Type III kerogen, is the ideal gas generator. Maturity is defined as 0.5 % vitrinite reflectance to about 2.0 % in the deeper part of the section. Maturity has also been defined in terms of the "Methylphenanthrene Index" which is based on aromatic hydrocarbons. Apparently the mature section is still in an active phase of hydrocarbon generation. Due to the tightness of the rock, the hydrocarbon transport mechanism seems to be dominated by diffusion processes. The light hydrocarbon distribution patterns observed throughout the wells suggest a dynamic trapping mechanism. Light hydrocarbons are lost at the top of the mature hydrocarbon-generating zone and are replenished in the middle part of the section where rich source rocks are found.

Based on this concept numerical treatment of gas diffusion with finite element computation is presented for well 6-28-68-13W6M of the Elmworth area. Using subsidence curves, time-temperature relationships, maturity-related methane generation data for source-rock intervals, and effective diffusion coefficients for methane (2×10^{-5} to 10^{-6} sq cm S^{-1}) concentration/depth curves were calculated as a function of geologic time. The results of the simulation for the present-day status compare remarkably well with the hydrocarbon distribution observed in this well today.

INTRODUCTION

The Deep basin in Western Canada is located east of the tectonically disturbed belt of the Rocky Mountains in the provinces of British Columbia and Alberta (Fig. 1). It represents the deepest part of the huge asymmetric Western Canada basin. The Deep basin is approximately 650-km (404-mi) long and reaches a width of about 130 km (81 mi). A schematic south-southwest to north-northeast cross section shows the principal geological features of the basin (Fig. 2). Paleozoic rock, mainly carbonates, rest unconformably on the Precambrian basement which dips gently to the southwest. They are overlain by a thick Mesozoic sequence largely consisting of Cretaceous dark shales with interlayered sandstones and conglomerates. Coal seams are frequent, particularly in the deeper part of the Lower Cretaceous. The thickness of the Mesozoic increases from approximately 300 m (984 ft) in Eastern Alberta to more than 4,000 m (13,123 ft) near the Foothills and the overthrust belt which forms the border of the basin to the west. The strata in general dip southwestward into the Deep basin where both porosity and permeability of the sandstones decrease significantly due to greater compaction, higher clay content, and more intense diagenesis. Gas occurs only in the deepest part of the basin, over an area of approximately 67,000 sq km, where almost the entire Mesozoic section at a depth level exceeding 1,000 m (3,281 ft) below the surface is gas-saturated as derived from resistivity logs (stippled area; Masters, 1979) and geochemical analyses of several wells (Welte et al, 1980; 1981). These gas-bearing zones correspond to the less porous and less permeable rock situated in a downdip position. The same strata in an updip position east of a transition zone, exhibiting higher porosities and permeabilities, are saturated with water. Thus, the situation is the reverse from a conventional gas field, where above the gas an impermeable seal would be expected; instead there are water-saturated reservoir-type strata. Throughout the Mesozoic, in the Triassic, the Jurassic and first of all in Cretaceous rock some 12 pay zones have been encountered with average porosities of 10 % and permeabilities of about 0.5 md (Masters, 1979). The better parts of these pay zones are often conglomeratic and exhibit permeabilities which range from

50 md to several darcys. To produce the gas the wells generally have to be stimulated by hydraulic fracturing techniques. Recoverable gas in the Deep basin may very well be around 50 tcf (50×10^{12} cu ft equivalent to $1,416 \times 10^9$ cu m, STP) or even more (Masters, 1983, personal communication).

This paper presents some results of combined geological and geochemical research that show that the gas occurrences in the Deep basin can best be explained as a dynamic situation between generation of gas from coals and carbonaceous shales on one hand, and losses to the shallower upper layers of the rock section and going updip on the other hand.

GAS GENERATION AND MIGRATION

The Mesozoic rock section in the area of the Elmworth gas field represents, due to its richness in organic carbon and type of organic matter, probably one of the finest source sections for gas anwhere. A profile monitoring the organic carbon content as derived from cuttings analyses of a typical Elmworth well is shown in Fig. 3. Down to 2,400 m (7,874 ft) depth, the organic carbon content averages around 1%. Then a very rich zone follows with values up to 80% which reflects, in essence, the presence of coal seams in this part of the section (2,400 to 3,200 m, or 7,874 to 10,499 ft). From 3,200 m to 3,555 m (10,499 to 11,663 ft) total depth (T.D.) the organic carbon level is around 2%.

The type of the organic matter was determined by ROCK-EVAL pyrolysis. The diagram showing the hydrogen index plotted versus oxygen index (Fig. 4) corresponds to the van-Krevelen diagram. The data of this well plot fairly close to the Type-III curve indicating a hydrogen-poor kerogen which is mainly derived from higher terrestrial plants. Microscopic investigation revealed low amounts of liptinite macerals, but high contents of vitrinite (up to 70%) which is known to be a good gas generator, and inertinite. Thus, the microscopic data agree with the results from ROCK-EVAL pyrolysis. Therefore, it is concluded that this kerogen is not able to generate large amounts of oil, but is a good source of gas if it is in the right stage of maturation.

Figure 5 shows vitrinite reflectance values (% R_m) for the same Elmworth well.

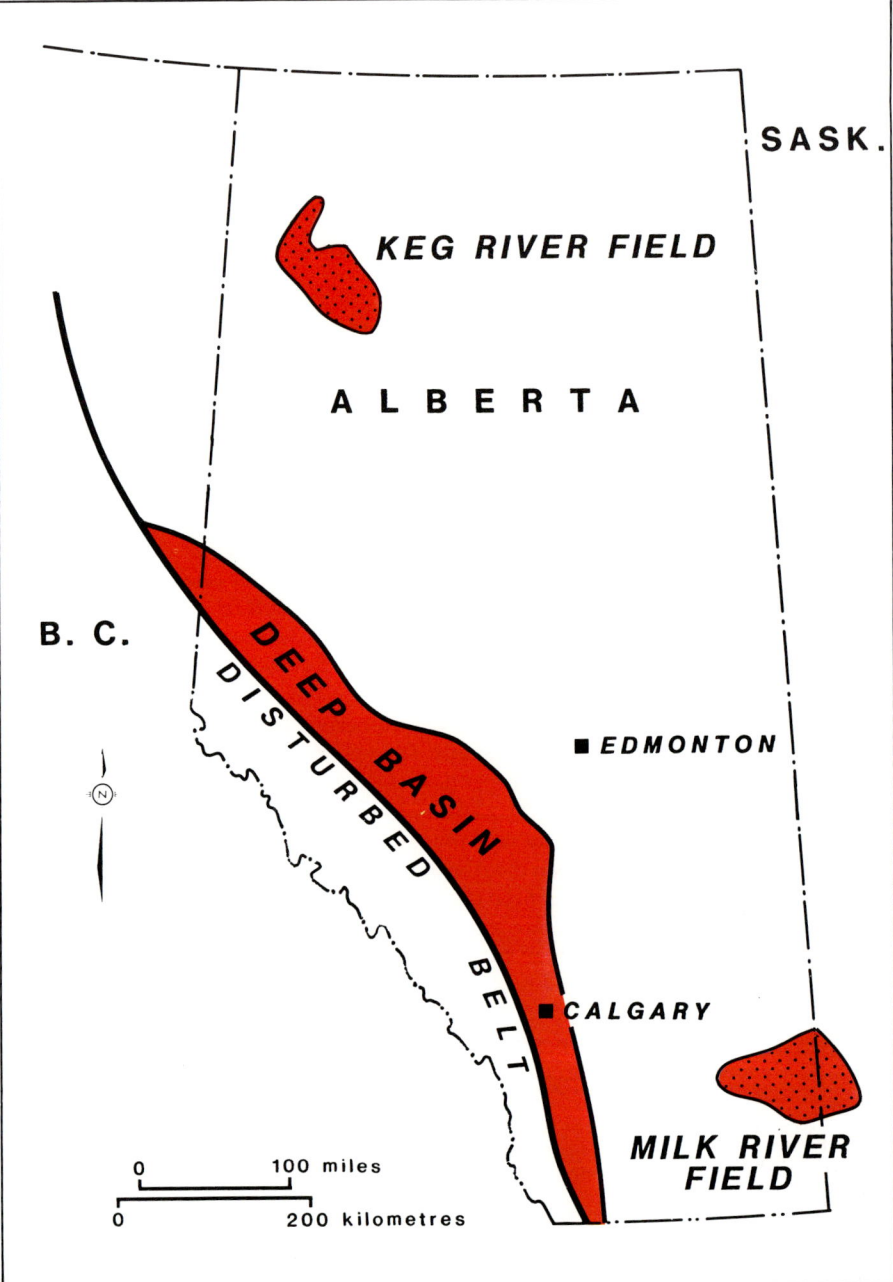

Figure 1. Location map of Western Canada Deep basin.

The maturity of the organic matter is approximately 0.7% mean vitrinite reflectance in the uppermost part of the well and shows only a slight increase down to 1,500 m (4,921 ft) where it reaches 0.8%. At this point there is a progressive change in the reflectance gradient and maturity increases quickly to 2.1% at total depth (T.D.). It is known that Type-III kerogen normally starts to produce liquid hydrocarbons at a maturity level around 0.7% R_m.

Oil generation reaches a maximum around 0.9% R_m and then decreases with further maturation toward the bottom of the "oil window" (at approximately 1.3 to 1.4% R_m). Significant amounts of gas are thought to be generated at maturities exceeding 0.9 to 1.1% R_m. The bottom of the "dry gas window" is not yet reached in this well at T.D.

The organic carbon-normalized concentrations of C_{15+} saturated hydrocarbons

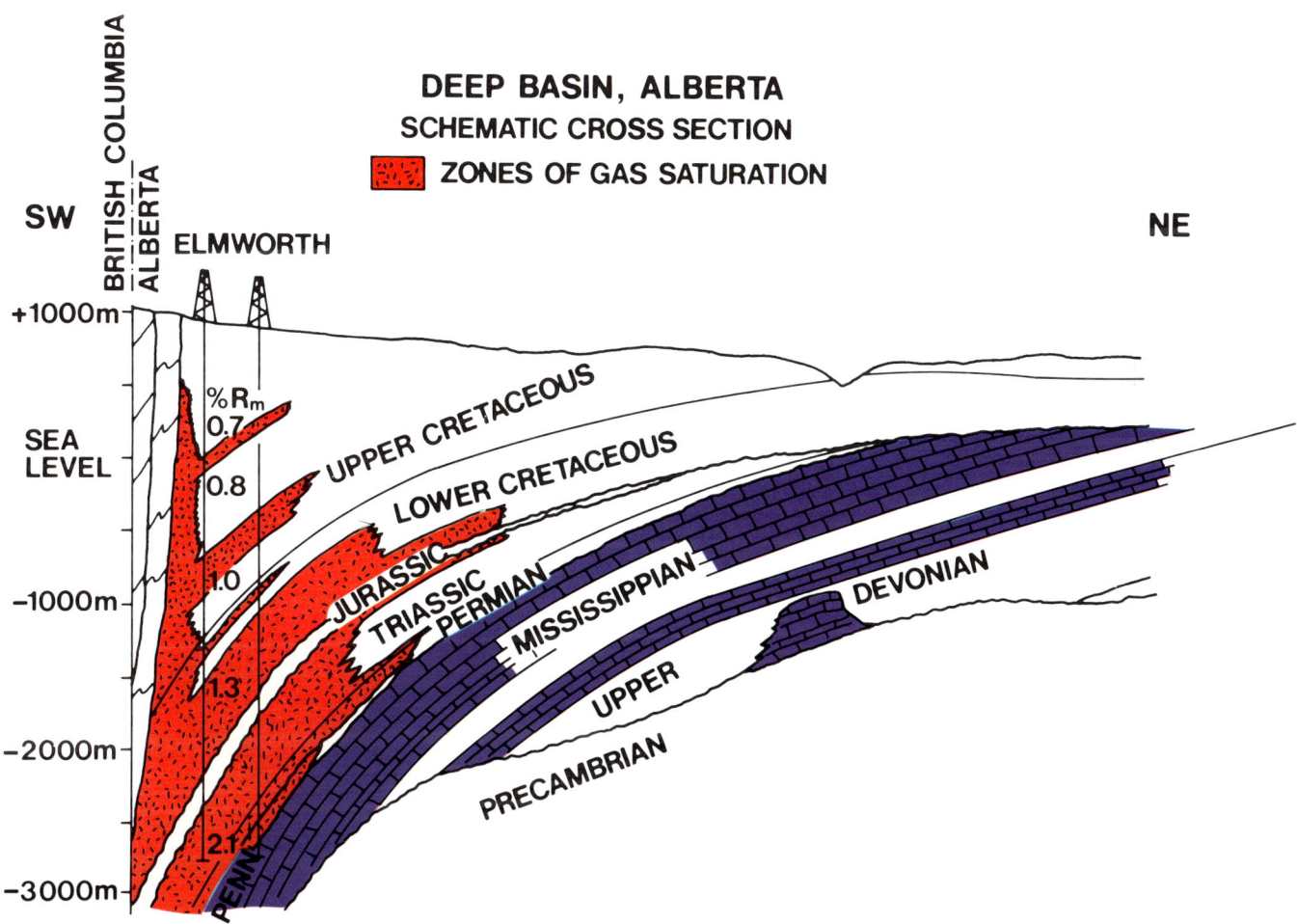

Figure 2. Schematic cross section through Deep basin showing zones of gas saturation.

and C_{15}-C_{35} n-alkanes, as determined for cuttings samples from the above well, when plotted versus depth (Fig. 6) exhibit the typical shape of a generation curve. This indicates that, in general, the C_{15+} hydrocarbons remained at the place where they were generated. There was no major redistribution (that is, migration) in a vertical direction. In this context, it is important to remember that the (liquid) C_{15+} hydrocarbons which were generated during maturation of the source rock at shallower depth are being cracked with further maturation progress and converted to smaller molecules (for example, gas) at greater depth. Therefore, hydrocarbon concentrations of the C_{15+} fraction pass through a maximum as shown in Figure 6. Similar curves exist for both the total extractable aromatics fraction and individual aromatic hydrocarbons. The aqueous solubility of the low-boiling aromatics is larger than that of the corresponding saturates. As these aromatic components also show a typical generation curve, it is suggested that no long-range vertical water flow occurred since the time of intense hydrocarbon generation. Such a water flow which would have blurred the original generation pattern of the low-boiling aromatics is also unlikely because of the very low porosity and permeability of the rocks.

Further evidence against a large vertical migration of the heavier hydrocarbons is derived from the application of the Methylphenanthrene Index (MPI). This chemical parameter permits the definition of the maturity of an oil or a rock extract in terms of calculated vitrinite reflectance (R_c) (Radke and Welte, 1983). In Figure 7 the hydrocarbon internal maturity (R_c) is compared with the mean vitrinite reflectance of the kerogen (% R_m). Obviously, there is an excellent agreement between the two curves. Any invasion of a more mature oil or condensation from greater depths would have resulted in a positive deviation of the R_c values from the vitrinite reflectance curve.

With decreasing carbon number (Fig. 8) the concentration profiles of hydrocarbons in the investigated Elmworth well get broader and eventually lose any similarity to a generation curve. Contrary to a previous figure (Fig. 6) depicting a typical generation curve in the depth profile of C_{15+} hydrocarbons in the shale, the propane concentration profile is rather broad and remarkably constant between 1,000 and 2,400 m (3,281 and 7,874 ft) depth, below which it decreases stepwise (Fig. 9). The propane distribution pattern as presented in this figure should show a definite trend toward lower concentrations going from 2,000 m (6,562 ft) depth upward if it would be influenced by generation rates.

The absolute concentration of total light alkanes (C_2-C_5) reaches a maximum

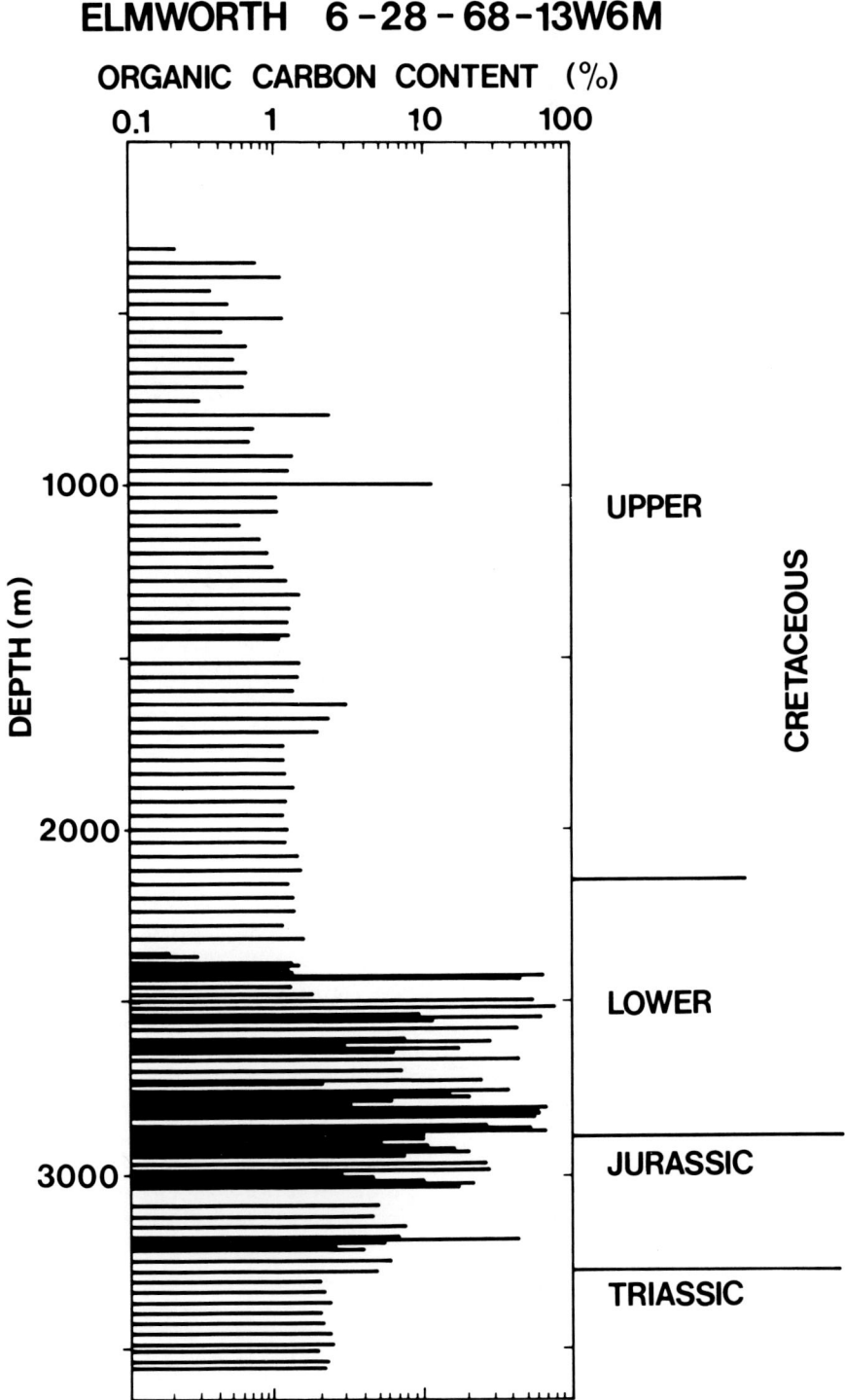

Figure 3. Organic carbon content (note logarithmic scale) plotted against depth for cuttings samples of well 6-28-68-13W6M.

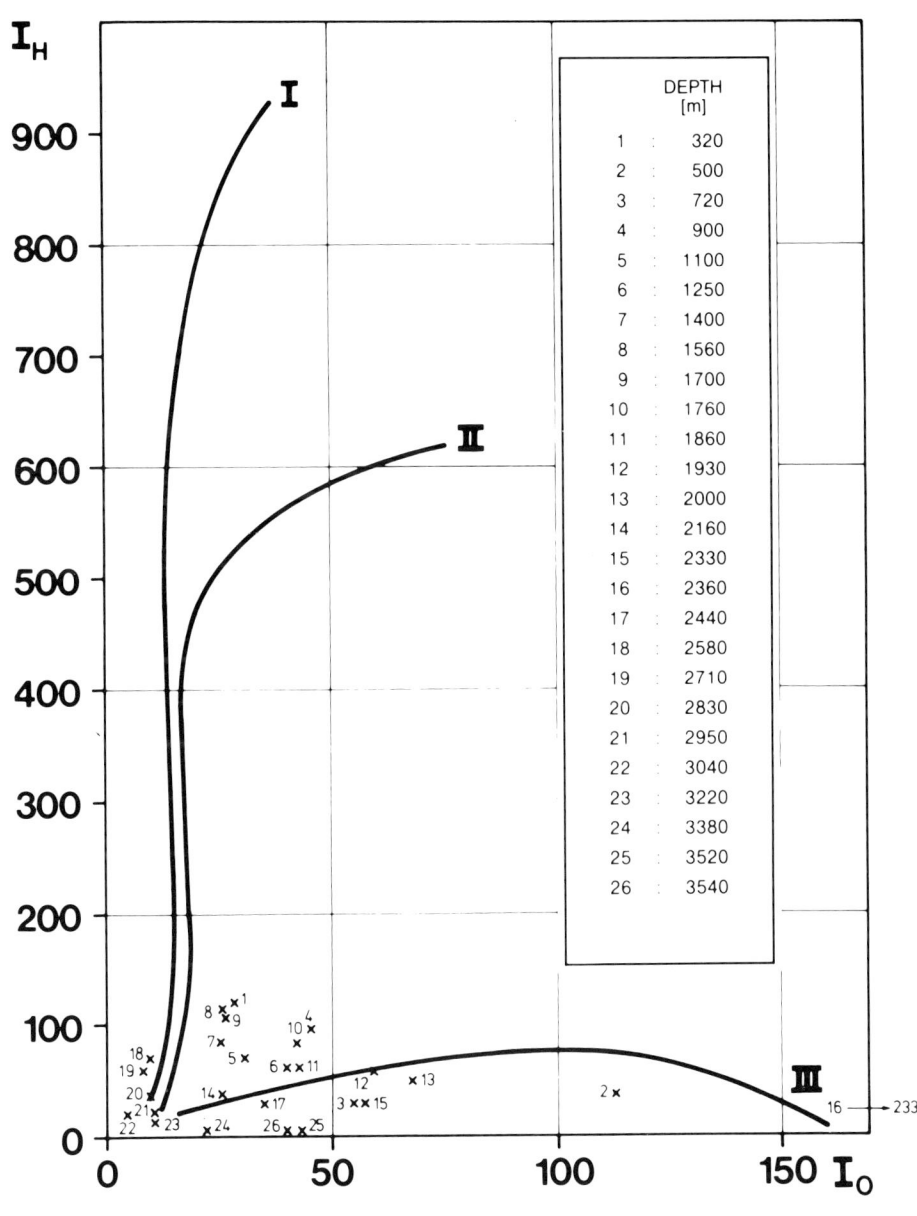

Figure 4. Van-Krevelen-type diagram derived from ROCK-EVAL pyrolysis data (hydrogen index I_H and oxygen index I_o) for selected rock samples of well 6-28-68-13W6M.

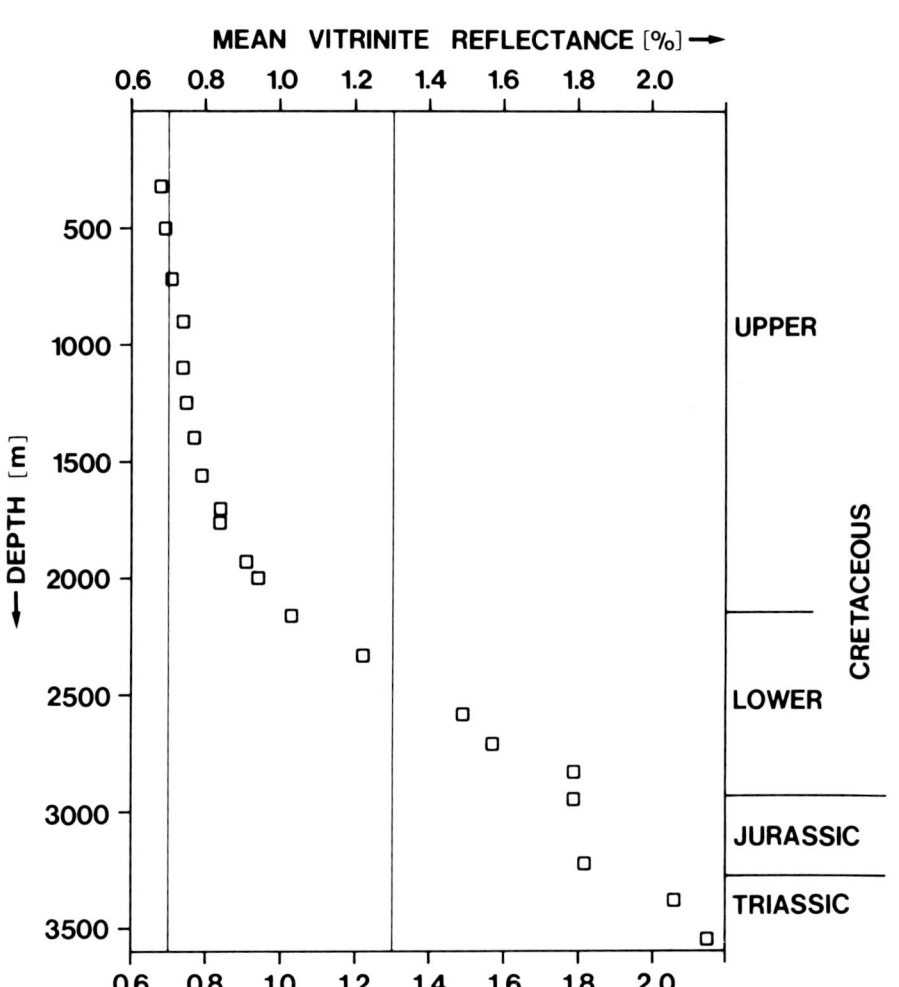

Figure 5. Mean vitrinite reflectance plotted against depth for rock samples of well 6-28-68-13W6M.

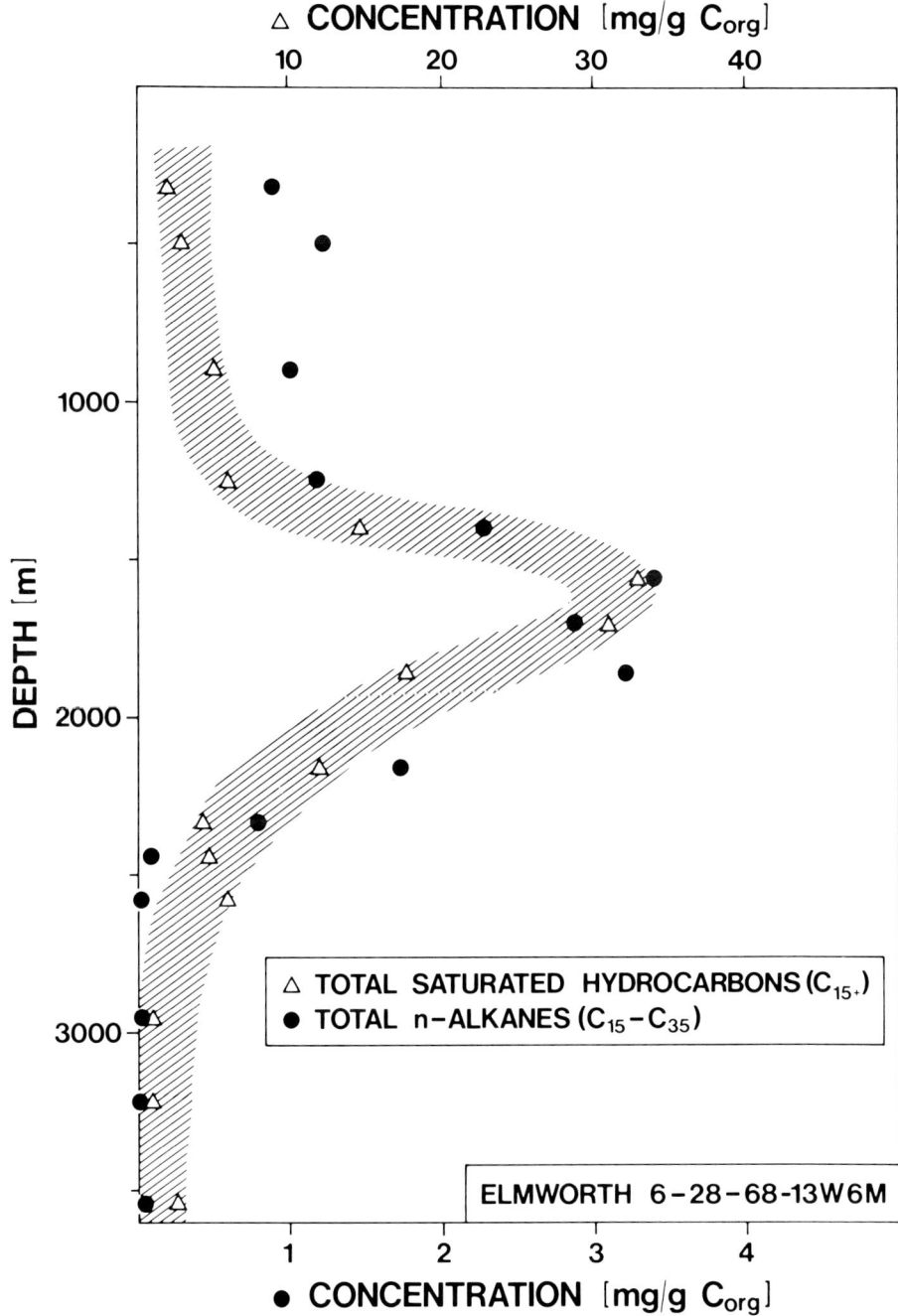

Figure 6. Total saturated hydrocarbon (C_{15+}) and total n-alkane (C_{15}-C_{35}) concentrations plotted against depth for selected rock samples of well 6-28-68-13W6M.

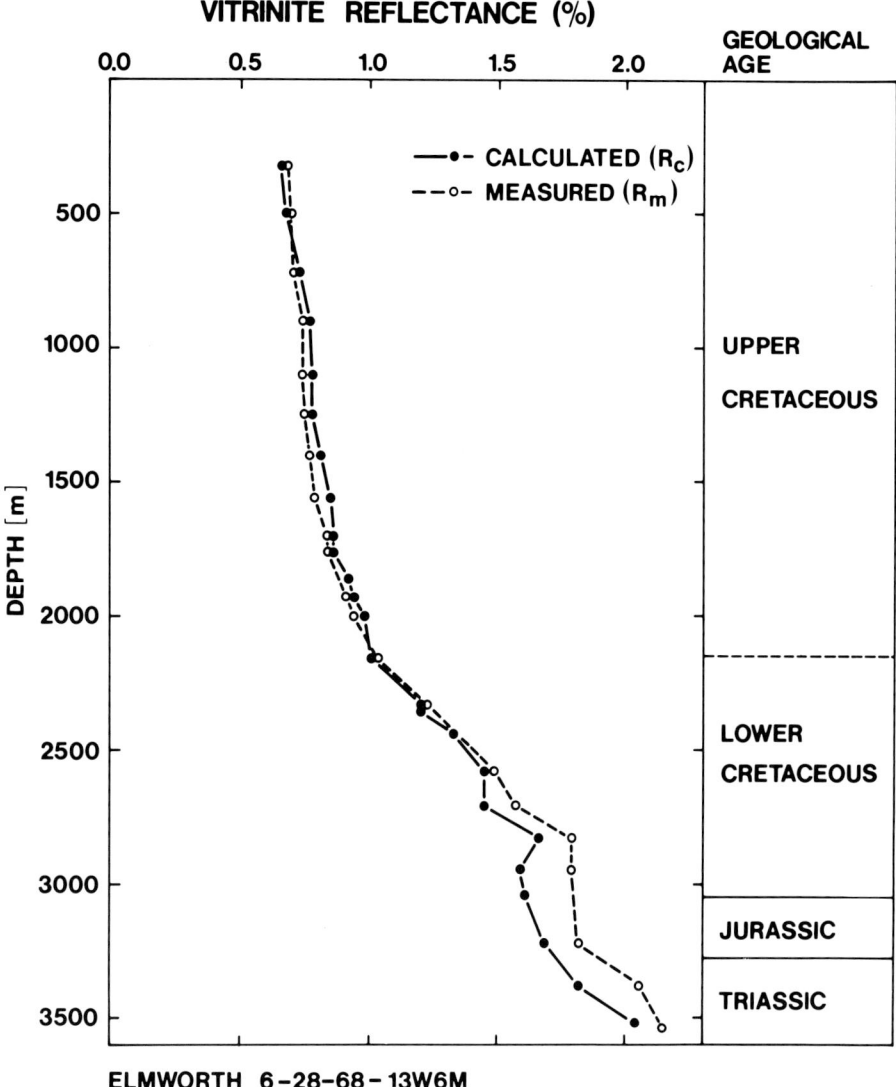

Figure 7. Calculated and measured vitrinite reflectance plotted against depth for selected rock samples of well 6-28-68-13W6M.

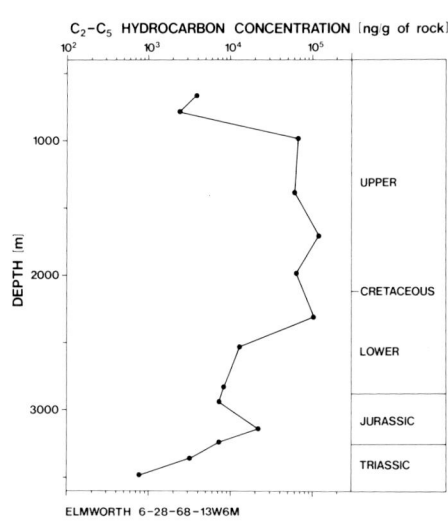

Figure 8. C_2-C_5 hydrocarbon concentration (absolute values obtained by combined thermovaporization/hydrogen stripping) plotted against depth for selected rock samples of well 6-28-68-13W6M.

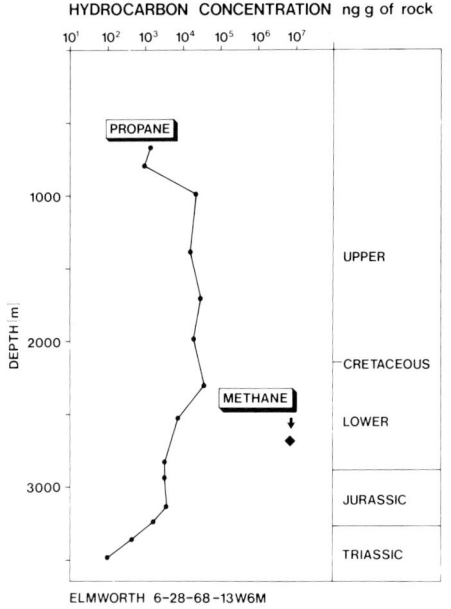

Figure 9. Propane concentration (absolute values obtained by combined thermovaporization/hydrogen stripping) plotted against depth for selected rock samples of well 6-28-68-13W6M. Also shown in this figure is the calculated methane concentration using the results of a pressure core-barrel gas analysis from another Elmworth well at the corresponding maturity level.

of 1.2×10^5 ng/g, or 120 g/metric ton of rock (Fig. 8), which is considered to be a very high value. It becomes even more spectacular when considering the fact that methane is not included. Hence, these geochemical analyses verify the conclusions based on resistivity log interpretations by Masters (1979) concerning the gas saturation. As the bulk of the light hydrocarbons is thought to be generated in maturity levels exceeding 0.9% R_m, corresponding to depth ranges below 2,000 m (6,562 ft), it has to be assumed that a considerable part of the light hydrocarbons (for example, C_1-C_3) has migrated over an appreciable distance into the overlying strata, contrary to the heavy C_{15+} hydrocarbons which apparently remain more or less at the site of their origin.

The concentrations displayed in Figures 8 and 9 were obtained using a combined thermovaporization/hydrogen stripping method developed by Schaefer et al (1978). These data are in good agreement with gas analyses including methane (Fig. 9) of pressurized core-barrel samples taken from another Elmworth well at about 2,600 m (8,530 ft) depth. The measured amount of methane and higher hydrocarbons from the pressurized core-barrel samples makes it possible to extrapolate the absolute methane yield (which cannot be measured from cuttings quantitatively) from the propane concentration at a given maturity. This calculated value is 6.7×10^6 ng/g or about 11 cu m (STP) methane/metric ton of coal, which is only the residual amount of gas present today. Yet, a coal is thought to have produced 8 to 10 times as much methane upon reaching this

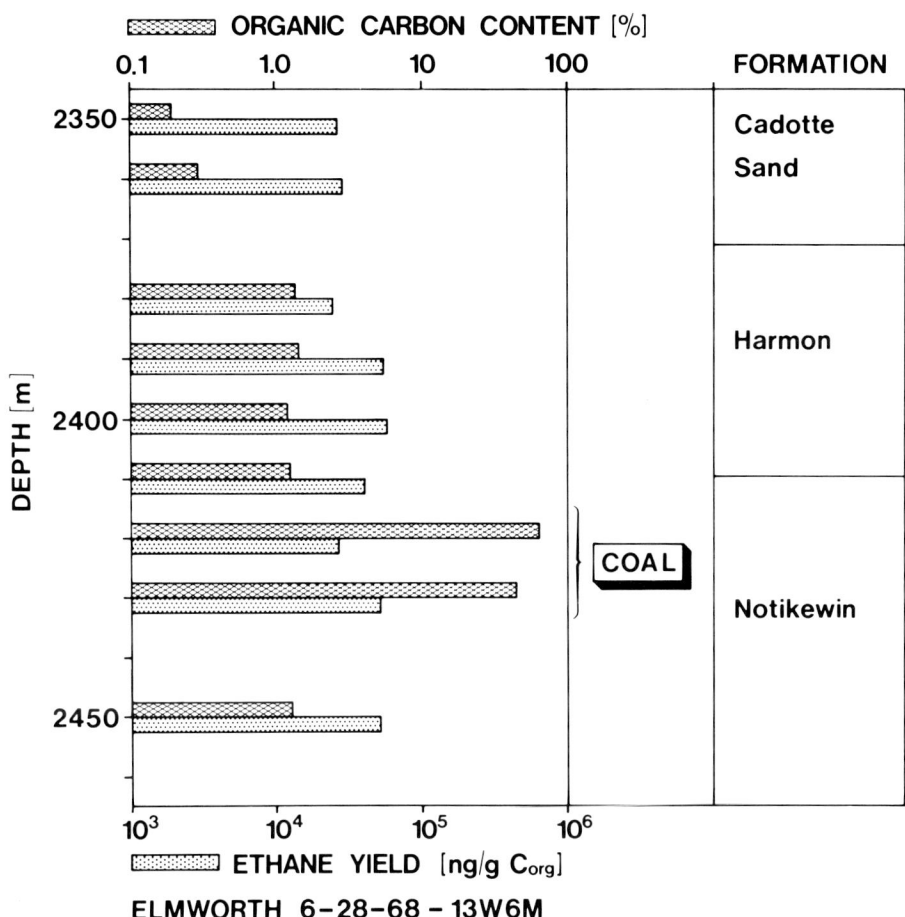

HYDROCARBON YIELD IN SAND/SHALE/COAL SEQUENCE

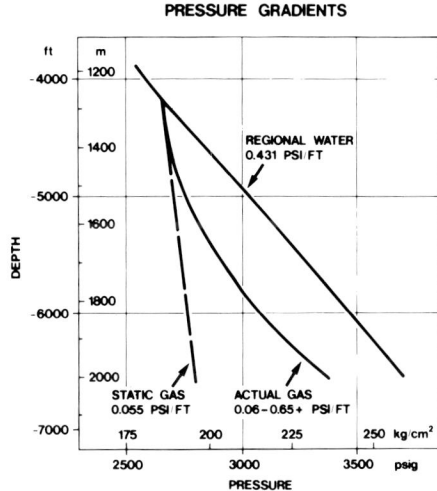

Figure 10. Organic carbon contents and ethane yield (obtained by hydrogen stripping) plotted against depth for a sand/shale/coal sequence of well 6-28-68-13W6M.

Figure 11. Pressure gradients for regional water, "static gas," and actual gas for the Elmworth area (after Gies, 1982).

maturation stage (approximately 1.3% R_m). This total volume generated cannot be stored by the coal itself and, therefore, migrated into adjacent strata.

Thus, the absolute concentration of methane and other gaseous hydrocarbons like ethane, propane, and butane per rock volume and available pore space of the different sedimentary units becomes an important parameter to understand the problem of gas generation and migration. Furthermore, knowing that this gas is generated from coals and Type-III kerogen, the methane concentrations normalized to content of organic carbon of these rocks inform us about the relative importance of the various gas generators (that is, the sources of the gas). Following this line of thought ethane concentrations were determined in detail in coal containing rock sections. Figure 10 may serve as an example where organic carbon content and ethane yields obtained by hydrogen stripping have been analyzed for a coal-containing 100-m (328-ft) interval.

The figure shows a section of the Lower Cretaceous in the deeper part of the well where the Notikewin coal seam is overlain by Harmon shales grading upward into shaly sands which are followed by the Cadotte sandstone. The organic carbon-normalized ethane content remains fairly constant over the whole depth interval. Therefore, a certain volume of coal contains much more ethane than the same volume of shale as it had to be expected. A similar curve has been found for propane. It can be assumed that methane concentrations follow the same trend (that is, that absolute methane concentration is highest in the coal). The fact that the organic carbon-normalized values for ethane are in the same order of magnitude for different rock sections with varying lithology and also for coals is interpreted as an indication that the gas generation process is still active and that concentration gradients for gas from the coals toward neighboring rocks are maintained. The concentration gradients would have been eliminated by diffusion if the coals had ceased to generate an appreciable amount of methane. These concentration gradients must be the driving force for an active gas diffusion going on today.

Hence, the conclusion is that in the Elmworth gas field there is, even at present, a zone of active gas generation mainly in the deeper part of the rock column exceeding 2,000 m (6,562 ft). In this zone, between about 2,000 and 3,500 m (6,562 and 11,483 ft) depth, temperatures range from about 80 to 120°C and maturities from 0.9 to 2.0% R_m.

Based on detailed pressure studies of the Lower Cretaceous Cadomin sandstone reservoir in the Elmworth gas field, Gies (1982) independently arrived at the same conclusion that the gas accumulation is not static equilibrium but is in a dynamic state with ongoing updip gas migration. The pressure gradients for regional water and the actual gas as determined by Gies (1983) are shown in Figure 11. It can be seen that the actual downdip gas pressures are greater than those predicted by a constructed hypothetical static gas pressure gradient based on a continuous gas saturation over a depth range of 800 m (2,625 ft). The conclusion, derived from these pressure data in the Cadomin sandstone

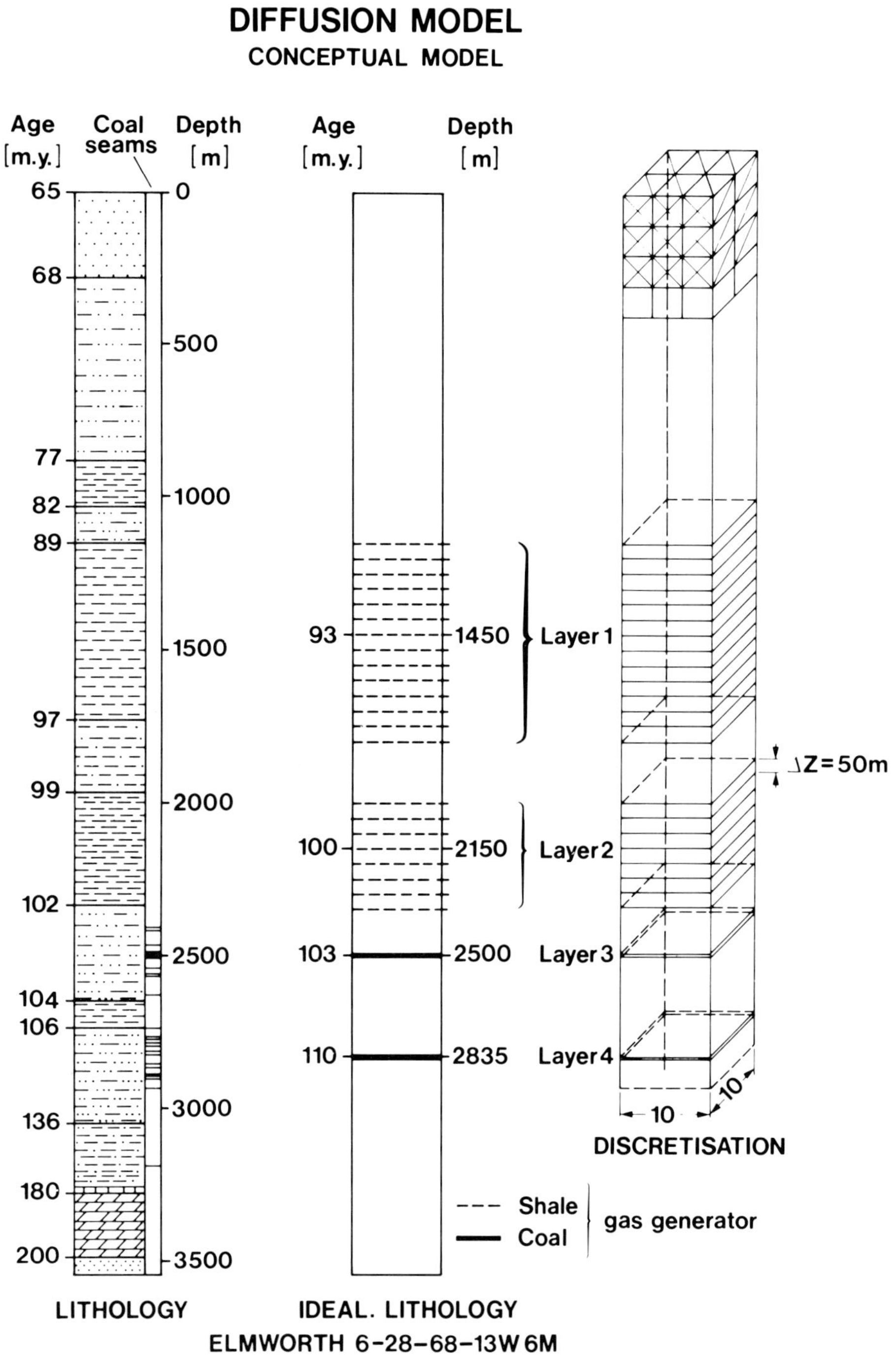

DIFFUSION MODEL
CONCEPTUAL MODEL

Figure 12. Conceptual model for gas generation and diffusion in well 6-28-68-13W6M.

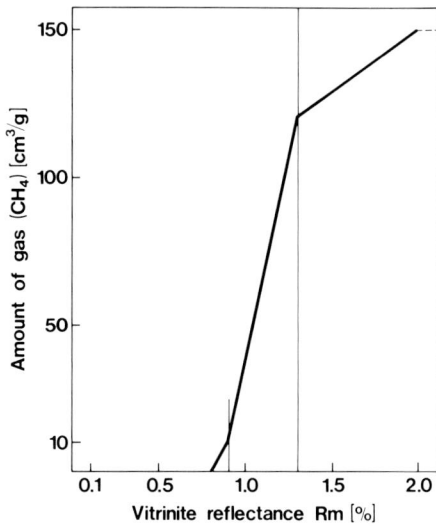

Figure 13. Integral amount of methane generated during coalification (modified after Jüntgen and Klein, 1975).

$$\frac{\partial c}{\partial t} - \frac{\partial}{\partial x_i}\left[(Dd)_{ij} \frac{\partial c}{\partial x_j}\right] = 0$$

where

c = concentration [M/L³]
$(Dd)_{ij}$ = components of the molecular diffusion [L²t]
x_i = components of the coordinates [L]
t = time [t]

The basic system, an idealized sedimentary section, as taken from an Elmworth well and the gas generators in that section, is shown in Figure 12. The left-hand column is a simplified lithologic log of this well, subdivided into individual lithologic units with their respective approximate ages. The column in the middle depicts the idealized system with four gas-generating sources. Layer 1 and layer 2 represent the two shale sequences whereas layer 3 and layer 4 represent the two groups of coal seams interlayered with shaly sands. The thick shales cover a wide maturity range. In order to improve the precision of the gas generation process these shales are subdivided into several "gas generating zones" with a depth interval of 50 m (164 ft). The right-hand column shows the discretized system under

reservoir, is that gas must be migrating upward in response to the pressure drop in an updip direction.

NUMERICAL MODEL OF GAS DIFFUSION

To understand the processes of gas generation and migration in the Elmworth area in a more quantitative manner a numerical model based on finite element discretisation with a moving interface was developed (Stoessinger, 1983). For a first simulation it was assumed that only molecular diffusion occurs within the system. Under these conditions the gas transport for unsteady state can be written in a tensor notation:

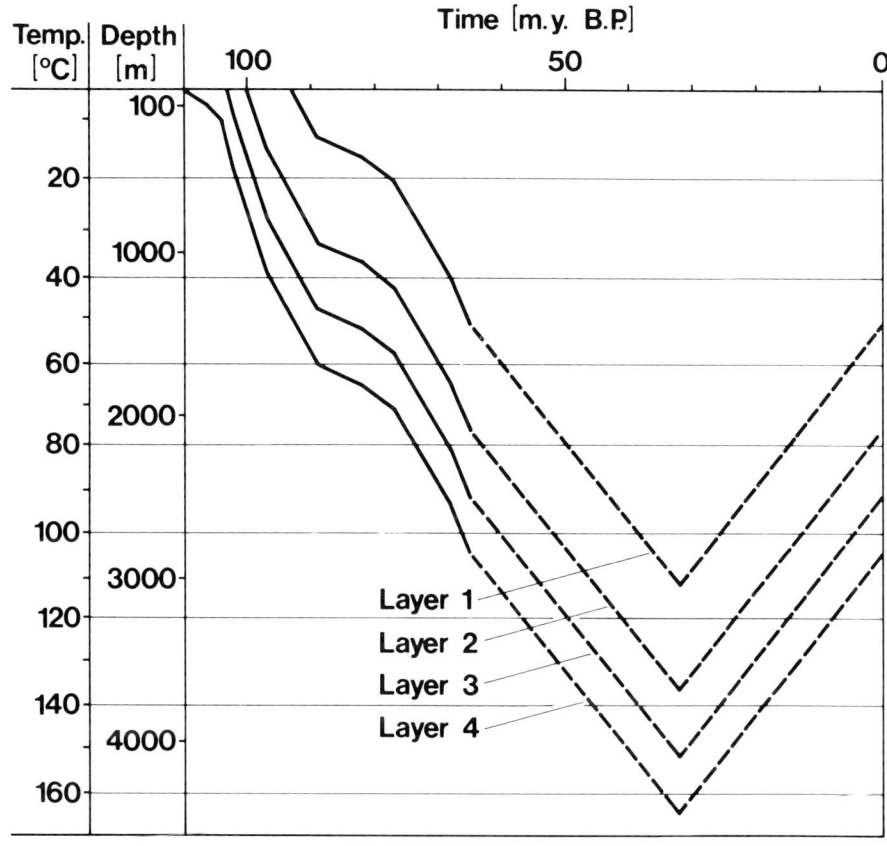

Figure 14. Depositional history for well 6-28-68-13W6M.

consideration. The boundary condition at the top of the system is a zero-concentration surface.

The numerical solution requires the definition of the total methane volume generated until today. The time dependent cumulative methane production can be calculated by using a modified curve from Juntgen and Klein (1975). It shows (Fig. 13) the cumulative amount of methane released from coal as a function of its maturity or degree of coalification. Methane generation begins around 0.8% vitrinite reflectance, then increases strongly between 0.9 and 1.3%; beyond that range it diminishes and levels off around 2.0% in the semianthracite stage.

More precise information on gas generation in natural maturation of organic matter taking into account the kind of source material and the kinetics of the gas generation process is not yet available.

Another prerequisite to calculate methane production is the knowledge of the burial history of the source rocks, which can be transformed into a thermal history by using a certain temperature-depth relationship. A temperature versus depth curve was kindly provided by T. Connolly, Canadian Hunter Exploration Ltd. (1981, personal communication) and was slightly modified (Fig. 14). The evolution of maturity can then be derived using a modified Lopatin (1971) or Waples (1980) approach.

No precise data are available on the thicknesses of the Tertiary and Upper Cretaceous rock which have been eroded since Oligocene (32 m.y.), the time of deepest burial (Hacquebard, 1977). For this reason the maximum burial depth is calculated and illustrated in Figure 14 from the recent mean vitrinite reflectance. In other words, that part of the curial curve corresponding to Tertiary and Quaternary (broken lines) is adjusted in such a way that calculated and measured maturity values (that is, vitrinite reflectance) could be matched.

With the above-mentioned boundary conditions the vitrinite-reflectance and the gas-generation rate as a function of time can be calculated. Both curves are presented in Figure 15. The upper part of the figure shows the evolution of the maturity of each source rock as it has been derived from the thermal history by Waples' relationship. Layer 1 has not yet reached the threshold of gas generation (0.8% R_m).

The time-dependent methane production can be calculated by taking into account the maturation, the thickness of the respective source rock, the organic carbon content and the modified production curve. The lower part of Figure 15 shows the unsteady production curve of the gas sources. The values are normalized to a 1.0 sq m base.

Because little is known about the actual tensor of molecular diffusion D_d and other petrophysical parameters throughout the whole sedimentary column, a sensitivity analysis was made in order to estimate the influence of the molecular diffusion on the concentration distribution. Figure 16 shows the methane concentration versus depth as calculated for present time for different diffusion tensors. The coefficients are assumed constant throughout the entire rock column. It becomes obvious from these curves that the unsteady concentration profiles are strongly influenced by the molecular diffusion. For instance, the bulk of the methane would still be at depths below 1,700 m (5,577 ft) if D is 10^{-7} sq cm s^{-1}. If D is assumed to be 10^{-5} sq cm s^{-1} the gas would have already reached the earth's surface and hence a major portion would have been lost.

Assuming that increasing temperature with mounting depth increases the coefficient of molecular diffusion more than the lower effect of decreasing porosity, a linear variation between $1.0\ 10^{-6} - 2.0\ 10^{-5}$ is considered reasonable. Following this line of thought concentration profiles are computed for different times in order to demonstrate the evolution of the gas concentration as a function of geologic history. The results are plotted in Figure 17. In the first stage, 44 m.y. ago, methane is generated only by the lower group of coal seams (layer 4). Concentration decreases quickly toward shallower depths and at 1,500 m (4,921 ft) above the seams it is virtually zero. The diffusion front has

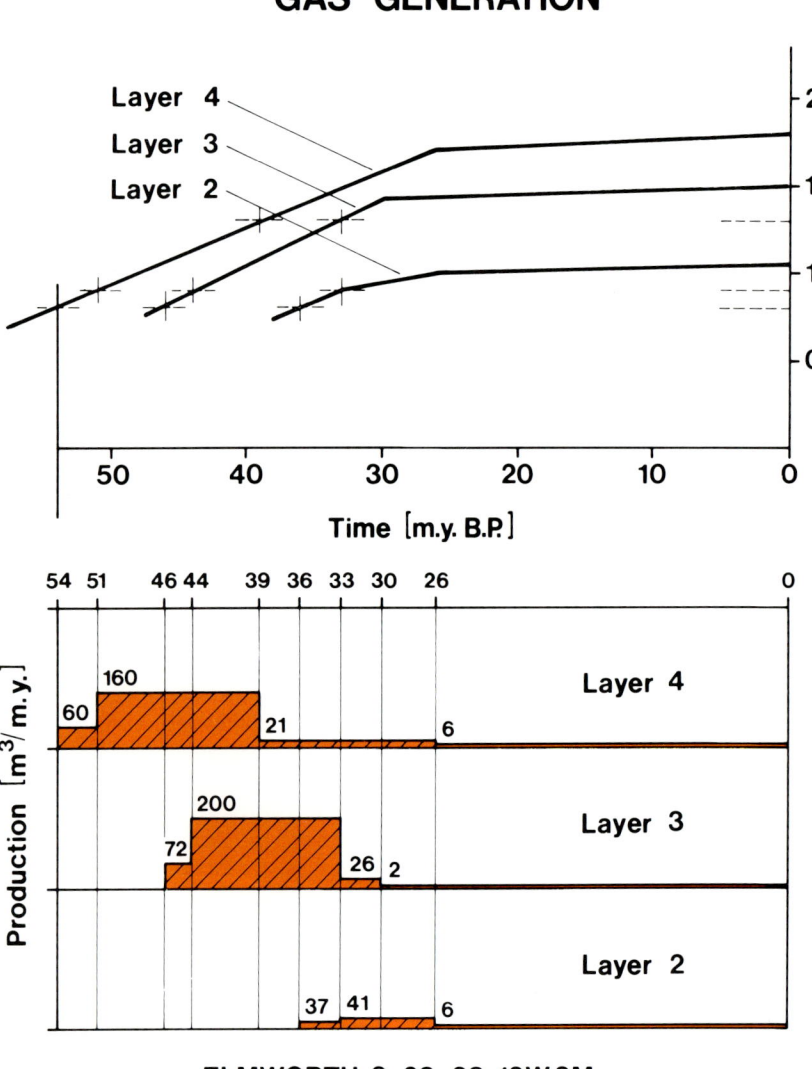

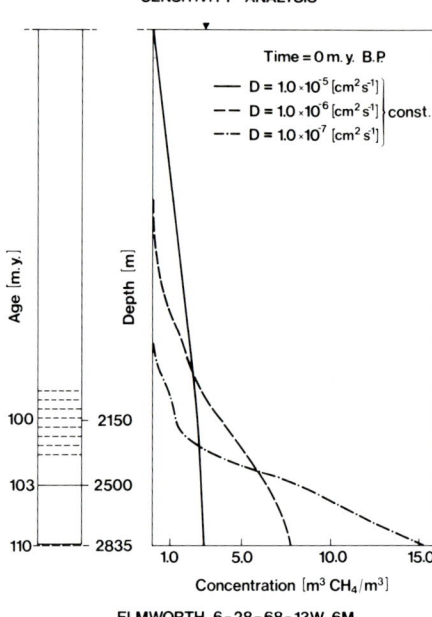

Figure 15. Time-dependent maturation (upper part of figure) and methane generation (lower part of figure) for different layers of well 6-28-68-13W6M (see Fig. 12).

Figure 16. Methane concentration plotted against depth calculated for three different diffusion coefficients in well 6-28-68-13W6M.

(that is, migration of gaseous hydrocarbons) and it can be shown that the main transportation mechanism must be diffusion inside the tight rock section.

Using the above concept, calculations show that present dry gas-saturation profiles in the center part of the Elmworth gas field can be simulated when assuming diffusion coefficients for methane in the range of 1.0×10^{-6} to 2.0×10^{-5} sq cm s^{-1} whereas increasingly higher diffusion coefficients have been adopted with increasing depth and temperature.

When compared with the gas analysis of the pressure core-barrel, the calculated concentration at the appropriate depth (2,500 to 3,000 m, or 8,202 to 9,842 ft) is only a factor of two higher than the actual methane concentration. As with any deterministic model the numerical solution requires accurate data. Considering the variation of all parameters which are difficult to measure in the field, this is a remarkably good result. Obviously the diffusional losses and other losses by conventional buoyancy-driven gas transport mainly toward more porous updip situated strata are compensated by a continuing gas generation from coals and organic rich shaly source rocks in the rock section. Temperatures between 80 and 120°C

migrated upward at 33 m.y., and the curve shows a sharp bend as layer 3 starts to produce large amounts of gas. Seven million years later (26 m.y.) the profile again shows a smooth shape which is caused by both decreasing gas production from layer 3 and beginning of gas generation from the shales (layer 2). Today the methane front has reached the earth's surface according to the model. Although models of this kind are very informative and highly desirable it must not be forgotten that they depend on the assumptions made, as for instance, that only diffusion controls gas migration in our example and other migration processes are negligible.

SUMMARY AND CONCLUSIONS

In summary, the following observations were made in the Elmworth gas field. In the tight, low-porosity center part of the Deep basin nearly the entire Mesozoic rock section is gas saturated. The gas saturation decreases rapidly in the shallower part of the rock column and updip toward the east where more porous and permeable rocks are found. Detailed geochemical analyses show that there is no major redistribution (that is, migration) of heavier hydrocarbons and no flow of water in the tight part of the rock section. However, there is evidence for massive redistribution

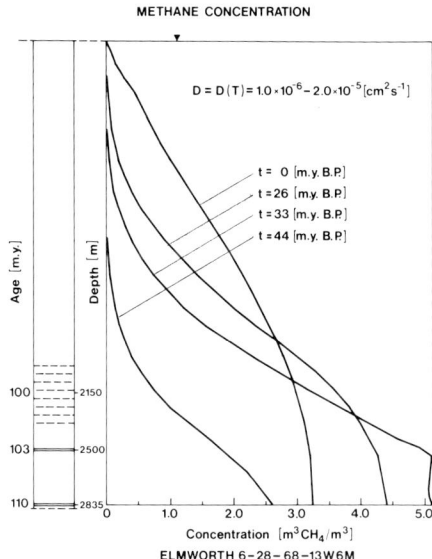

Figure 17. Methane concentration plotted against depth for different time stages assuming a linearly increasing diffusion coefficient with depth for well 6-28-68-13W6M.

seem to be high enough to guarantee an ongoing coalification process and hence gas generation. Balance calculations show the coal measures to be important sources for the gas. Cumulative coal thicknesses of Jurassic to Lower Cretaceous coals in the Elmworth region may range up to 70 m (230 ft) and more (Masters, 1983, personal communication).

All in all the Elmworth gas field is a dynamic situation where gas is continuously being generated in the center part of the Deep basin, and lost toward the surface and the more porous edge. In the inner core of the gas-generating rock column diffusion processes seem to be the predominating mode of transportation.

From the numerical model and its results it can be concluded that given a database similar to the one used in this example, it seems to be feasible in the future to predict gas distributions with much higher accuracy.

ACKNOWLEDGMENTS

We acknowledge the excellent cooperation with Canadian Hunter Exploration Ltd., Calgary, and the stimulating discussions with Mr. John Masters, President of Canadian Hunter Exploration.

We are indebted to Dr. J. Gormly (organic carbon values and ROCK-EVAL data), Dr. P. K. Mukhopadhyay (vitrinite reflectance measurement and maceral analysis), and Mr. H. M. Weiss (subsidence curves, time-temperature calculations), all at KFA-Julich.

Assistance with the experimental work by Mrs. M. Weiner and B. Winden, as well as Messrs. P.W. Benders, U. Disko, W. Laumer, F. Leistner, H. Pooch, F. Schlosser, B. Schmidl, K. Schmitt, H. W. Schnitzler, H. G. Sittardt, R. Weckheuer, and H. Willsch, all at KFA-Julich, is gratefully acknowledged.

REFERENCES

Gies, R.M., 1982, Basic physical principles of conventional and Deep basin gas entrapment (abs.): AAPG Bulletin, v. 66, p. 572.

Hacquebard, P.A., 1977, Rank of coal as an index of organic metamorphism for oil and gas in Alberta, *in* G. Deroo et al, eds., The origin and migration of petroleum in the western Canadian sedimentary basin, Alberta; a geochemical and thermal maturation study: Geological Society of Canada Bulletin, v. 262, p. 11–22.

Juntgen, H. and J. Klein, 1975, Entstehung von erdgas aus kohligen sedimenten: Erdol und Kohle, Erdgas, Petrochemie, Erganzungsband, v. 1, p. 52-69.

Leythaeuser, D., R. G. Schaefer, and A. Yukler, 1980, Diffusion of light hydrocarbons through near-surface rocks: Nature, v. 284, p. 522-525.

Lopatin, N.V., 1971, Temperature and geologic time as factors in coalification: Akademiya Nauk, Uzb. SSSR, Ser. geologicheskaya, Izvestiya, no. 3, p. 95-106. (Translation by N.H. Bostick, Illinois State Geological Survey, February 1972).

Masters, J.A., 1979, Deep basin gas trap western Canada: AAPG Bulletin, v. 63, p. 152–181.

Radke, M. and D. H. Welte, 1983, The Methylphenanthrene Index (MPI); a maturity parameter based on aromatic hydrocarbons, *in* Advances in Organic Geochemistry 1981: Bergen, Proceedings of the 10th International Meeting on Organic Geochemistry, p. 504–512.

Schaefer, R.G., B. Weiner, and D. Leythaeuser, 1978, Determination of sub-nanogram per gram quantities of light hydrocarbons (C_2-C_9) in rock samples by hydrogen stripping in the flow system of a capillary gas chromatograph: Analytical Chemistry, v. 50, p. 1848-1854.

Stoessinger, W., 1983, Numerical treatment of gas diffusion with finite element computation: Internal Report of KFA Julich, unpublished.

Waples, D.W., 1980, Time and temperature in petroleum formation; application of Lopatin's method to petroleum exploration: AAPG Bulletin, v. 64, p. 916–926.

Welte, D.H., et al, 1980, Organic geochemistry of well Canadian Hunter, Elmworth 10-35-71-13W6M, Canada, and its implication for an "unconventional" gas accumulation: Internal Report of KFA-Julich, unpublished.

———, et al, 1981, Organic geochemistry of well Canadian Hunter, Elmworth 6-28-29-13W6M, Western Canada: International Report of KFA-Julich, unpublished.

Paleogeography of the Lower Cretaceous Mannville Group of Western Canada

Paul C. Jackson
*Mannville Oil and Gas Ltd.
Calgary, Alberta, Canada*

Great volumes of oil and gas have been generated, migrated and trapped in the clastic sediments of the Lower Cretaceous Mannville Group (Kootenai Formation equivalent) of the Western Canada basin. Aptian to middle Albian depositional environments have been summarized in a series of paleogeographic maps which show the many potential reservoir sands in their geological setting. The maps illustrate the important phases, or "events," of a general sea-level rise in the basin.

Lower Mannville continental sediments of Aptian age deposited on a mature, erosional surface of Early Cretaceous to Paleozoic beds were inundated by a southward transgression of the Boreal Sea. Major northwest-trending fluvial systems were encroached and ultimately an interior seaway was extended into Montana. In Albian time, a terrigenous influx of Cordilleran origin prograded upper Mannville coastal and continental deposits more than 300 mi (480 km) from southern to northern Alberta. A multitude of shoreline sand trends which span Alberta in an east-west direction were deposited. The Mannville Group was capped by marine shales of the middle Albian Boreal and Gulfian (Skull Creek) seas which joined and inundated the western interior basin of North America.

Mannville sand trends contain an array of hydrocarbon traps including western gas-saturated "Deep basin" reservoirs, conventional stratigraphic oil and gas traps, and structural-stratigraphic traps in the heavy oil region of eastern Alberta and western Saskatchewan.

INTRODUCTION

Industry and academic interests have been focused for many years on the Lower Cretaceous Mannville Group of the Western Canada basin (Figs. 1, 2), but few comprehensive regional maps have been published illustrating facies of the principal "events" of the Mannville. In this paper, a series of Mannville paleogeographic maps, based on a regional study of over six thousand well logs, is presented. This basin-wide mapping allows the illustration of important continental and marine reservoir sand trends.

The initial goal of the study was to map Mannville sequences within the area of the "Deep basin" (Masters, 1979) which is identified on Figure 3, a map of Mannville Group oil and gas fields. The decision to include all of Alberta, northeastern British Columbia, and a portion of western Saskatchewan was based on a desire to understand the deposits involved in the generation, migration, and accumulation of vast Cretaceous gas reserves in the Deep basin (Masters, 1984) and the giant heavy oil fields in eastern Alberta and western Saskatchewan (Fig. 3).

Exploration of the area between the Deep basin and the Heavy Oil Trend has resulted in the discovery of several thousand fields, concentrated primarily in southern and eastern Alberta (Fig. 3). Gas fields generally of small to medium size predominate but nonetheless the total recoverable Mannville gas reserve of this area is in excess of 17 tcf (ERCB, 1981). Considerable potential for both oil and gas remains, particularly in the more lightly explored Mannville sand trends of central and northern Alberta, within and updip of the Deep basin. In the eastern Heavy Oil Trend, application of Mannville environments of deposition and associated reservoir sand parameters will become increasingly necessary to complement engineering-oriented heavy oil exploration.

METHODS OF STUDY

Well logs were studied to determine the principal environments of deposition and lithofacies of significant Mannville sequences. The gamma ray – sonic log combination was most commonly used after calibration by means of core and sample log study. There are, of course, limitations to using log profiles to always correctly identify coarsening-up, progradational marine and lacustrine units and fining-up, channel or transgressive marine units. Statistically, however, the method will provide reasonable definition of regional marine to continental relationships provided enough well control is utilized. In the Deep basin area, nearly all wells were studied, elsewhere an average of one well per township was used.

To identify regionally important Mannville sequences, a basin-spanning grid of cross sections of gamma ray – sonic logs was constructed (Fig. 4). These stratigraphic sections were hung on the post-Mannville Base of Fish Scales Zone, an excellent basin-wide lithologic marker. They cover all of the Lower Cretaceous and several hundred feet into the pre-Mannville section. Major transgressive and regressive successions were identified before a finer subdivision was attempted using more local, transgressive marine

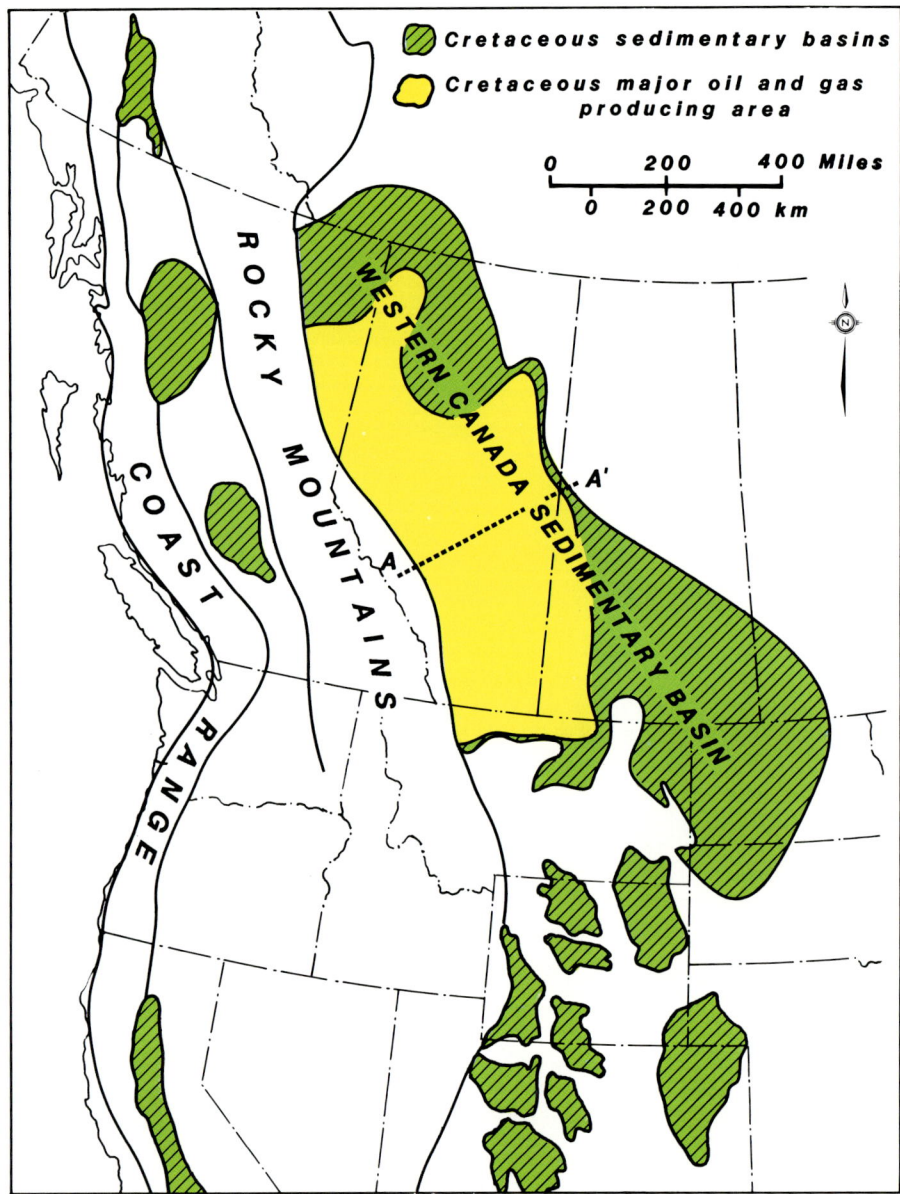

Figure 1. Sedimentary basins of western North America.

shales and extensive coastal plain coal beds.

Each paleogeographic map summarizes the principal continental, coastal, and offshore facies of a significant, regionally identifiable stage of Mannville deposition. Depositional environments of Mannville transgressive and regressive systems varied with time, therefore only the dominant paleoenvironment interpreted for each log interval was mapped in order to illustrate "events." Land-sea relationships are clarified by showing the maximum onlap position of transgressions and the maximum seaward position of coastal progradations at each stage of deposition.

REGIONAL GEOLOGICAL SETTING

The basin developed as a southwest-dipping asymmetrical trough (Fig. 5) in Early Cretaceous time, as an overthrusted, early Cordillera loaded the western margin of this foreland basin (Eisbacher et al, 1974) to create the Deep basin (foredeep) area of maximum subsidence. The pre-Mannville surface, the basin floor, comprised Devonian, Mississippian and lower Mesozoic formations (Fig. 6) and was a mature, erosional surface of subdued, cuesta-like relief formed prior to a Late Jurassic - Early Cretaceous orogeny and early Cordilleran uplift. The rising Cordilleran Belt to the west was a major source of Early Cretaceous terrigenous clastics and pulses of sediment into the subsiding basin reflected renewed periods of structural activity. The Mannville Group sequence was deposited in a lower trans-

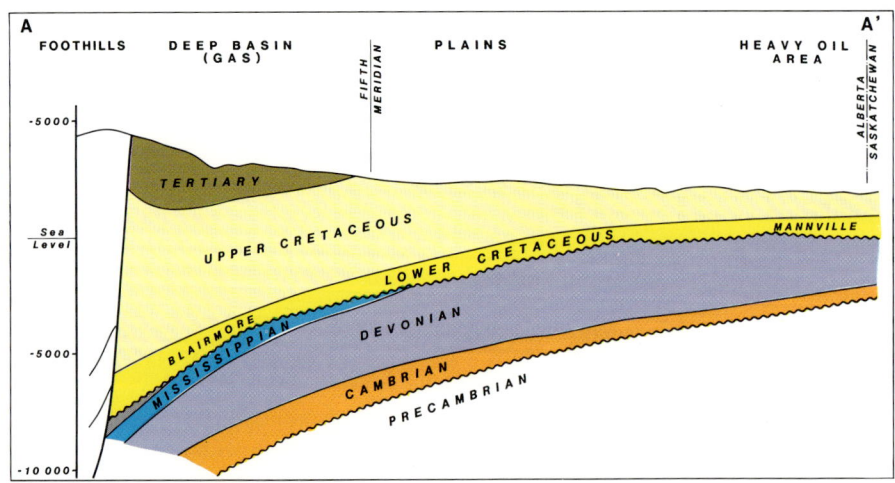

Figure 2. Regional basin cross section A-A', showing the Mannville Group as the lower part of a westward-thickening Cretaceous section.

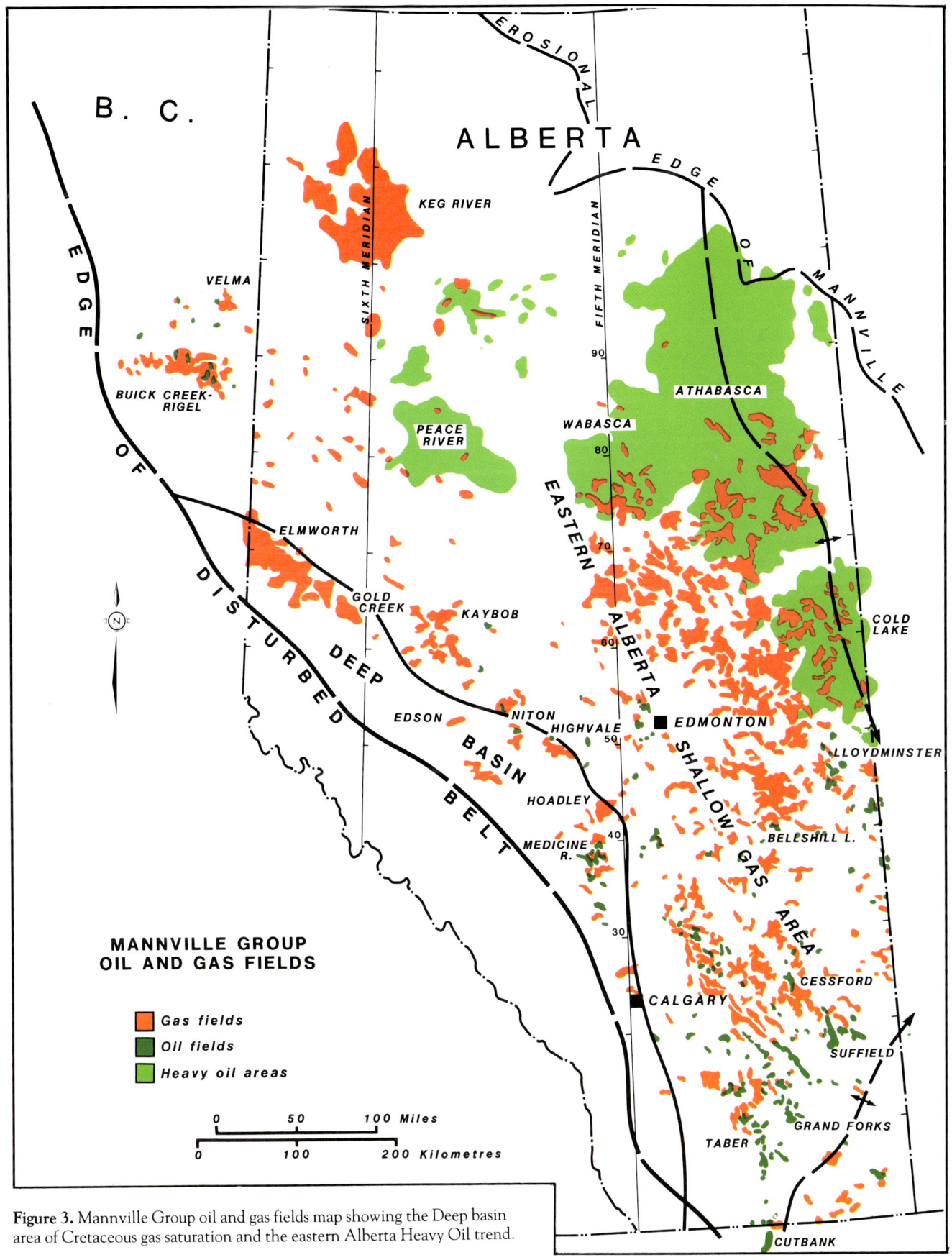

Figure 3. Mannville Group oil and gas fields map showing the Deep basin area of Cretaceous gas saturation and the eastern Alberta Heavy Oil trend.

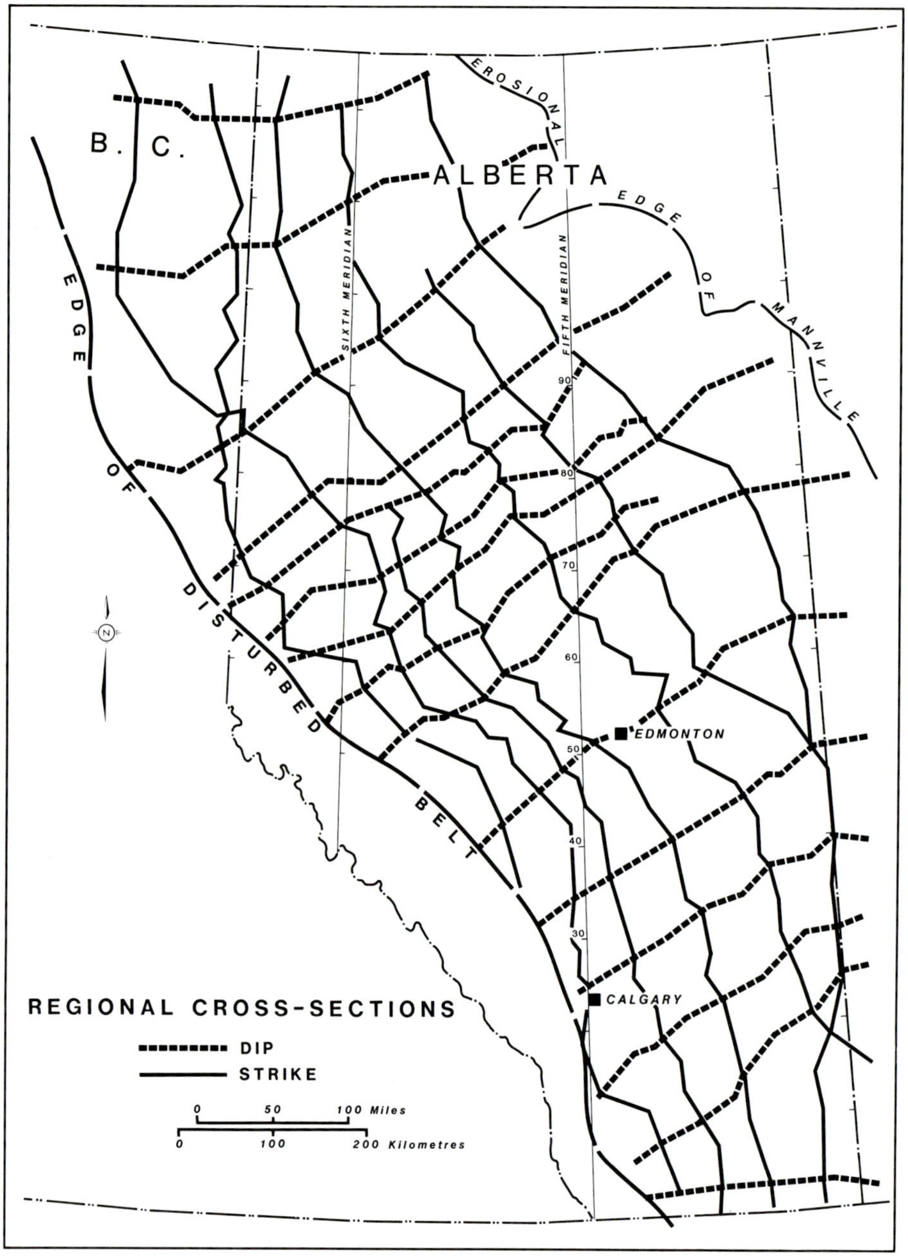

Figure 4. Regional grid of gamma ray-sonic log cross sections.

gressive phase and then, as a major influx of Cordilleran sediment entered the basin, in an upper regressive phase. Subsidence of the basin continued through the Tertiary Laramide orogeny to produce the present-day structural attitude as shown in Figure 7, a structure map of the erosional surface underlying the Mannville Group. Generally, the rate of subsidence increased downdip to the southwest, but variations which occurred in some areas caused distinct "hingelines" (Fig. 8). These variable rates of subsidence, as well as some shifts of depocenters, were critical to Mannville Group sedimentation.

Structural Influences on Sedimentation

Structural changes, which altered basin morphology and influenced Mannville sedimentation (Figs. 9a, b), were determined from analysis of sequential basin fill patterns shown by the lower Mannville and total Mannville Group isopach maps (Figs. 10, 11) and from the regional cross sections (Fig. 4).

Figure 9a illustrates the structural features which significantly influenced lower Mannville sedimentation. They were the development of the foredeep trough with a distinct hingeline on its eastern flank, an axial high region and an eastern regional low. Lower Mannville section thickness varies from in excess of 500 ft (150 m) in the foredeep; to zero in areas of nondeposition along the axial high trend; to over 300 ft (90 m) in the low along the Alberta-Saskatchewan border (Fig. 10).

The axial high (Figs. 7, 8, 9a), which extended from northeastern British Columbia to the Swift Current platform (Christopher, 1974) in southwestern Saskatchewan, was created by development of

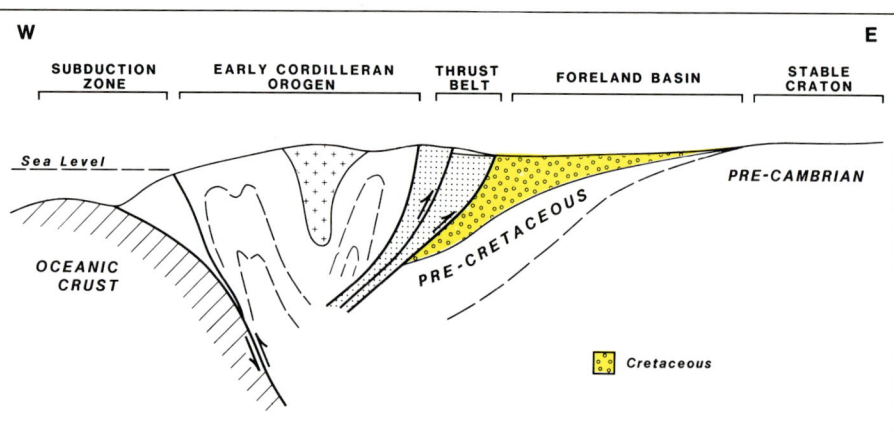

Figure 5. Schematic cross section, illustrating the relationship of the uplifted early Cordillera to the Foreland basin during the Cretaceous (after Weimer, 1983; Eisbacher et al, 1974).

the foredeep trough and an eastern regional low caused by subsurface leaching of Devonian salt. In effect, two subbasins for lower Mannville deposition were formed.

Paleotopography of the pre-Mannville surface, including the axial high, is generally considered to have been the result of early Mannville erosion of variably resistant lithologies of pre-Mannville formations. However, it is noted that many of the topographic features trend obliquely to pre-Mannville formations (Fig. 6). Erosional patterns, although significantly influenced by lithology of the pre-Mannville formations, are suggested here to have been primarily determined by structural developments with the axial high, for example, having directed early Mannville drainage to adjacent structural lows. Paleotopography of the pre-Mannville surface was locally enhanced by incision of broad valleys (Fig. 6) up to several hundred feet in depth.

The axial high played a major role in a marine phase of lower Mannville sedimentation, by separating the transgressing Boreal Sea, which encroached in a southeasterly direction in late Aptian time into two major lobes. An onlapping marine sand and shale sequence was extended southward into Montana by inundation of fluvial and lacustrine environments of the foredeep low trend by the western lobe.

During Albian time, and deposition of the upper Mannville, rapid subsidence of the Peace River area expanded a broad trough to the northeast allowing development of the Clearwater Formation marine shale basin (Figs. 9b, 11). This structural low, with a prominent hingeline on its southern flank, halted the northward progradation of upper Mannville coastal and continental sediments. Formation of this new subbasin was indicative of renewed subsidence of an area which had been an arch during the Early Paleozoic, and had commenced subsiding in late Devonian time.

Continued influence of the axial high on early phases of upper Mannville sedimentation is evident from minor section thinning which may have been due to compaction and drape over relief on the underlying Paleozoic surface or further subsidence of adjacent areas. Salt collapse of the eastern area (Fig. 9b) probably continued during deposition of the upper Mannville as section thickens in that direction (Fig. 11).

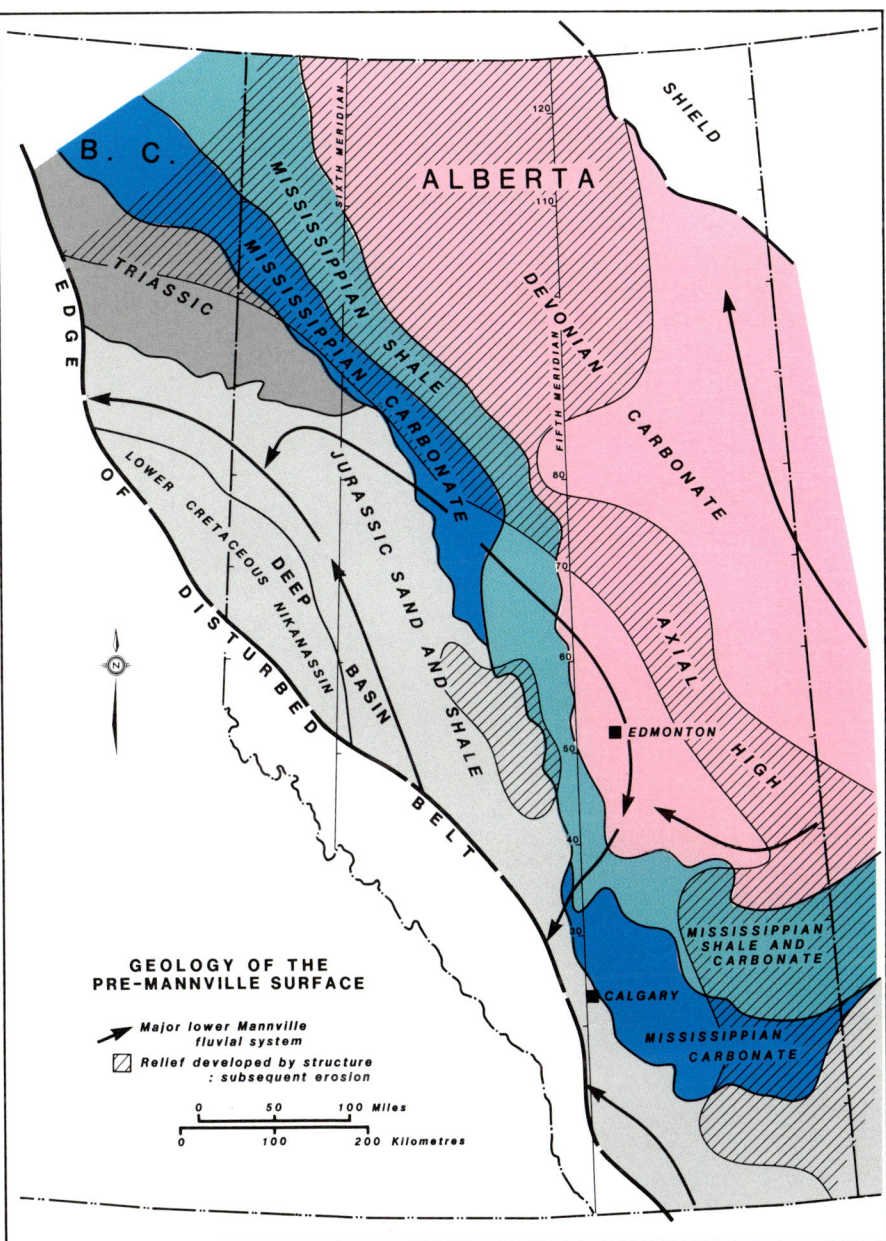

Figure 6. Geology of the pre-Mannville surface; routes of major lower Mannville drainage systems on each side of the axial high trend are shown.

Mannville sedimentation in late Albian time ended with a cessation of the clastic supply, continued subsidence of the Clearwater shale depocenter, and creation of a new depocenter in southern Alberta. Marine transgressions were able to encroach the Mannville surface as illustrated by Figure 12, an isopach of the post-Mannville lower Colorado Group. Continued subsidence of the Clearwater subbasin allowed the Boreal Sea to transgress southward and cap the Mannville Group with Harmon Member marine shale. A younger northward transgression of the Gulfian (Skull Creek) Sea resulted in deposition of Joli Fou Formation marine shale over most of the remaining continental Mannville surface of Alberta and the creation of the late Albian western interior seaway.

MANNVILLE GROUP STRATIGRAPHY

Correlations and principal facies of the Lower Cretaceous, and most specifically of the Mannville Group, are summarized in Figure 13, while the areas referred to in the stratigraphic columns are shown in Figure 14. The stratigraphic chart is adapted from the work of numerous authors (Glaister, 1959; Rudkin, 1964; Mellon, 1967; Stott, 1967; McLean and Wall, 1981; and others).

Continental, shoreline, and offshore marine facies relationships of the major transgressive-regressive cycle forming the Mannville Group are depicted by three diagrammatic cross sections (Fig. 15). Sections N^1-S^1 and N^2-S^2 follow basin strike along the western Deep basin and the eastern Heavy Oil Trend respectively. These are intersected by basin-dip section SW-NE as shown on the cross section location map. All sections are hung on the Base of Fish Scales marker, a good regional lithologic unit at the top of the Lower Cretaceous, and the entire Mannville Group stratigraphic section and pre-Mannville erosional surface are presented. Depositional environments of the major sequences are color indexed in the same manner as the correlation chart (Fig. 13) and the paleogeographic maps to follow.

To present a finer breakdown of the cycles composing the Mannville transgressive-regressive major cycle, two more-detailed well log sections, A-A' and B-B' (Figs. 17, 18), and an index map (Fig. 16), are provided.

As illustrated by the cross sections, the significant regional events or phases of Mannville deposition during a period of rising sea-level in the basin were as follows:

1. An early Mannville continental phase in which coarse clastics were shed eastward from the rising Cordillera to form a piedmont alluvial plain (Stott, 1973) along the western foredeep. Northwest trending drainage systems developed basin wide on the pre-Mannville erosional surface and extensive fluvial deposits accumulated. The Boreal Sea (Jeletzky, 1971), containing marine shale, reached northeastern British Columbia and possibly the northeast corner of Alberta.

2. A continuation of the lower Mannville transgression of the Boreal Sea from the northwest, in which fluvial and lacustrine systems were inundated and onlapping sequences of marine sand and shale were deposited southward into Montana. Eastward-thinning wedges of fluvial clastics derived from the rising Cordillera spread over large areas of the western basin margin.

3. A middle Mannville continuation of the transgression over the northern half of Alberta, as a progradation infilled the shallow marine basin in southern Alberta. This is the turn-around or transition from the transgressive lower part to the regressive upper part of the Mannville sequence. The Bluesky, Wabiskaw,

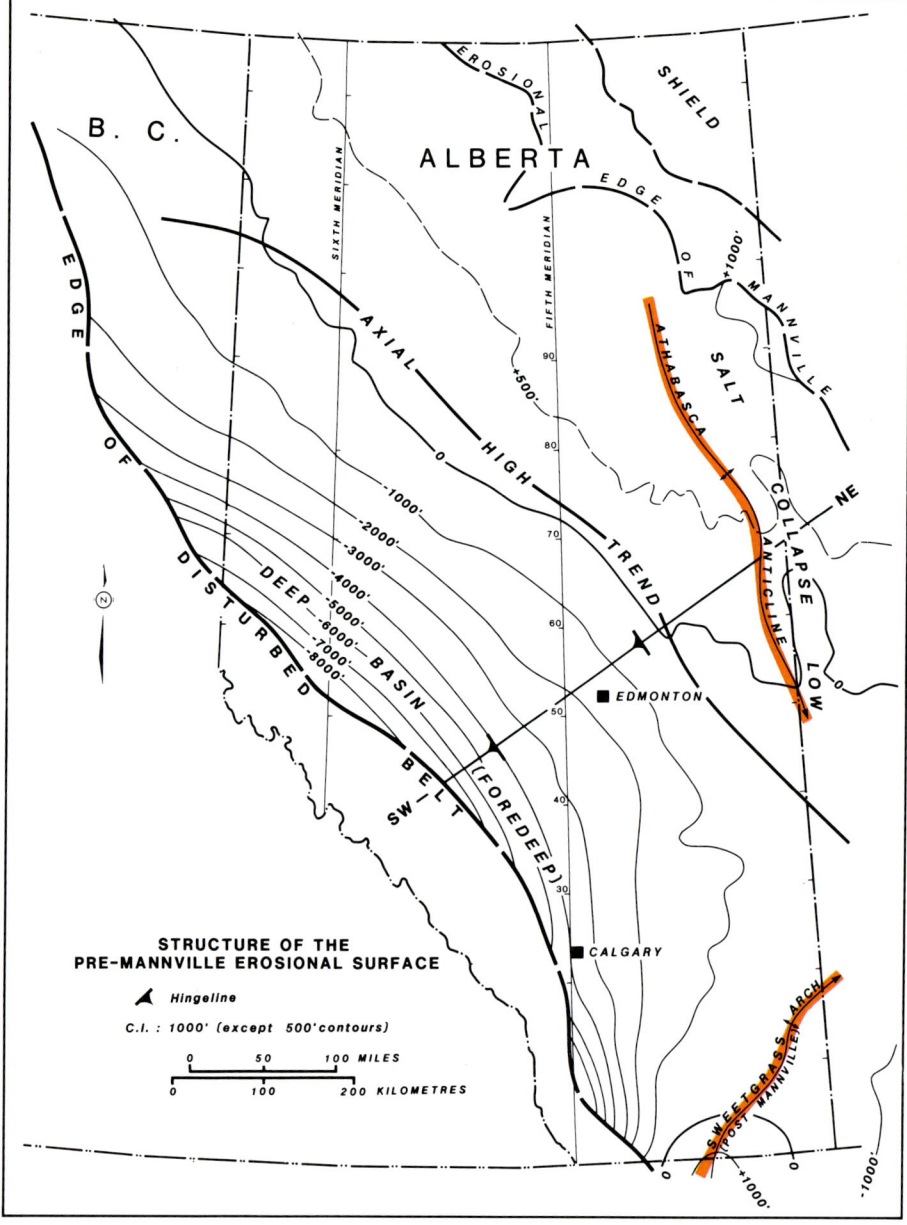

Figure 7. Structure map of the pre-Mannville surface.

and Cummings formations and the Glauconite informal member are traditionally included in the upper Mannville on the basis of previous lithologic distinctions (Rudkin, 1964). They are informally designated middle Mannville herein to more accurately portray their transitional position in the major transgressive-regressive Mannville cycle.

4. An upper Mannville regression, in which shorelines were prograded, in a series of pulses, more than 300 mi

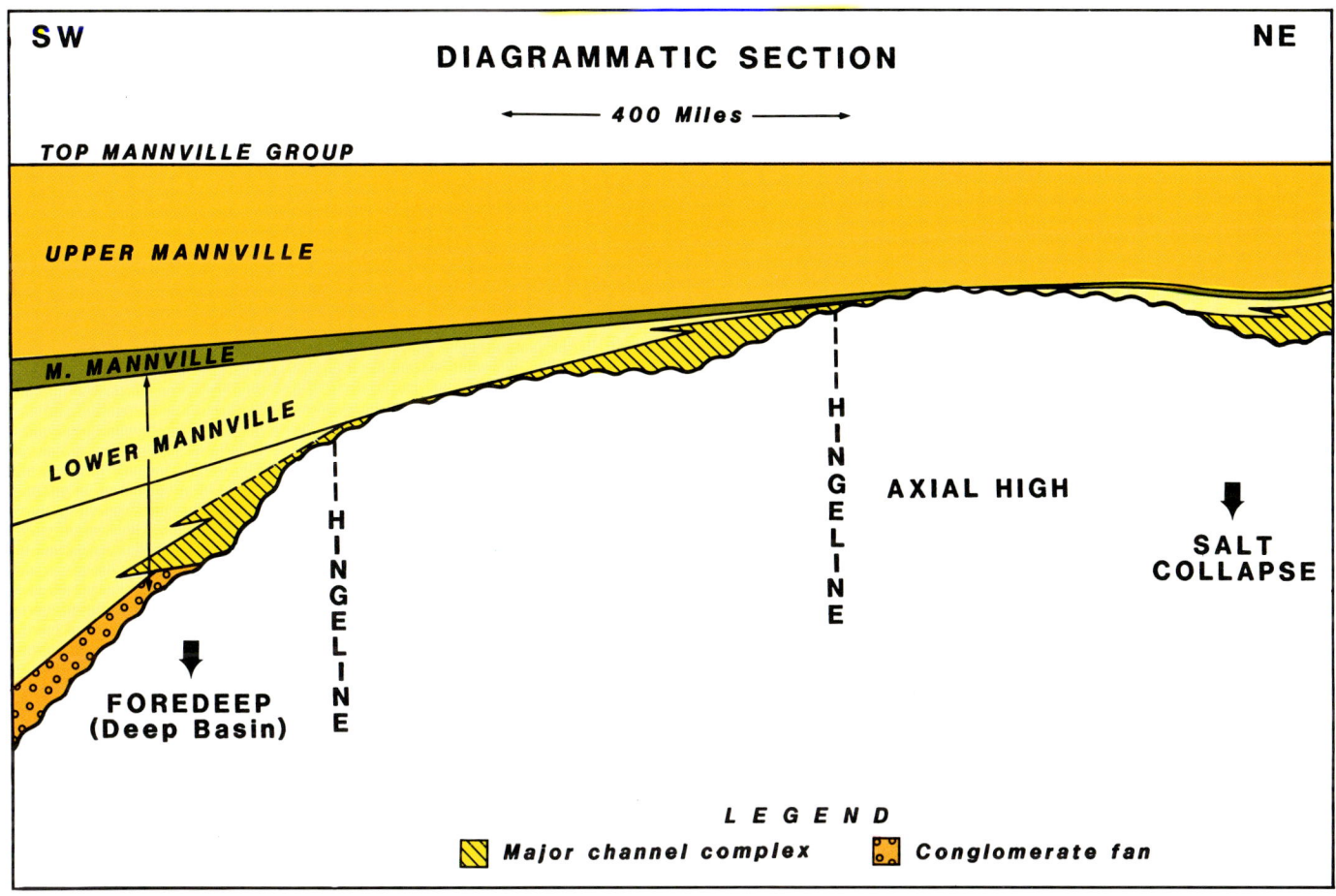

Figure 8. Diagrammatic cross section, SW-NE, illustrating the important hingelines and subsidence areas affecting the shape of the early Lower Cretaceous basin and Mannville deposition.

(480 km) from south-central to northern Alberta into the subsiding Clearwater Formation marine shale basin.

5. A brief drop in relative sea level and minor erosion of the upper Mannville continental surface. Capping marine shales of the Harmon Member and Joli Fou Formation were deposited in a second major southward advance of the Boreal Sea and a younger northward advance of the Gulfian Sea respectively. These seas eventually joined, inundated nearly all the Mannville continental surface, and formed the western interior seaway of North America.

LOWER MANNVILLE PALEOGEOGRAPHY

The first three "event" maps (Figs. 19, 20, 21) illustrate, in order, the initial continental phase of the lower Mannville, an early onlap stage of the transgression of the Boreal Sea, and the maximum onlap stage. These are the principal stages of an overall transgression during Aptian time which culminated in the development of a shallow seaway over a large area of the Western Canada basin.

Lower Mannville Continental Phase

The depositional environments and lithofacies of the Cadomin Formation, the lower Gething Formation, and equivalents are shown in Figure 19. Extensive Cadomin Formation conglomerate fans were initially shed eastward from the rising Cordillera into the subsiding, foredeep trough. This event was followed by deposition of a thick wedge of Gething Formation fluvial deposits eastward over the Cadomin Formation to the position of the foredeep hingeline (Fig. 9a). Here drainage entered a major northwest-trending "channel" system, the Spirit River Channel (McLean, 1977), in a broad valley incised several hundred feet into Jurassic sand and shale. This valley, or meander belt, was contained by the eastward slope of Gething deposits from the Cordillera and by an eastern escarpment known as the Fox Creek Escarpment in Central Alberta (McLean, 1977). The Fox Creek Escarpment parallels the foredeep hingeline and likely has structural connotations. A belt of fluvial sand up to 200 ft (60 m) thick is mapped from Montana (Cutbank Formation) to northeastern British Columbia, where Gething Formation deltaic and marine equivalents occur in the position of the present day overthrust belt (Stott, 1973). Major tributaries, the Edmonton channel system (Williams, 1954) and the Peace River channel system, are shown draining from the west flank of the axial high into the Spirit River channel. Of note is the interpretation of a southward flow direction for the Edmonton "channel" system, opposite to other interpreta-

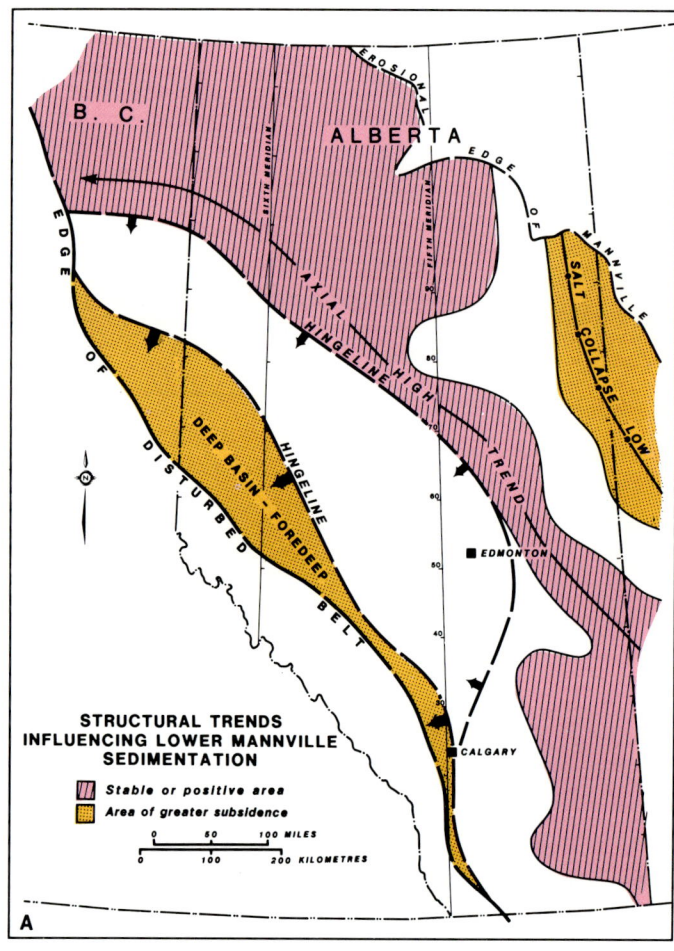

Figure 9a. Structural trends that influenced lower Mannville Group sedimentation.

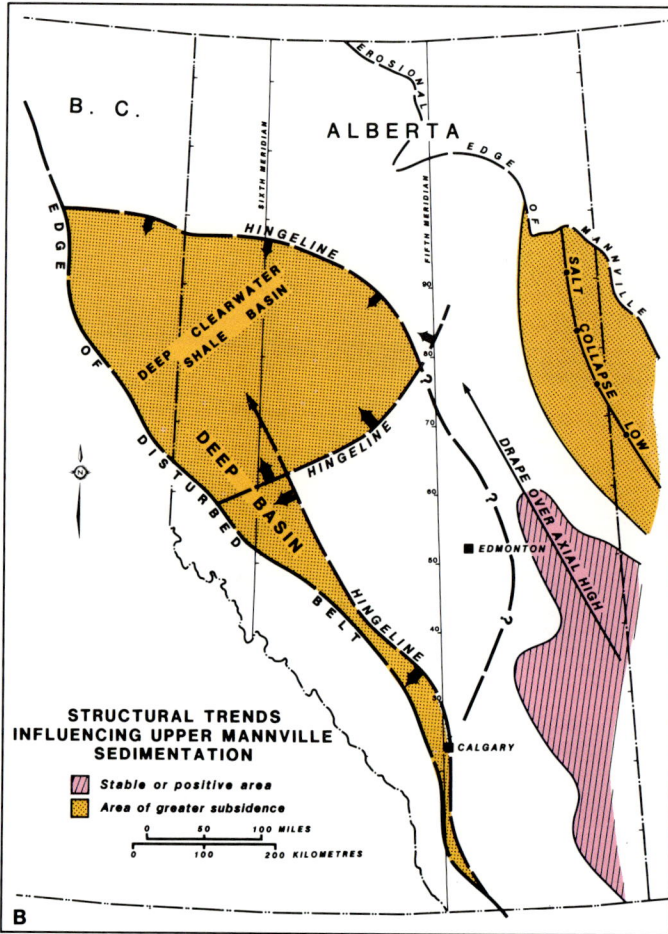

Figure 9b. Structural trends that influenced upper Mannville Group sedimentation.

tions (McLean and Wall, 1981) which link the Edmonton and Peace River systems in a northward direction. Well logs indicate a minor high (lower Mannville thin, Fig. 10) on the pre-Mannville (Mississippian) surface which was interpreted as a watershed separating the two systems.

The Edmonton and Peace River channel systems incised wide valleys parallel to a basin hingeline (Fig. 9a) on the western flank of the axial high trend. To the west, the Pembina erosional high was formed of Jurassic and Mississippian beds. Clastic sediments were likely shed from this high into both the Spirit River and Edmonton Channels to contribute to the Gething Formation and Basal Quartz Formation respectively. Broad belts of Basal Quartz Formation fluvial sand up to 200 ft (60 m) in thickness were deposited in the Peace River, Edmonton, and Bellshill Lake channel systems of central and southern Alberta. Many other smaller channels dissect the axial high and its southern extension, the Jurassic highland of southeastern Alberta. Fluvial sands of the lower Sunburst Member of southern Alberta are in many cases difficult to distinguish from underlying Jurassic sands from which they were sourced.

To the east of the axial high, which extended from southwestern Saskatchewan to northeastern British Columbia, drainage flowed into another major channel system known as the St. Paul channel and deposited the Dina and lower McMurray Formations. This channel system occupied a regional collapse low created by Devonian salt solution (Fig. 9a) and flowed northward to a postulated marine area in extreme northeastern Alberta (Mossop and Flach, 1981). In places, the St. Paul channel system incised the Paleozoic surface to a depth of several hundred feet, and contains like thicknesses of quartzose sand.

Erosion of the axial ridge in southern Alberta provided mature, quartzose clastics, probably from Jurassic section, to the early Mannville fluvial systems depositing the lower Sunburst Member and the Basal Quartz and Dina formations. Some of the coarser, quartzose sands of the Dina Formation in northeastern Alberta, suggest a contribution from fluvial systems, now eroded, which flowed off the Canadian Shield (de Wit, personal communication). Cadomin and Gething formation sediment was largely derived from the western Cordillera and is characterized by "salt and pepper" sandstones containing chert and volcanic detritus and by chert conglomerates.

Hydrocarbons

Major hydrocarbon accumulations along the eastern updip side of the major channel systems and their tributaries include

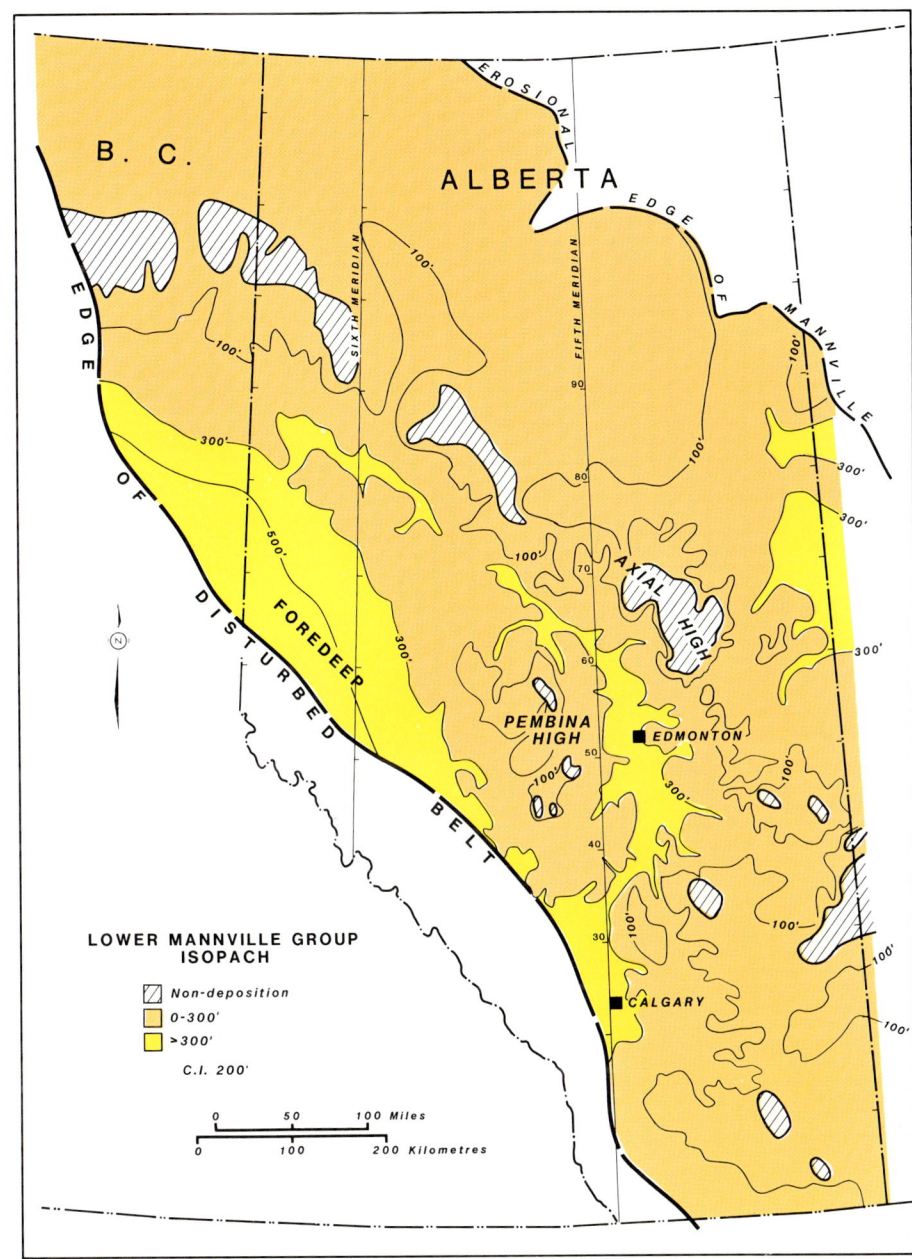

Figure 10. Lower Mannville isopach, illustrating thicker basin fill in areas of greater subsidence, and in channels incised into the pre-Mannville surface.

the Cutbank oilfield (over 200 million barrels recoverable) in northern Montana, the Taber oilfields (over 14 million barrels recoverable), in southern Alberta, and the Grand Forks oilfield (over 50 million barrels recoverable) in southern Alberta. In northern Alberta the massive Peace River heavy oil deposit contains over 90 billion barrels in place, and in British Columbia, the Buick Creek – Rigel fields have recoverable gas reserves of over 1 tcf. The Bellshill Lake field in central Alberta with 60 million barrels recoverable is a stratigraphic trap within a major channel complex. A large proportion of the 990 billion barrels of oil in place in the Athabasca tar deposit occurs in saturated McMurray Formation channel sands.

In the Deep basin area, the gas-saturated Cadomin Formation (Gies, 1984) contains over 1 tcf of recoverable gas in a piedmont alluvial plain conglomerate facies.

In view of the great quantities of oil and gas that migrated northeast and updip along the pre-Cretaceous unconformity surface and lower Mannville path (Masters, 1984), and the complexity of the fluvial trends, a large number of plays remain to be made in basal Mannville continental clastics.

Early Marine Onlap

Figure 20 shows the paleogeography of an initial transgression of the Boreal Sea which progressed in a southeasterly direction over continental lowlands. Fluvial systems of the basal Mannville occupying the western foredeep trough and the regional northeastern Alberta salt collapse low were first inundated. Upper Gething Formation marine and deltaic facies are reported from surface and subsurface mapping (Stott, 1973) in northeastern British Columbia and marine sand units occur in the upper McMurray Formation of northeastern Alberta (Stewart and MacCallum, 1978). The marine incursion was separated into two lobes by the low-lying land mass of the axial high trend.

In the Edmonton and Calgary regions, lacustrine areas are postulated to have developed in lowlands previously occupied by the Edmonton and Spirit River channel systems. These lakes may have eventually become joined with the Boreal Sea although marine connections (Fig. 20) are tenuous. The upper Gething Formation and upper Basal Quartz Formation in these "lake" areas exhibit a number of correlative, coarsening-up, progradational sand sequences on well logs. Many fining-up sequences, interspersed with correlative lacustrine or marine units in the upper part of the section, are interpreted to be tidal channels and suggest development of estuarine complexes adjacent to fluvial systems of the Basal Quartz Formation and the upper Sunburst Member (Fig. 20).

In western Alberta, continued deposition of upper Gething fluvial sands derived from the Cordillera infilled the foredeep trough and maintained the continental area of the Deep basin. Continental conditions also persisted in southern Alberta along the northwest-trending axial high and over a wide area of northwestern Alberta.

Hydrocarbons

Hydrocarbon occurrences of note in the lacustrine and marine sequence are the

Niton oil and gas field of central Alberta and a proliferation of potentially sizable gas and gas condensate fields in the Ricinus – Medicine River area of south-central Alberta. In northeastern Alberta a large amount of heavy oil is trapped in thin marine sand units, interbedded with delta-plain sand and shale (Stewart and MacCallum, 1978) in the upper McMurray Formation at Cold Lake and Athabasca, although the amount presently recoverable is not known.

Maximum Lower Mannville Marine Onlap

Continued basin subsidence allowed marine onlap of most of the basin by late Aptian time, as depicted in Figure 21. Upper Gething Formation fluvial clastics continued to be deposited in the foredeep area to maintain much of it as a continental lowland. The Boreal Sea continued its southward transgression in two lobes separated by emergent parts of the axial high trend. The eastern lobe extended into central Saskatchewan, the western lobe into Montana following inundation of fluvial systems and lacustrine areas of the foredeep low. Southeastern Alberta was a channelled, continental plain where fluvial sands of the upper Sunburst Member were deposited.

Generally, fairly thin offshore bar and tidal bar sands characterize the marine Ostracod member in central and southern Alberta. A distinctive thin limestone or calcareous shale unit, informally known as the "Calcareous member," occurs at the top of the Ostracod member over a wide area from central Alberta southward into Montana (Burden, 1982). This lithologic unit is an easily identifiable marker on well logs and is often used as a datum in subsurface work.

No such regional marker has been identified over the northern part of the marine basin. The outline of marine areas in Figure 21 is quite speculative, as marine upper Gething or Ostracod equivalents cannot be distinguished from younger marine Bluesky and Wabiskaw Formations. It is quite likely that the early phase of Bluesky and Wabiskaw sand deposition (Fig. 22) in northern Alberta and British Columbia is time equivalent to late Ostracod member deposition in southern Alberta.

Hydrocarbons

Oil and gas fields of the Ostracod member are generally small, most of them with

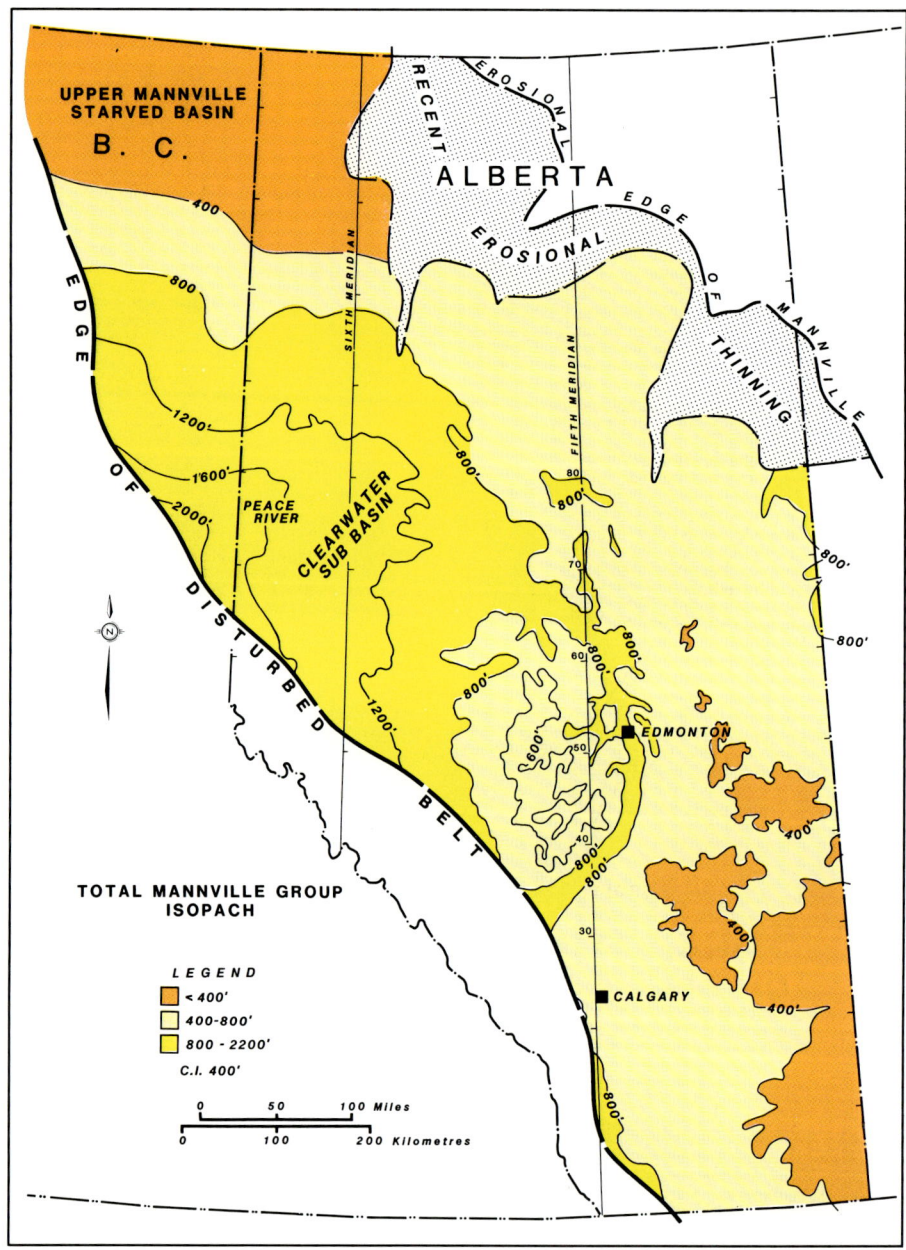

Figure 11. Total Mannville Group isopach, illustrating thicker basin fill into areas of greater subsidence. Of note is the development of the Clearwater Formation shale depocenter in northwestern Alberta during deposition of the upper Mannville.

reserves of less than three million barrels of oil recoverable or 50 bcf of gas. However, an indication of what might be found is the Highvale field in the Pembina region where over 5 million barrels of recoverable oil are contained in marine sand bars. Numerous small oil and gas fields occur throughout the southern Alberta portion of the marine basin and in channel sands of the equivalent upper part of the continental Sunburst Member.

Although these fields are small, they make attractive targets because of their potentially large numbers and relatively shallow drilling depths.

MIDDLE MANNVILLE PALEOGEOGRAPHY

The informal "middle" Mannville designation refers to the sequence result-

Lower Cretaceous Mannville Group 59

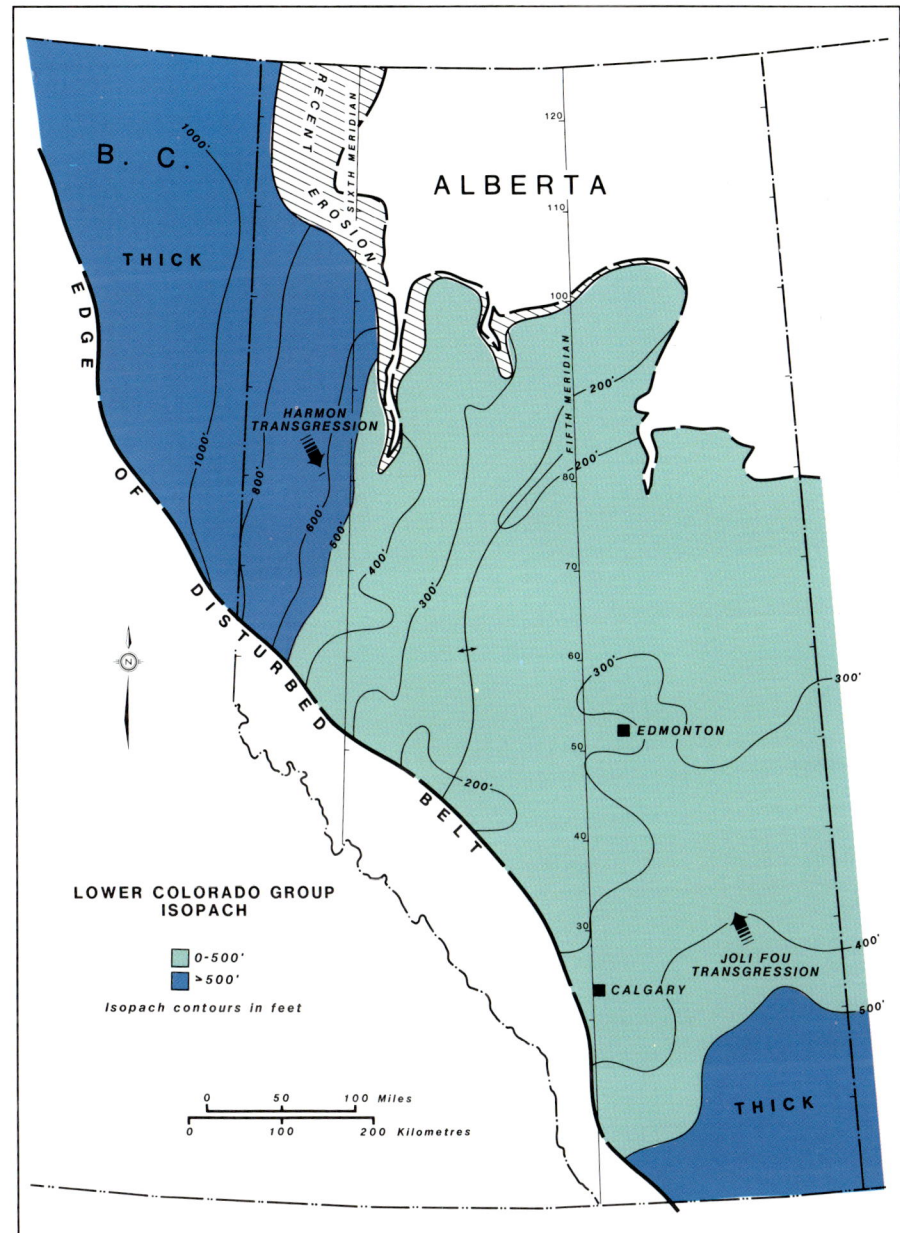

Figure 12. Lower Colorado Group isopach, illustrating the shale depocenters following transgressions of the Boreal (Harmon Member) and Gulfian (Joli-Fou Formation) seas.

progradation which returned shorelines to central Alberta. The final beach and barrier island sequences of the Glauconite member faced open-marine conditions and deposition of basal-marine shales of the Clearwater Formation to the north.

Bluesky Formation and Equivalents

In northern Alberta and British Columbia, continued transgression inundated large areas of Gething Formation continental deposits and pre-Mannville formations of the axial high trend. Shoreline sands and tidal-bar sequences of the Bluesky Formation developed around a portion of the axial high trend, a low-relief landmass straddling the British Columbia-Alberta border (Fig. 22). Both Triassic and Mississippian Banff Formation sands of this landmass were likely contributing clastic sediments to bordering marine areas. The Bluesky Formation is represented in many areas by widespread transgressive lag deposits, while coarsening-up beach and offshore bar cycles are indicative of localized, regressive pulses into the deepening basin. At Elmworth, several stacked progradational cycles, the upper capped by coastal plain deposits near the British Columbia border, are interpreted as part of a large delta complex. A late phase of transgressive reworking of the upper Bluesky Formation section and evidence of tidal channelling indicates cessation of clastic supply from the Cordillera and transgression over most of the delta.

Beach, barrier island, and offshore bar trends of the Cummings and Wabiskaw formations of eastern Alberta represent a coastal progradation northward from the Lloydminster region. A subsequent transgression returned shallow marine conditions southward to Lloydminster where a large number of tidal and post-depositional fluvial channels dissect beach and nearshore bar facies.

Glauconite Informal Member

Progradational marine sand units of the Glauconite informal member were mapped from immediately north of the Montana border to the Hoadley-Pembina region of south-central Alberta. These marine sands are characterized by a high glauconite content similar to that of the Bluesky Formation of northern Alberta and British Columbia.

Beach and deltaic sequences were capped by a thick coal formed in a coastal marsh environment associated with the

ing from the turn-around of the southward transgression of the Boreal Sea and the initiation of a major northward regession. Lower Mannville and upper Mannville designations are used to identify the transgressive and regressive parts of the major Mannville cycle.

Sand trends illustrated in the paleographic map of the Bluesky Formation and "equivalents" (Fig. 22) appear to fall into two groups, from detailed study of the regional cross sections (Fig. 4).

Sands of the northern basin area, the Bluesky, Wabiskaw, and Cummings Formations, were deposited as transgressive lag deposits or as regressive pulses in a continuing transgression of the Boreal Sea. As previously suggested, early phases of sand deposition associated with these formations are likely time equivalent to the Ostracod member of southern Alberta.

The Glauconite informal member of southern and central Alberta was deposited in a somewhat younger northward

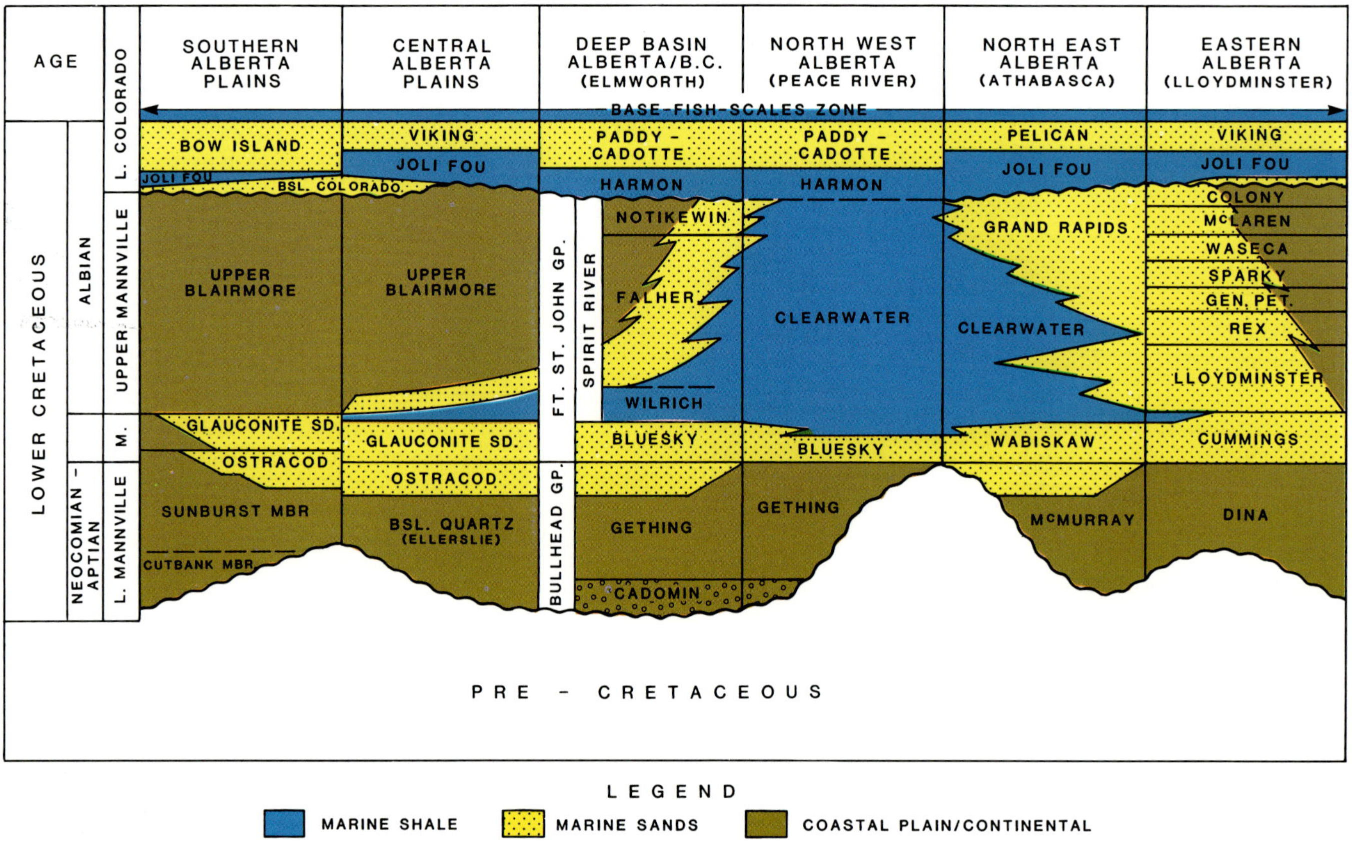

Figure 13. Lower Cretaceous correlation and general facies chart (Jackson, 1983).

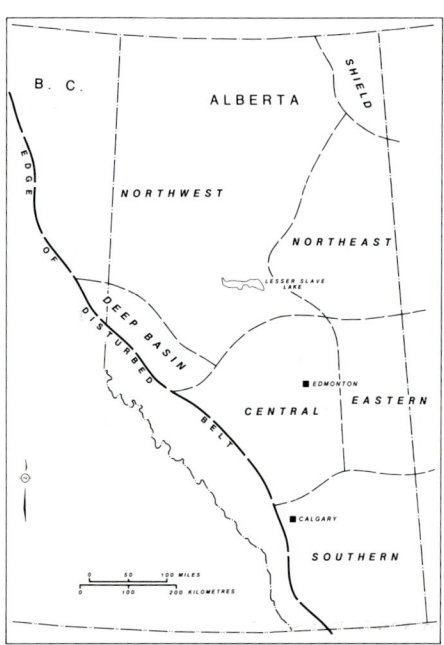

Figure 14. Stratigraphic reference areas for Figure 13, the correlation and general facies chart.

development of a widespread coastal plain in southern Alberta. The coastal plain was extensively dissected by fluvial channels and delta distributaries of a general northwest orientation. At some point during the northward coastal progradation, a barrier island trend developed in the Hoadley region of south-central Alberta. This regressive sequence reached thicknesses in excess of 75 ft (23 m) and extended nearly 150 mi (240 km) to the northeast across the basin (Chiang, 1984). As the coastal progradation reached the Hoadley trend, an offshore bar complex formed to the north at Bigoray (Fig. 22) and developed into another large northeast-trending barrier island complex. The coastal progradation evidently did not extend northward to it as there is an absence of capping coal on the Bigoray sand trend. Both the Hoadley and Bigoray trends are cut by numerous shale-filled channels, possibly of tidal origin, which are of major importance in trapping hydrocarbons. At its full development, the Bigoray barrier trend faced open-marine conditions to the north as transgression continued, and Clearwater Formation marine shales were deposited over the Bluesky Formation and equivalents.

Hydrocarbons

Within the area of the western "Deep basin" (Masters, 1984), sands and conglomerates are totally gas-charged and a number of large gas and condensate fields (100 to 300 bcf recoverable) have been discovered in the Bluesky Formation in the Elmworth-Edson region. Other large "Deep basin" type fields include the Pembina-Bigoray field (over 400 bcf recoverable) and the Hoadley field (6 to 7 tcf potential; Chiang, 1984), both in the

Figure 15. Diagrammatic regional cross section panel. Strike sections 1-1′ and 2-2′ and basin dip section SW-NE, cross section index map.

Lower Cretaceous Mannville Group

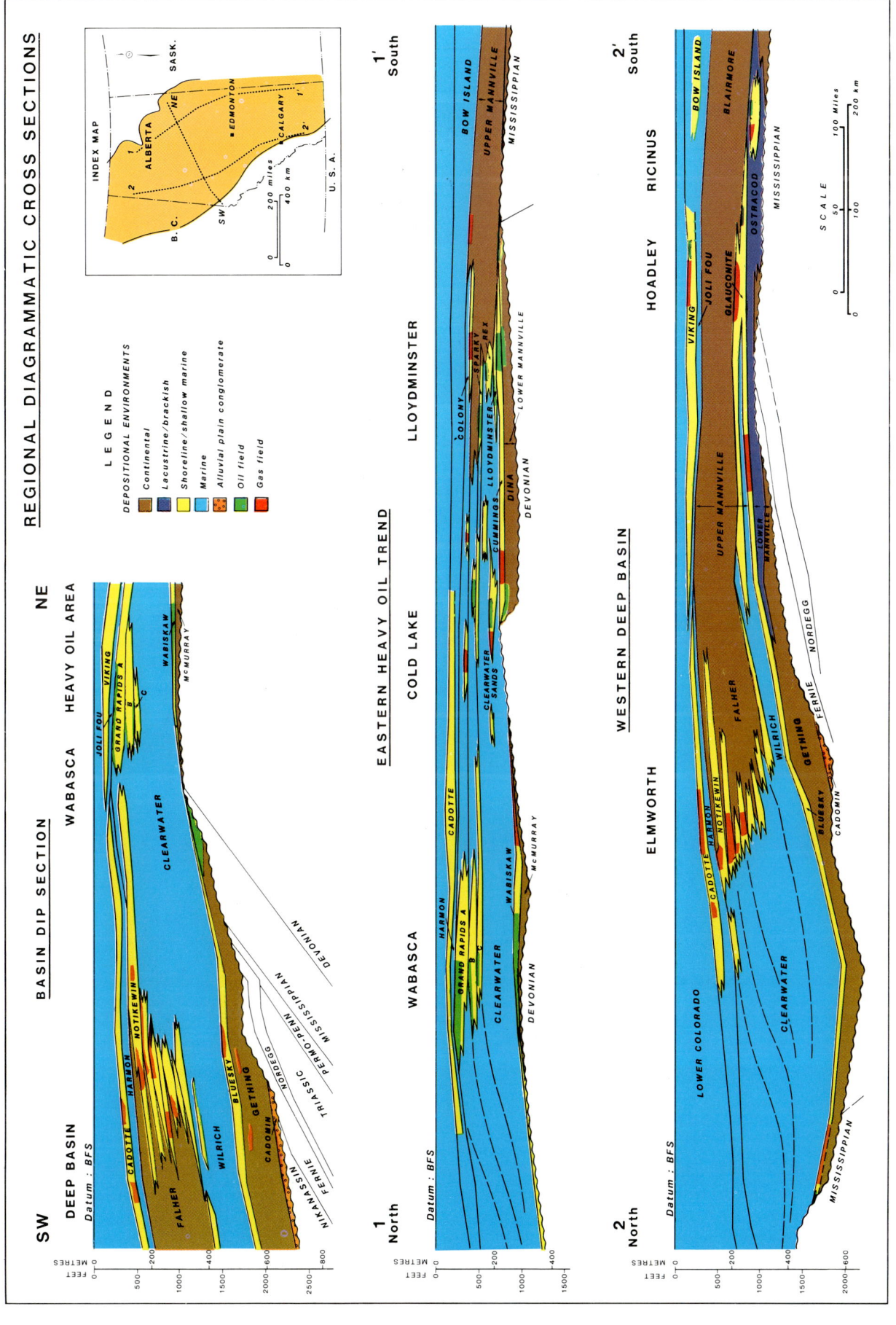

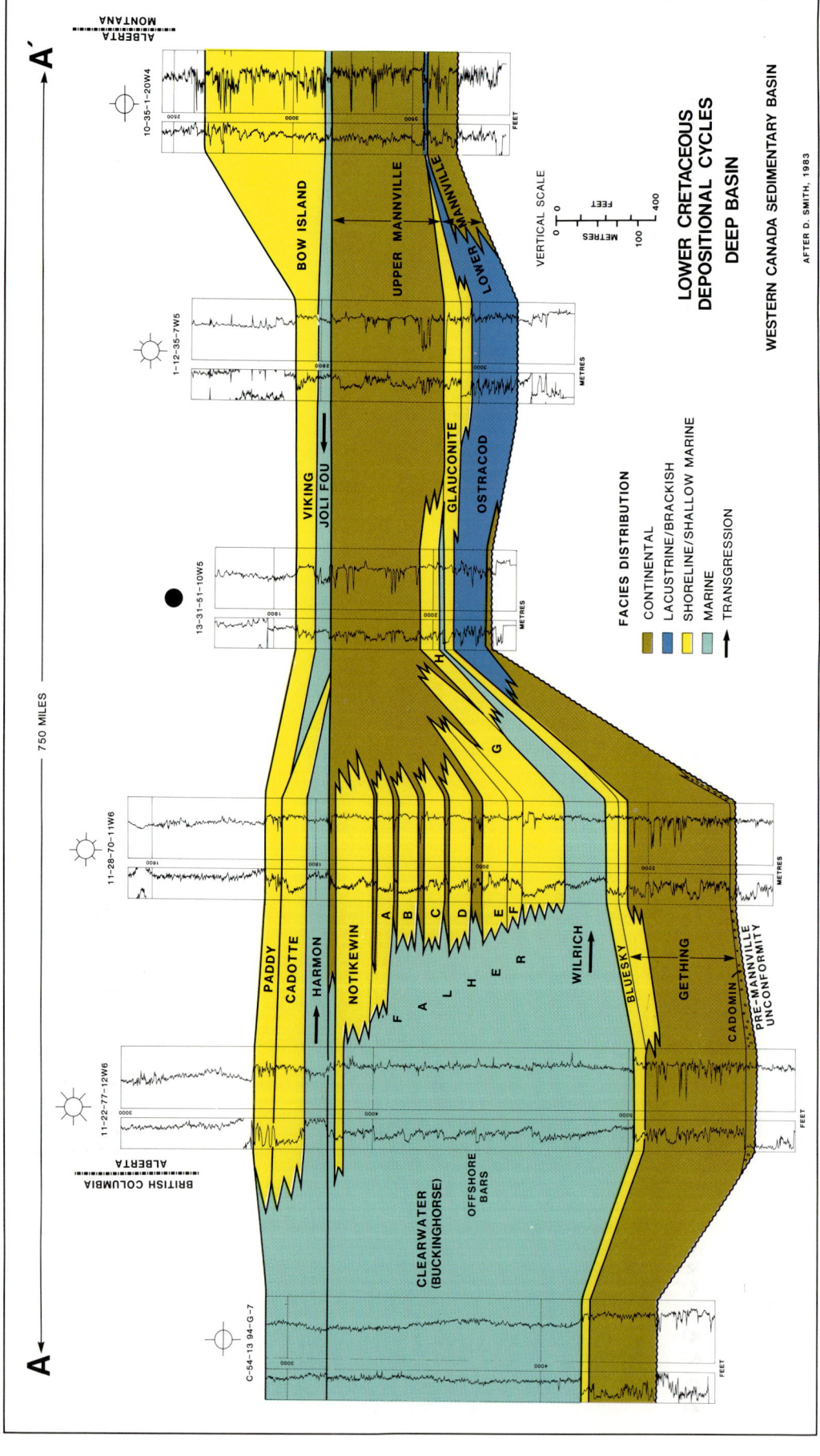

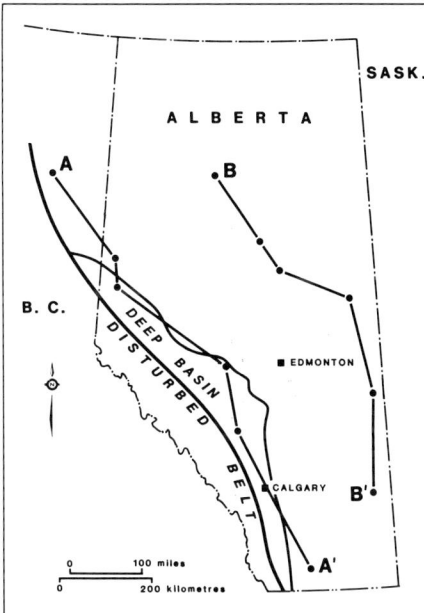

Figure 16. Detail cross section (A-A', B-B') index map, for Figures 17, 18.

Glauconite Member. Both the Pembina-Bigoray and Hoadley gas fields terminate updip against shale-filled channels cutting across otherwise continuous reservoir conduits and more such traps are envisioned in these and other barrier sand trends.

A large number of small to intermediate-sized oil and gas fields have been found in the Glauconite informal member of southern Alberta in a complex array of marine and fluvial traps. Oil discoveries include the Jenner field (over 7 million barrels recoverable) in a barrier-island facies (Holmes and Rivard, 1976) and the Bantry field (over 40 million barrels recoverable) in fluvial facies.

Heavy oil traps occur in the Cummings Formation in the Lloydminster region of eastern Alberta where marine beach and bar sands terminate updip against shale-filled tidal and fluvial channels.

In northwestern Alberta, the shallow Keg River gas field (over 700 bcf recoverable) in the Bluesky Formation covers an area of nearly 3,000 sq mi (7,769 sq km).

Marine sand trends of the Wabiskaw Formation in northeastern Alberta form major heavy oil and tar traps at updip stratigraphic terminations and in structural traps where Cretaceous beds cross a Devonian salt-collapse front and drape over the Athabasca anticline (Fig. 3).

Figure 17. Regional detail gamma ray - sonic log cross section A-A' (see Fig. 16).

THE UPPER MANNVILLE SEQUENCE

The upper Mannville summary map (Fig. 23) shows the extent of a great coastal progradation from approximately the position of the Hoadley barrier (Fig. 22), in central Alberta, to the Clearwater shale basin, a distance of over 300 mi (480 km).

A large influx of western-sourced clastic sediments overcame basin subsidence to build the northward thinning wedge of coastal and continental sediment in a series of progradational pulses. Successive shoreline sand belts, many individually exceeding 50 ft (15 m) in thickness, are mapped from west to east across Alberta, a distance of over 350 mi (560 km). Although not mapped in this paper, these trends continue into a marine embayment which occupied west-central Saskatchewan.

Previously described as a region of greater subsidence (Fig. 9b) during the upper Mannville, the Clearwater shale basin accumulated over 900 ft (274 m) of marine shales and silts (Fig. 24). Isopach thinning toward the northern boundary of British Columbia, seen in Figure 24, is due to clinoform beds sloping northward into the basin. The south flank of the shale basin is marked by a prominent hingeline beyond which coastal sequences thicken, then terminate against offshore sands and shales. These features are seen adjacent to wellbore 11-28-70-11 W6 on cross section A-A' (Fig. 17).

Upper Mannville Regressive Cycles

Figure 25 illustrates and compares the successive cycles of progradation on the east and west sides of the basin. Based on lithology and log correlation, many of the major regressive pulses are considered correlative basinwide as indicated in Figure 25 and as presented in the upper Mannville paleogeographic map series.

Although the upper Mannville section thins from over 600 ft (180 m) to less than 400 ft (120 m) in an eastward direction, a reasonable correlation of progradational sequences was determined from the regional cross section grid (Fig. 4).

Similarity of lithofacies suggests a Cordilleran provenance for upper Mannville clastics deposited across the basin at least to the Saskatchewan border. Lithic sandstones containing chert and volcanic grains characterize the Fahler cycles of the Deep basin as well as the sand units of the Clearwater Formation (Putnam and Pedskalny, 1983) in eastern Alberta. The eastern Grand Rapids Formation (Figs. 13, 18), although a more quartzose sand to suggest a possible Canadian Shield provenance, contains chert grains which indicate a western contribution. More quartzose sand, even if of Cordilleran origin, is to be expected in eastern Alberta considering that content of lithic material is reduced the greater the transport distance from the Cordillera. For example, the amount of chert present in the Falher Member either as capping conglomerate, or as discrete grains, diminishes rapidly east of the Deep basin.

Some individual cycles, particularly in eastern Alberta, show profiles characteristic of progradation into relative sea-level stillstand (Fig. 26a) but the majority indicate relative sea-level rise (Fig. 26b). The terms relative sea-level stillstand and relative sea-level rise are used in the context of a structurally stable depositional surface and a subsiding surface respectively. As progradations neared the more rapidly subsiding Clearwater Formation shale basin, shoreline deposits became stacked within single cycles as well as in multi-cycle sequences. A number of Grand Rapids Formation and Fahler Member shoreline sands attain thicknesses of 100 ft (30 m) seaward of the basin hingeline (Fig. 24).

In the Deep basin area, many coarsening-up Fahler beach sand and conglomerate sequences are capped by coal deposits formed in extensive coastal lowland swamps developed with each progradation. Minor transgressions periodically returned areas of these western coastal lowlands to marine conditions, however, in eastern Alberta, more extensive transgressions repeatedly inundated as far south as the Lloydminster area. Thick coastal plain coals of the western basin were usually preserved during these transgressions while thinner coals of the eastern area were more often destroyed. Preserved top coal distribution is shown in the upper Mannville paleogeographic map series.

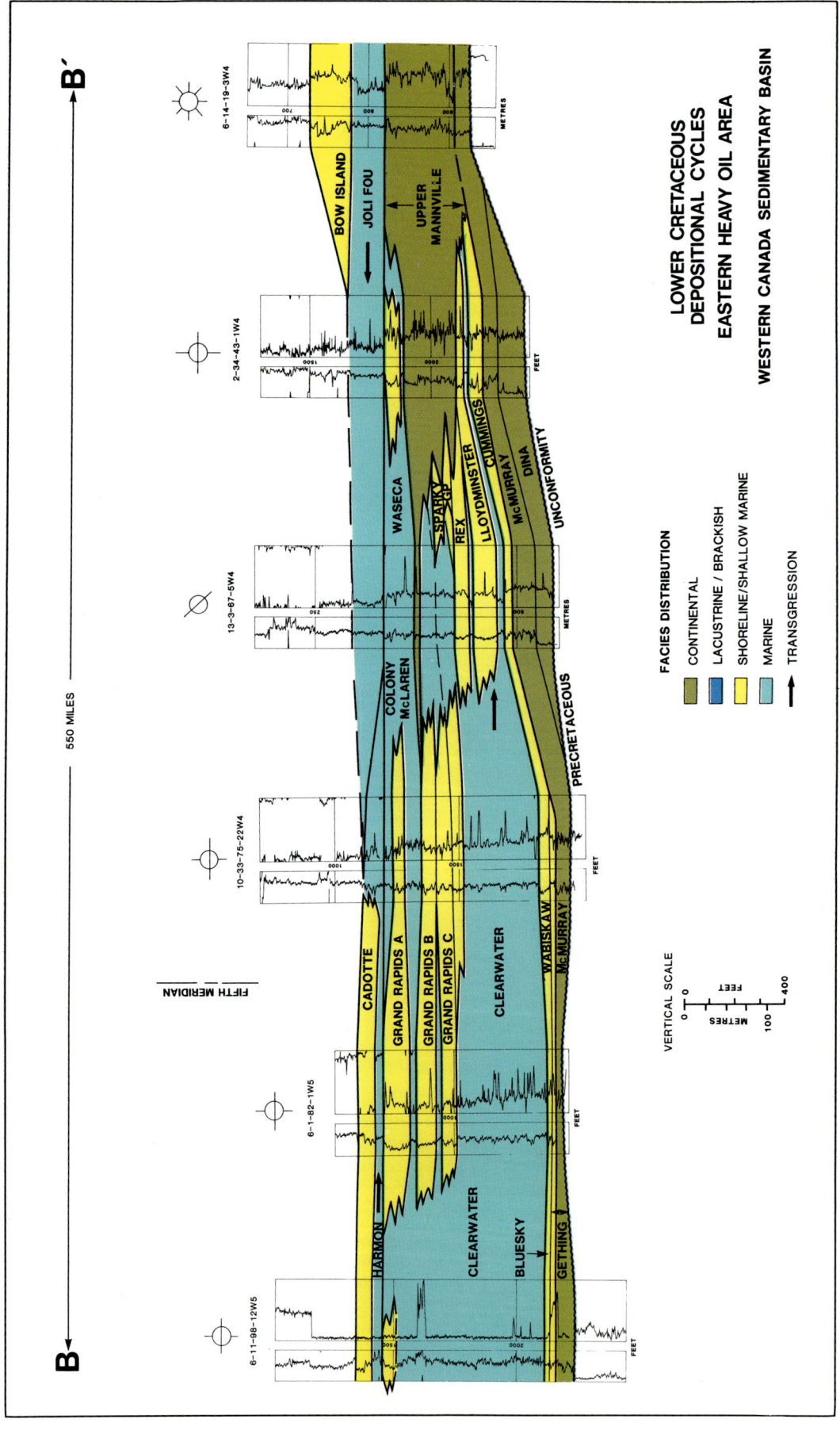

Lower Cretaceous Mannville Group 65

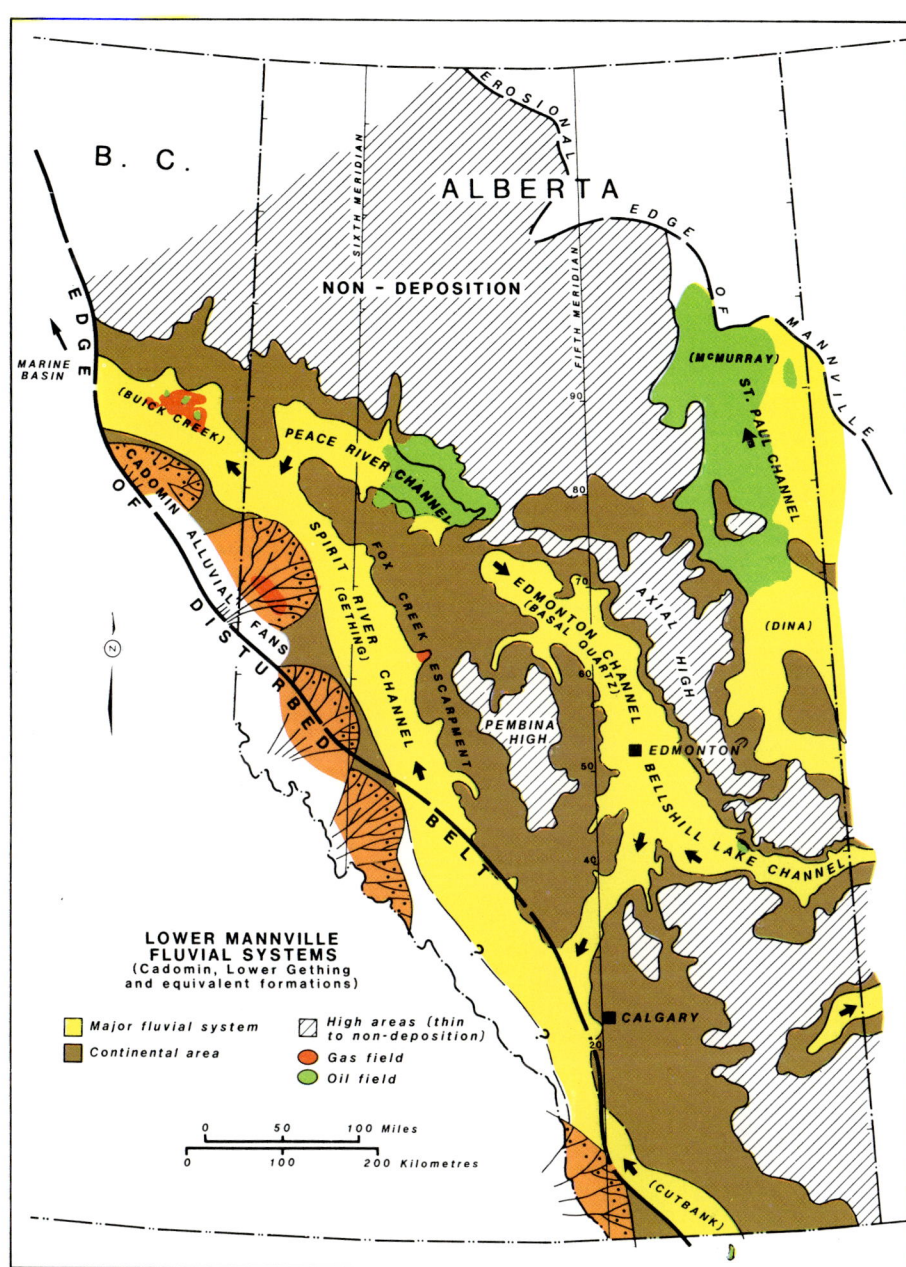

Figure 19. Map of lower Mannville fluvial systems, the first phase of Mannville Group deposition.

UPPER MANNVILLE PALEOGEOGRAPHY

The following series of paleogeographic maps illustrate the successive shoreline and offshore sand trends of major progradations comprising the upper Mannville

Figure 18. Regional detail gamma ray - sonic log cross section B-B' (see Fig. 16).

regression. The maps show the southern starting position of each coastal progradation, shoreline and offshore sand trends in which sands of 25 ft (7.6 m) or more in thickness are highlighted, the seaward limit of capping coastal-plain coals, and finally, the distribution of basinal shales. The sand belts consist of barrier beach, deltaic, barrier island, and offshore bar facies and trend in a general west to east direction for over 350 mi (560 km) across the basin. Specific formation and member names are shown on the appropriate maps and also on well log cross sections A-A' and B-B' (Figs. 17,18).

Falher H and Equivalents

Figure 27 maps the paleogeography of the earliest Falher Member sand and its approximate equivalent, the Lloydminster Member in eastern Alberta. These first progradations into the Clearwater basin built out over a very shallow shelf, where an increased supply of terrigenous clastics produced rapid seaward advances.

The Falher H unit, as yet without any significant hydrocarbon production, contains some conglomerate-capped beach ridges to the west and certainly has exploration potential in the gas-saturated Deep basin. The eastern Lloydminster Member and its northern equivalent, the Clearwater B Member, both contain extensive heavy oil fields and some gas fields in updip barrier-trend terminations or where sands drape over the Athabasca anticline. Shale-filled tidal channels in the Lloydminster area are important trapping mechanisms that give rise to a multitude of traps which now contain heavy oil. The presence of numerous channels, both tidal and fluvial, is characteristic of almost every cycle in the Lloydminster region as transgression repeatedly returned this area to shallow nearshore marine conditions.

Falher G and Equivalents

The overlying Falher G cycle with its approximate eastern basin equivalents, the General Petroleum and Rex Members (Fig. 28), continued the northward progradation. In western Alberta only a minor transgression over the Fahler H sequence interrupted the rapid progradation of the Falher G shoreline sands. In eastern Alberta, progradation continued following extensive transgression over the Lloydminster Member (Fig. 27) as far south as the Lloydminster area. The repeated return of eastern Alberta to shallow marine conditions may indicate continued regional subsidence due to Devonian salt solution and removal. Development of a large marine embayment southward to the Lloydminster area is a feature of all major upper Mannville cycles.

Western Falher G sands and conglomerates are gas-saturated in the Deep basin as is the entire Mannville Group. To date, no significant commercial fields have been discovered in the Deep basin or in the updip central area of the basin although

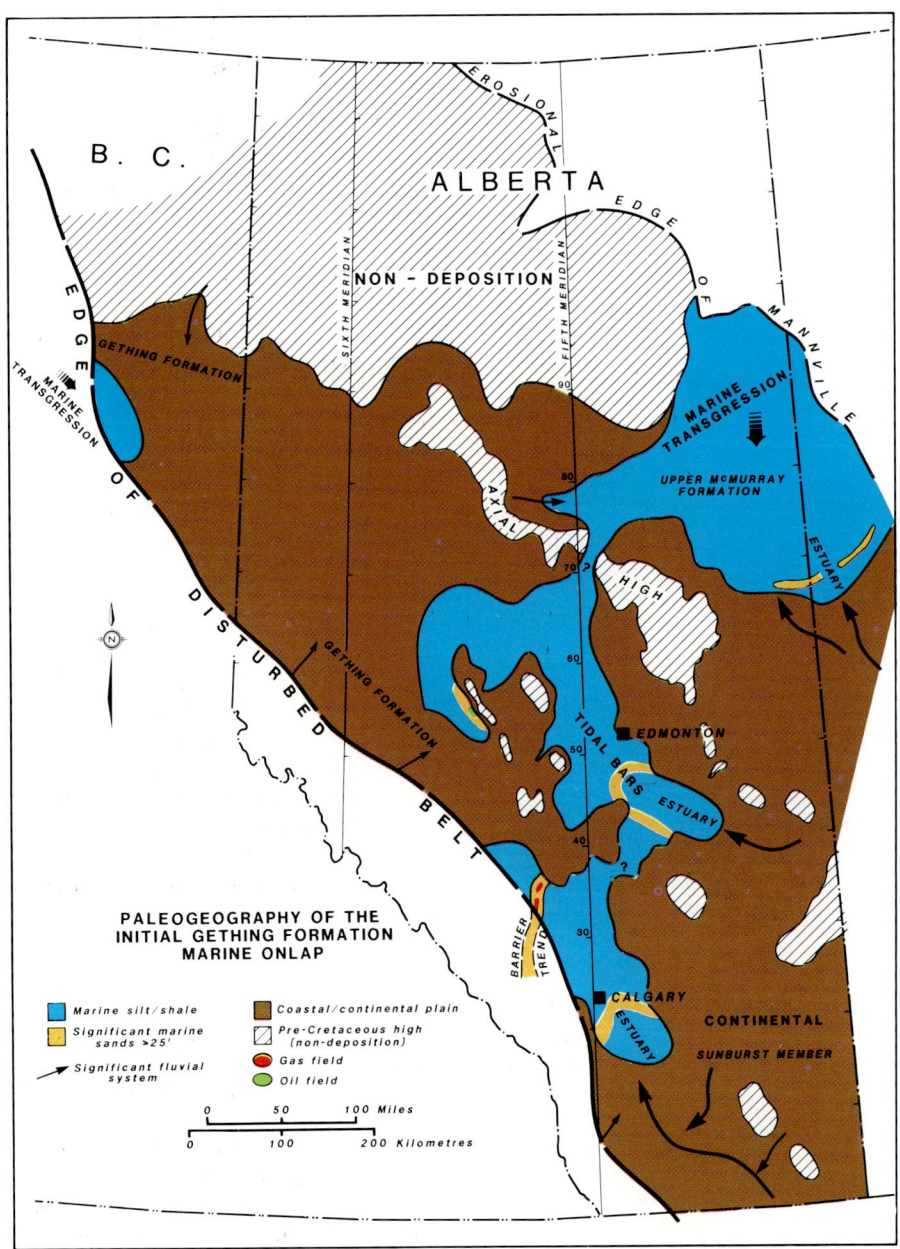

Figure 20. Paleogeography of the initial phase of Gething Formation (and equivalents) marine onlap.

the former, in particular, has good potential.

Heavy oil fields and gas fields occur in the Rex, General Petroleum, and Clearwater A members in eastern Alberta.

Fahler C, D, E, F, and Equivalents

Figure 29 illustrates the stacked Fahler Member C, D, E, and F cycles in western Alberta and the much thinner eastern equivalents comprising the Sparky Member. This section of the upper Mannville posed the greatest correlation difficulty for two reasons. A pronounced eastward thinning of the entire section as well as of individual cycles occurs and there is not much coastal plain coal preserved over large areas of the eastern embayment to help identify separate cycles.

The western shorelines were deposited out to a steepening slope of the Clearwater shale basin past the shale-basin hingeline which controlled the position of these and younger Fahler Member shoreline sequences. Here sediment supply and basin subsidence reached a balance and the shorelines stacked.

The eastern basin equivalent, the multicycle Sparky Member, was deposited in a pulsing sequence of prograding shorelines and offshore bars which accreted on a very shallow shelf. Once again extensively dissected by tidal and postdepositional fluvial channels, these inner shelf and shoreline sands contain a large number of traps.

In the Wapiti-Elmworth region of the Deep basin, several gas fields in these intervals of the Falher Member (R. Smith, 1984) individually exceed 1 tcf of recoverable gas. Extensive, heavy oil production occurs in the Sparky Member of the Lloydminster region in eastern Alberta.

Upper Falher Cycles and Equivalents

The western Fahler Member A and B cycles (Figs. 30, 31) are thick, stacked beach and nearshore sand sequences (D. Smith, 1984) deposited adjacent to the Clearwater deep shale basin. The overlying coastal deposits of the Notikewin Member (Fig. 32) prograded further northward than the Fahler Member as clastic influx was apparently great enough to overcome subsidence of the Clearwater subbasin. A Notikewin-equivalent turbidite deposit is mapped in the deep Clearwater shale basin in northeastern British Columbia.

The eastern Grand Rapids Member sequences (Grand Rapids A, B, and C) prograded out into the Clearwater deep shale basin, where stacked beach and barrier sequences in places exceed over 100 ft (30 m) in thickness. Massive barrier island and barrier beach complexes repeatedly formed across the seaward end of the eastern Alberta embayment (Fig. 18). These barriers, although breached by tidal channels, effectively cut off wave energy to the embayment and created lagoonal conditions. Coastal lowlands formed over wide areas only to be repeatedly inundated by southward transgressions as the embayment was re-established. The shape of this large embayment would suggest focusing of tidal energy prior to development of the Grand Rapids A, B, and C barrier island systems. A similar but much smaller area of likely tidal focusing persisted through a number of depositional cycles in the Kaybob region (Fig. 32) and numerous (tidal?) channels are evident on well logs. In both areas channels have a general northward orientation. Most of the upper Mannville shoreline to shallow-marine sands are east to west trending linear barrier beach, barrier island, and offshore bars deposited in a wave-dominated environment.

The upper Falher A and B sequences (Figs. 30, 31) in the Deep basin contain the giant Elmworth and Wapiti gas fields, with over 5 tcf recoverable in beach conglomerate trends (R. Smith, 1984). Many more conglomerate-capped beaches will be identified as exploration continues in the Deep basin.

The Grand Rapids sequence contains the bulk of the giant Wabasca heavy oil deposits (120 billions barrels in place), where Grand Rapids A and B (Figs. 31, 32) beaches drape over a Devonian salt-collapse front. Grand Rapids heavy oil fields and gas fields occur in both structural and stratigraphic traps from Lloydminster to Cold Lake.

Although numerous shows and some small fields exist in the upper Mannville of the area between the Deep basin and the Heavy Oil Trend, there is only one large field of note. That is the Kaybob gas field (over 500 bcf recoverable) in a beach deposit of the Notikewin Member adjacent to an updip shale-out. The many coastal and offshore sand trends of this central area are bound to yield discoveries as more stratigraphic breaks in these reservoir conduits become identified.

The Mannville Surface

A brief drop in relative sea level, coupled with a cessation of clastic influx, allowed minor erosion of the continental upper Mannville surface. This erosion was followed by a second major transgression of the northern Boreal Sea to deposit capping Harmon Member marine shale over the Mannville Group. A later northward transgression of the Gulfian Sea encroached the remaining Mannville continental area and joined the Boreal Sea to form the western continental interior seaway. As these transgressions inundated the Mannville surface, both shoreline and continental sediments were reworked to form a number of thin basal clastic zones in the Harmon Member and Joli-Fou Formation. These zones are not mapped in this paper, but include the lower Colorado sands of southern Alberta, an upper Colony sand in eastern Alberta, and upper units of the Notikewin Member in western Alberta. All contain generally thin but areally significant gas reservoirs.

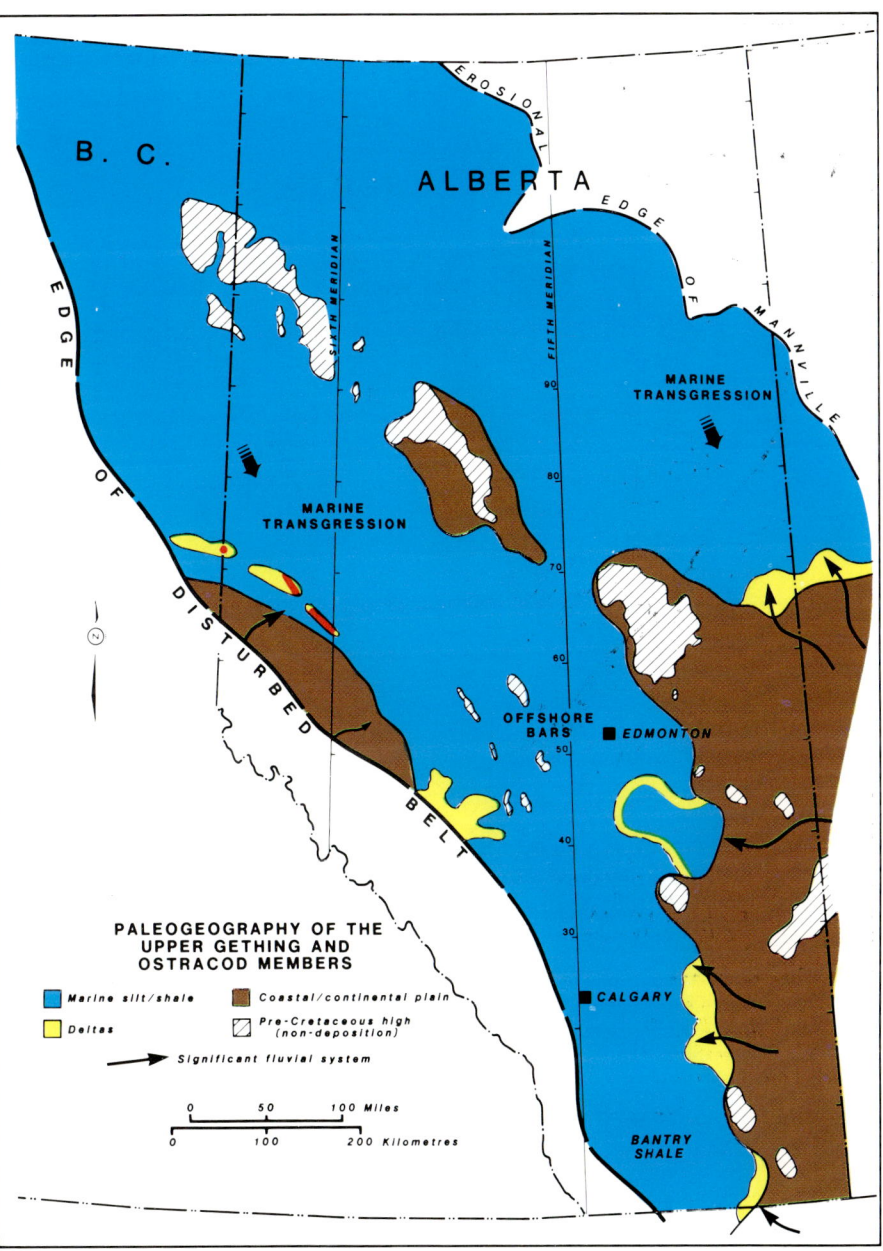

Figure 21. Paleogeography of the upper Gething Formation and Ostracod member. The maximum southward extent of transgression of the Boreal Sea is shown.

SUMMARY

The depositional history of the Mannville group has been defined in a series of paleogeographic maps. These maps have illustrated the significant stages of the lower Mannville transgression, the upper Mannville regression, and most importantly, the major sand trends deposited.

The Mannville Group conglomerate and sand belts are the principal reservoirs of the western Deep basin gas accumulations. In eastern Alberta and western Saskatchewan, they are also the reservoirs of the vast heavy-oil fields. In between, they offer an intricate network of sands through which huge volumes of hydrocarbons migrated, particularly in the lower Mannville section.

A large proportion of Alberta's future gas reserves can already be recognized in Mannville trends within the Deep basin. The general location of important new oil and gas fields updip from the Deep basin can be postulated as well.

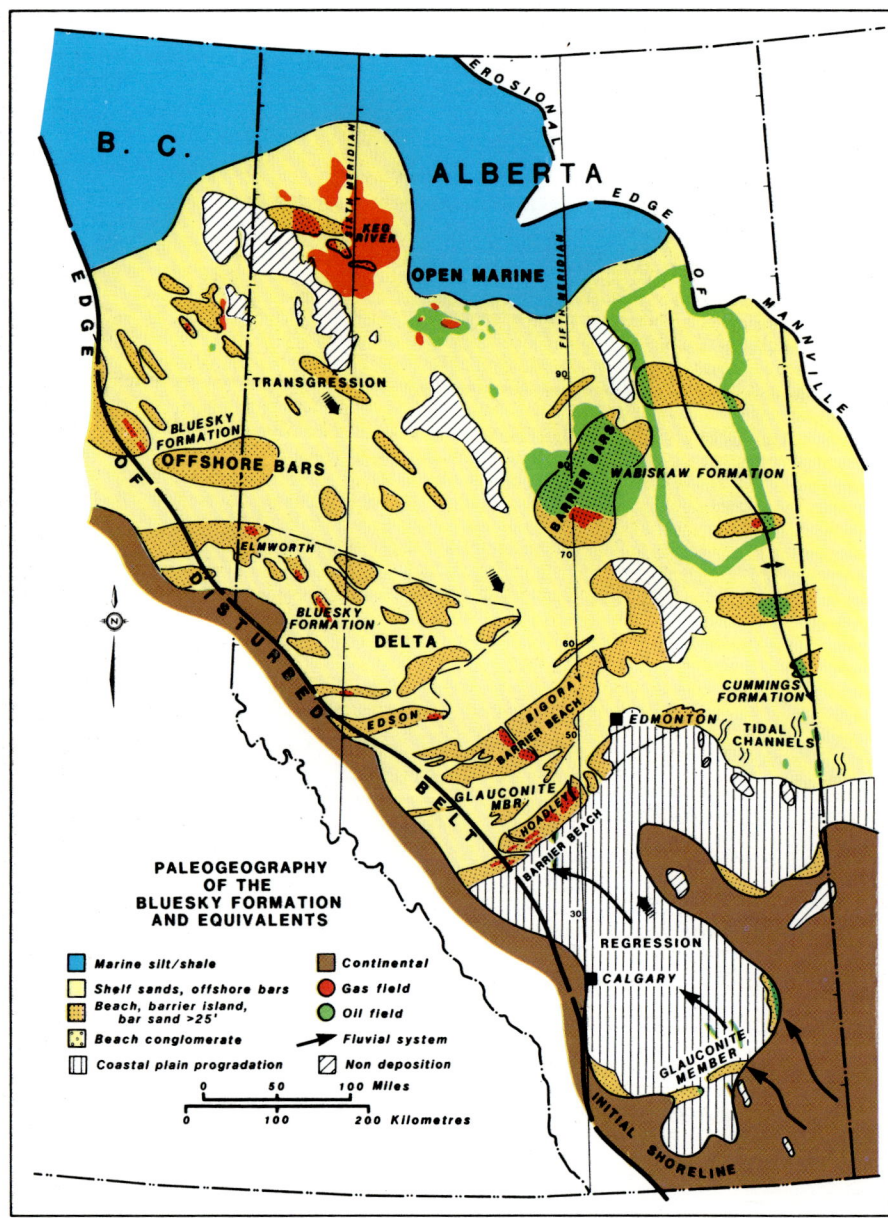

Figure 22. Paleogeography of the Bluesky Formation and equivalents.

ACKNOWLEDGMENTS

Gratitude is expressed to D. Yakimishyn for assistance in geological mapping and to G. Cone for drafting. The many suggestions and constructive criticisms of Dr. R. de Wit, J. Masters, and Dr. W. Brideaux are gratefully acknowledged.

REFERENCES CITED

Burden, E. T., 1982, Lower Cretaceous terrestrial palynomorph biostratigraphy of the McMurray Formation, northeastern Alberta: University of Calgary, Ph.D. Thesis, 59 p.

Chiang, K. T., 1984, The giant Hoadley gas field, south-central Alberta, in J. A. Masters, ed., Deep basin gas: AAPG Memoir 38, this volume.

Eisbacher, G. H., M. A. Carrigy, and R. B. Campbell, 1974, Paleodrainage pattern and late-Orogenic basins of the Canadian Cordillera: Society of Economic Paleontologists and Mineralogists, Special Publication #22, p. 143-166.

ERCB, 1981, Alberta's reserves of crude oil, gas, natural gas liquids and sulfur at December 31, 1981: Calgary, Alberta, Energy Resources Conservation Board.

Gies, R. M., 1984, Case history for a major Alberta Deep basin gas trap; the Cadomin Formation, in J. A. Masters, ed., Deep basin gas: AAPG Memoir 38, this volume.

Glaister, R. P., 1959, Lower Cretaceous of southern Alberta and adjoining areas: AAPG Bulletin, v. 43, p. 590-640.

Holmes, I. G., and Y. A. Rivard, 1976, A marine barrier island bar, Jenner Field, southeastern Alberta, in The sedimentology of selected clastic oil and gas reservoirs in Alberta: Canadian Society of Petroleum Geologists, p. 44-61.

Jeletzky, J. A., 1971, Marine Cretaceous biotic provinces and paleogeography of Western and Arctic Canada; illustrated by a detailed study of ammonites: Geological Survey of Canada, Paper 70-22, 92 p.

Masters, J. A., 1979, Deep basin gas trap, Western Canada: AAPG Bulletin, v. 63, p. 152-181.

——, 1984, Lower Cretaceous oil and gas in western Canada, in J. A. Masters, ed., Deep basin gas: AAPG Memoir 38, this volume.

McLean, J. R., 1977, The Cadomin Formation; stratigraphy, sedimentology and tectonic implications: Bulletin of Canadian Petroleum Geology, v. 25, p. 792-827.

Mellon, G. B., 1967, Stratigraphy and petrology of the Lower Cretaceous Blairmore and Mannville Groups, Alberta Foothills, and Plains: Research Council of Alberta Geological Division, Bulletin 21, 270 p.

Mossop, G. D. and P. D. Flach, 1981, Oil sands geology and technology; field guides to geology and mineral deposits: Calgary, GAC, MAC, CGU, p. 143-154.

Putnam, P. E., and M. A. Pedskalny, 1983, Provenance of Clearwater Formation reservoir sandstones, Cold Lake, Alberta, with comments on feldspar composition: Bulletin of Canadian Petroleum Geology, v. 31, p. 148-160.

Rudkin, R. A., 1964, Lower Cretaceous in the geological history of Western Canada: Alberta Society of Petroleum Geologists, p. 169-189.

Smith, D. G., C. E. Zorn, and R. M. Sneider, 1984, The paleogeography of the Lower Cretaceous of western

Alberta and northeastern British Columbia in and adjacent to the Deep basin of the Elmworth area, *in* J. A. Masters, ed., Deep basin gas: AAPG Memoir 38, this volume.

Smith, R., 1984, Gas reserves and production performance of the Elmworth/Wapiti area of the Deep basin, *in* J. A. Masters, ed., Deep basin gas: AAPG Memoir 38, this volume.

Stewart, G. A., and G. T. MacCallum, 1978, Athabasca oil sands guidebook: Calgary, Canadian Society of Petroleum Geologists, 33 p.

Stott, D. F., 1967, Jurassic and Cretaceous stratigraphy between Peace and Tetsa rivers, Northeastern British Columbia: Geological Survey of Canada, Paper 66-7, 73 p.

—— , 1973, Lower Cretaceous Bullhead Group between Bullmoose Mountain and Tetsa River, Rocky Mountain foothills, northeastern British Columbia. Geological Survey of Canada Bulletin 219, 228 p.

Vail, P. R., R. M. Mitchum, and S. Thomson III, 1977, Seismic stratigraphy and global changes of sea level, part 3; relative changes of sea level from coastal onlap, *in* Seismic stratigraphy – applications to hydrocarbon exploration: AAPG Memoir 26, p. 63–81.

Weimer, R. J., 1983, Relation of unconformities, tectonics, and sea level changes, Cretaceous of the Denver Basin and adjacent areas, *in* Mesozoic paleogeography of the west-central United States: Rocky Mountain Section, Society of Economic Paleontologists and Mineralogists, Rocky Mountain Paleogeography Symposium 2, p. 359–376.

Williams, G. D., 1963, The Mannville Group (Lower Cretaceous) of central Alberta: Bulletin of Canadian Petroleum Geology, v. 2, p. 350–368.

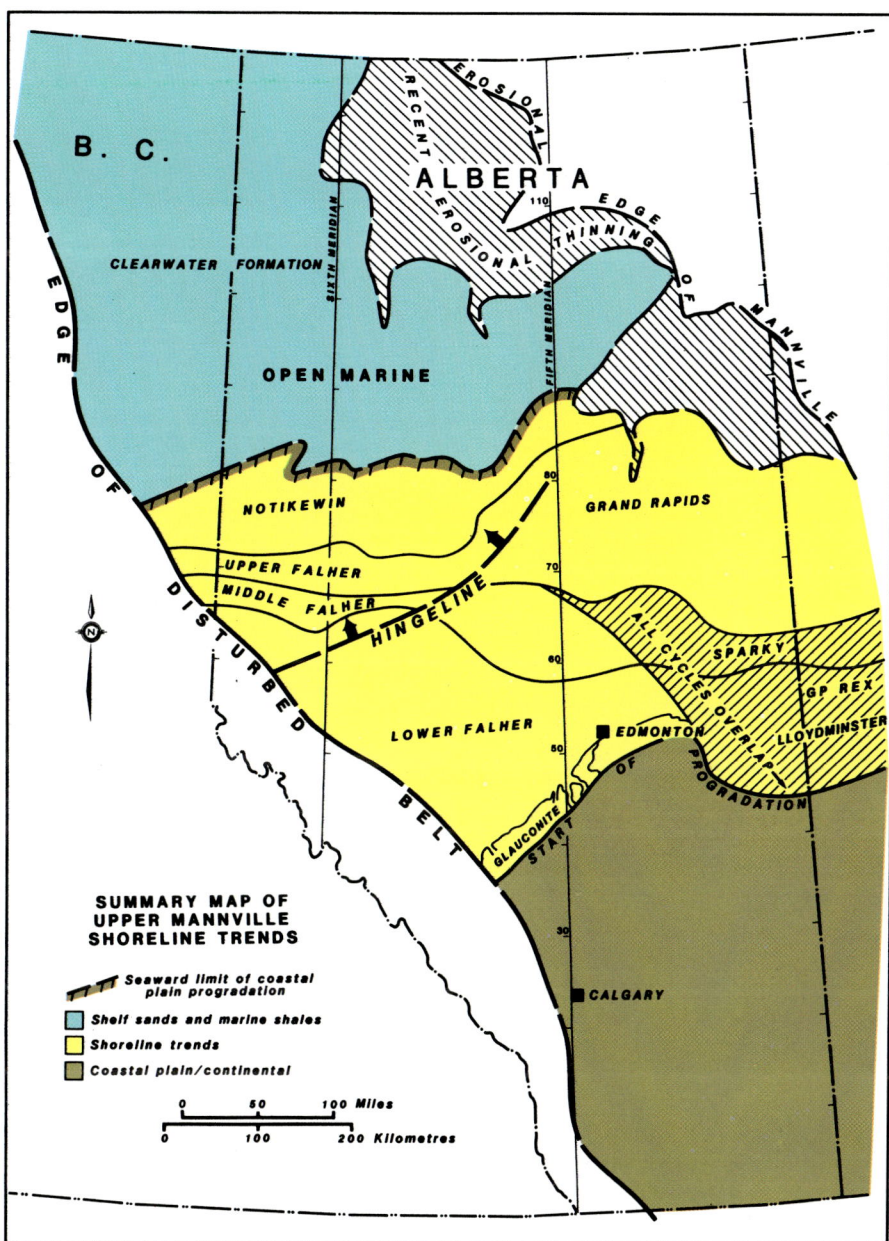

Figure 23. Summary map of upper Mannville shoreline trends from the initiation point of progradation in south central Alberta northward to the Clearwater Formation marine shale basin.

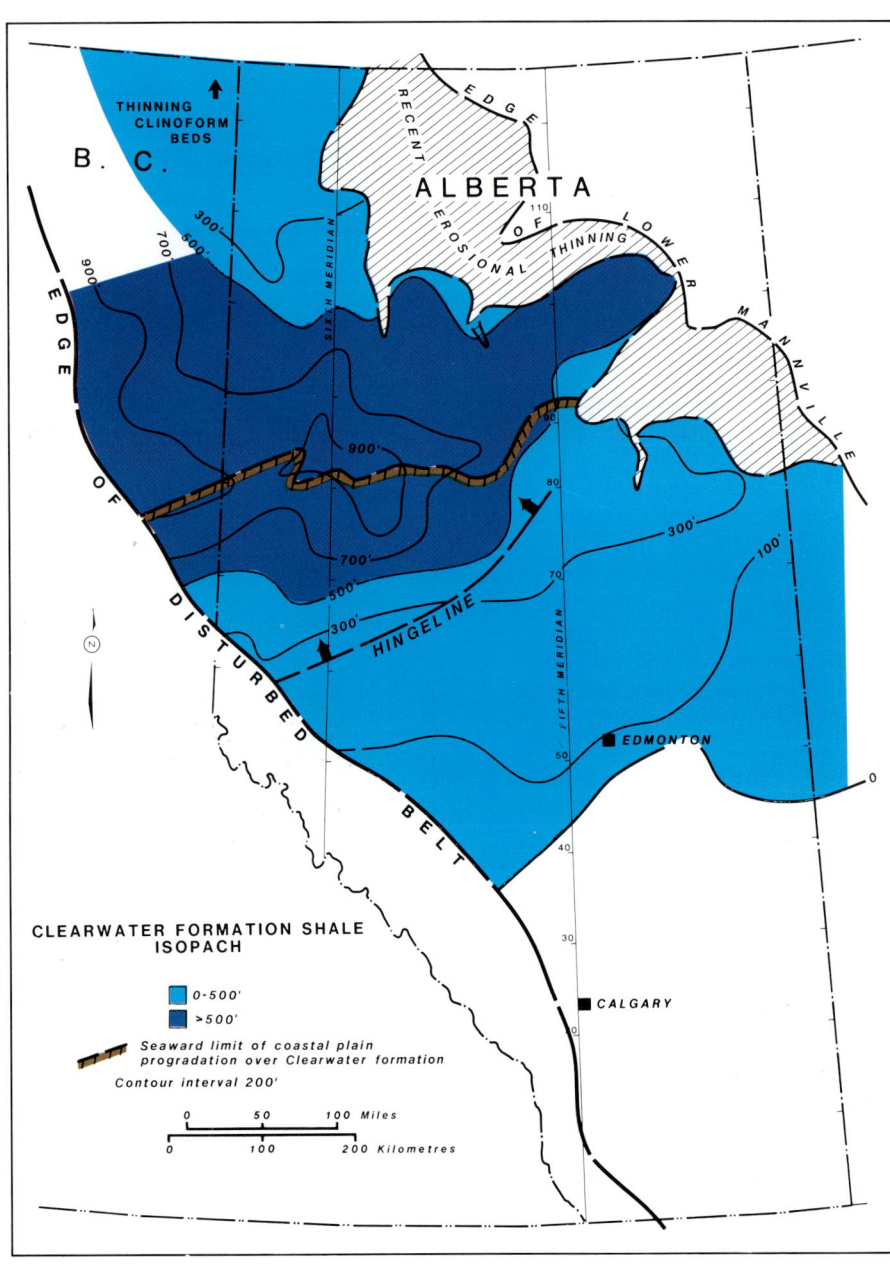

Figure 24. Clearwater Formation marine shale isopach, showing the basin hingeline and northward thinning shale clinoform beds into the basin.

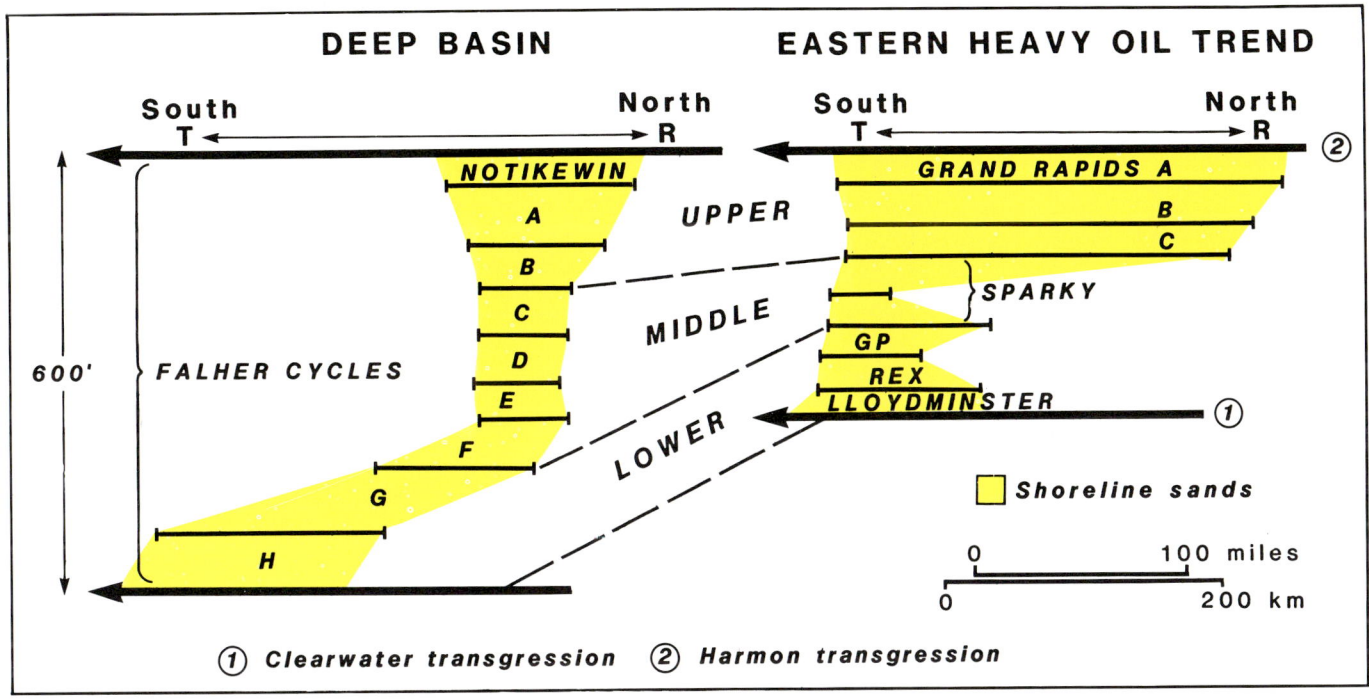

Figure 25. Comparison of depositional cycles in the western Deep basin and eastern Heavy Oil areas.

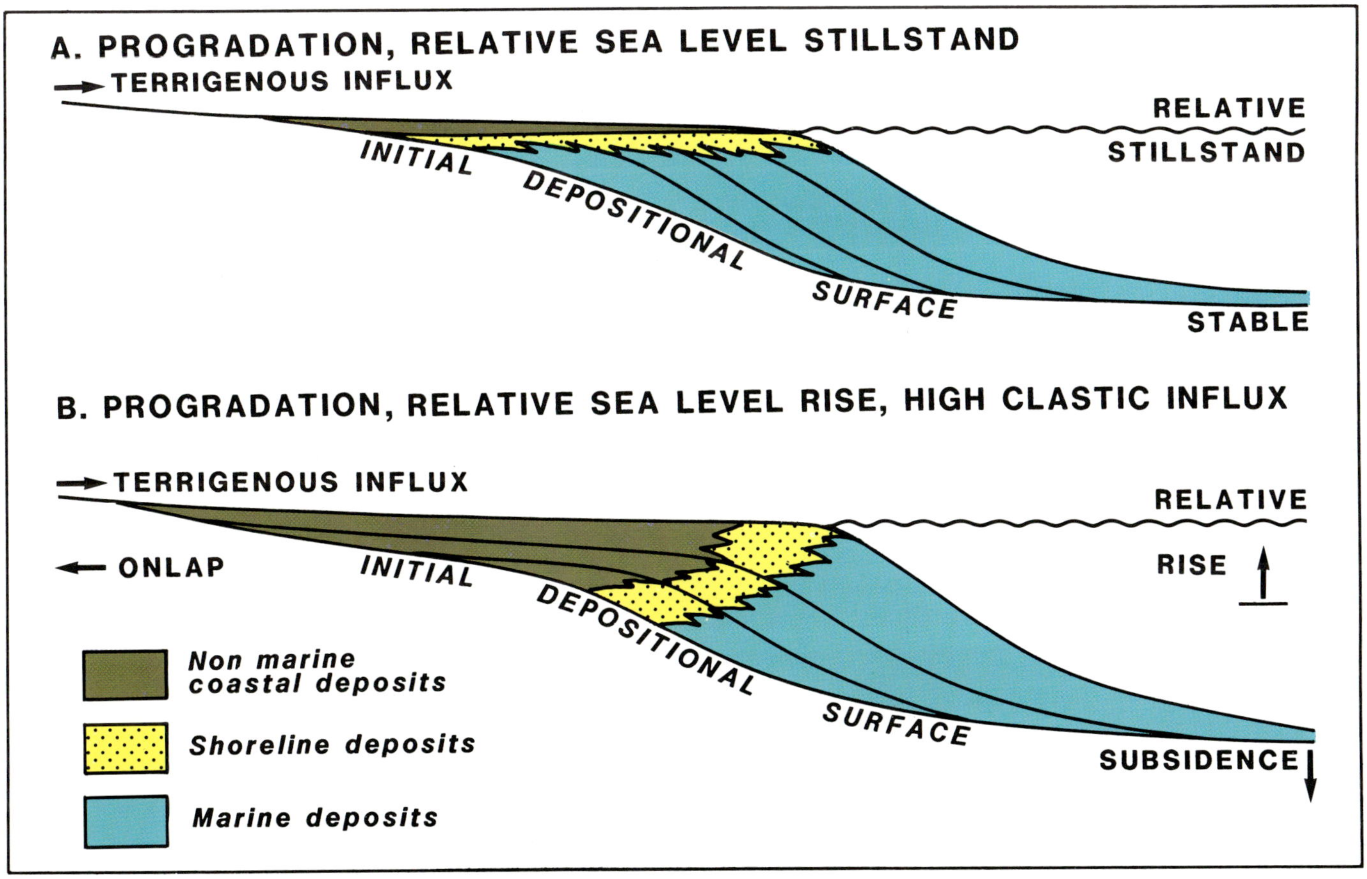

Figure 26a. Progradational profile with a relative sea-level stillstand, a stable basin.
Figure 26b. Progradational profile with a relative sea-level rise, a subsiding basin.

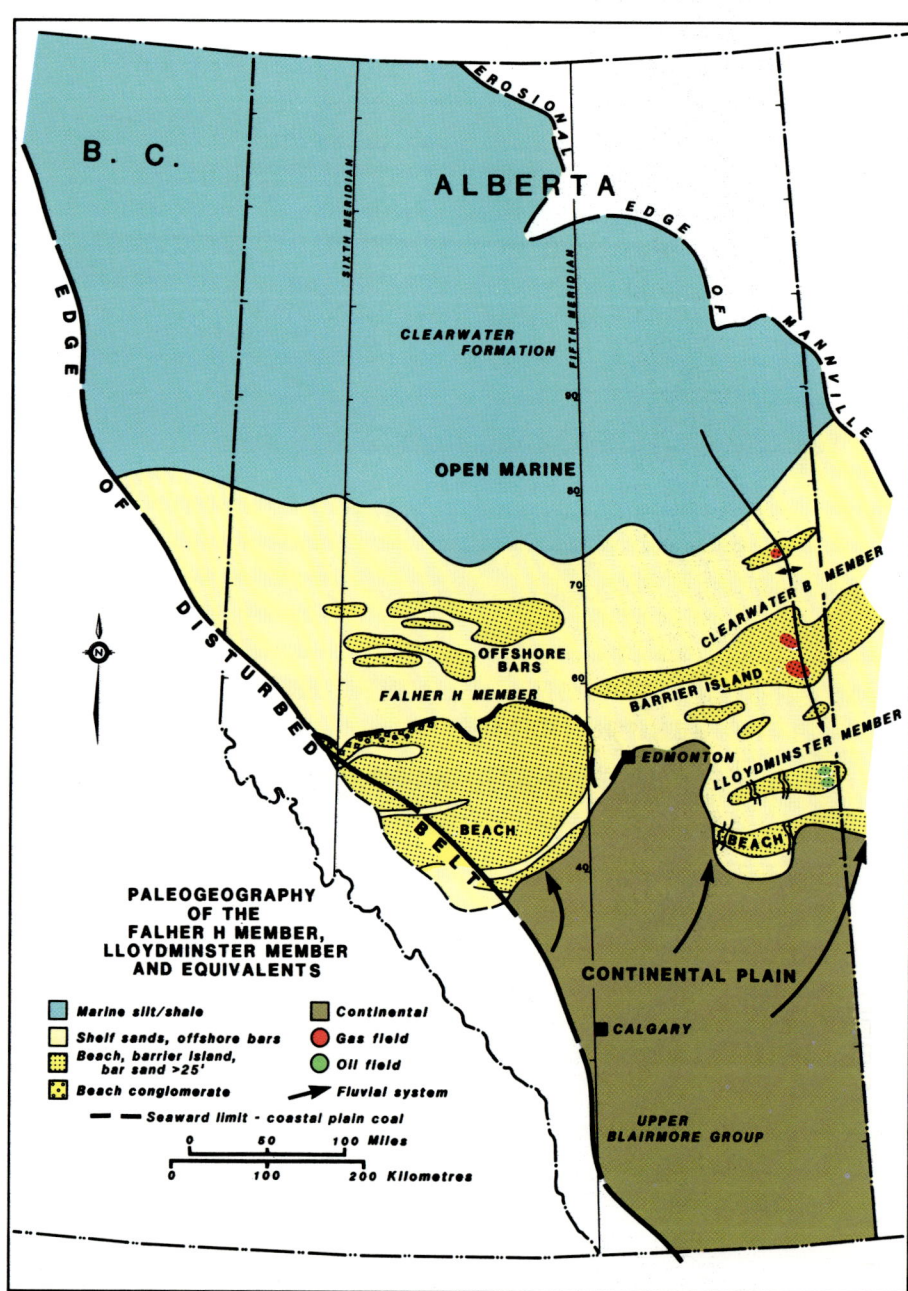

Figure 27. Paleogeography of the Falher H Member, Lloydminster Member, and equivalents.

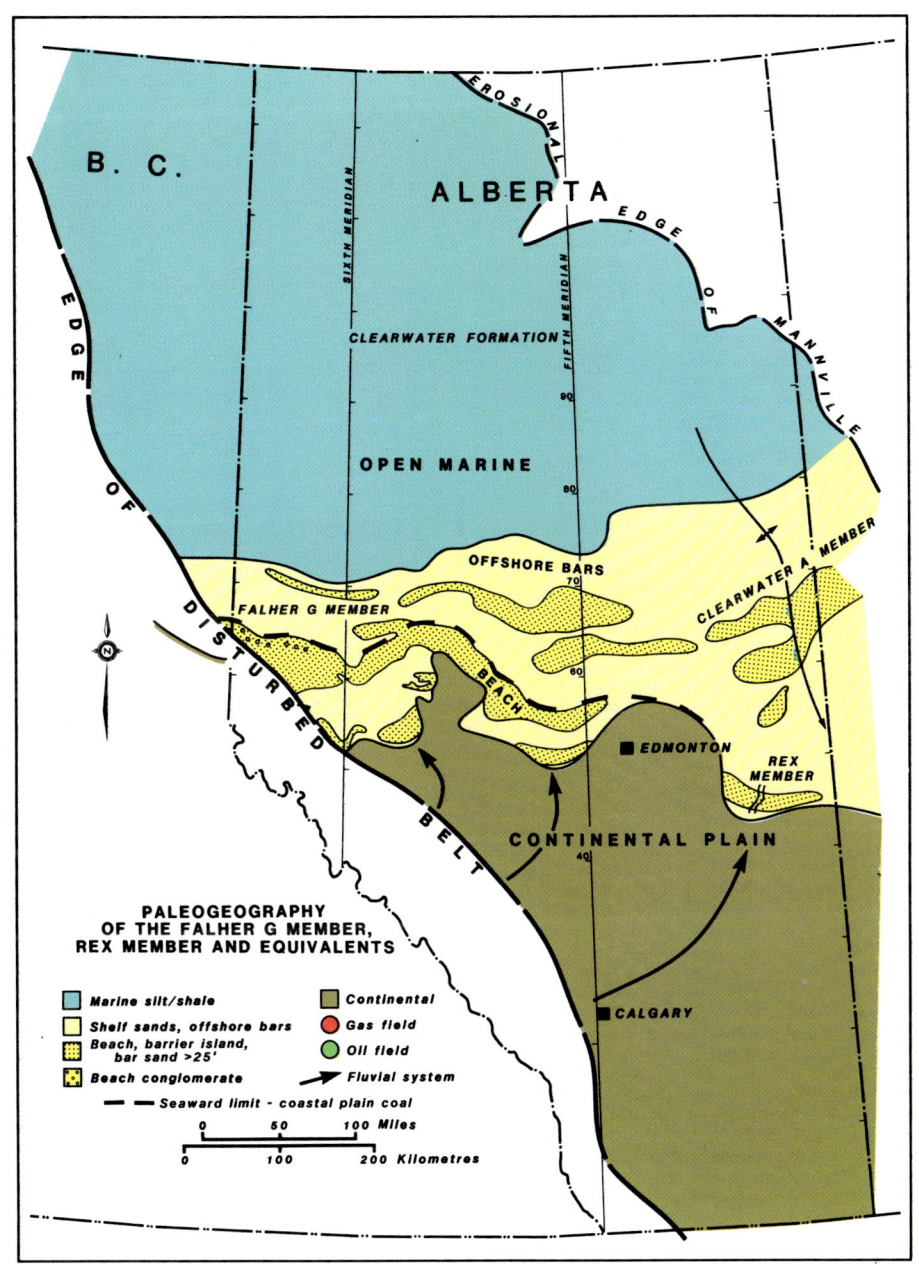

Figure 28. Paleogeography of the Falher G Member, Rex Member, and equivalents.

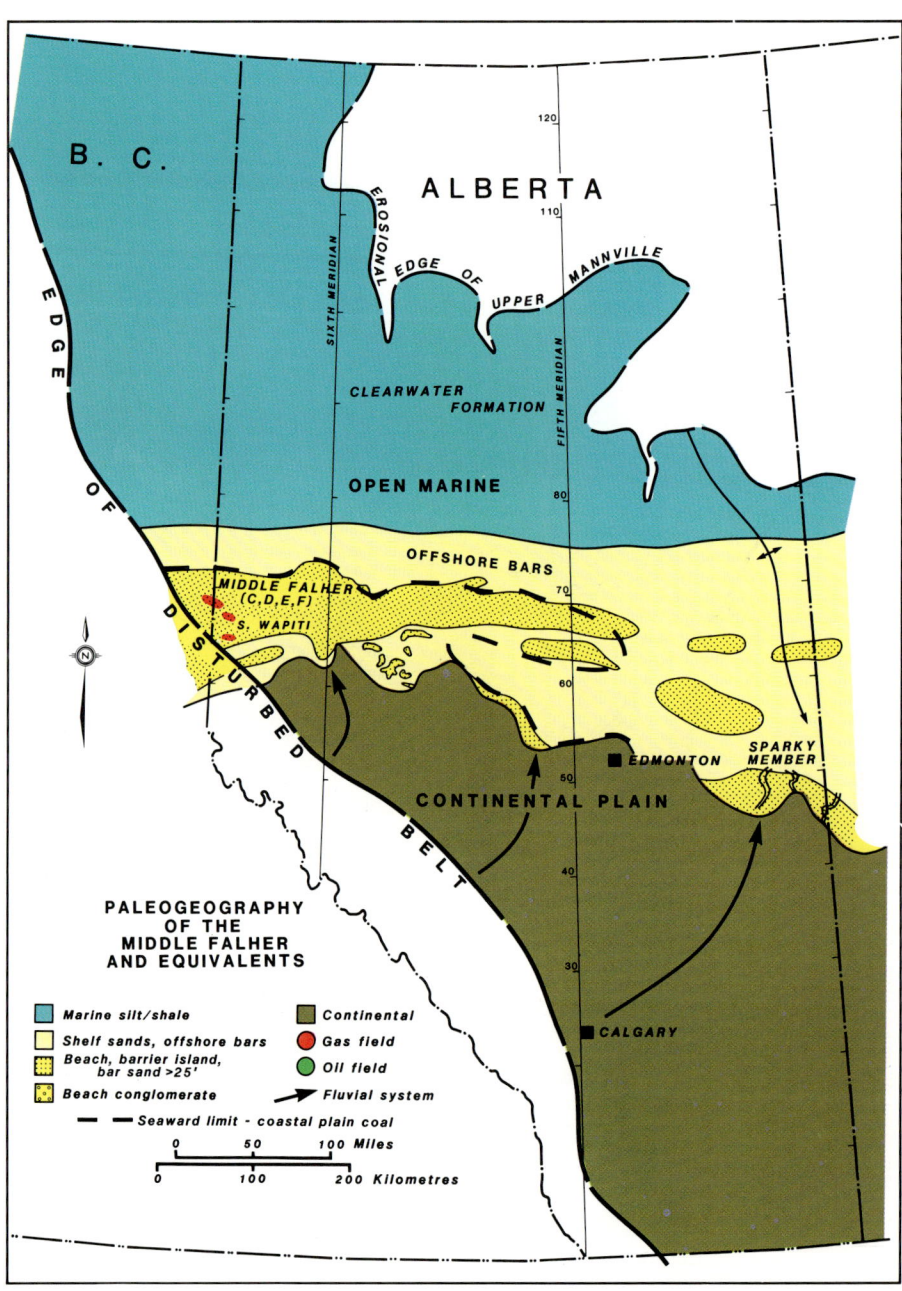

Figure 29. Paleogeography of the Middle Falher (C, D, E, and F cycles) and equivalents.

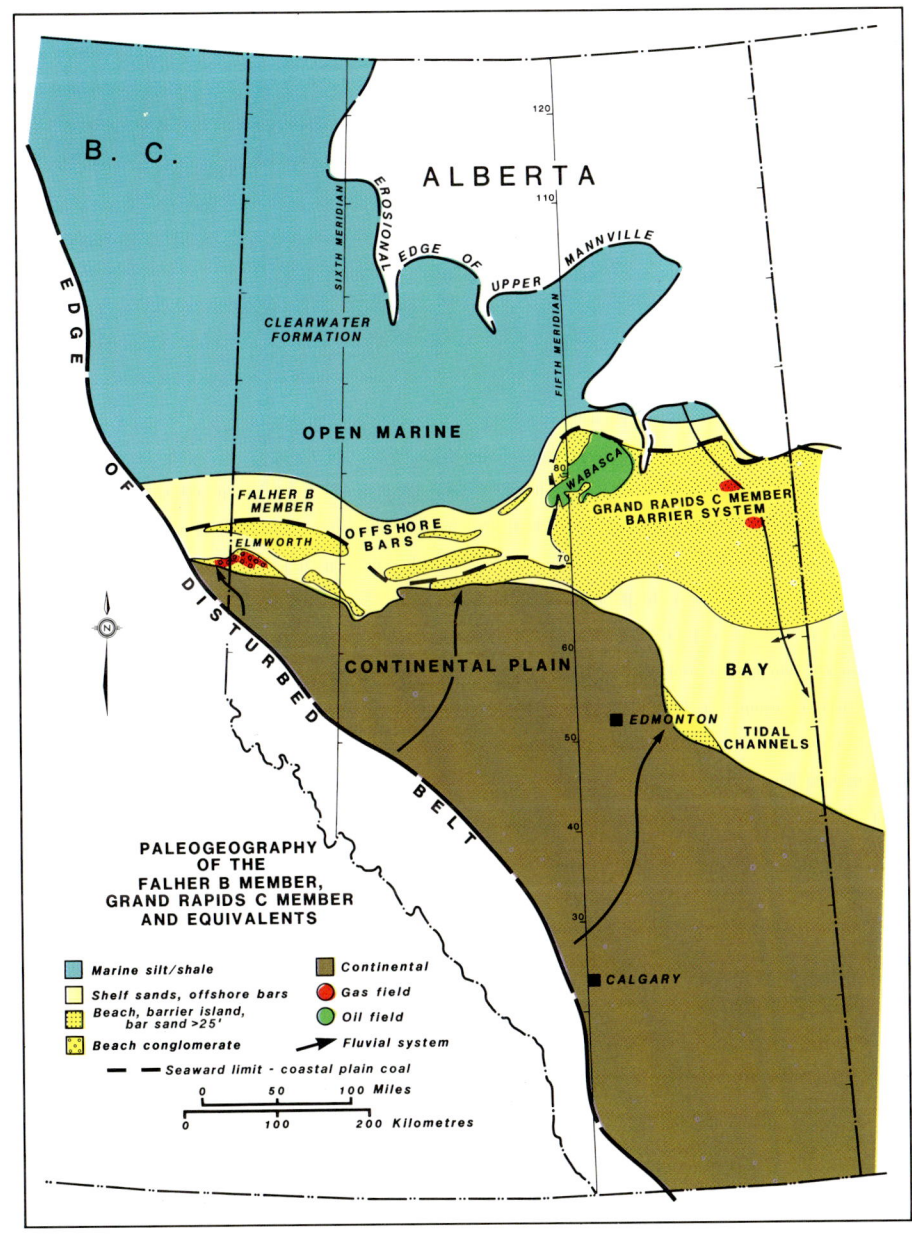

Figure 30. Paleogeography of the Falher B Member, Grand Rapids C Member, and equivalents.

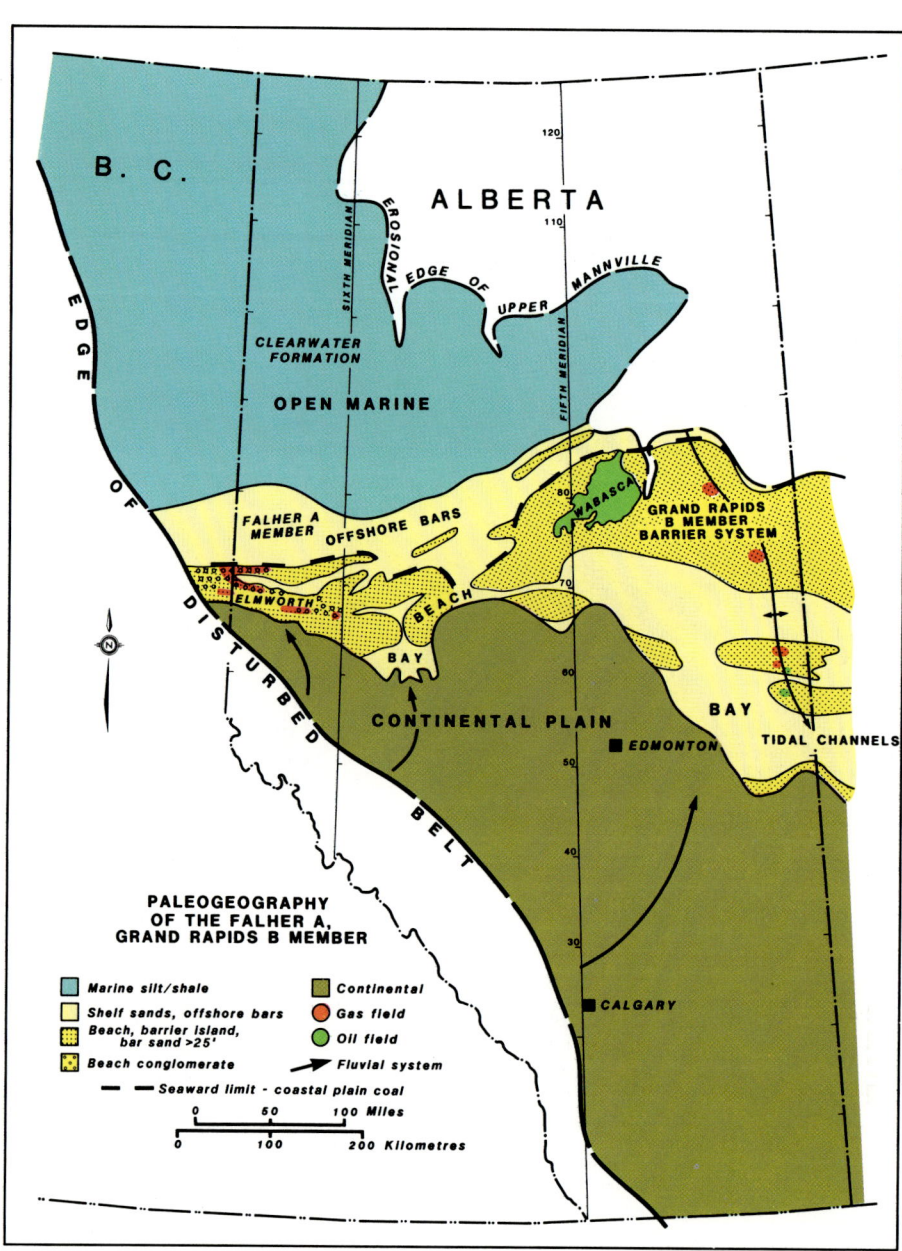

Figure 31. Paleogeography of the Falher A Member, Grand Rapids B Member, and equivalents.

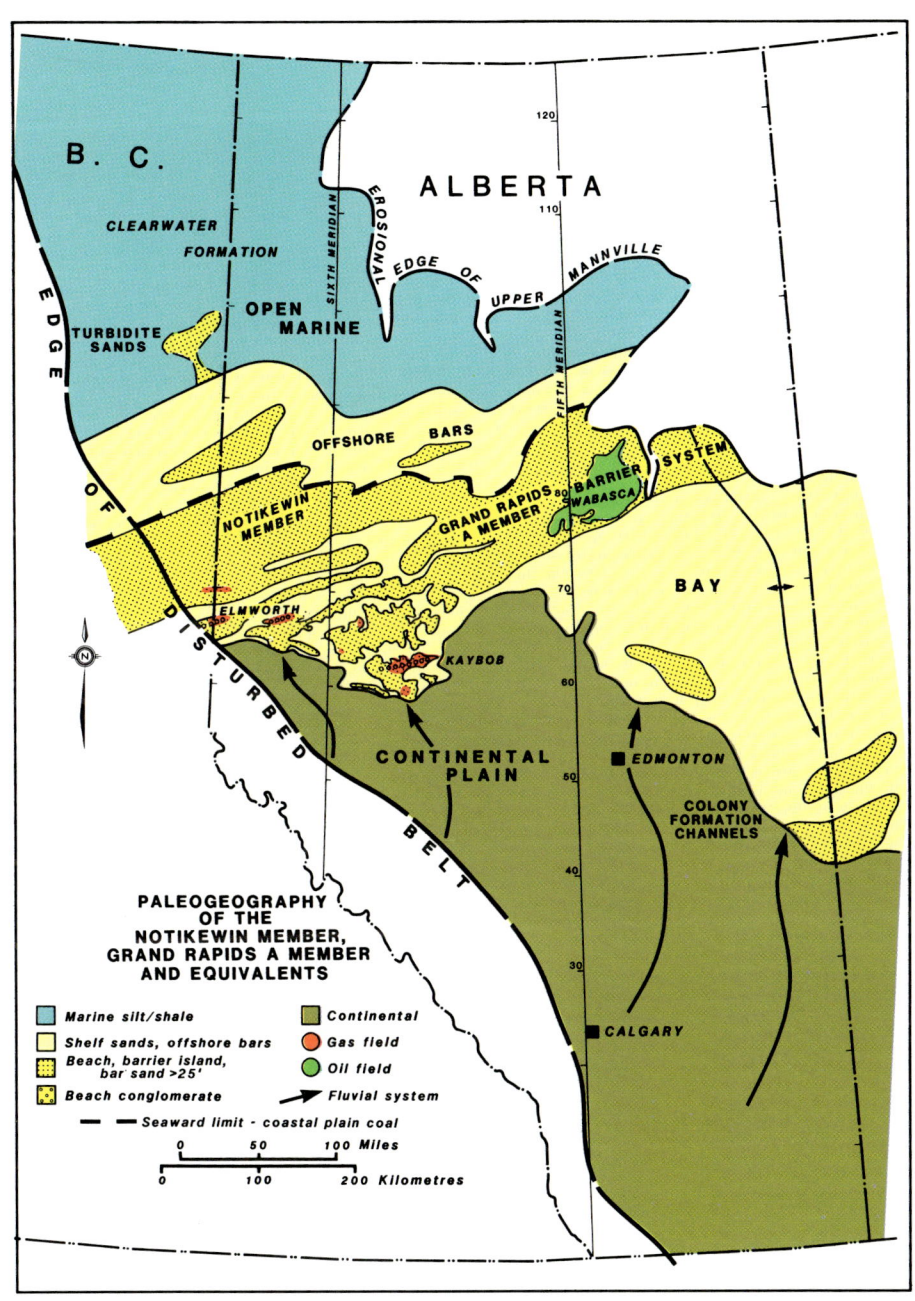

Figure 32. Paleogeography of the Notikewin Member, Grand Rapids A Member, and equivalents.

The Paleogeography of the Lower Cretaceous of Western Alberta and Northeastern British Columbia in and Adjacent to the Deep Basin of the Elmworth Area

David G. Smith
Carl E. Zorn
*Canadian Hunter Exploration, Ltd.
Calgary, Alberta*

Robert M. Sneider
*Robert M. Sneider Exploration, Inc.
Houston, Texas*

Following a period of Early Cretaceous Hauterivian uplift and erosion, the Deep basin area of western Alberta and northeastern British Columbia began to subside and receive sediments from the rising Cordillera to the west.

During Barremian time, the alluvial fan and braid plain conglomerates of the Cadomin Formation were deposited in a belt flanking the eastern margin of the Cordillera.

In Aptian time, the developing trough continued to deepen resulting in the accumulation of the fluvial and delta plain sediments of the Gething Formation.

In early Albian time, continued subsidence of the trough coupled with a eustatic rise in sea level resulted in a major transgression of the Deep basin from the north by the Boreal sea. This event is represented by the coastal and shallow marine sandstones of the Bluesky Member. The Bluesky was capped by the marine shales of the Moosebar Formation/Wilrich Member as the Boreal sea continued to deepen and advance southward.

Due probably to increased Cordilleran tectonism, marine conditions did not persist. During middle- to late-early Albian time, a major flood of sediment restored the Deep basin to continental conditions. The regressive, coastal/deltaic sandstones and conglomerates of the Falher and Notikewin members represent the northward advance of the coastline during this period of marine retreat.

A second major marine transgression advanced across the Deep basin during early-middle Albian time. This event is represented by the marine shales of the Hulcross/Harmon members which cap the Notikewin Member. A well-developed regressive cycle, represented by the coastal plain to shallow marine sediments of the Paddy and Cadotte members occurs within this transgressive pulse, which continued until at least the end of Albian time.

Lower Cretaceous sandstone reservoirs in the Deep basin exhibit average porosities around 8.0% and average permeabilities near 0.001 millidarcys, reflecting the "tight sand" nature of the Deep basin. In coarse-grained sandstone and conglomerate reservoirs however, average permeabilities are much higher, ranging from 20 to 80 millidarcys. As a result, in areas of the Deep basin containing such reservoir rock, gas productivity is quite high.

To provide an understanding of the depositional framework of Lower Cretaceous sandstones of the Deep basin, the paleogeography of all Lower Cretaceous sandstone units has been mapped. This will aid in defining hydrocarbon exploration fairways as well as in outlining potential areas of coarser grained sediments where higher permeabilities should be present.

INTRODUCTION

The Deep basin of western Alberta and northeastern British Columbia is now well established as one of North America's giant gas fields with recoverable gas reserves somewhere in the 50 to 150 tcf range (Masters, 1979). However, development of this huge field continues to be a major challenge to the energy industry. Much of the gas is found in low porosity, low permeability reservoir rocks. This problem coupled with present economics makes commercial development of many of the gas zones extremely difficult.

Paramount to the successful exploitation of this huge natural gas resource is the ability to identify areas of good porosity and permeability which are usually associated with chert pebble conglomerates and coarse-grained sandstones.

This paper is presented as an aid in understanding the depositional framework of Lower Cretaceous sandstones in and adjacent to the Deep basin of the Elmworth area (see study area Figure 1). The purpose of the paper is to identify and map all potentially commercial Lower Cretaceous depositional systems, and to provide a spatially accurate model of sandstone distribution to assist in the identification of areas of good porosity and permeability.

METHODS OF STUDY

During the exploration and development of the Deep basin play, approximately 27,000 ft (8,200 m) of Lower Cretaceous rocks were cored as operators sought to understand lithology, reservoir

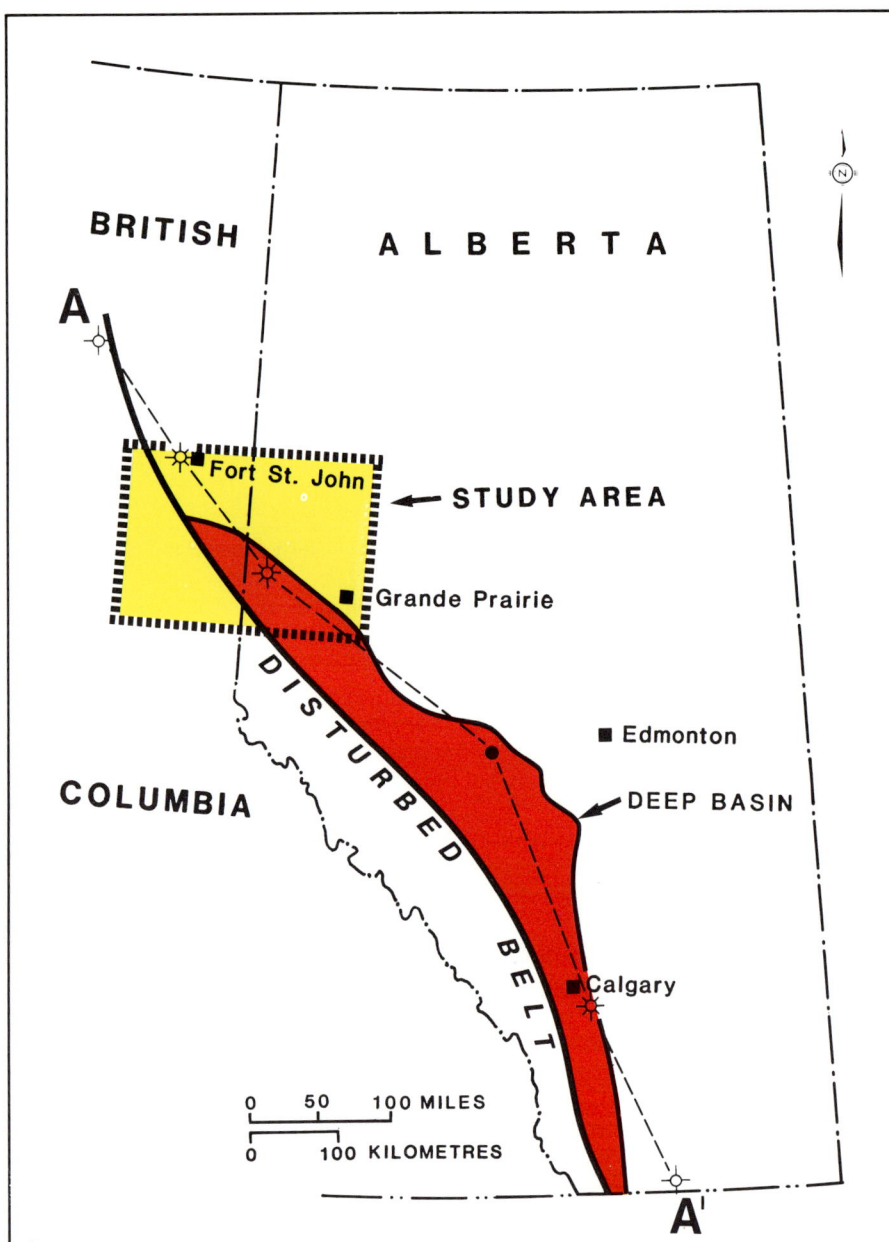

Figure 1. Location map showing the position of the study area relative to the Deep basin of Alberta and the trend of regional cross section A-A'(Fig. 3). The segment of the Deep basin which occurs within the study area (Alberta only) is essentially the Elmworth gas field.

depositional environments are based on gamma ray – sonic log responses. All available well control within the study area was used in defining the paleogeography of each Lower Cretaceous system.

Understandably, in any prograding system, the depositional environment changes with time within the mapped zone. In the subsequent series of paleogeographic maps, the paleoenvironment depicted is the dominant facies represented by the well bore.

GEOLOGIC SETTING

By Early Cretaceous Neocomian time, the Cordillera of western North America had become essentially continuous from Mexico to Alaska (Williams and Stelck, 1975; Jeletzky, 1971). Subsequent periodic uplifts within the Cordillera throughout Cretaceous time resulted in the shedding of clastics westward into the Pacific Ocean and eastward into the North American interior. The physiography of the interior during this time varied between a low-relief, near sea-level plain and a shallow epicratonic sea in which sedimentation essentially kept pace with basin subsidence.

The changing relationship between land and sea in the interior during the Early Cretaceous is best understood by examining the various lithologies and environments of deposition of Lower Cretaceous sediments. Figure 3, a cross section stretching from the U.S. border to northeastern British Columbia, ties the various Lower Cretaceous formations to regional facies patterns in the basin.

Although not covered in this paper, some Early Cretaceous (Valanginian) clastics are found beneath the sub-Cretaceous unconformity in the otherwise Jurassic Nikanassin Formation of the British Columbia foothills (Stott, 1973).

The sub-Cretaceous unconformity was developed during Early Cretaceous Hauterivian time as a result of a pause in Cordilleran orogenic activity. Minimal basin subsidence or uplift, combined with

quality and environments of deposition within this complex geological system. Borehole logs, when calibrated with cores, allow numerous inferences to be made regarding lithology, reservoir quality and environments of deposition from log response. The gamma ray – sonic log combination is particularly useful in that it readily identifies coarsening upward, prograding sequences (barrier bars, offshore bars, mud- or sand-dominated deltas), fining upward sequences (tidal or fluvial channels), and coals or carbonaceous beds (lagoonal, delta plain or alluvial plain facies). See the example log in Figure 2.

In this paper, all interpretative core data with respect to environments of deposition have been taken from the files of Canadian Hunter Exploration Ltd. Within the greater Elmworth field, cores have been calibrated with logs to match environments of deposition with log signatures. Throughout the remainder of the study area, all interpretations regarding

Figure 2. Lower Cretaceous correlation chart for the Foothills outcrop belt and the adjacent Deep basin of the study area. Type log is from the Elmworth field.

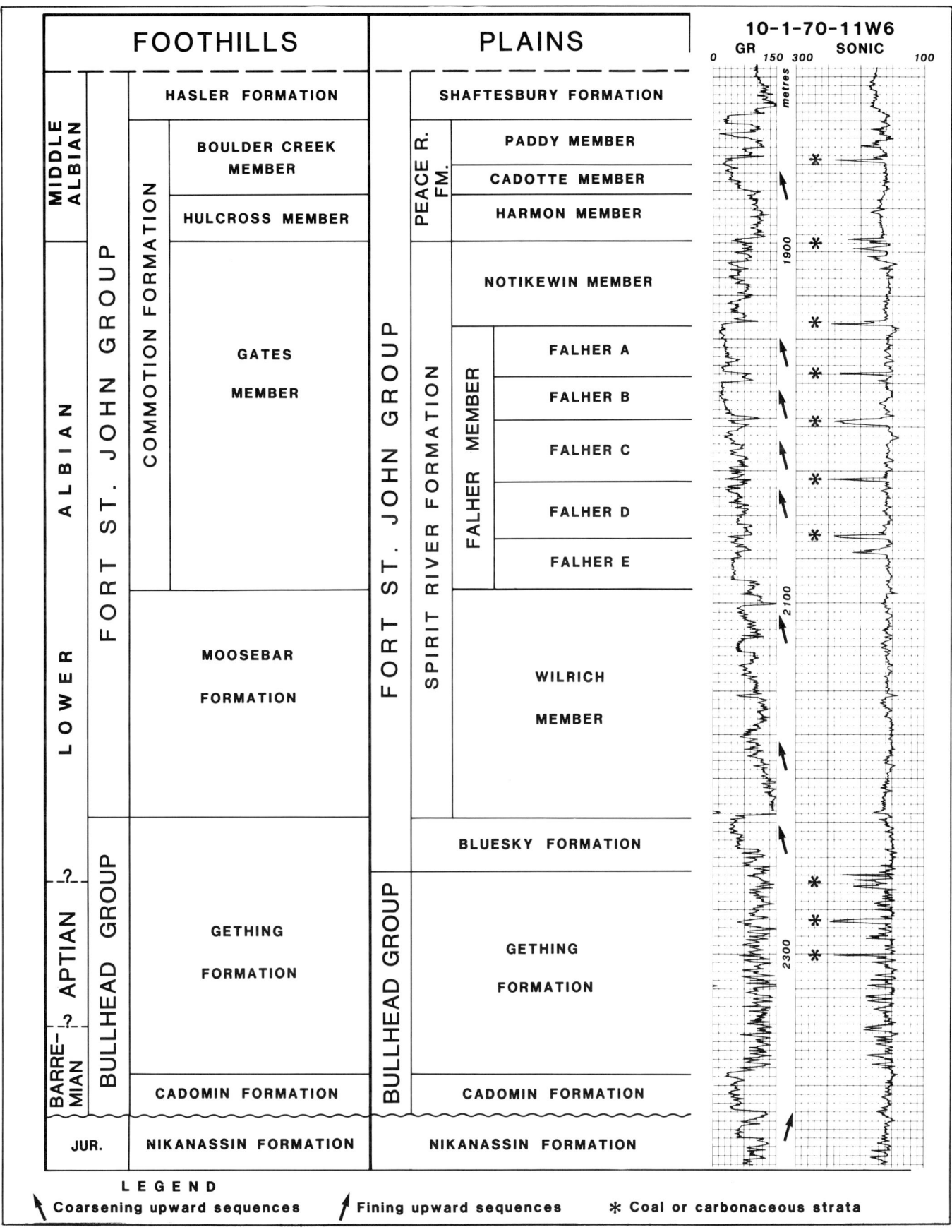

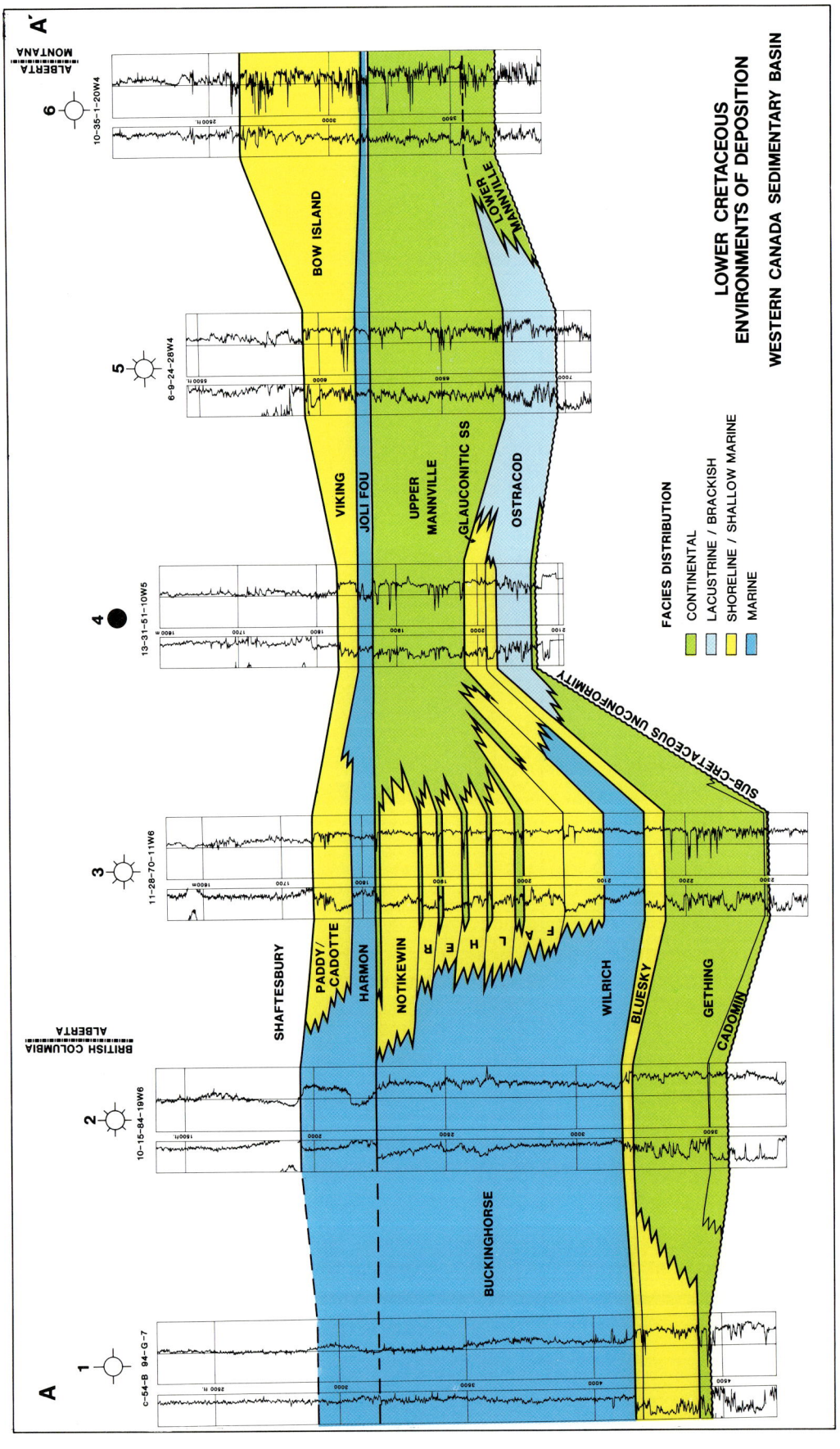

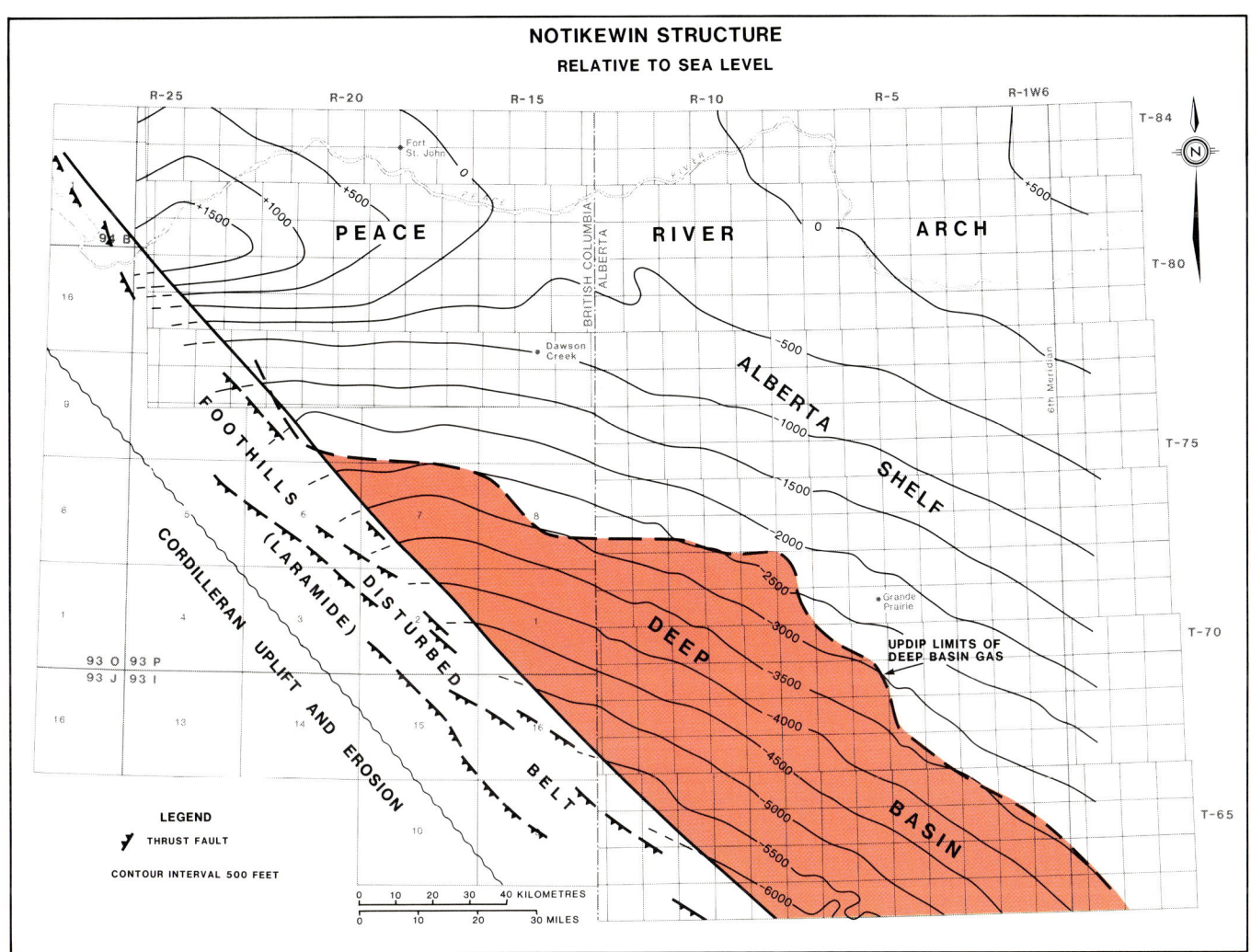

Figure 4. Structure map of the study area at the Lower Cretaceous Notikewin level showing configuration of the Deep basin and the adjacent Peace River Arch.

lack of sediment supply, caused erosion across the bulk of the interior. Truncation of pre-existing sediments increased from west to east across the basin.

A strong Cordilleran uplift during Barremian time once again caused shedding of clastics (probably from the Main Ranges of the Rocky Mountains) into the interior. This event is represented by the alluvial fan and braid-plain conglomerates of the Cadomin Formation, found along the western edge of the Alberta basin (Schultheis and Mountjoy, 1978; McLean, 1977).

The developing trough east of the Cordilleran uplift (Alberta basin) continued to deepen during Aptian time resulting in the accumulation of the incised valley, fluvial sediments of the Lower Mannville Formation in southern Alberta, and the fluvial and delta plain sediments of the Gething Formation in northern Alberta. Rivers flowed eastward from the Cordilleran uplands, then northward across a broad low-relief plain to the Boreal sea, probably joining en route other fluvial systems with sources on the Canadian Shield. Position of the Boreal coastline in northeastern British Columbia during Gething/Lower Mannville time is documented by Stott (1973). He describes Gething coastal/deltaic sediments as far south as the Peace River and classifies the entire Gething, north of the Sikanni Chief River as marine. In the subsurface of Alberta, Gething equivalent (McMurray Formation) coastal/deltaic and marine sediments are documented near Fort McMurray by Nelson and Glaister (1978).

Although precise age relationships have not yet been established in areas of fluvial/marine transitions (Stott, personal communication), it is clear that the Boreal sea advanced into northeastern British Columbia and northern Alberta during Aptian time. This is much further south than many previous authors had assumed. This transgression, which marks the beginning of a much more widespread invasion of the interior basin during Albian time,

Figure 3. Regional cross section A-A′ depicting Lower Cretaceous environments of deposition from north to south along the axis of the Western Canada Sedimentary Basin. Plains nomenclature is indicated as it relates to the various environments of deposition.

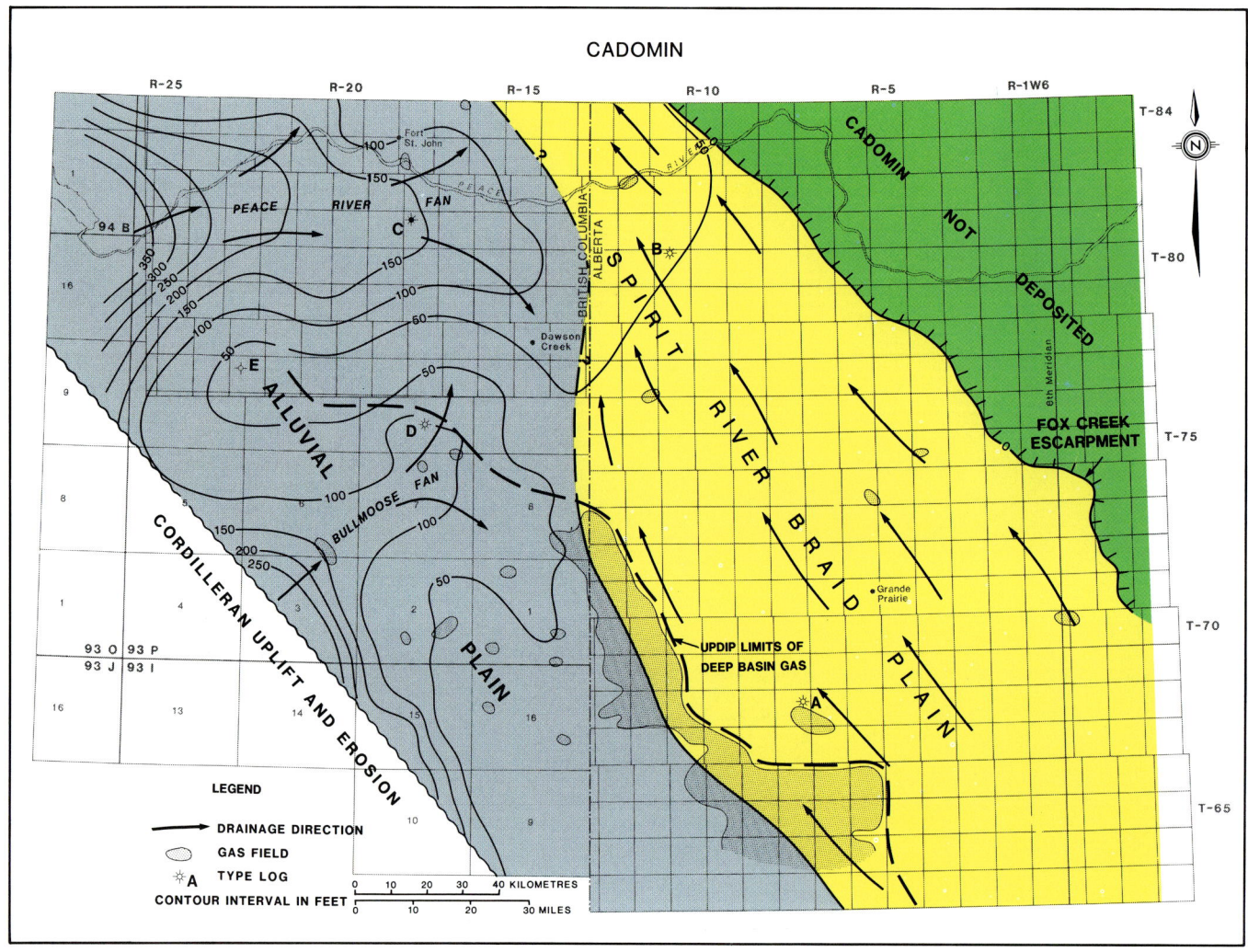

Figure 5. Conglomerate isopach and paleogeography of the Cadomin Formation.

correlates closely in time with similar advances on the Atlantic, Gulf of Mexico and Pacific margins, and probably represents an eustatic rise in sea level as well as basin subsidence (Vail et al, 1977, Hancock and Kauffman, 1979).

The strong southern advance of the Boreal sea across the Gething delta plain probably began in early Albian time as marked by the occurrence of lower Albian fauna in the marine shales of the Moosebar Formation of the British Columbia foothills and in the Clearwater Formation (Moosebar equivalent) in east-central Alberta (Stelck and Kramers, 1980). This marine invasion of the interior plains is represented by the coastal and shallow-marine sandstones of the Bluesky Member in northern Alberta, the barrier bars of the Glauconitic Sandstone in central Alberta, and the brackish water sediments of the Ostracod Zone (Calcareous Member) in southern Alberta. These sediments are, in turn, overlain in central and northern Alberta by the marine shales of the Moosebar/Wilrich/Clearwater. This transgressive event, commonly called the Moosebar or Clearwater sea, flooded much of Alberta and Saskatchewan and appears to have reached as far south as Montana, as indicated by the presence of brackish water dinoflagellates and acritarchs in the uppermost Sunburst Member near Great Falls, Montana (Burden, 1982, p. 59).

Marine conditions did not prevail in the south, however. Probably because of intensified tectonism in the Cordillera in middle- to late-early Albian time, a major flood of sediment inundated the interior, quickly restoring south and south-central Alberta to continental conditions to the approximate latitude of Edmonton. From then until the end of Notikewin/Gates time (late Early Albian) shorelines moved steadily northward. The coastal/deltaic sandstones and conglomerates of the Falher Member of the Elmworth gas field are part of this overall regressive cycle. This regressive pulse reached its maximum northward extent during Notikewin time when coastal/deltaic sediments are documented as far north as Dawson Creek, British Columbia.

A second major marine transgression inundated much of the interior during early-middle Albian time. This cycle is represented by the Hulcross/Harmon marine shales of northern Alberta and British Columbia and the Joli Fou shales of southern Alberta and Saskatchewan. This overall transgressive pulse, which continued until at least the end of Albian time (Shaftesbury Formation) and which ulti-

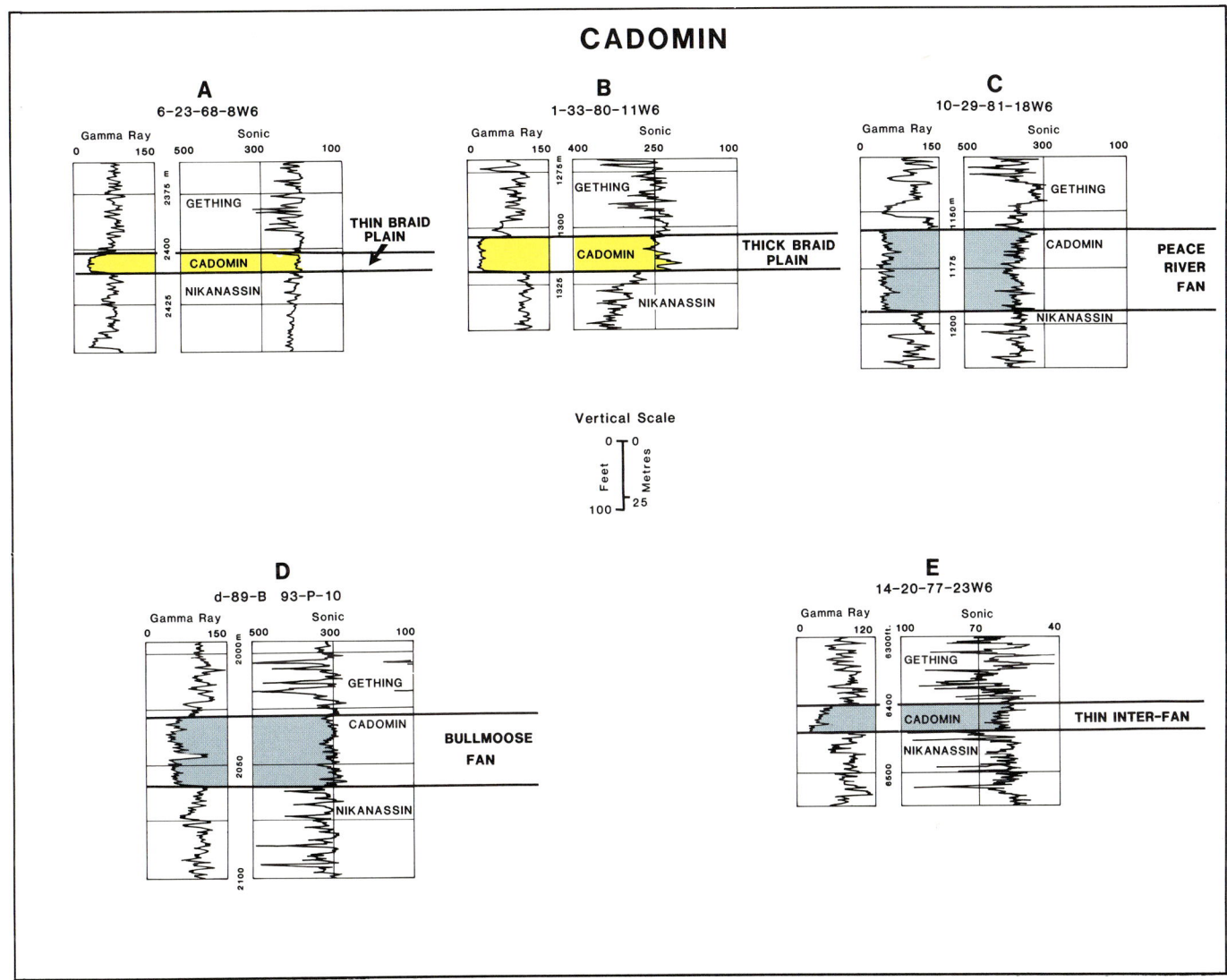

Figure 6. Type logs of the various facies of the Cadomin Formation.

mately connected the Gulfian and Boreal seas, had two separate and distinct regressive pulses within it. The Paddy/Cadotte regression, the older of the two, is a lobe of shallow-marine to coastal-plain clastics essentially localized over the Peace River Arch. It originated from a sediment source in the Cordillera, southwest of Fort St. John, British Columbia. The younger Viking regression (Stelck, 1958; Oliver, 1960) is a series of barrier- and shallow-marine bars in central and southern Alberta. These sands were derived primarily from the Cordillera by fluvial transport across a delta plain which is represented by the Upper Blairmore continental sediments of the southern Rocky Mountain foothills.

The middle and late Albian transgressive events of Alberta are not isolated events in the geologic record. Rather they are part of a system of Albian transgressions which are recognized worldwide and as such are almost certainly the result of a global eustatic rise of sea level (Vail et al, 1977; Hancock and Kauffman, 1979).

Tectonic elements affecting Lower Cretaceous deposition within the Alberta basin were minimal. Apart from the Cordilleran uplift, the only tectonic feature to affect Lower Cretaceous sediments within the study area was the Peace River Arch (Fig. 4). Even this structure, previously very active during Paleozoic time both as a positive and negative feature (De Mille, 1958; Lavoie, 1958), had limited effect on the Lower Cretaceous. The only evidence of activity is a gentle thickening of Lower Cretaceous strata along the arch axis, particularly southwest of Fort St. John. Subsequent to Lower Cretaceous deposition, however, probably during the Laramide orogeny of the Cordilleran area (Late Paleocene to Early Eocene, 55 m.y. ago), the Peace River Arch became active once again as a positive feature. This is evident from the present-day structural configuration of Lower Cretaceous strata (Fig. 4), which shows major uplift along the axis of the arch.

DEPOSITIONAL MODELS AND PALEOGEOGRAPHY

Cadomin Formation

The conglomerates of the Cadomin Formation are considered to have formed

Figure 7. Artist's concept of the landscape of the study area during Cadomin time. View is to the north. (Compare with Fig. 5).

in a piedmont alluvial plain environment (McLean, 1977; Stott, 1968; Gies, 1984). The terrestrial origin of these conglomerates is also supported by their intertonguing with coal-bearing Gething sediments in foothills outcrops (Stott, 1968, p. 110).

At various places along the Cordilleran uplift, Cadomin rivers debouched onto the interior plains. Upon emerging from the confines of the mountain canyon system, these streams abruptly deposited their bed-load material to build large alluvial fans. With continued water flow and redistribution of sediment, these fans expanded away from their source area and ultimately merged to form a single widespread alluvial plain covering much of western Alberta (Figs. 5, 6 and 7).

In the area of study, two piedmont source areas are readily observable by examining an isopach of net Cadomin conglomerate (Fig. 5). The smaller of the two, named the Bullmoose Fan in this paper, appears to have originated somewhere southwest of Bullmoose Mountain. Further north a much larger fan, centered on the Peace River, forms a prominent lobe extending eastward essentially to the eastern limits of Cadomin deposition. Stott describes over 700 ft (213 m) of Cadomin strata exposed in the vicinity of the W.A.C. Bennett Dam on the Peace River (Stott, 1968, sections 61-22, 61-23).

Stott records a third fan centered near Mount Belcourt where he logs 530 ft (161.5 m) of Cadomin in outcrop (Stott, 1968). Little evidence of this fan could be found by the authors in the subsurface. In the well b-27-I/93-I-8, located in the foothills thrust belt just 14 mi (22.5 km) east of Mount Belcourt, two fault slices containing Cadomin strata are noted. In the lower slice only 30 ft (9 m) of Cadomin conglom-

Figure 8. Net sand isopach and paleogeography of the Gething Formation. Contour interval is in feet.

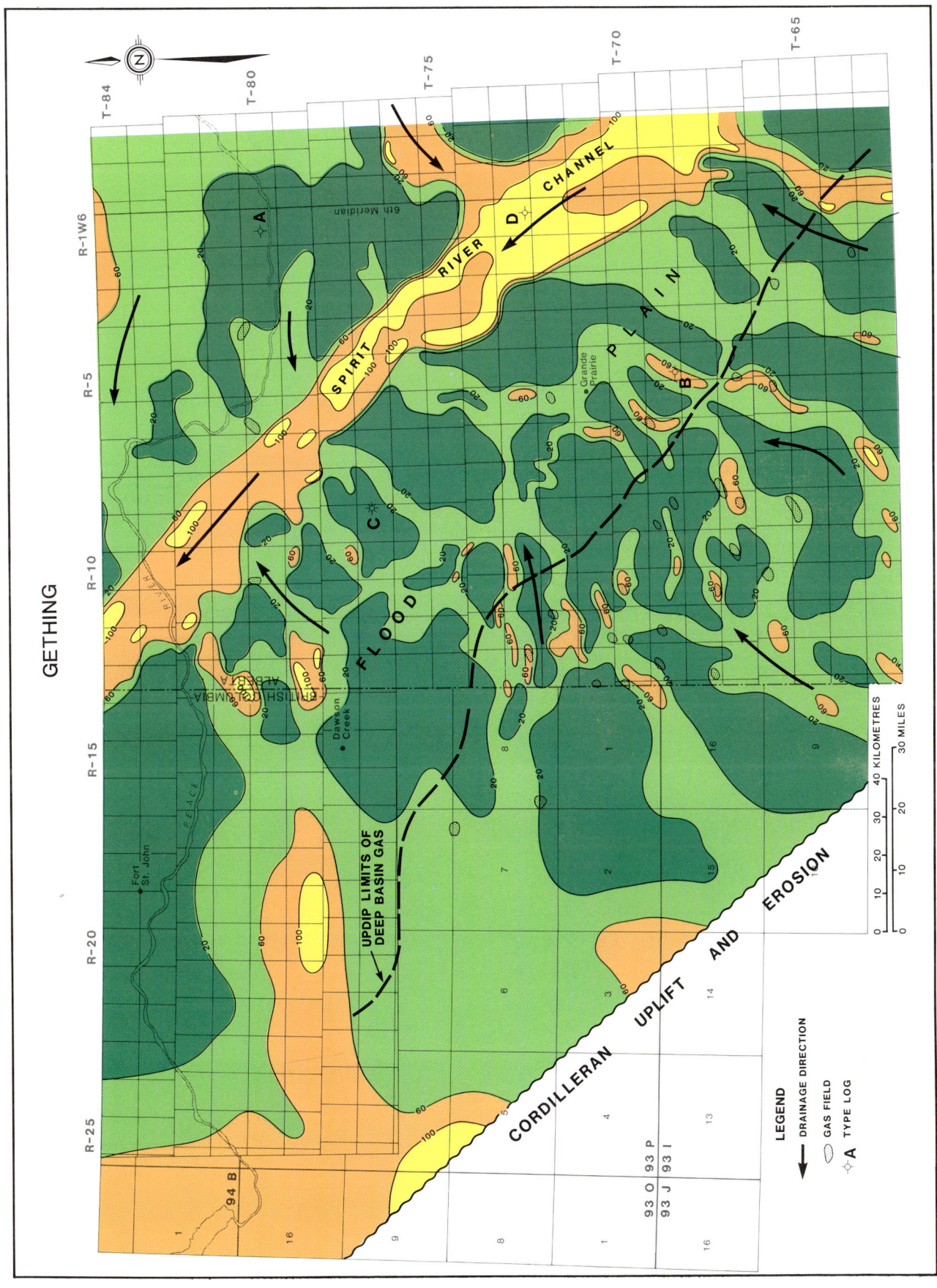

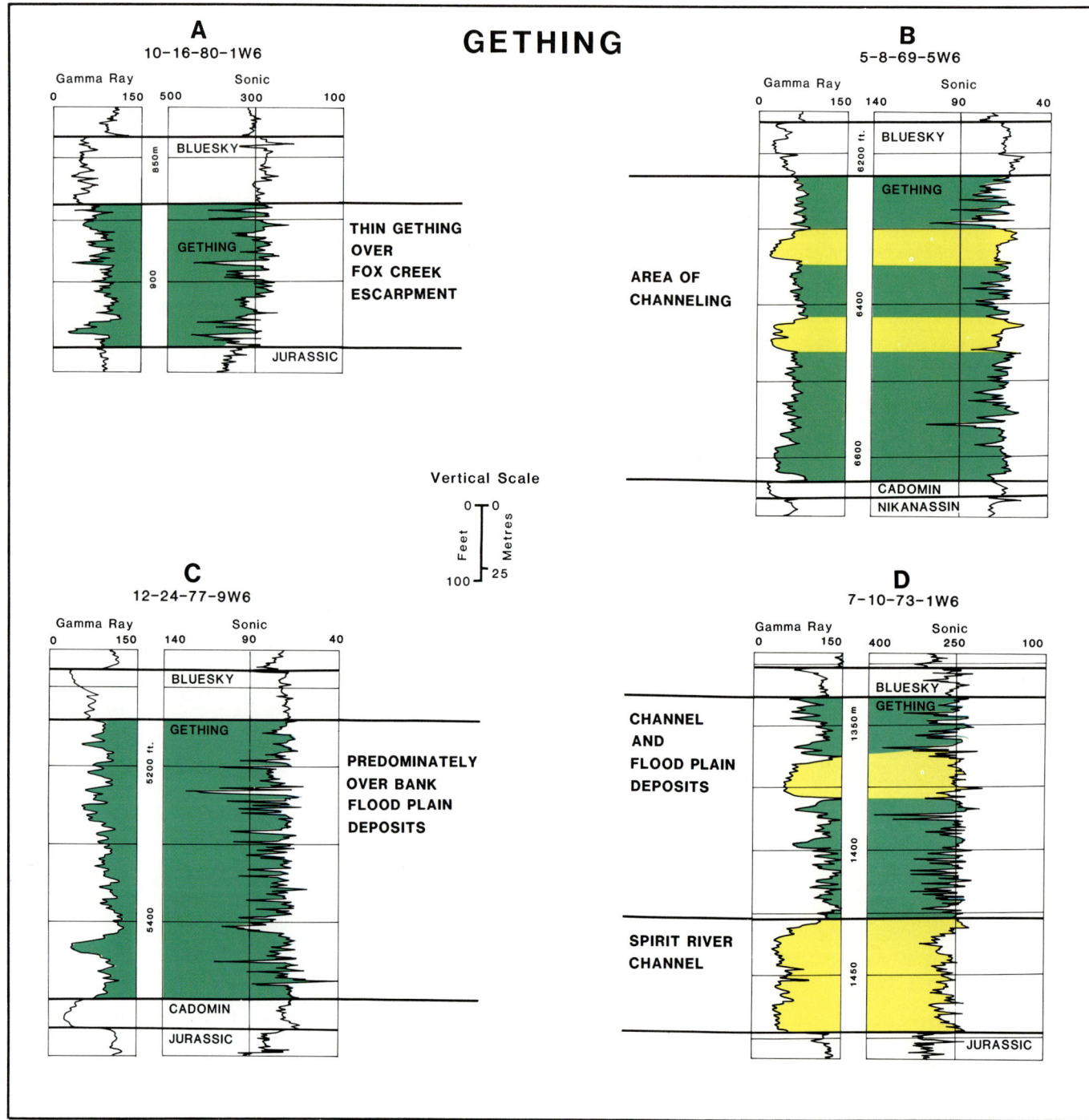

Figure 9. Type logs of the various facies of the Gething Formation.

erate is observed from logs and samples. In the upper slice, 145 ft (44 m) of conglomerate is logged in samples, but it is not known whether all or part of this section is Cadomin, with the remainder being the Nikanassin Formation (commonly conglomeratic in this area). It is postulated that Laramide thrusting displaced the Mount Belcourt section eastward relative to its original position during Cadomin deposition, and Mount Belcourt is in reality either part of the Bullmoose fan or a more localized fan not widely expressed in the subsurface.

The eastern limit of Cadomin deposition is marked by the Fox Creek Escarpment, a gentle range of hills during Barremian time east of the Cadomin catchment basin. This paleotopographic high is expressed in the subsurface by a thin Gething, no Cadomin (although basal Gething channels may be present and can be confused with the Cadomin) and an abnormally thick Jurassic section pre-

Figure 10. Artist's concept of the landscape of the study area during Gething time. View is to the northeast. (Compare with Fig. 8).

served beneath the sub-Cretaceous unconformity.

Between the Fox Creek Escarpment and the Cadomin alluvial plain facies lies a broad Cadomin braid-plain formed by a north-flowing trunk river system called the Spirit River Channel (McLean, 1976, 1977). This facies is separate and distinct from the alluvial plain facies to the west (Gies, 1984; Varley, 1983). Alluvial plain deposits consist of poorly-sorted, sandy, dark chert pebble conglomerates, while Spirit River fluvial braid plain conglomerates are better sorted, consist of lighter-colored chert pebbles that are generally smaller. Quartzite pebbles are common in the alluvial-plain facies, but absent in the braided-river facies. Detrital clays are found in cores of the alluvial-plain facies whereas shaly interbeds and associated clay rip-up clasts are common in the Spirit River Channel. The western limit of the braided river facies is outlined in Figure 5 (after Gies, 1984). It should be noted, however, that this represents the maximum westward extent of the Spirit River Channel. There is much inter-tonguing of alluvial plain and braided river sediment along this trend as each system, in turn, dominated in the area.

Gething Formation

Sediments of the Gething Formation in the study area (see Figs. 8, 9 and 10) consist primarily of interbedded fine- to medium-grained sandstones, siltstones, mudstones and coal. The sequence is clearly terrestrial. Sandstones are fining upward or thin-bedded. Trough and planar crossbedded, ripple-bedded and parallel-laminated sandstones are common. Plant material including fossil leaves, stems, logs, stumps, and carbonaceous debris is common. Coalbeds occur throughout the Gething Formation; even dinosaur footprints have been documented in the Peace River canyon (Sternberg, 1932).

The Gething can best be described as a low relief interior drainage plain on the eastern flank of the Lower Cretaceous Cordillera. During early Gething time,

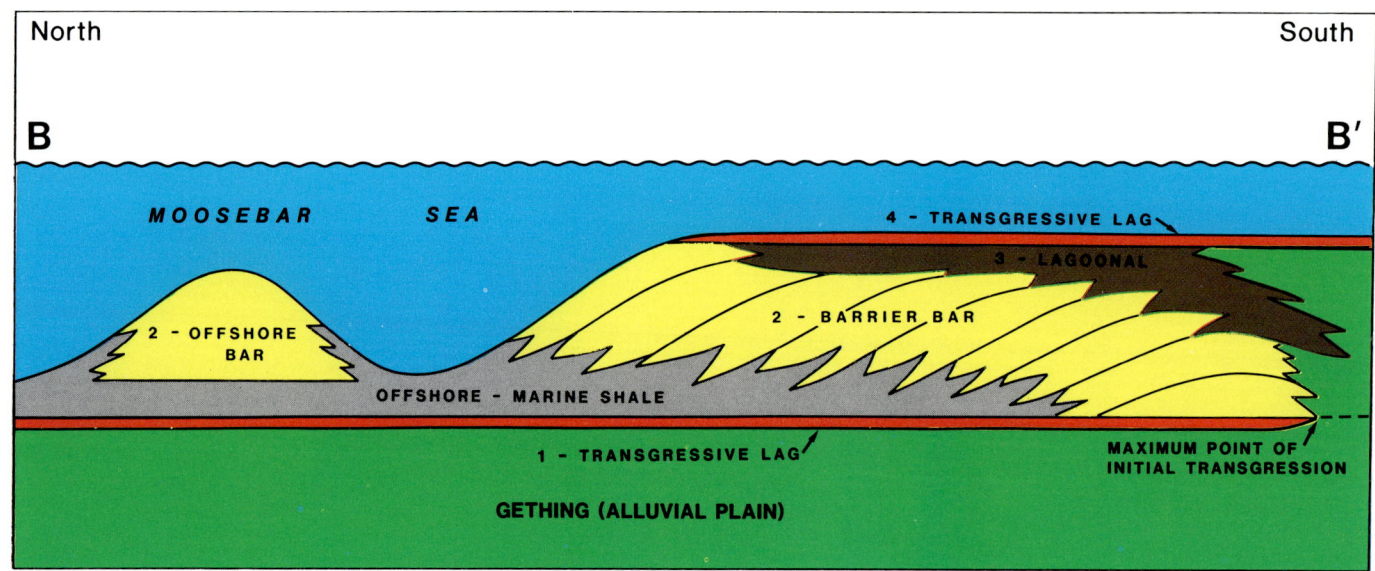

Figure 11. Schematic cross section B-B' illustrating the sequence of events that occurred in the study area during the transgression of the Moosebar sea (see Fig. 12 for cross section location). *(1)* A "Bluesky" transgressive lag is formed as the sea advances across the Gething alluvial plain. *(2)* An increase in sediment supply results in the development of a prograding "Bluesky" barrier bar/offshore bar system. *(3)* As the barrier progrades seaward, a bay/lagoon develops landward of the barrier. *(4)* The barrier system is ultimately overpowered by the sea and is once again transgressed. A transgressive lag caps the system.

rivers flowed northeast from the Cordilleran region, then northwest parallel to the mountain front along a major trunk system (Spirit River Channel; Fig. 10). Figure 8, an isopach of net Gething sandstone demonstrates the drainage pattern. Gething sandstone thick spots, representing belts of active tributary channel systems, clearly show a northeast drainage pattern. The Spirit River Channel has narrowed from Cadomin time, but is still a major northwest-trending trunk system. Smaller, westward-flowing tributaries also entered the Spirit River Channel from the east along its course.

It should be noted that the sandstones of the Spirit River Channel are primarily restricted to the lower Gething. During late Gething time, the drainage pattern changed significantly. The Fox Creek Escarpment became buried by Gething sedimentation and the drainage plain expanded eastward. With the disappearance of the Fox Creek Escarpment, there appears to have been no local constraints on drainage so that river flow expanded randomly across the drainage plain.

The physiography of this large drainage basin was a low-lying, swampy plain with numerous lakes, and the area was heavily forested with conifers, cycads and ferns. It is probable that streams were braided to the west near the Cordillera and meandering to the east in lower gradient areas of the drainage plain.

No evidence of marine sedimentation was observed in the study area, although the coastline was not far away. Stott (1973, p. 29) reports in his Bullhead study of the British Columbia foothills, "the Gething Formation north of Peace River, in contrast to the coal-bearing succession of the type section [Peace River Canyon] is dominantly fine-grained sandstone of deltaic to marine origin ... The formation grades laterally northward into thinly interbedded sandstone and shale, and presumably into shale of the Buckinghorse Formation."

Bluesky Formation

The Bluesky Sandstone is often considered to be a transgressive lag deposit representing the reworking of the top few feet of the Gething alluvial plain by the advancing Boreal sea. Although locally this is true, in many areas the Bluesky is a coarsening-up, marine, regressive sandstone. This may at first seem unusual when considering that during early Albian time, the interior basin was undergoing overall transgression. But it is important to remember that the sediment supply to the interior was still more or less continuous throughout this period and as a result, many regressive pulses are seen in this overall transgressive environment, particularly in areas adjacent to sediment sources. One such regressive cycle is present in the study area. It is represented by a major barrier bar sequence that prograded northward and eastward into the Moosebar sea from a sediment source southwest of Fort St. John. Cross section B-B¹ illustrates this system (Fig. 11).

Following an initial transgression, usually represented by a glauconite-rich sandstone and/or conglomerate lag, a major barrier bar/offshore bar system built seaward (Figs. 12, 13 and 14). This trend is represented by coarsening-upward sandstone. Bioturbation is very common in the finer-grained lower shoreface facies. Crossbedding is common in the coarser middle to upper shoreface. Most sandstones are slightly to moderately glauconitic.

Lagoonal and bay sediments, consisting of siltstone, shale, and occasionally coal, cap most of the barrier system. South and east of the main barrier trend in the south-central part of the study area, low energy bay sediment predominates. In this area, the Bluesky is represented by thin, highly-burrowed sandstones, siltstones and shales with occasional thin molluscan shell banks. Further offshore, in the east and southeast part of the study area, a strong northwest-southeast offshore bar trend is

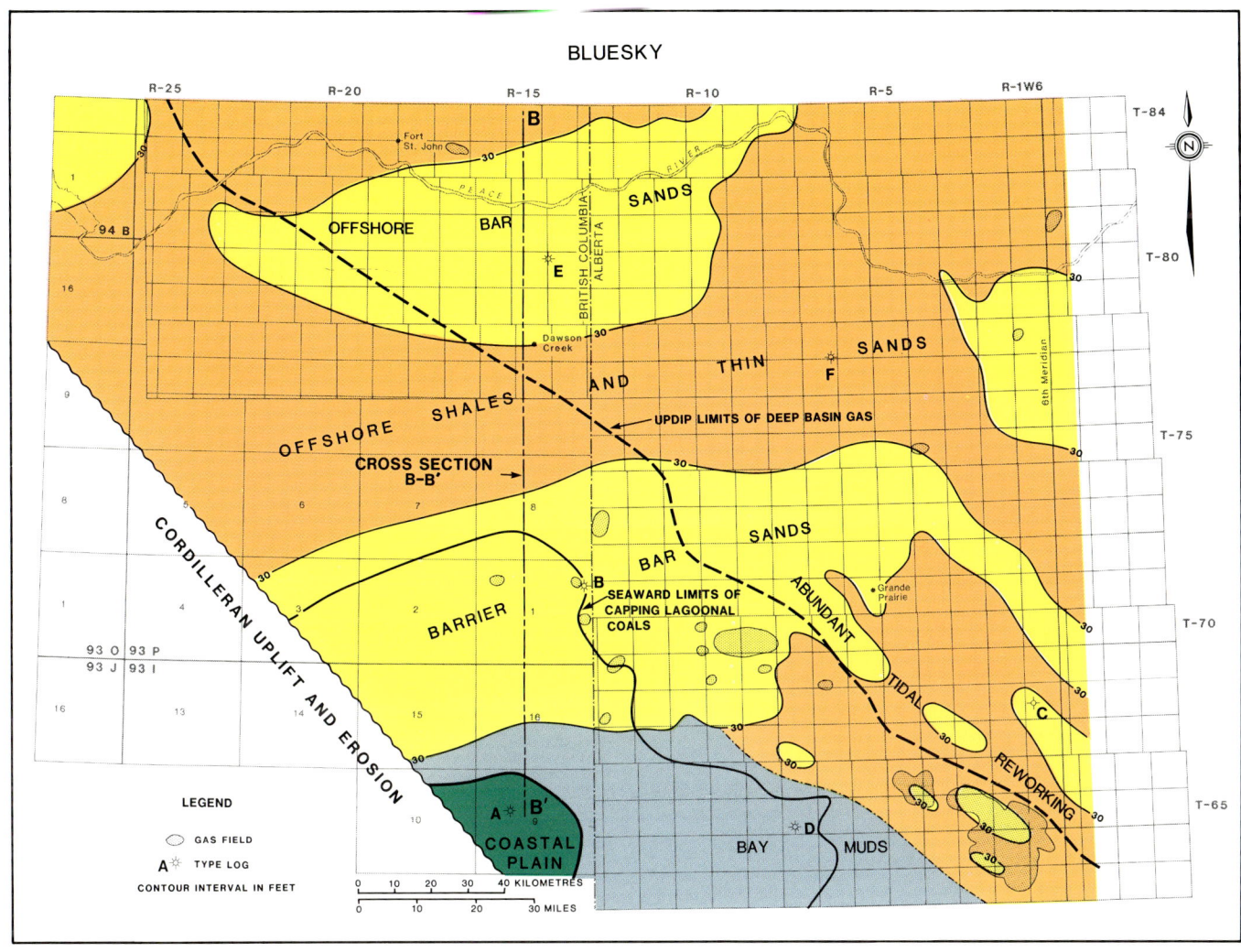

Figure 12. Net sand isopach and paleogeography of the Bluesky Formation.

developed. Many of the sandstones in this area exhibit blocky profiles on logs suggesting tidal reworking of the sands.

In the extreme northwest corner of the study area, north of Peace River (Block 94-B-1), a thick, coarsening-up Bluesky Sandstone is observed in well logs. This appears to be another prograding barrier system with a western sand source that resisted the advance of the Moosebar sea. Although no connection is evident from well control, it may have been the sediment supply for the large offshore bar system in the north-central part of the study area that trends east-to-west in the vicinity of the Peace River (Fig. 12).

Falher Member

Following the inundation of the interior by the Moosebar sea, intensified Cordilleran tectonism once again provided a strong sediment supply to the interior. Sedimentation began to exceed relative sea-level rise and shorelines prograded northward. Probably by middle-early Albian time, Moosebar shales within the study area were being overlain by coastal/deltaic sediments of the Falher Member (see Stelck and Kramers, 1980). This strong progradational system continued through Falher and Notikewin deposition so that by the end of Notikewin deposition, the Boreal sea had retreated to the latitude of Dawson Creek, British Columbia.

Five distinct Falher cycles are mappable in the study area (Figs. 15 to 25; also see correlation chart, Fig. 2). Each cycle represents a rapid transgression followed by a slow regression as the coastline moved back and forth across the study area. This oscillation of the marine/non-marine interface is due to the combined effect of continuing but varying subsidence of the basin, probable rising sea level and intermittent interruptions of the sediment supply to the area.

On each of the five Falher paleogeography maps (Figs. 15, 17, 19, 21, 23) the "landward limits of transgression" is indicated. This is identified in the subsurface by: (1) the disappearance of the basal transgressive lag of fining-upward sandstones and conglomerates; and (2) the replacement of the coarsening-upward sandstones and conglomerates of the marine system with the thin-bedded fine-grained sandstones, siltstones, shales, coals, and channel sandstones of the lower delta-plain facies.

Also indicated on each of the five maps is the "seaward limits of capping lagoonal coals." This line defines the seaward

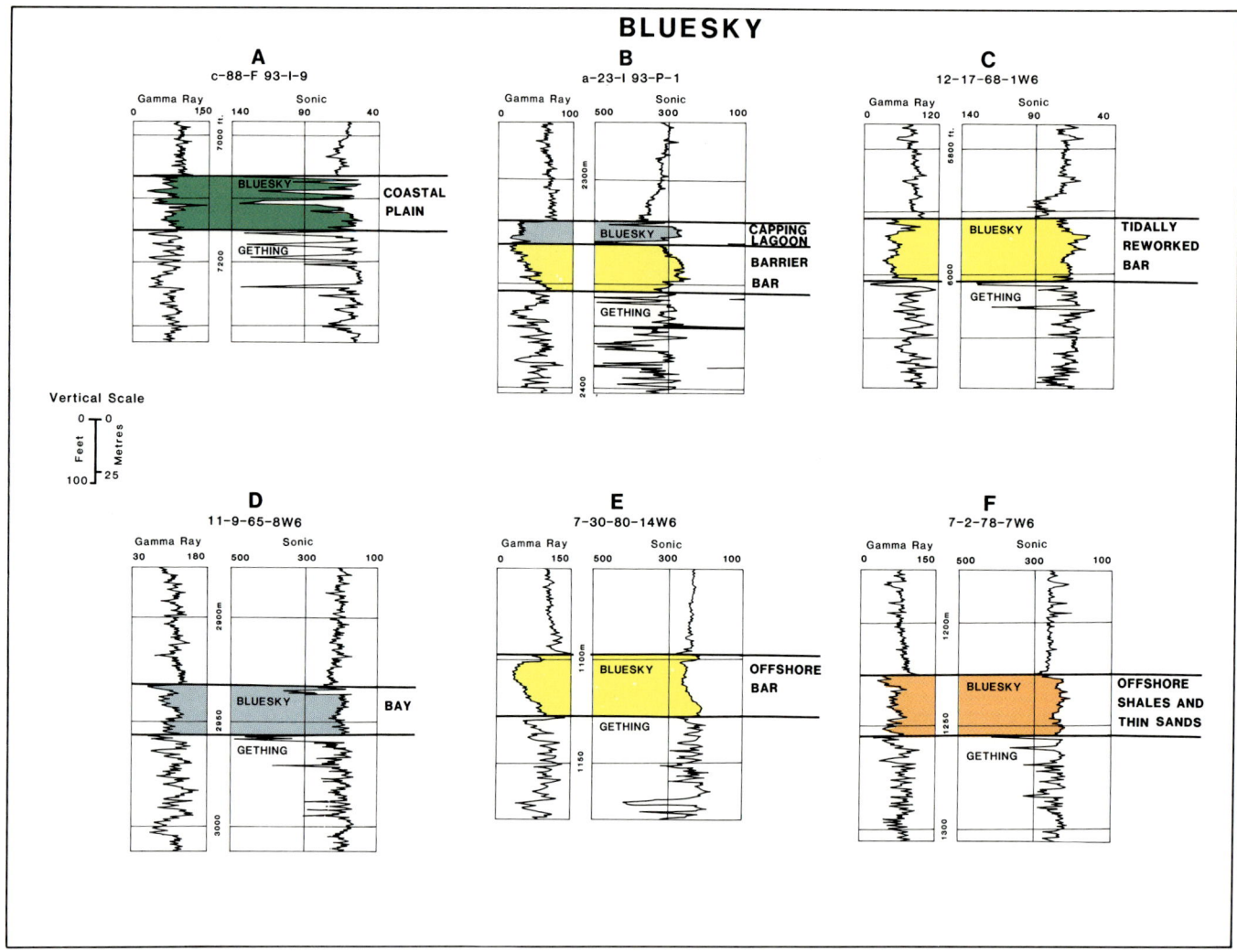

Figure 13. Type logs of the various facies of the Bluesky Formation.

extent of coal and thus indirectly maps the maximum progradational limits of the back-barrier lagoonal facies. Seaward of this point only barrier bar and offshore sediments should occur.

The "seaward limits of conglomerates and clean sandstones" serves two purposes. First, where this line extends seaward of the capping lagoonal coals, it marks the progradational limits of the barrier-bar facies. Second, where it is found landward of the limits of capping lagoonal coals, it indicates that beyond the line, the coarsening-up cycles beneath the capping lagoonal coals contain little or no clean sandstone or conglomerate. A close examination of well logs in such areas reveals a log profile much more indicative of a river-dominated "mud" delta. In addition, correlation of individual sandstones from well to well in such areas is much more difficult, as is expected in a mud delta.

All fluvial channels identified from well logs and cores are plotted on each of the five maps. Although the current well control can identify only a fraction of the channels that must exist in this deltaic system, it does indicate that channels are common throughout the length of each Falher coastline trend (Cant, 1983; Youn, 1983). This indicates that all five cycles are mainly deltaic in the study area. The occurrence of highly correlative, clean sandstones and conglomerates in the west in cycles A, B, C, and D, however, indicate a destructive or wave-dominated delta (Elmworth Field) while the "dirty," poorly correlative sandstones to the east indicate a constructive or river-dominated delta. The Falher E cycle, like the Notikewin (see below for a detailed description of the processes involved), is different in that it consists of an initial barrier trend followed by a progradational river-dominated delta system.

Although no attempt has been made to map the occurrence of sandstone in the offshore marine facies, sand is abundant in the offshore. It is common to see coarsening-up sandstones in the offshore facies over 40 mi (64 km) from the mapped coastline. This occurs because we are dealing with a relatively shallow epeiric sea wherein offshore transport of sediment by storm action was significant (Leckie and Walker, 1982).

Notikewin Member

The Notikewin Member is another regressive pulse similar to the various cycles of the Falher Member (Figs. 26 and 27). It consists of continental sediments in

Figure 14. Artist's concept of the landscape of the study area during late Bluesky time just prior to the final transgression of the Moosebar sea. View is to the northeast (Compare with Fig. 12).

the south, a prograding delta facies trending east to west across the central study area, and an offshore facies to the north. There are, however, important differences in the lithology and distribution of Notikewin strata that point to changing sedimentation conditions. The "landward limits of transgression" mark the point of maximum southern advance of the Boreal sea at the close of Falher "A" deposition. It is common to see a transgressive lag at the base of the Notikewin in wells seaward of this line. Immediately seaward of this point of maximum transgression, in the regressive cycle of the Notikewin, is a very well-developed, coarsening-upward, fine-grained sandstone to conglomerate barrier bar that stretches with occasional breaks across the entire study area. This barrier clearly developed during the initial stages of the regressive cycle and was part of a wave-dominated coastal system. Suddenly, however, things changed. The barrier bar, after prograding seaward as much as 15 mi (24 km) in some places, was abruptly overrun by a massive influx of sediment that engulfed the entire coastal margin changing the coastline from a wave-dominated to a river-dominated system. The evidence for this is found in the lithologic character of the Notikewin. The barrier bar is a clean, highly-correlative, coarsening-upward sandstone, usually with well-sorted conglomerate at the top of the cycle. The Notikewin throughout the remainder of the area is generally argillaceous with numerous shale interbeds, contains little conglomerate, correlates poorly from well to well and often displays incising fluvial channel profiles. In addition, the width of the progradation measured from point of maximum transgression to the seaward limit of capping lagoonal coals is over twice the width of the average Falher regressive cycle, indicating a strong, continuous sediment supply relative to relative sea-level change.

Notikewin sandstones continue many miles seaward in the offshore facies. At "The Gates" on Peace River, which is the type section for the Gates Formation of

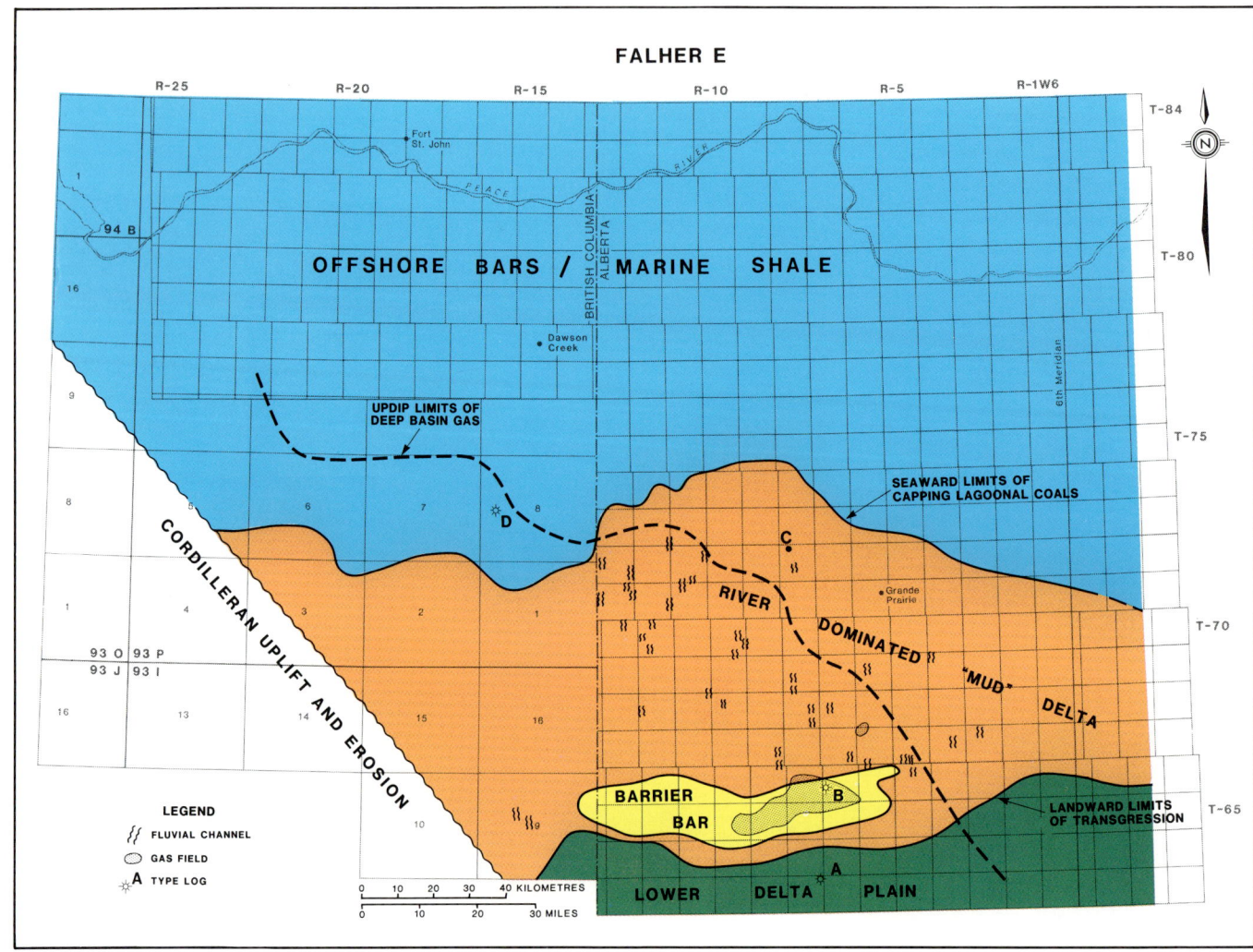

Figure 15. Paleogeography of the Falher "E" submember.

the British Columbia foothills (Sec. 14, Twp 82, Rge 25W6) the upper 50 ft (15 m) of the formation is exposed. This section, lying directly beneath the shales of the Hasler Formation (Hulcross equivalent), is almost certainly the stratigraphic equivalent of the Notikewin. "The Gates" on Peace River, which is formed by the resistive sandstones and siltstones of this unit, is over 30 mi (48 km) beyond the seaward limits of coal as mapped in the subsurface. North of Peace River, however, the Gates Formation (Notikewin) grades into marine shale as it has not been recognized in outcrop north of the river (Stott, 1968). This belt of offshore sand is similar in width to the offshore sand belts in the Falher cycles.

Paddy and Cadotte Members

Early in middle Albian time, the Boreal sea again advanced across much of the interior, with the deposition of the Harmon (Hulcross) marine shales in northern Alberta, and somewhat later the Joli Fou shales of central and southern Alberta. This transgressive event occurred during the period of major global rise in sea level during Albian time postulated by Vail et al (1977). It is clearly the harbinger of the major inundation of the interior which intensified during late Albian time with the spread of the Shaftesbury/Colorado sea across much of western North America.

Intruding into this major transgression is the Paddy/Cadotte regressive pulse (Figs. 28 to 32) which developed as a result of a strong and continuing sediment source southwest of Fort St. John. Throughout the study area, the Cadotte is a coarsening-upward sequence ranging in grain size from very fine sandstone to conglomerate. It represents a prograding barrier sequence that initiated beyond the southern limits of the study area and built northward and eastward with time until it reached its maximum seaward extent, a northeast-southwest trend just south of Fort St. John (Figure 28). Seaward of this position for another 20 mi (32 km) or so to the north, offshore silts and thin sands accumulated, probably as a result of seaward sediment transport due to storm and tidal action.

Many subsurface cores in the Elmworth Deep basin exhibit an apparent erosional surface within the Cadotte Member (Ethier, 1982). This scour separates underlying very fine- to fine-grained, coarsening-upward sandstones of the lower to middle shoreface from overlying coarser, parallel and cross-bedded sandstones and conglomerates of the upper shoreface and foreshore

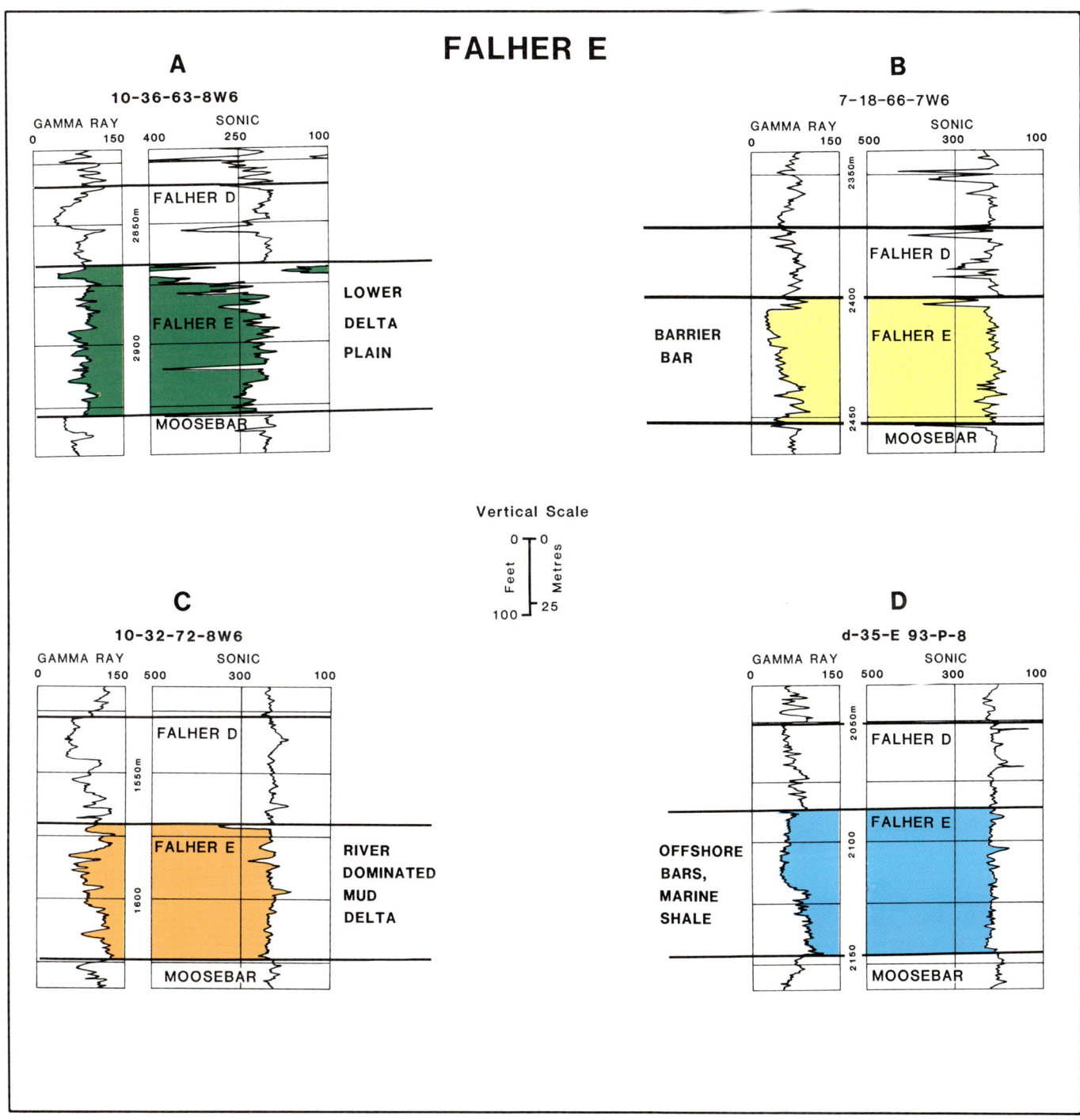

Figure 16. Type logs of the various facies of the Falher "E" submember.

(Fig. 29-D). Stott appears to have observed this same phenomena in several Boulder Creek (Paddy-Cadotte) outcrop sections in the British Columbia Foothills, where he describes massive conglomerates lying with sharp or unconformable contact on fine-grained sandstones (Stott, 1968; p. 83).

This rather unusual vertical sequence probably represents a prograding, high-energy, barred coastline. Hunter et al (1979) in their SCUBA analysis of the nearshore systems of the southern Oregon coast (a high-energy, barred coastline) predict a vertical profile for the prograding model of that system that is virtually identical to the above-described sequence.

Normal, coarsening-upward Cadotte sections, without erosional breaks, observed in other Deep basin cores, very closely resemble the prograding model predicted by Clifton et al (1971) for high-energy, non-barred portions of the Oregon coast.

Average Cadotte grain size increases westward toward the sediment source area.

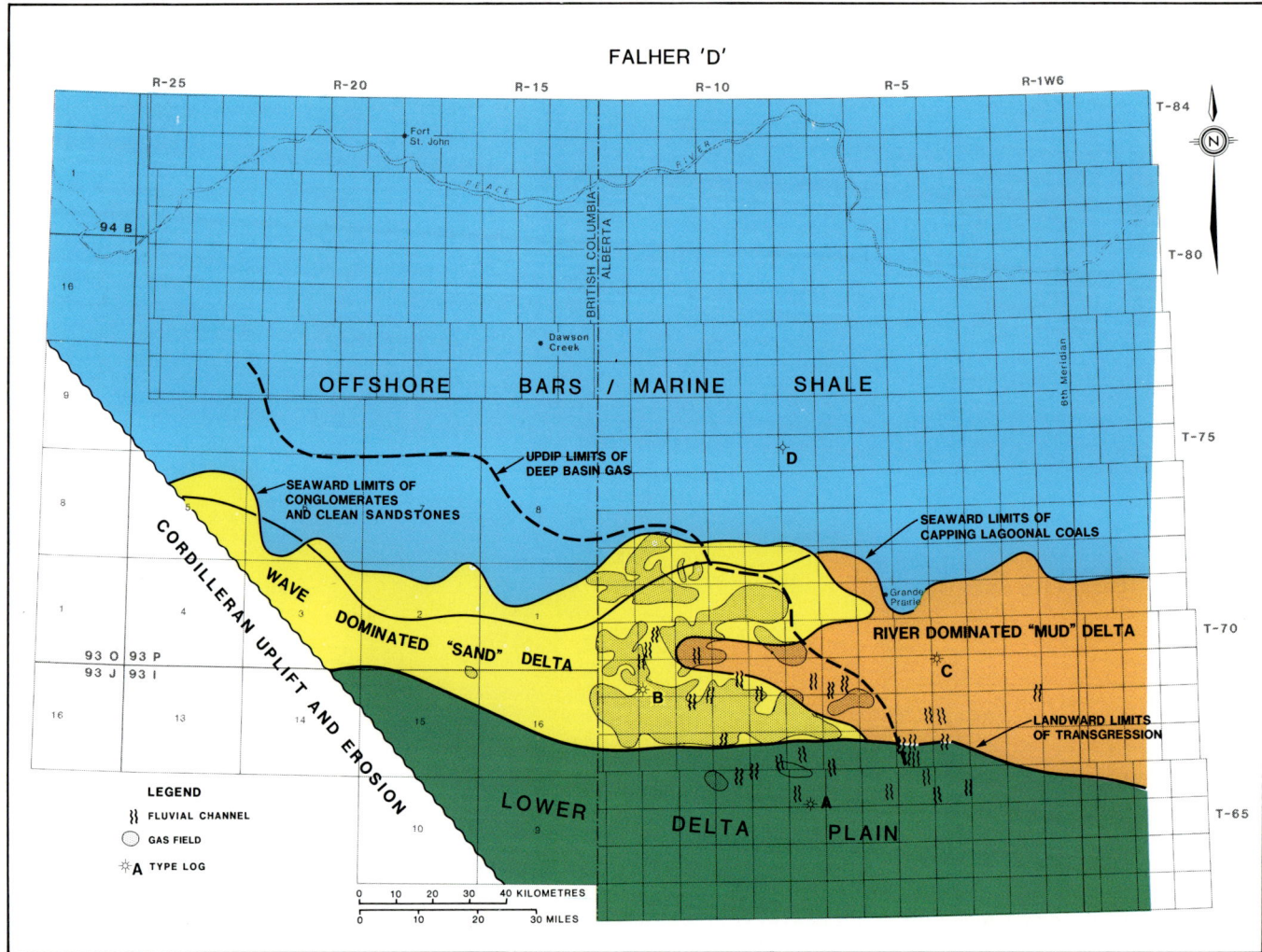

Figure 17. Paleogeography of the Falher "D" submember.

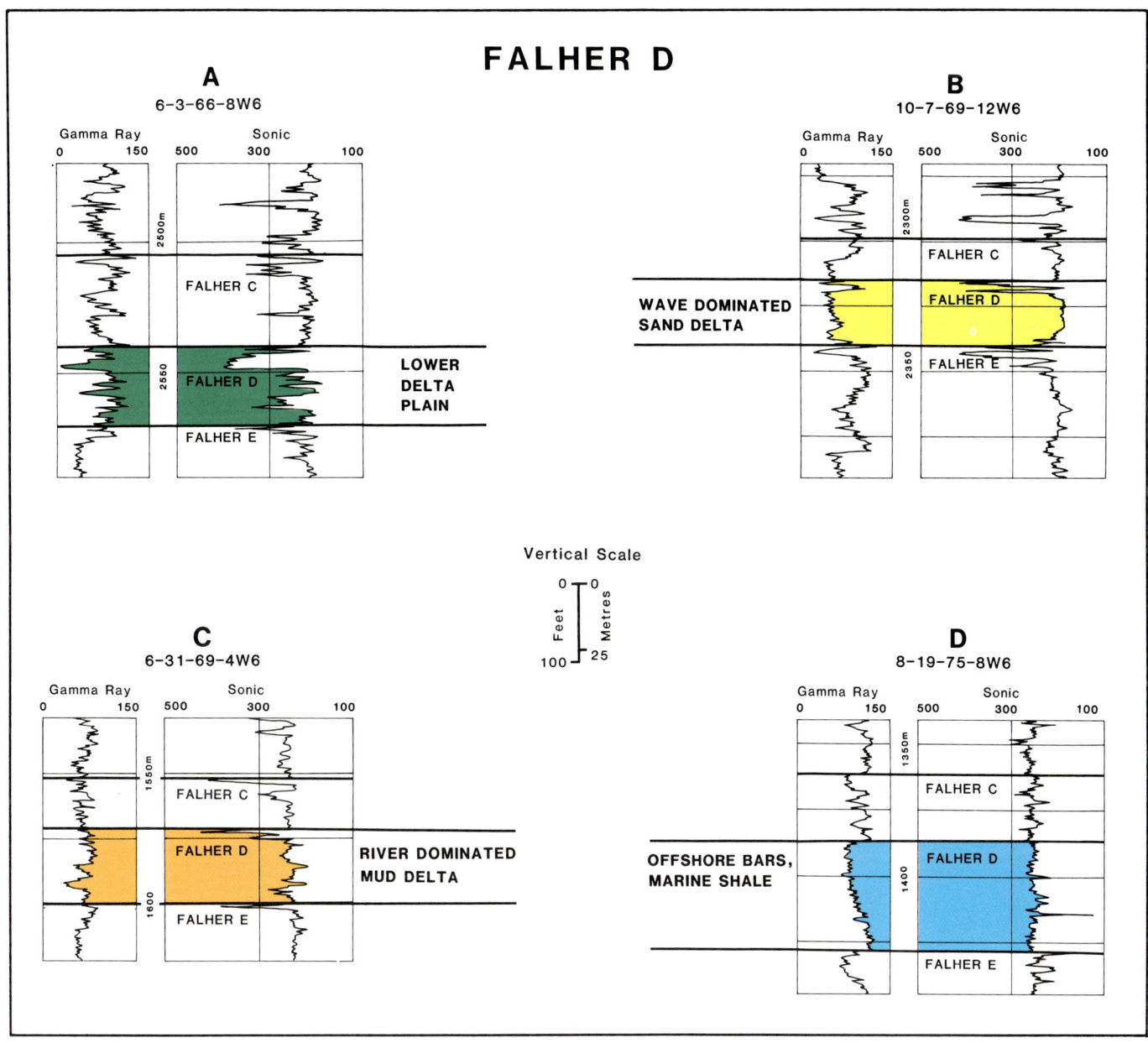

Figure 18. Type logs of the various facies of the Falher "D" submember.

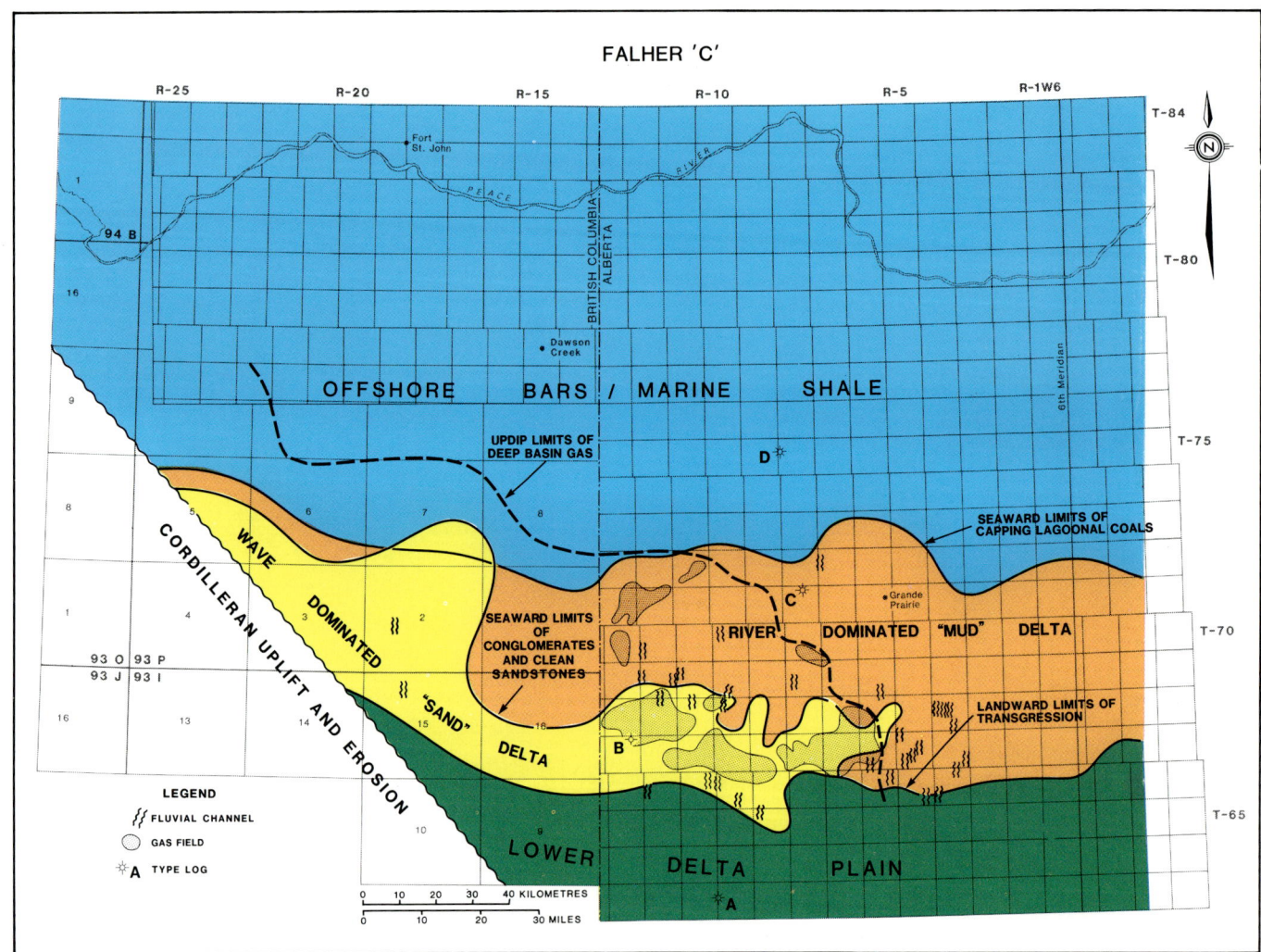

Figure 19. Paleogeography of the Falher "C" submember.

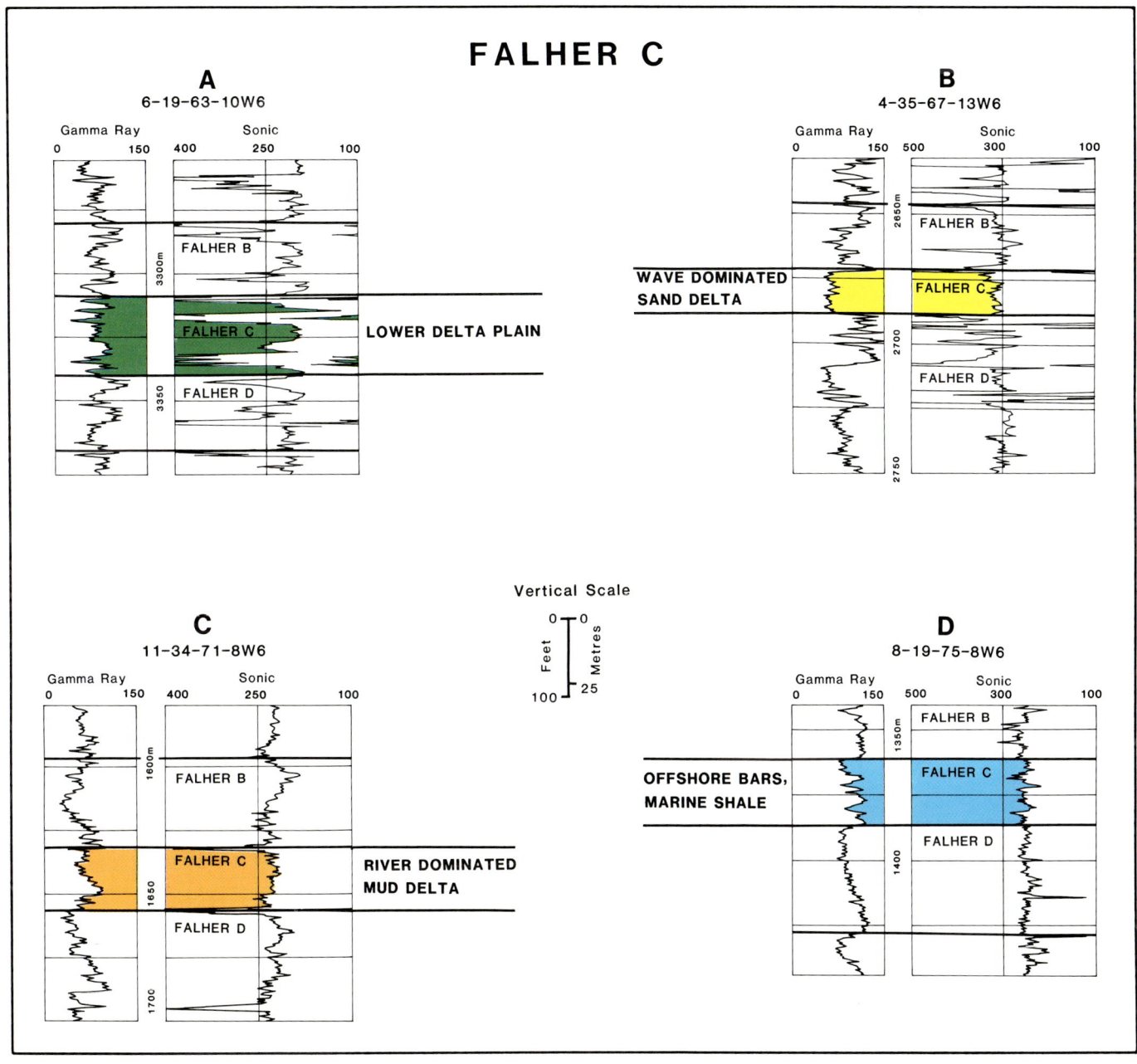

Figure 20. Type logs of the various facies of the Falher "C" submember.

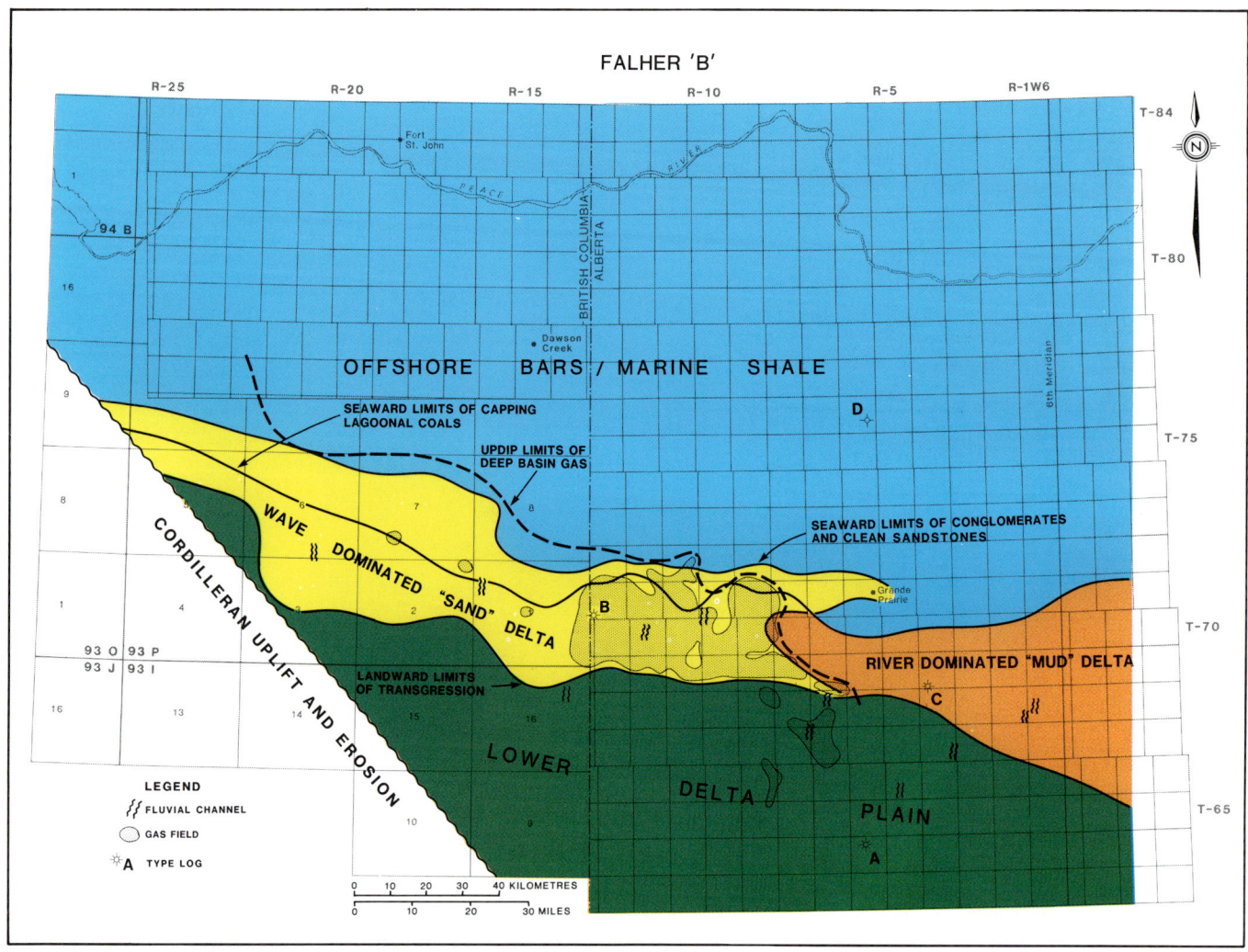

Figure 21. Paleogeography of the Falher "B" submember.

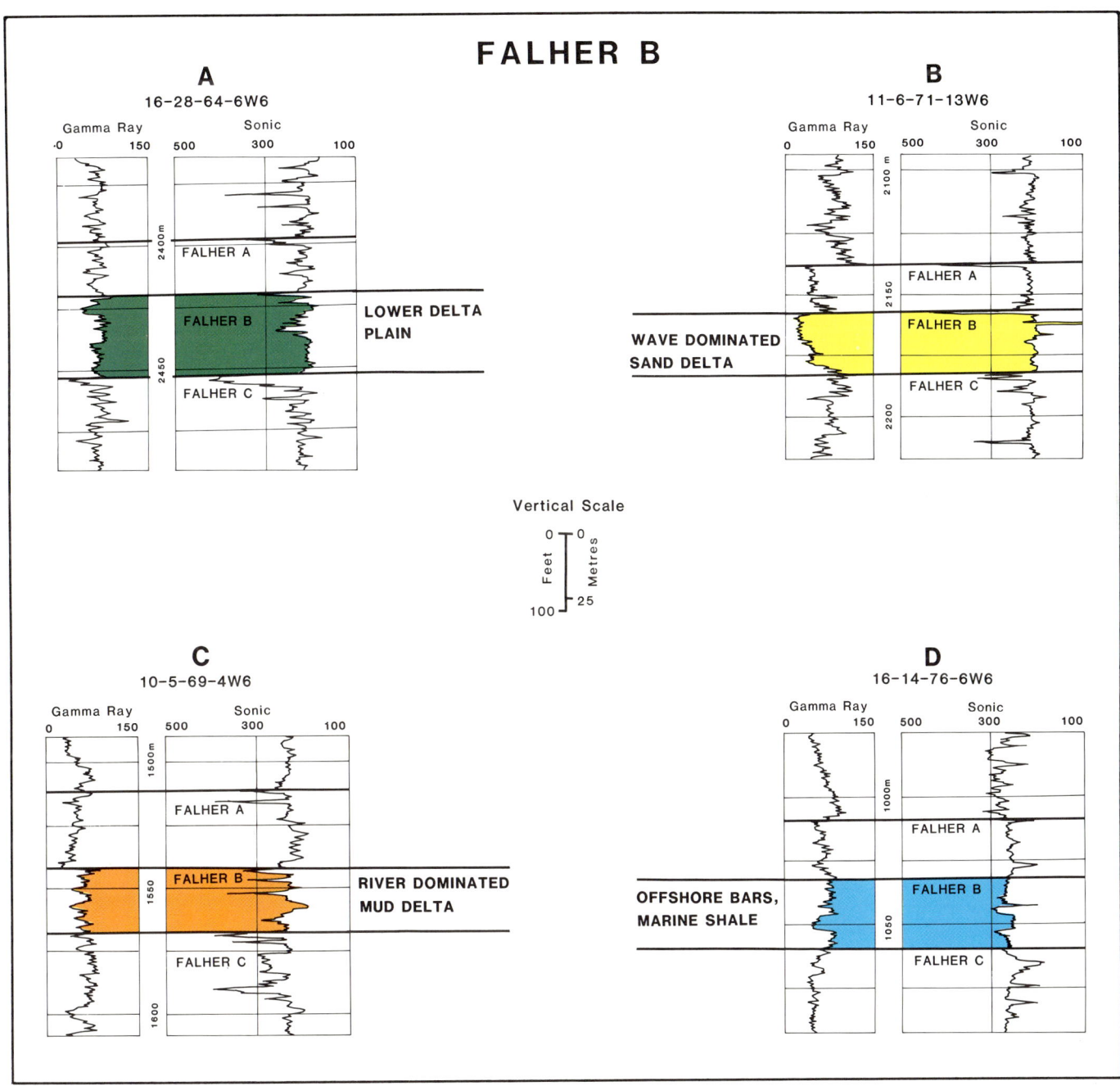

Figure 22. Type logs of the various facies of the Falher "B" submember.

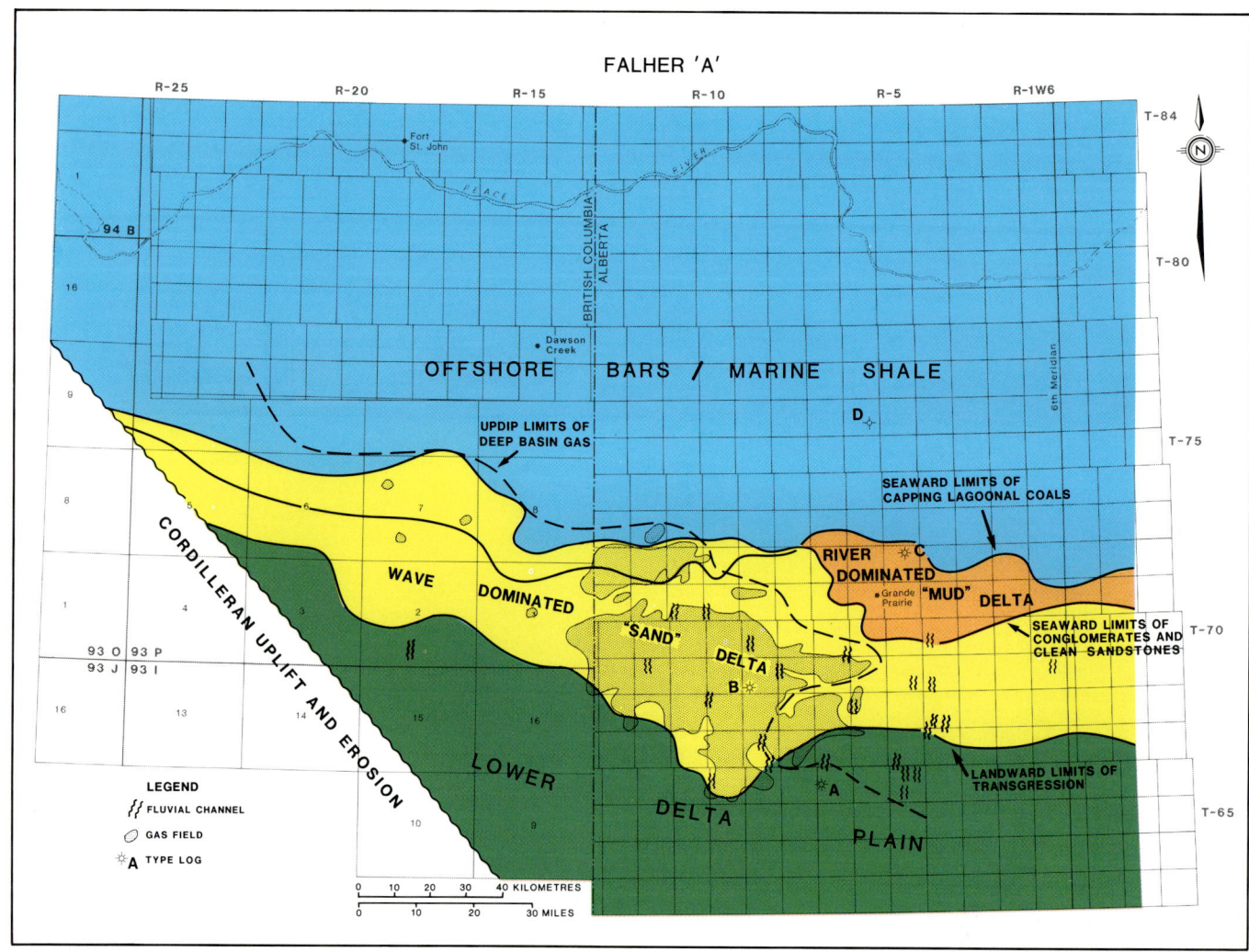

Figure 23. Paleogeography of the Falher "A" submember.

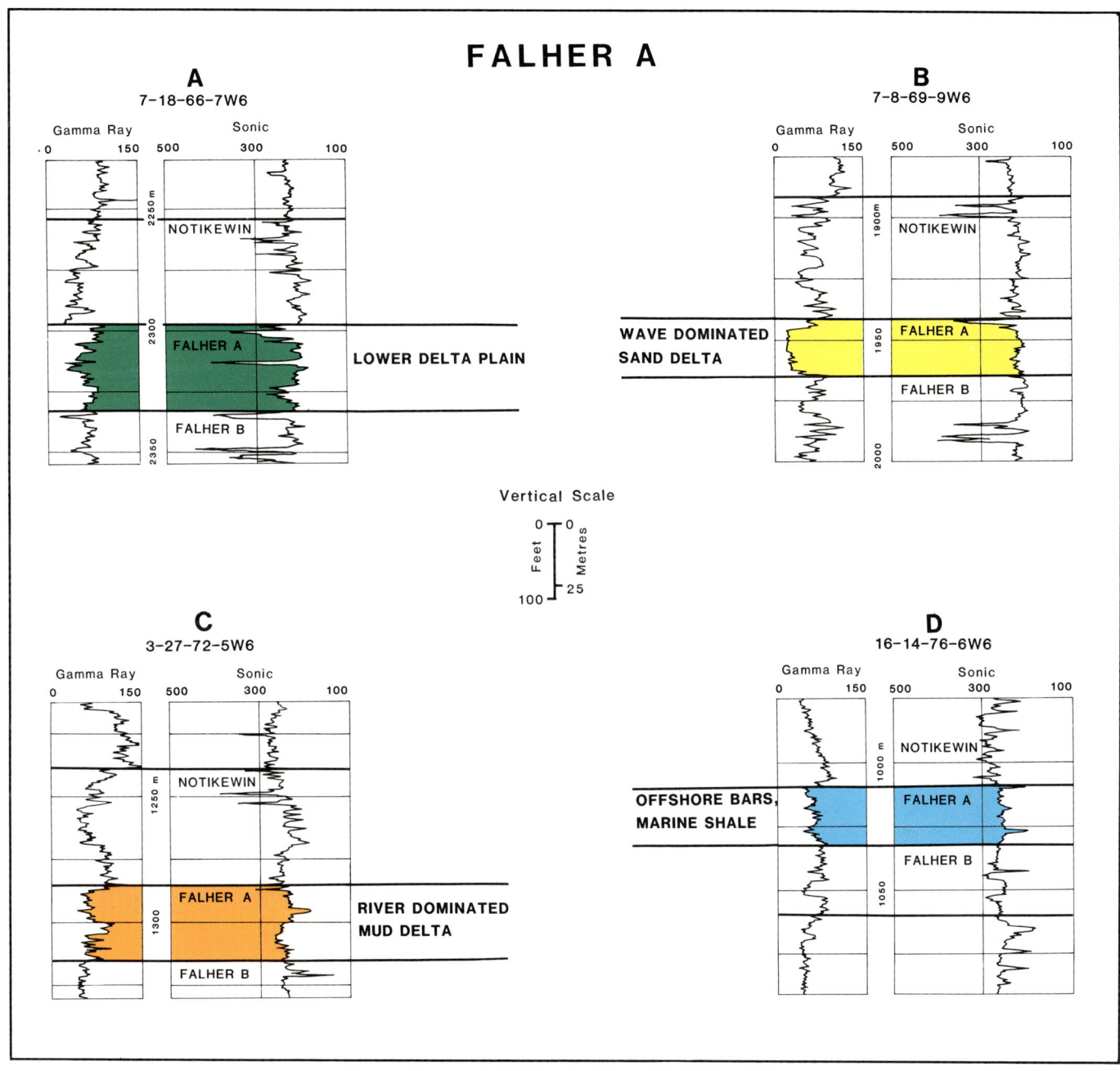

Figure 24. Type logs of the various facies of the Falher "A" submember.

Figure 25. Artist's concept of the landscape of the study area during Falher time. View is to the north. Note the sand-dominated coastline in the west and the mud-dominated, highly irregular coastline in the east.

Sandstones near the eastern edge of the study area are generally very fine- to medium-grained whereas in the foothills, the Cadotte generally consists of coarsening-upward fine- to coarse-grained sandstones overlain by conglomerates. The conglomerates, as well as the lithic grains in the sandstones, are predominantly chert.

Foothills nomenclature (Fig. 2) does not distinguish between Paddy and Cadotte, but rather combines them as the Boulder Creek Member. Genetically, this may be more accurate since the Paddy, or the "upper carbonaceous facies" of the foothills (Stott, 1968), is probably a nonmarine time equivalent of a Cadotte barrier bar further seaward.

The dominant Paddy lithofacies are illustrated in Figure 30 and essentially represent paleogeographic conditions close to the end of Paddy deposition.

A large northward-draining delta plain (the upper carbonaceous facies of the foothills) occupied the western sector of the study area. Sediments include thin sandstones, siltstones, shales, thin coal beds and channel sandstones, often containing conglomerate.

East of the delta plain facies, the section is represented by thin laminated siltstones, mudstones, carbonaceous shales grading to thin coals, and occasional thin fining-upward sandstones. Channel sandstones are notably missing. Bioturbation is common. This profile is interpreted as an intertidal flat facies. The thin fining-up sandstones are interpreted as tidal creek deposits, the coaly layers as swamp and salt marsh deposits, the laminated siltstones and mudstones as the mud flats of the intertidal zone.

East of the intertidal facies is a vertical sequence consisting primarily of stacked, thin (20 to 30 ft; 6 to 9 m) coarsening-upward cycles. Each cycle typically contains in ascending order: (1) platy shale with occasional storm lags; (2) highly burrowed, interbedded ripple-laminated sandstone and shale; and (3) horizontally laminated sandstone occasionally cut by cross-bedded sandstone. The capping sandstone unit rarely exceeds 10 ft (3 m) in thickness. This sequence is interpreted as a back-barrier subtidal facies consisting of quiet-water bay muds grading to wave-dominated (horizontally laminated) and tidally-reworked (cross bedded) sand shoals.

Occasionally, the subtidal facies described above is cut by thick cross-bedded sandstones, often occupying the entire Paddy section. These sandstones contain shale rip-up clasts, exhibit a thin fining-upward sequence at the top, rest with a scoured contact on the subtidal facies and appear to correlate closely in time with a thin transgressive lag often found at the top of the Paddy Member. This thick sandstone facies is interpreted as representing very late Paddy time tidal channels incising pre-existing subtidal bay sediments.

In the east-central part of the study area, additional intertidal deposits are found similar to those of the intertidal facies straddling the British Columbia border. Again siltstones, mudstones and carbonaceous layers constitute the bulk of the section.

Trending diagonally across the study area is a major Paddy barrier bar trend consisting of coarsening-upward sandstone as indicated by "cone-shaped" log profiles. Occasional blocky log profiles within this trend suggest tidal channels incising the barrier system. This barrier is essentially part of the same prograding system that advanced northward across the bulk of the study area during Cadotte deposition. The barrier continued its advance during Paddy deposition to its ultimate point of advance, a northeast-southwest trend coincident with the course of the present-day Peace River. The same offshore silty apron that existed seaward of the Cadotte barrier is also recognized in Paddy sediments.

The end of Paddy deposition is marked by a major transgression which ultimately inundated the entire North American interior basin from Alaska to the Gulf of Mexico. The advance of this transgression across the study area usually can be seen as a thin transgressive sandstone at the top of the Paddy Member capped by the marine shales of the Shaftesbury Formation.

Figure 32 is an artist's conception of the geography of the study area near the end of Paddy time. In the west, rivers flowed northward across a low-relief delta plain transporting sediment to the coastline to feed the advancing barrier system. Long shore drift then moved beach sands northeastward along the barrier trend. To the east, tides moved through major breaks in the barrier system into a large subtidal bay bordered on the east and west by intertidal mud flats and salt marshes.

During the early stages of the Shaftesbury transgression, as the sea deepened and the Paddy barrier trend foundered, major tidal currents, accelerated by the movement of water from the deeper offshore onto the very shallow back-barrier shelf, cut deep tidal channels across the subtidal areas of the shelf. As the sea continued to spread across the Paddy shelf, the upper few feet of Paddy sediments were reworked by wave and tidal energy, creating the widespread lag sandstone usually found at the top of the Paddy Member.

HYDROCARBON POTENTIAL

The concept of widespread, pervasive hydrocarbon saturation within the Deep basin is now widely accepted by the oil industry. The problem of low permeability in Deep basin reservoirs, however, continues to be a major economic barrier to commercial development in spite of demonstrable, very large reserves potential. Table 1 summarizes, within the study area, porosity and permeability averages and ranges (where known) for Lower Cretaceous Deep basin reservoirs and their counterparts on the updip Alberta shelf or Peace River Arch. The Deep basin permeability problem can be readily seen by a comparison of reservoir characteristics of updip fields with Deep basin pools. Most of the Deep basin reservoirs exhibit average permeabilities around 0.001 millidarcys, while updip shelf fields are generally well above 50 millidarcys. It should be noted, however, that the average permeability and permeability range in some Deep basin reservoirs is relatively high. This is due primarily to the preservation and/or diagenetic enhancement of permeabilities in the well-sorted coarser-grained sandstones and conglomerates present in some reservoirs. The extremely high permeability range (0.003 to 850 millidarcys) of the Falher foreshore (beachface) conglomerates is an excellent example of this phenomena.

Deep basin exploration then must rely heavily on the accurate delineation of depositional trends where high permeability reservoirs such as foreshore conglomerates or tidally reworked sandstones are known to exist.

By comparing Deep basin productive areas and fields outside the Deep basin with the various environments of deposition of Lower Cretaceous reservoirs (refer to paleogeography maps), it can be seen

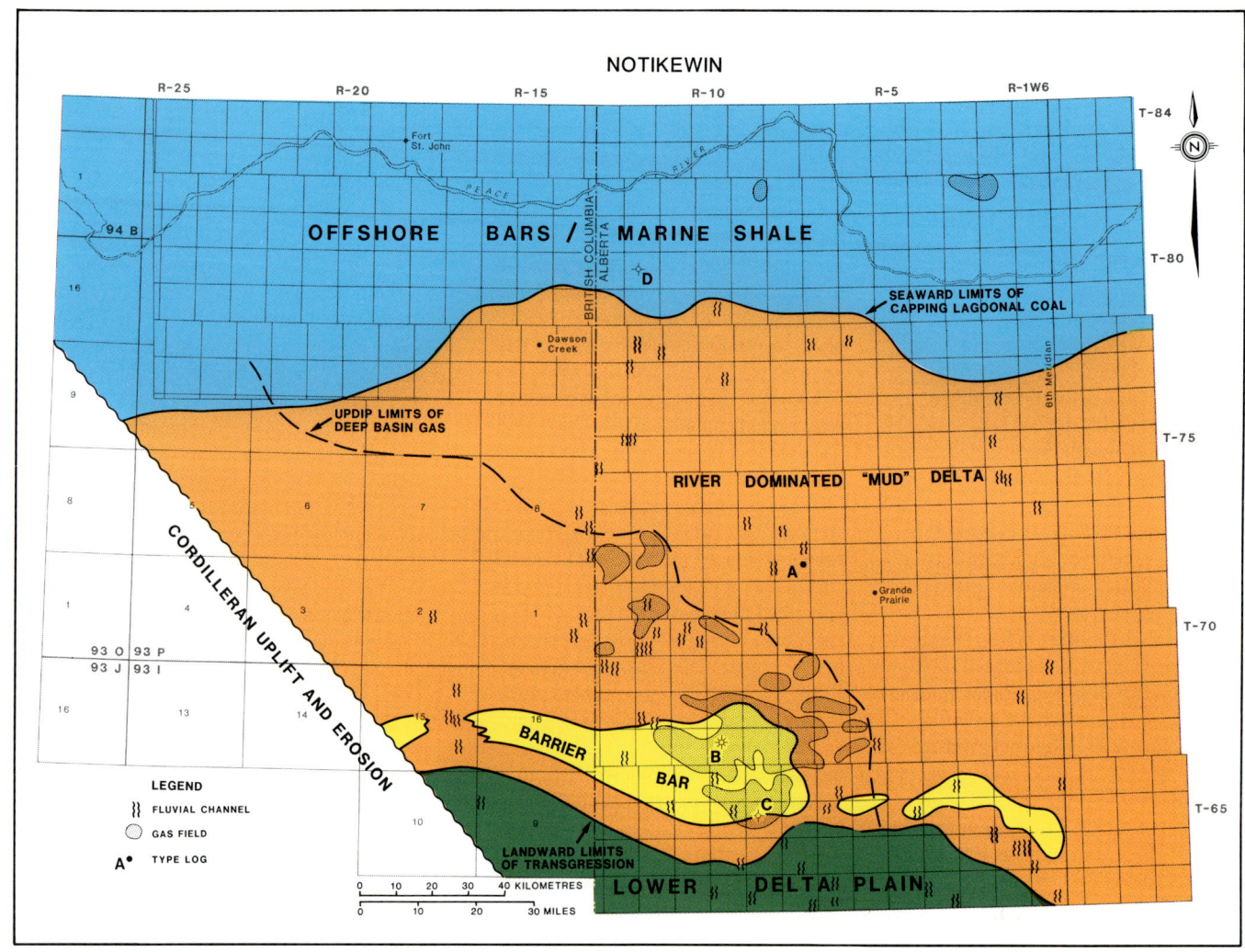

Figure 26. Paleogeography of the Notikewin Member.

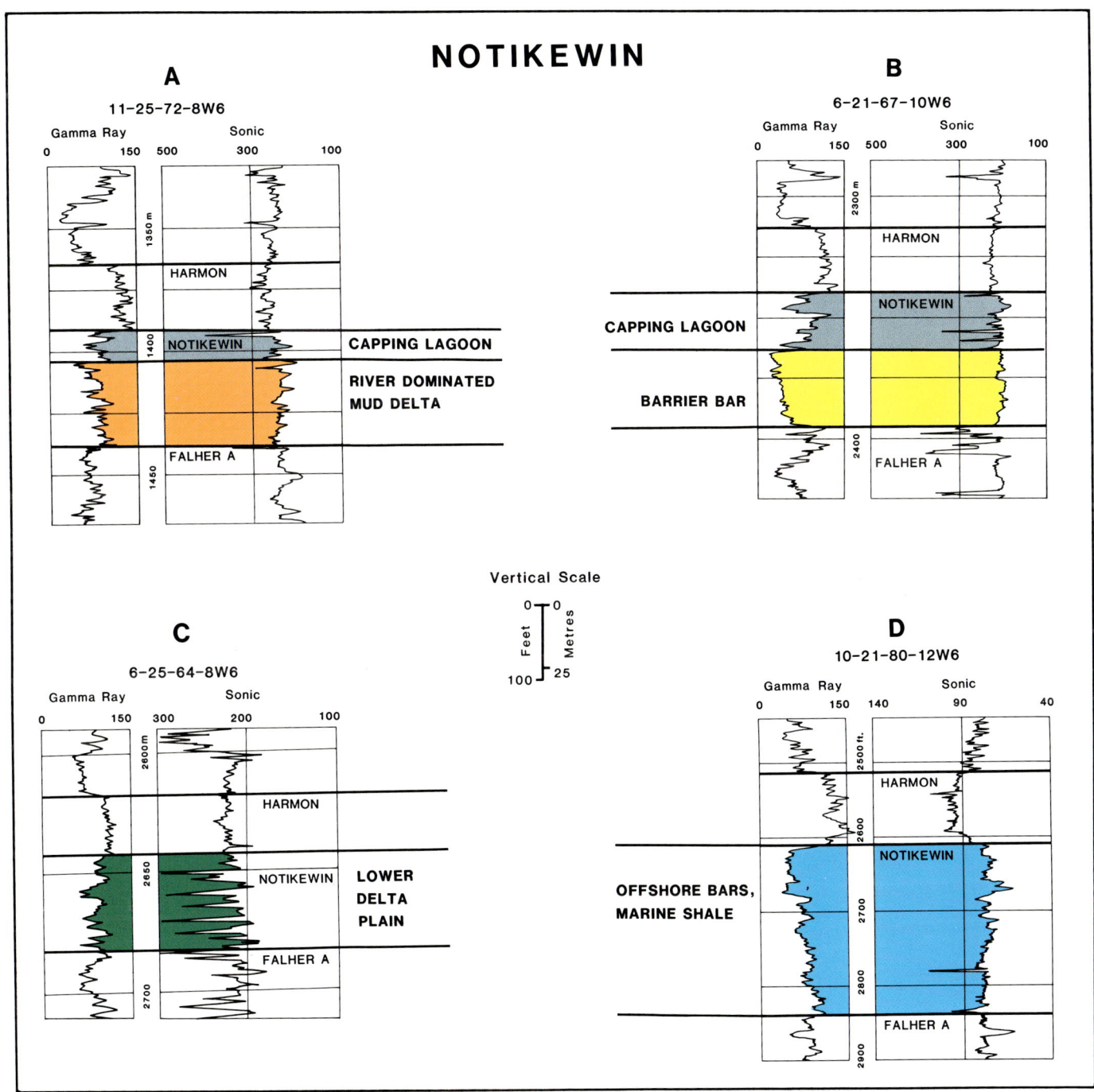

Figure 27. Type logs of the various facies of the Notikewin Member.

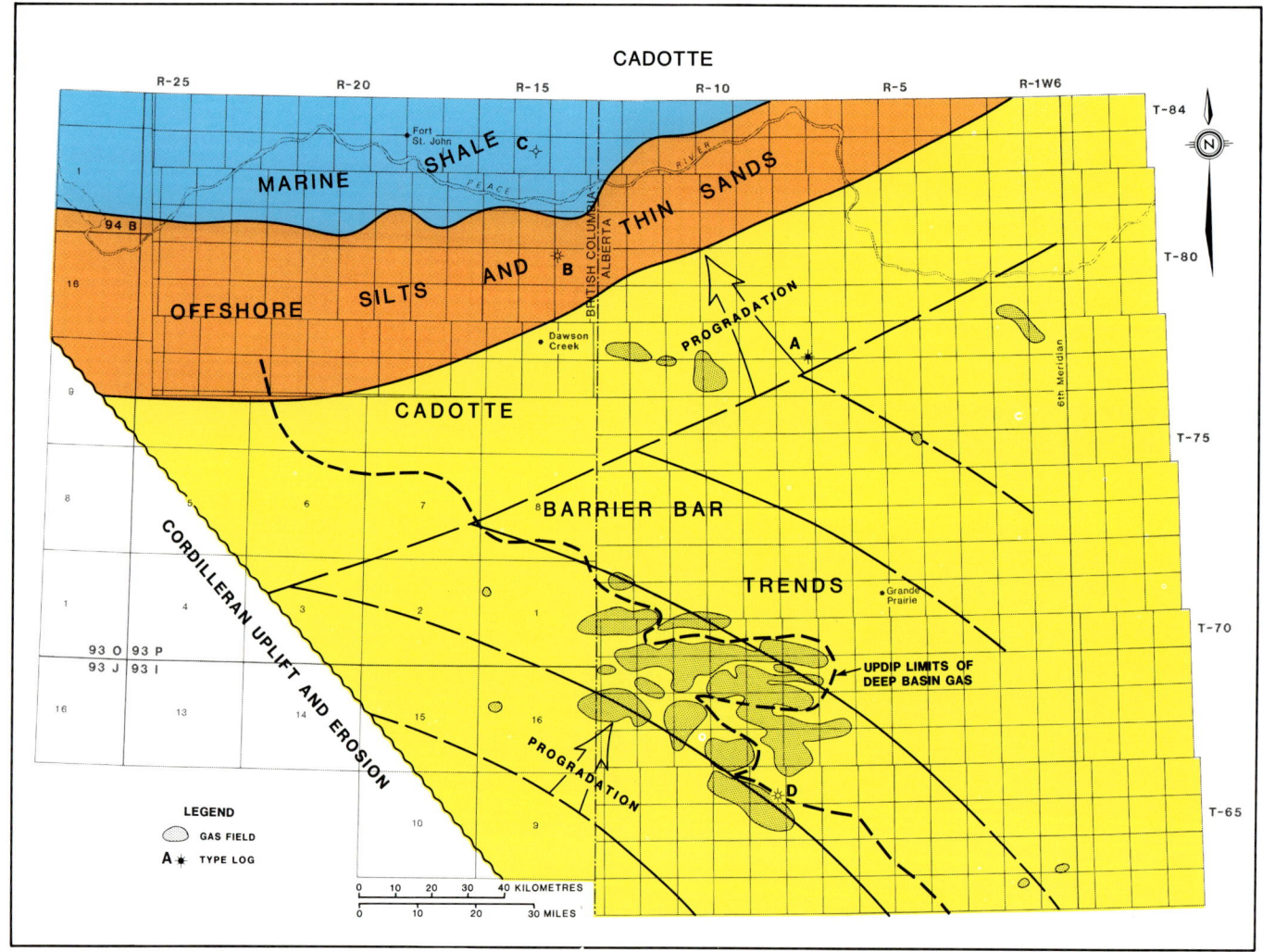

Figure 28. Paleogeography of the Cadotte Member.

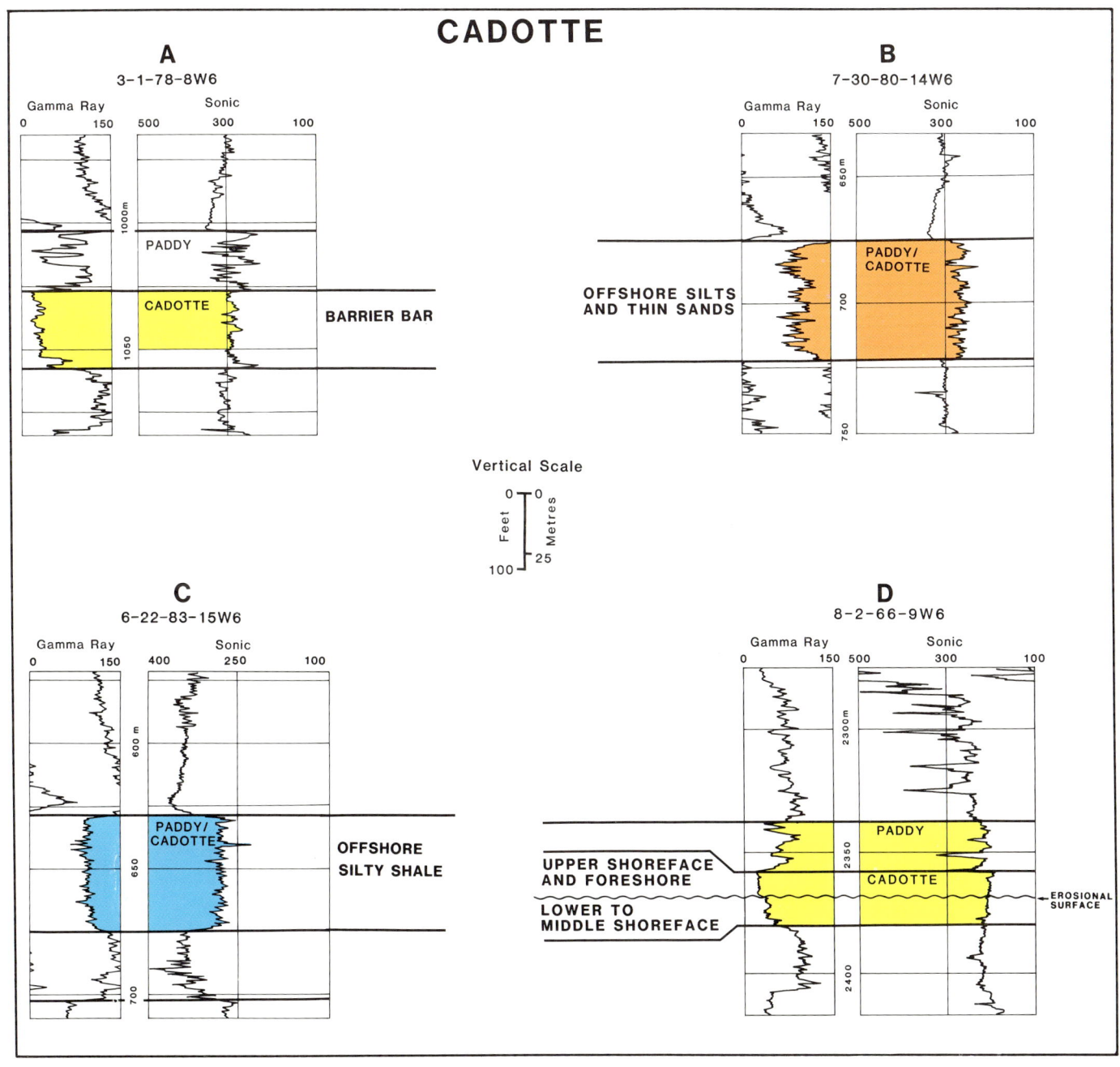

Figure 29. Type logs of the various facies of the Cadotte Member.

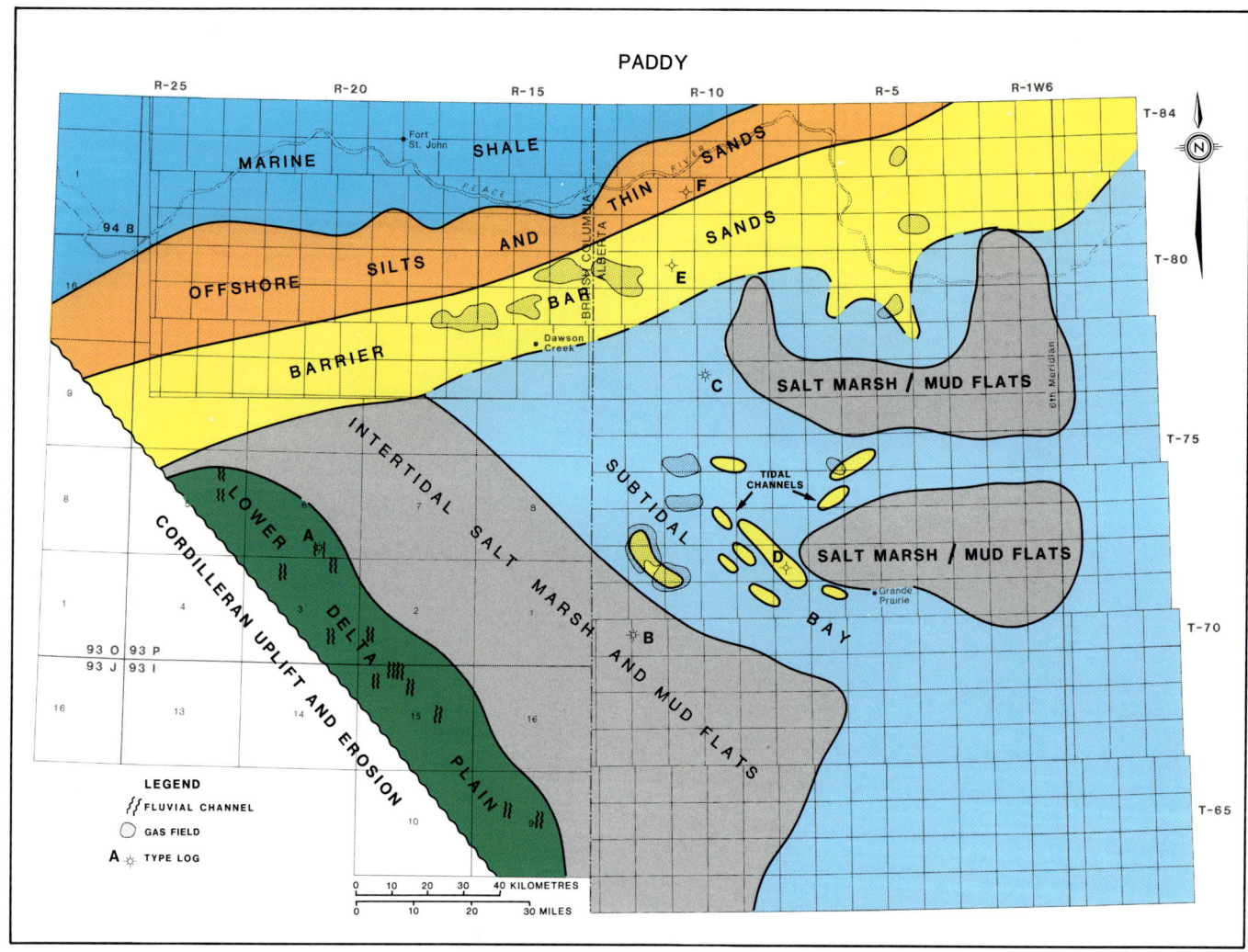

Figure 30. Paleogeography of the Paddy Member.

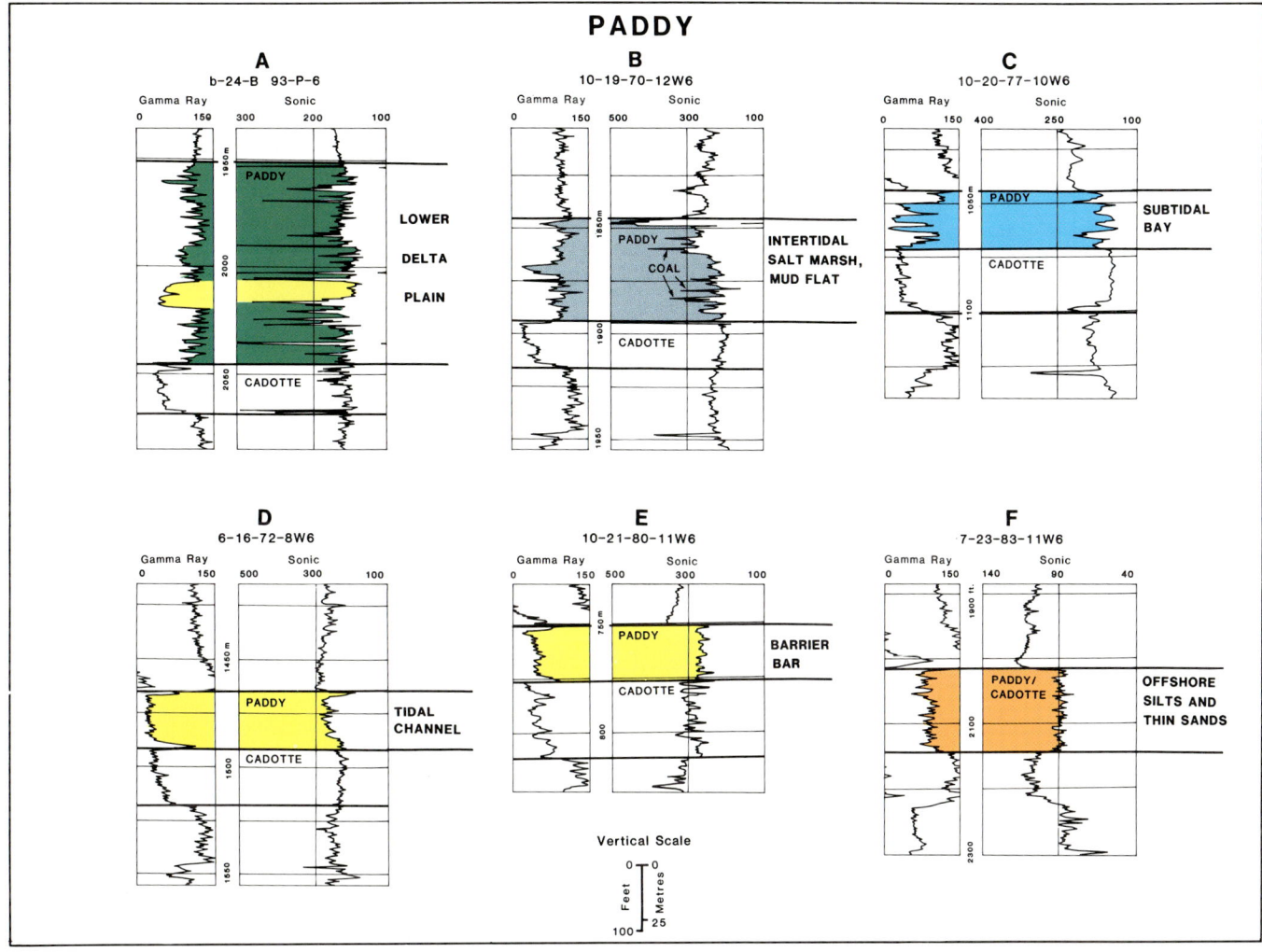

Figure 31. Type logs of the various facies of the Paddy Member.

Figure 32. Artist's concept of the landscape of the study area during late Paddy time just prior to the transgression of the Shaftesbury sea. View is to the northeast (Compare with Fig. 30).

that the most attractive facies of each zone in terms of reservoir quality are as follows: Cadomin, fluvial braid plain; Gething, fluvial channels; Bluesky, tidally-reworked barrier bars; Falher E, barrier bars; Falher D, wave-dominated delta; Falher C, wave-dominated delta; Falher B, wave-dominated delta; Falher A, wave-dominated delta; Notikewin, barrier bars; Cadotte, barrier bars; and Paddy, tidal channels (subtidal facies) and barrier bars.

Within the confines of the Deep basin, the explorationist, in defining areas of interest, would need to supplement the above-described facies information with regional data on grain size and sorting trends, and diagenesis (Sneider and King, 1984). Outside the Deep basin, tactics must of necessity differ because each trap is limited by downdip water. On the Peace River Arch, hydrocarbons are trapped structurally, forcing the explorationist to blend facies-related porosity trends with seismic data. On the Alberta shelf, monoclinical dips preclude structural traps, but set up the opportunity for stratigraphic updip porosity pinchouts forcing the explorationist to pay very close attention to the geometries of facies-related sandstone trends.

CONCLUSIONS

The Cadomin is a blanket deposit of piedmont alluvial plain and fluvial-braid plain conglomerates derived from the Cordillera during the initial stages of Lower Cretaceous deposition. Sediment was deposited from the source area to the Fox Creek Escarpment, a gentle range of hills during Cadomin deposition, which trended northwest to southeast across the eastern part of the study area. Fluvial braid plain conglomerates, which are better sorted and contain less clay than the alluvial plain conglomerates, constitute the bulk of commercial Cadomin reservoir rock.

Gething sediments represent channel and overbank deposition across a wide, low-relief drainage plain located on the eastern flank of the Cordilleran uplift. Rivers flowed east from the Cordillera

Table 1. A comparison of the porosity and permeability of Lower Cretaceous reservoirs in the Deep basin and on the Peace River Arch/Alberta Shelf.

Zone	Average Por. (%)	Porosity Range (%)	Average K (md)	K Range (md)	Established Reserves (Bcf)	Reservoir Facies
Deep Basin						
Paddy	11.7	9.7–13.5	81.19	0.98–237.99	216	Tidal channels
Cadotte	7.8	5.1–10.4	2.45	0.001–20.7	473	Barrier bars
Notikewin	8.7	7.8–9.6	81.82	0.004–775.3	167	Barrier bars
Falher A – Congl.	7.5	5.7–10.8	34.59	0.002–434.5	959	Wave dominated delta
– Sand	8.6	7.8–9.3	0.001	0.0001–0.01	1434	Wave dominated delta
Falher B – Congl.	7.4	4.7–11.0	62.56	0.003–853.17	692	Wave dominated delta
– Sand	8.8	7.8–10.2	0.001	0.0001–0.01	587	Wave dominated delta
Falher C – Congl.	9.1	7.5–11.1	69.31	0.01–381.81	128	Wave dominated delta
– Sand	8.0	5.5–8.8	?	?	137	Wave dominated delta
Falher D – Congl.	6.2	4.6–7.1	55.47	0.002–388.06	256	Wave dominated delta
– Sand	7.3	5.7–8.7	?	?	373	Wave dominated delta
Falher E – Congl.	8.4	7.3–9.4	21.125	0.004–206.0	52	Barrier bars
– Sand	7.9	7.8–8.0	?	?	91	Barrier bars
Bluesky	7.4	6.2–9.1	8.34	0.06–59.48	98	Tidally reworked barrier bars
Gething	6.9	5.6–8.3	2.8	0.009–55.77	117	Fluvial channels
Cadowin	6.1	4.8–8.9	1.22	0.004–20.99	1160	Fluvial braid plain
Peace River Arch/Alberta Shelf						
Paddy	24	13–30	300	50–1000+	203	Barrier bars
Cadotte	22	15–30	250	75–1000+	135	Barrier bars
Notikewin	19	9–27	?	?	47	Offshore bars
Falher – Undifferentiated	22	12–27	?	?	39	Offshore bars
Bluesky	18	14–24	200	20–1000+	64	Offshore bars
Gething	17	10–24	10	0.1–1000+	245	Fluvial channels
Cadowin	16	11–23	75	0.1–4000+	125	Fluvial braid plain

Data sources: – Deep Basin: Canadian Hunter Exploration Ltd.
 – Peach River Arch/Alberta Shelf: Alberta's Reserves of Crude Oil, Gas, Natural Gas Liquids and Sulphur, ERCB Report, Dec. 31, 1982.
 – Deep basin permeabilities were calculated from field well flow tests and therefore represent in situ conditions.
 – Peace River Arch/Alberta Shelf permeabilities were taken from field core analyses.

across the plain to join a major north-flowing trunk system, the Spirit River Channel. Sediments include fining-upward fluvial channel sandstones, overbank thin sandstones, siltstones, shales, and coal. Reservoir rock is restricted primarily to channel sandstones.

The Bluesky Formation represents sediments deposited during the transgression of the Moosebar sea. During this period, a strong sediment source southwest of Fort St. John supplied sediment to the area resulting in a prograding barrier bar and offshore bar sequence that advanced across much of the study area during early Moosebar time. Extensive tidal reworking of the barrier sands, indicated by a blocky gamma-ray log profile is found throughout the barrier system. Typically, the Bluesky Formation consists of either a basal lag sandstone overlain by coarsening-upward sandstone or a thick, massive sandstone (tidally reworked). In British Columbia, this sequence is capped with lagoonal, thin sandstones, shales, and coal. Behind the barrier bar trend, in the extreme southern part of the study area, bay shales with occasional molluscan shell banks are found. Reservoir rock is common in the barrier bar (particularly when tidally reworked) and offshore bar facies.

The Falher Member, within the study area, contains five stacked, mappable cycles, each consisting of lower delta plain sediments in the south and a regressive deltaic sequence followed by shallow marine offshore sedimentation to the north. Lower delta plain sediments include fining-upward channel sandstones, and lagoonal/overbank siltstones, shales, and coal. The prograding delta sequence contains transgressive lag sandstones and conglomerates, coarsening-upward sandstones (often capped by conglomerate), and occasional fining-upward fluvial channel sandstones. The above sequence is often capped with lagoonal shales and coal. Offshore sediments include shales and coarsening-upward sandstones. Coastlines were wave-dominated to the west (clean sandstone/conglomerate) and river-dominated to the east (muddy, discontinuous sandstones, shales). Reservoir rocks in the Deep basin are restricted to the clean sandstones and conglomerates of the wave-dominated systems. On the Alberta Shelf/Peace River Arch, gas is found in coarsening-upward, offshore sandstones.

The regressive cycle of the Notikewin Member, with the exception of a widespread sandstone and conglomerate barrier-bar trend near the point of maximum transgression of the Notikewin sea, consists entirely of river-dominated delta sediments. Argillaceous, coarsening-upward sandstones and shales are the dominant strata. Conglomerates are rare. Numerous fining-upward channel sandstones incise the system. A capping lagoonal facies of fine-grained sandstone, shale and coal is found across the width of the transgression. The offshore facies, found in the vicinity of the Peace River, consists of marine shales and coarsening-upward sandstones. The bulk of commercial Deep basin gas in this member is found in the barrier-bar trend. Alberta Shelf/Peace River Arch fields are primarily restricted to offshore bar sandstones.

The Paddy and Cadotte members represent a local regressive pulse within the transgressive Shaftesbury/Colorado sea. A strong sediment source southwest of Fort St. John supplied the regression. A prograding barrier-bar trend built eastward and northward, capped in the west by delta plain sediments and in the east by intertidal and shallow subtidal shelf strata. The Cadotte consists of coarsening-upward barrier bar sandstones often grading in the west, near the sediment source, to conglomerate. The Paddy in the west contains shales, coals, and fining-upward channel sandstones (lower delta plain). In the east, it consists of shales (intertidal, subtidal), thin coals (salt marsh), thin coarsening-upward sandstones (subtidal), and occasionally, thick, fining-upward sandstones (tidal channels). Reservoir rock is found throughout the Cadotte barrier bar system. In the Paddy, the tidal channels and occasionally the thin coarsening upward subtidal sandstones, provide commercial production.

REFERENCES CITED

Burden, E. T., 1982, Lower Cretaceous terrestrial palynomorph biostratigraphy of the McMurray Formation, northeastern Alberta: University of Calgary, Ph.D. thesis, 59 p.

Cant, D. J., 1983, Spirit River Formation – a stratigraphic-diagenetic gas trap in the deep basin of Alberta: AAPG Bulletin, v. 67, p. 577–587.

Clifton, H. E., R. E. Hunter, and R. L. Phillips, 1971, Depositional structures and processes in the non-barred high-energy nearshore: Journal of Sedimentary Petrology, v. 41, n. 3, p. 651–670.

De Mille, G., 1958, Pre-Mississippian history of the Peace River Arch: Journal of the Alberta Society of Petroleum Geologists, v. 6, p. 61–69.

Ethier, V., 1982, Channel and shoreline sequences in Paddy and Cadotte members of the Peace River Formation, Deep basin of Alberta, in J. Hopkins, ed., Depositional environments and reservoir facies in some western Canadian oil and gas fields: University of Calgary Core Conference, p. 29–41.

Gies, R. M., 1984, Case history for a major Alberta Deep basin gas trap; the Cadomin Formation, in J. A. Masters, ed., Deep basin gas: AAPG Memoir 38, this volume.

Hancock, J. M., and E. G. Kauffman, 1979, The great transgressions of the Late Cretaceous: Journal of the Geological Society of London, v. 136, p. 175–186.

Hunter, R. E., H. E. Clifton, and R. L. Phillips, 1979, Depositional processes, sedimentary structures, and predicted vertical sequences in barred nearshore systems, southern Oregon coast: Journal of Sedimentary Petrology, v. 49, n. 3, p. 711–726.

Jeletzky, J. A., 1971, Marine Cretaceous biotic provinces and paleogeography of western and arctic Canada; illustrated by a detailed study of ammonites: Geological Society of Canada Paper 70-22.

Lavoie, D. H., 1958, The Peace River Arch during Mississippian and Permo-Pennsylvanian time: Journal of the Alberta Society of Petroleum Geologists, v. 6, p. 69–74.

Leckie, D. A., and R. G. Walker, 1982, Storm- and tide-dominated shorelines in Cretaceous Moosebar – Lower Gates interval – outcrop equivalents of Deep basin gas trap in western Canada: AAPG Bulletin, v. 66, p. 138–157.

Masters, J. A., 1979, Deep basin gas trap, western Canada: AAPG Bulletin, v. 63, p. 152–181

McLean, J. R., 1976, Cadomin Formation; eastern limit and depositional environment: Geological Survey of Canada Paper 76-1B, p. 323–327.

——— , 1977, The Cadomin Formation; stratigraphy, sedimentology and tectonic implications; Bulletin of Canadian Petroleum Geology, v. 25, p. 792–827.

Nelson, H. W., and R. P. Glaister, 1978, Subsurface environmental facies and reservoir relationships of the McMurray oil sands, northeastern Alberta: Bulletin of Canadian Petroleum Geology, v. 26, p. 177–207.

Oliver, T. A., 1960, The Viking-Cadotte relationship: Journal of the Alberta Society of Petroleum Geologists, v. 8, p. 247–253.

Schultheis, N. H., and E. W. Mountjoy, 1978, Cadomin conglomerate of western Alberta – a result of Early Cretaceous uplift of the Main Ranges: Bulletin of Canadian Petroleum Geology, v. 26, p. 297–342.

Sneider, R. M., and H. R. King, 1984, Integrated rock-log calibration in the Elmworth field area, Canada, in, J. A. Masters, ed., Deep basin gas: AAPG Memoir 38, this volume.

Stelck, C. R., 1958, Stratigraphic position of the Viking Sand: Journal of the Alberta Society of Petroleum Geologists, v. 6, p. 2–7.

——— , and J. W. Kramers, 1980, Freboldiceras from the Grand Rapids Formation of north-central Alberta: Bulletin of Canadian Petroleum Geology, v. 28, p. 509–521.

Sternberg, C. M., 1932, Dinosaur tracks from Peace River, British Columbia: National Museum of Canada Annual Report, 1930, Bulletin 68, p. 59–86.

Stott, D. F., 1968, Lower Cretaceous Bullhead and Fort St. John groups between Smoky and Peace rivers, Rocky Mountain Foothills, Alberta and British Columbia: Geological Survey of Canada Bulletin, no. 152, 279 p.

——— , 1973, Lower Cretaceous Bullhead Group between Bullmoose Mountain and Tetsa River, Rocky Mountain Foothills, northeastern British Columbia: Geological Survey of Canada Bulletin, no. 219, 228 p.

Vail, P. R., R. M. Mitchum Jr., and S. Thompson III, 1977, Seismic stratigraphy and global changes of sea level, part 4; global cycles of relative changes of sea level: AAPG Memoir 26, p. 83–97.

Varley, C. J., 1983, The sedimentology and diagenesis of the Cadomin Formation, Elmworth area, northwestern Alberta: University of Calgary, Master's Thesis, 173 p.

Williams, G. D., and C. R. Stelck, 1975, Speculations on the Cretaceous palaeogeography of North America, in W. G. E. Caldwell, ed., The Cretaceous system in the western interior of North America: Geological Association of Canada Special paper no. 13, p. 1–20.

Youn, S. H. 1983, Depositional environments and their significance on diagenetic processes of Falher Member (Lower Cretaceous), Spirit River Formation, Elmworth area, Alberta: University of Calgary, Ph.D. thesis.

Case History for a Major Alberta Deep Basin Gas Trap: The Cadomin Formation

Robert M. Gies
Canadian Hunter Exploration Ltd.
Calgary, Alberta

Abundant, high-quality geological data from the gas productive Elmworth region of the Alberta Deep basin reveal the existence of enormous gas accumulations found in an unconventional form of trap. This special form of "deep basin" gas trap defies conventional concepts of gas entrapment by turning them virtually upside-down. Gas is trapped in the deepest part of the basin, rather than on the flank of the basin. The gas/water contact occurs at the updip end of the accumulation, rather than the downdip end. Original gas accumulation pressures lie below the regional formation water pressure gradient, rather than above it as in conventional traps. Gas in at least one giant accumulation is in a dynamic state of updip migration, rather than in static equilibrium as found in conventional traps. The Lower Cretaceous Cadomin formation in the Alberta Deep basin provides some of the best available information on basic characteristics of deep basin gas traps.

Physical principles behind this form of gas entrapment were confirmed in fluid flow models designed to simulate the Cadomin gas-trapping conditions.

INTRODUCTION

In February 1976, a commercial gas discovery was made when a well was drilled near the small town of Elmworth, Alberta. As a consequence of that discovery, 700 follow-up wells were drilled throughout the surrounding region during the next 6½ years. The discovery well helped to confirm suspicions, based on log evaluations for a number of abandoned wells scattered throughout the area, that the Mesozoic section in the deepest portion of the Alberta sedimentary basin was virtually gas saturated, whereas the adjacent updip regions were mainly water bearing.

The follow-up drilling confirmed this unexpected geological phenomenon with the establishment of almost 6 tcf (170×10^9 cu m) of proven gas reserves. Ten gas plants have been built with a combined raw sweet gas processing capacity of 1.2 billion cu ft (34×10^6 cu m) per day, or the energy equivalent of 200,000 barrels of oil (31,800 cu m) per day.

Gas production is obtained from numerous formations extending throughout an elongate region covering several thousand square miles situated in the deepest part of the basin just in front of the northwest-trending disturbed belt (Fig. 1). This is not a unique gas entrapment situation. Most of the principal features are similar to the long established San Juan basin of New Mexico which contained an estimated 25 tcf (708×10^9 cu m) of gas in the deepest part of that basin.

In the beginning it was difficult to accept early indications of possible large gas accumulations trapped downdip from water, but as development progressed it became apparent from the results that, for reasons unknown at that time, our conventional concepts of gas entrapment, (that is, gas updip from water), were being turned virtually upside down. Gas/water contacts, for example, are developed at the updip termination of the deep basin gas accumulations whereas in conventional traps, the gas/water contacts are found at the downdip end. Most deep basin gas accumulations have original gas pressures that lie below the regional formation water pressure gradient, rather than above it as in conventional gas accumulations (Fig. 2). In at least one case, a deep basin accumulation appears to exist in a dynamic state of updip gas migration, while most conventional gas accumulations are found in a state of static equilibrium. Today we have developed an understanding of the basic physical principles behind this important form of hydrocarbon entrapment and as time goes on, we continue to learn new and important characteristics of deep basin traps.

Elmworth development has provided a vast quantity of high-quality, modern technical data of all kinds. Much of it is publicly available.

One of the best examples for illustrating the features of deep basin entrapment is found in the widespread Cadomin formation of Lower Cretaceous age (Fig. 3). An estimated 15 tcf (425×10^9 cu m) of gas in place are contained in the 3,800 sq mi (9,840 sq km) Cadomin gas accumulation. This paper presents a brief description of the geological setting for the Cadomin formation followed by a description and discussion of the gas and water it contains. The two-part approach emphasizes and demonstrates the physical compatibility that must exist between reservoir geology and reservoir fluid distribution (which includes fluid pressures). Experience has shown that failure to consider and integrate relevant data of all kinds can result in a seriously inaccurate concept of the trap. The geological setting of the reservoir involves not only structural considerations, but the equally important aspects of sediment types, sedimentary conditions, sediment distribution and diagenesis. The latter processes combine to establish present day reservoir "plumbing." The reservoir plumbing is a combination of variables, like porosity and permeability, together with reservoir continuity trends and fluid barriers (reservoir discontinuities) and seals. Such geological conditions in turn exert major controls over the distribution of gas and water throughout the reservoir formation. Mutual dependence between reservoir fluid distribution and reservoir rock geometry means that we can use fluid data to provide valuable additional information about the reservoir's plumbing; especially in areas somewhat remote from the well bore.

DEPOSITIONAL SETTING OF THE CADOMIN

By early Cretaceous time an extensive mountain system had formed in the region of the province of British Columbia (Fig. 4). The climate must have been humid with abundant rainfall. At numerous points along the east- and northeast-facing mountain front, rivers emerged out onto the eastern plains. The sand and pebble debris they carried was deposited abruptly as the rivers left the confines of their mountain canyon channels. Near the mountains, alluvial fans of great thickness developed. Water flow across the fans redistributed some of the sediments downslope toward the northeast to develop broad alluvial plains. Water runoff from the alluvial plains eventually joined a large northwestward-flowing braided river system which transported both water and sediments derived from mountain sources, and from eastern drainage areas as well, in a direction more or less parallel to the mountain front. This large braided trunk river is named the Spirit River (McLean, 1977).

The overall drainage system, therefore, consisted of two complementary but distinctively different fluvial types. In this case the braided trunk river system was necessary to carry away water derived from the mountain rivers. The river was, no doubt, controlled in its position and direction of flow by topography and structural activity of the basin at the time. Recognition of the two interacting fluvial systems is fundamental to developing a proper understanding of the resultant reservoir plumbing patterns and consequent entrapment of the deep basin gas accumulation in the Cadomin.

As time passed, the two great fluvial systems competed for dominance of the region bordering the mountain front. Alluvial plains sediments, spread by stream flow downslope to the northeast, eventually were captured and transported northwestward by the Spirit River. The Spirit River channel migrated laterally across the region beginning perhaps at its farthest eastward position, which is defined by a gently rising erosional feature called the Fox Creek Escarpment (Fig. 5). These and other geological interpretations of the Cadomin history throughout the region are founded largely on subsurface well control supplemented by outcrop investigations. Lateral channel migration may have been promoted by gradual basin subsidence in late Cadomin time which

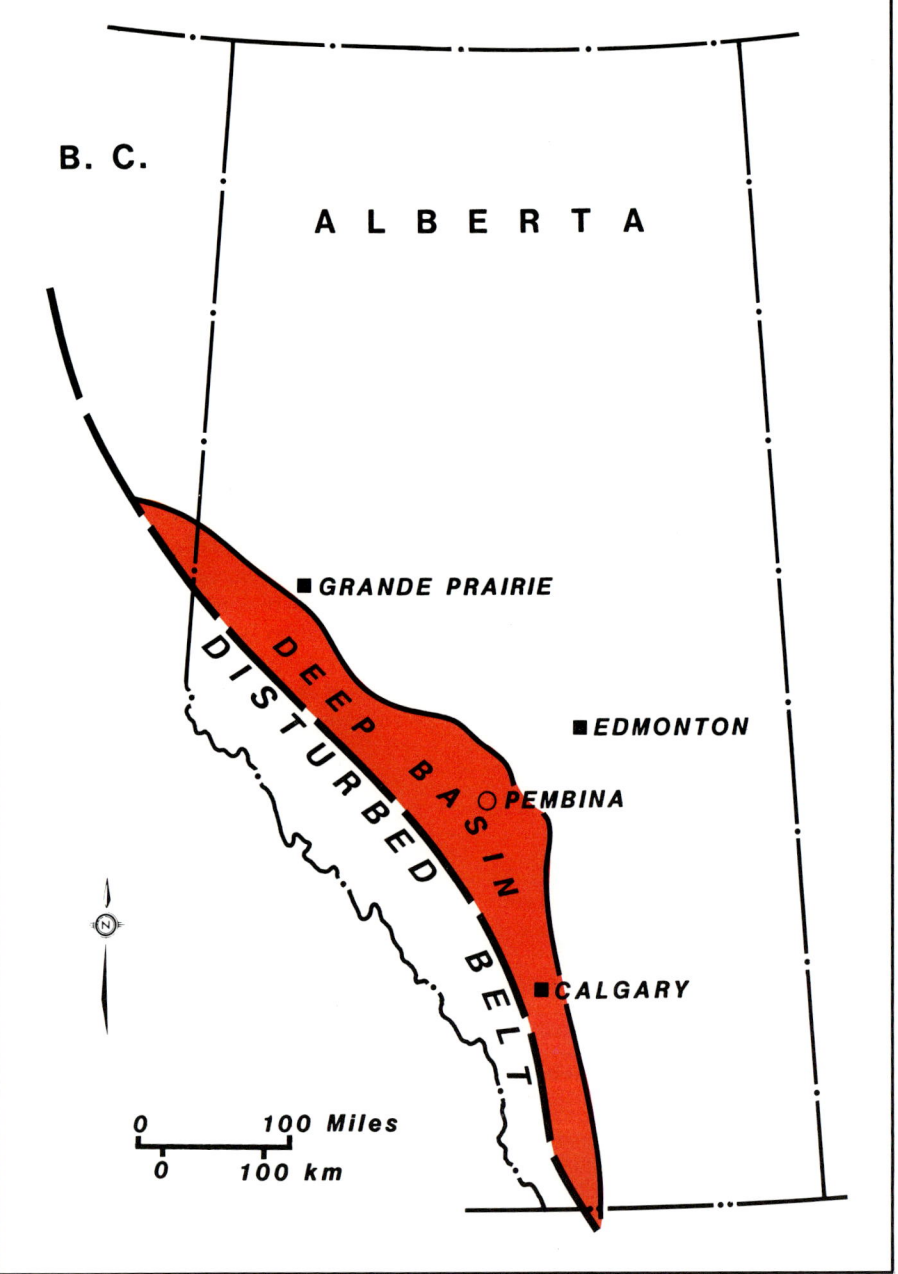

Figure 1. Index map showing location of the Elmworth deep basin.

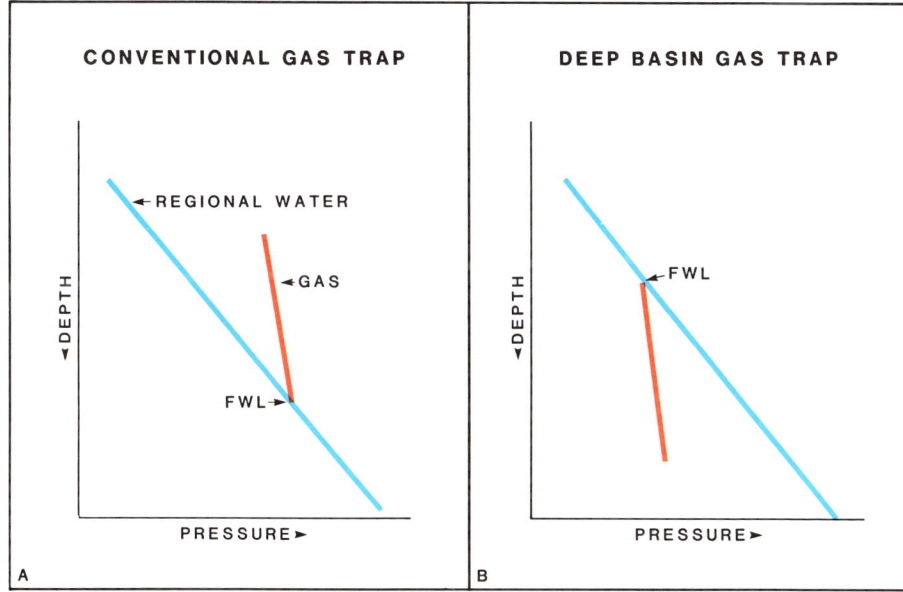

Figure 2. Pressure-depth relationships between conventional and deep basin gas traps.

saw the basin tilt to the southwest with the consequent lateral shifting of the Spirit River closer to the mountain front (Fig. 6). Remnant outliers of former alluvial plains sediments were cut off or became interbedded with Spirit River sediments throughout an extensive region. At the river's closest position to the mountain front, braided river sediments are overlain by a thin layer of alluvial plains sediments derived from the fans to the southwest.

In places, the Cadomin has been found from well control to be very thin or absent along northwest-trending belts parallel to the Spirit River channel courses. There is some evidence of clay channel fills and shale infilling, of possible late channel courses, by the basal Gething formation. This resulted in reservoir discontinuities or barriers along the Spirit River channel course. Such barriers act to influence present-day distribution of gas and water in the Cadomin reservoir.

Figure 7 is an isopach map of the total Cadomin interval. The map illustrates important Cadomin thickness trends throughout the region of interest. Thick isopachs in British Columbia reflect proximity to major alluvial fan sediment sources which spread north and eastward from the disturbed belt. These thick trends are dissected by a linear, isopach thin trend oriented northwestward which marks the position of the late Spirit River channel system.

Examination of Cadomin sediments in cores and drill cuttings permits the identification of the original type of fluvial conditions during deposition, that is, alluvial plain or braided river (Spirit River). The alluvial plains sediments consist of sandy conglomerates containing mainly chert and some quartzite pebbles that are characteristically poorly sorted with maximum pebble diameter of about 3 in (75 mm). Dark pebble colors predominate. The associated matrix sands are similarly dark in color and usually fine-grained chert sands with about 10% quartz grains (Fig. 8). Spirit River sediments also consist of sands and sandy conglomerates. However, the pebbles tend to be smaller, better sorted and light colors (white and pale apple green) predominate (Fig. 9). The associated matrix sands are also light colored, well sorted, coarse grained and mainly chert with up to 20 percent quartz. Quartzite pebbles are absent. Alluvial plains sediments that became reworked by the Spirit River are better sorted with coarser sand grains, although typical Spirit River light-colored clastics are the most abundant (Fig. 10).

Detrital clays have been identified in cores from the alluvial plains sediments. They line pore walls or partly infill some pores. These clays are thought to have been deposited in the pore system by muddy mountain river waters which infiltrated the alluvial plains sediments (sieve deposits). Clay rip up clasts, thin coaly beds and shale beds tend to be characteristic of the Spirit River braided channel deposits.

Although the usual processes of burial diagenesis greatly reduced original porosity and permeability, distinct differences in the magnitudes of permeability are still governed by original sedimentary processes. As expected, the permeability of alluvial plains sandy conglomerate is very low (often less than a millidarcy at in situ conditions) due to small pore throat sizes of the fine-grained sand matrix. Even the occasional well-sorted, sand-free conglomerate beds may show reduced permeability due in part to clay sieve deposits. The well-sorted, coarse-grained braided river deposits, on the other hand, have permeabilities ranging up to several hundred millidarcys because of larger pore throats and less interstitial clays.

RESERVOIR FLUIDS

The second half of the Cadomin story concerns the reservoir fluids. Figure 11 is a map of the Elmworth region showing the gas and water distribution throughout the Cadomin. Geological control and reservoir fluid interpretations are based on examination of data from the 650 wells scattered throughout the area. Note the Cadomin structure contours which depict a very gently southwestward-dipping monocline. Cadomin sandy conglomerate was deposited throughout the entire map region. Updip areas (shown in white) are water bearing.

The Cadomin deep basin as accumulation (red) extends across 3,800 sq mi (9,840 sq km) in the downdip region of the map. There are also a few small, conventional type gas accumulations (orange), thought to be both structurally and stratigraphically trapped in the updip, predominantly water saturated region of the Cadomin. About 100 wells capable of initial gas flow rates in excess of 500,000 cu ft (14,160 cu m) per day have been completed to date. There are also a similar number of potential gas wells waiting on completion, or wells with lesser flow rates.

An important feature of the main gas accumulation is that it occupies a large proportion of the downdip, low-permeability alluvial plains sediments. The more permeable braided river deposits predominate throughout the updip region which is shown to be mainly water bear-

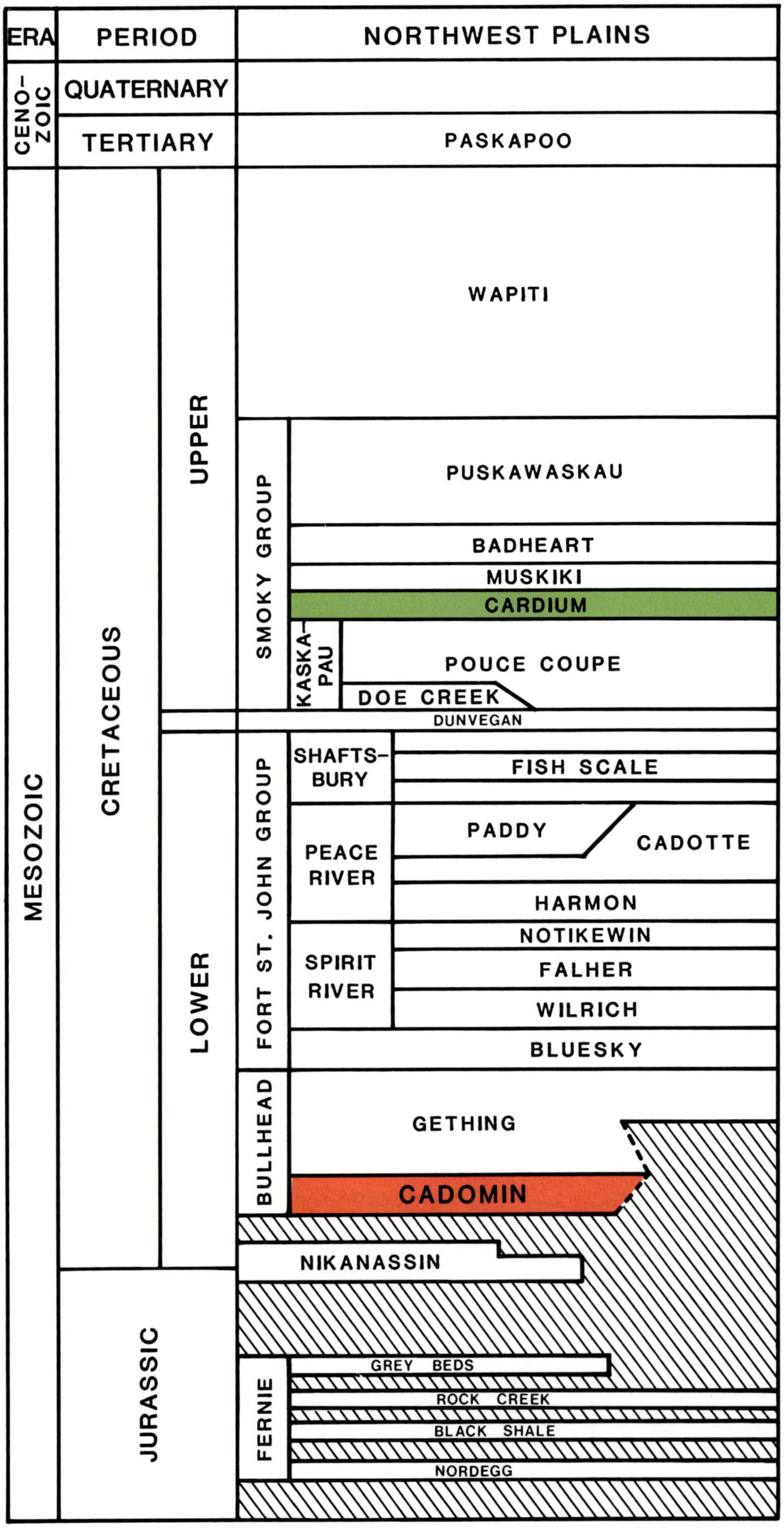

Figure 3. Table of stratigraphic names in the Elmworth deep basin-Mesozoic section.

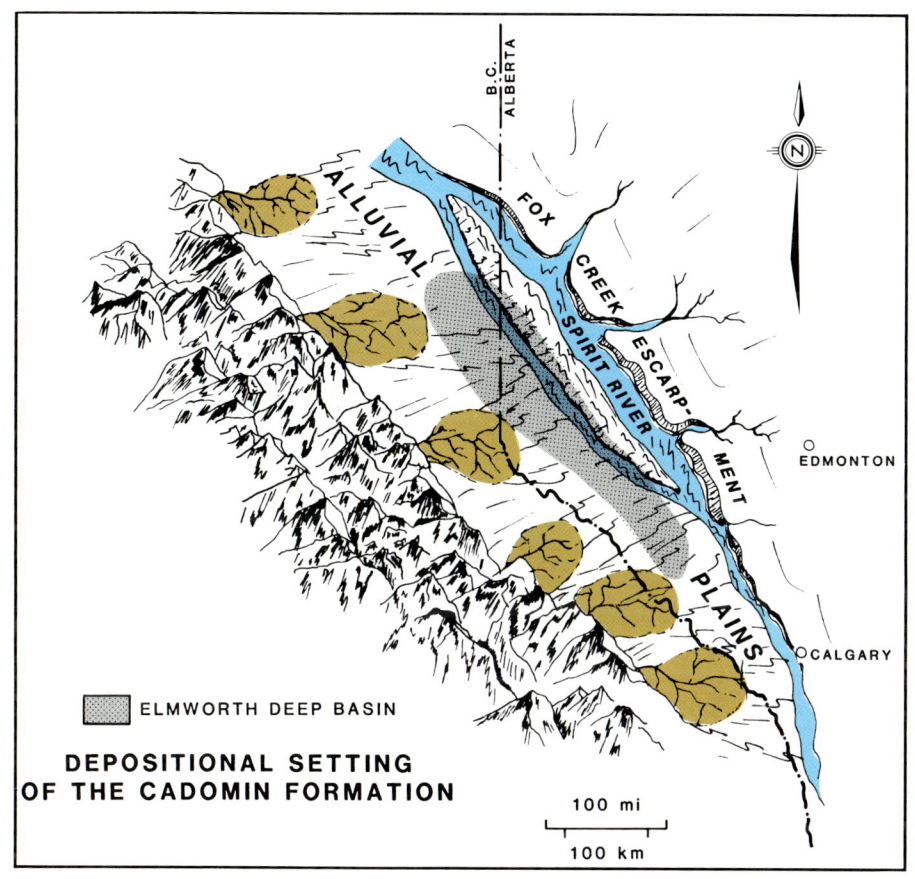

Figure 4. Sandy conglomerate derived from alluvial fans was deposited across broad alluvial plains to the east where the conglomerate was reworked by the northward flowing Spirit River system.

Figure 5. Model of early Cadomin depositional setting.

The Cadomin Formation

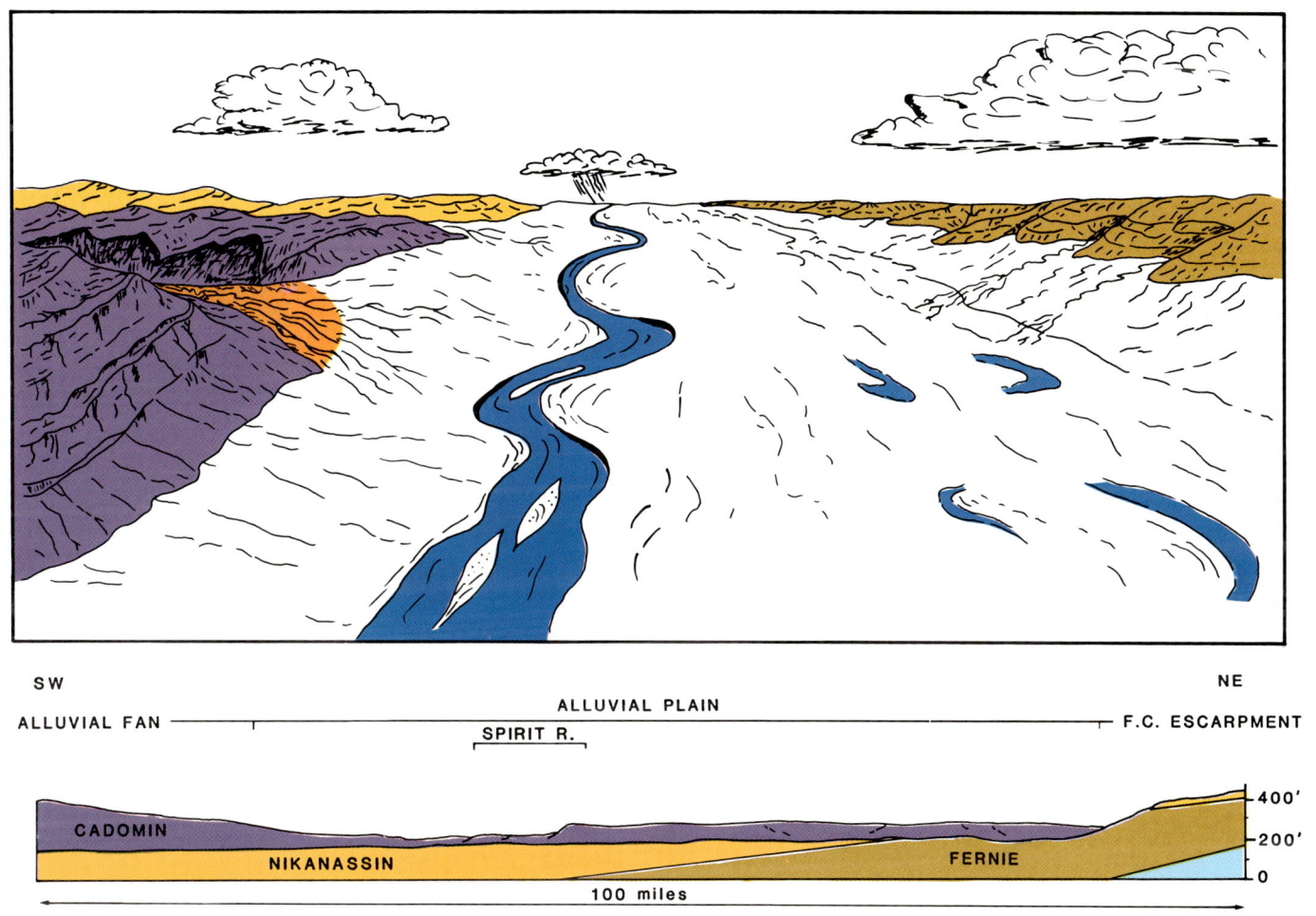

Figure 6. Model of late Cadomin depositional setting.

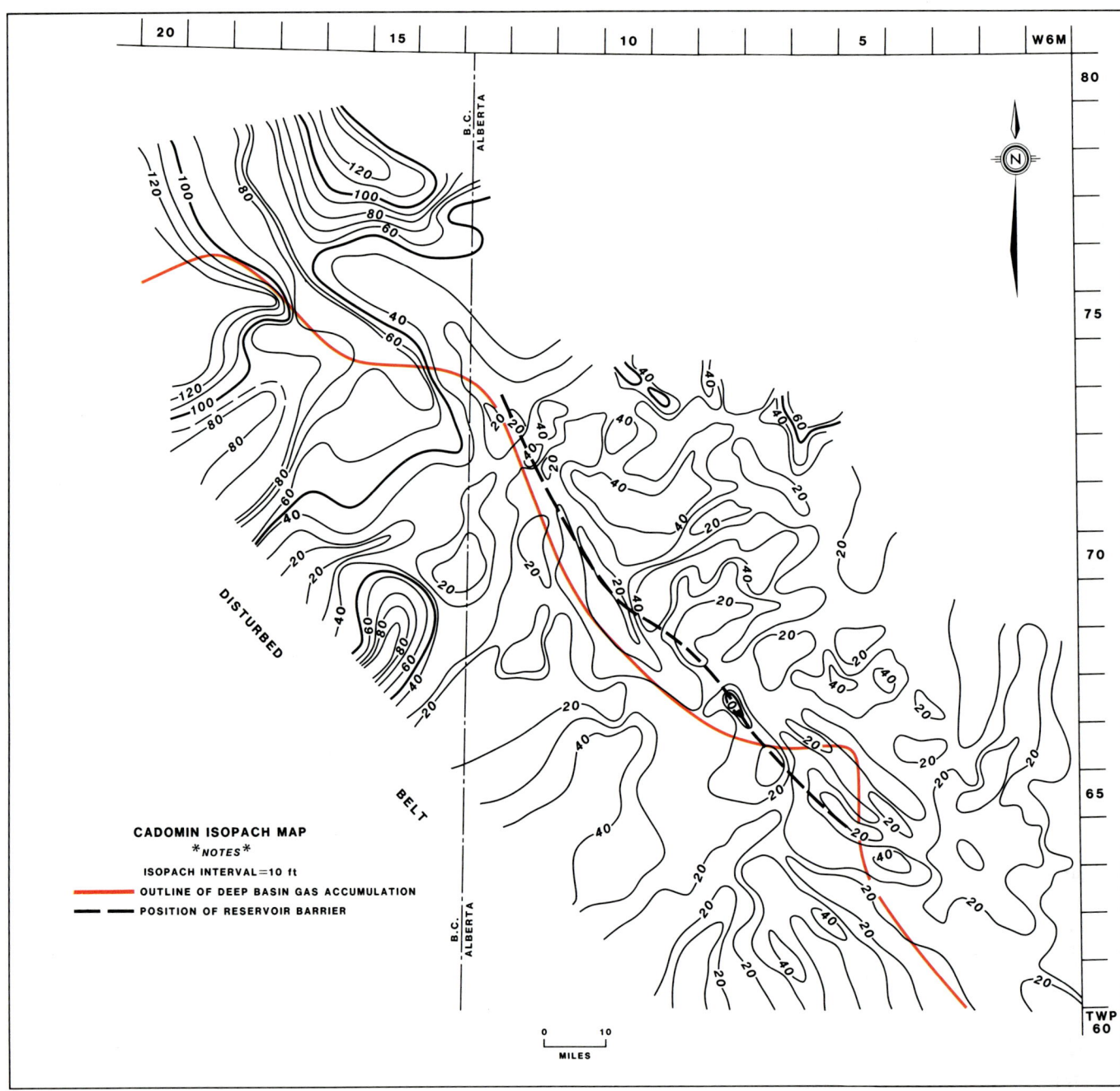

Figure 7. Isopach map of total Cadomin interval. Note northeast to east oriented "thicks" which reflect alluvial plains sedimentary trends versus linear northwestward "thin" along the course of a late Cadomin Spirit River channel system.

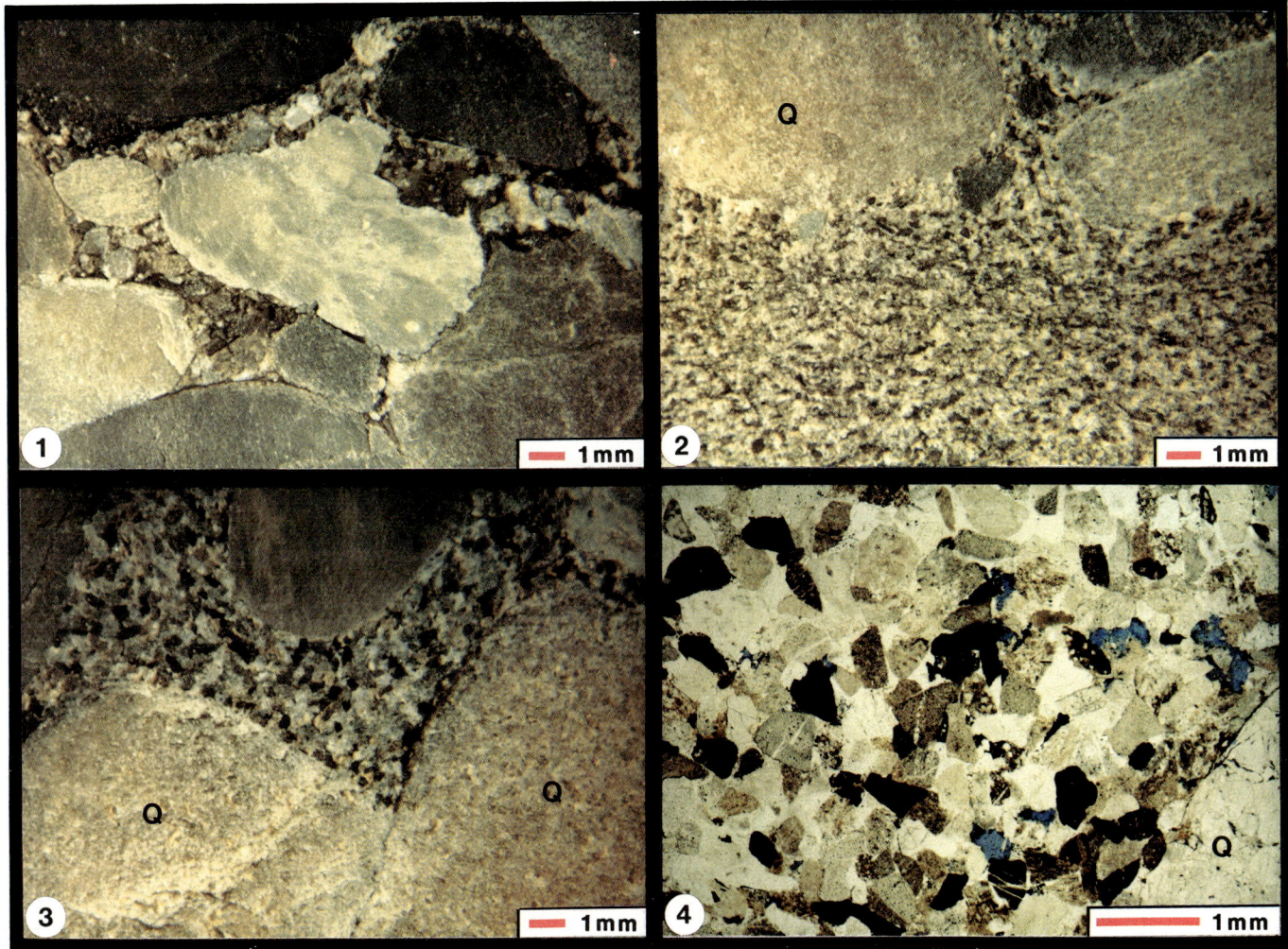

Figure 8. Microscopic views of Cadomin alluvial plain sandy conglomerate facies.

(1) 10-8-70-9W6, Cadomin Formation, 2,203m, 6.8×, ϕ = 1.2%, k = 1 md.
Typical dark-colored chert pebble conglomerate with medium grained, dark-colored chert and quartz sand matrix as seen through the binocular microscope with reflected light.

(2) 11-28-70-11W6, Cadomin Formation, 2,301m, 6.8×, ϕ = 2.2%, k = 0.065 md.
This binocular microscope view illustrates the fine-grained, tight-sand matrix frequently found in the alluvial plain conglomerate facies. Note the pink quartzite pebble (Q).

(3) 11-7-68-12W6, Cadomin Formation, 2,830.5m, 6.8×, ϕ = 1.9%, k = >0.01 md.
This binocular microscope view also illustrates the typical, dark colored, tight, fine sand matrix found in many alluvial plain conglomerates. Note pink quartzite pebbles (Q).

(4) Some visible porosity (blue color) can be seen in thin section through the polarizing microscope within the quartz-rich sand matrix of the same sample described in (3) above. Additional, unseen microporosity also is present among and within some of the sand grains. Note pervasive porosity infilling quartz overgrowth cement, and quartzite pebble (Q).

Figure 9. Binocular microscope views of Cadomin Spirit River sand conglomerate facies.

(1) 10-8-70-9W6, Cadomin Formation, 2,195.6m, 6.8×, ϕ = 4.5%, k = 20 md.
This view and photos 2 and 3 are from the same well which also contains alluvial plain facies as seen in Figure 7. Note that light chert colors predominate (white, pale green, and light brown), the sand matrix is coarse grained and matrix porosity is obvious. Quartzite pebbles are very rare or absent.
(2) 10-8-70-9W6, Cadomin Formation, 2,197m, 6.8×, ϕ = 3.6%, k = 18 md.
The porous, light-colored chert grain texture of this sandy conglomerate is very distinctive.
(3) This is an enlarged view (10.2×) of photo 2 illustrating the well sorted, essentially silt and clay free texture of the porous Spirit River facies.
(4) 7-11-65-2W6, Cadomin Formation, 2,342m, 6.8×, ϕ = 2.2%, k = 70 md.
Similar chert mineralogy and clean, porous texture are seen here in this example of a coarser-grained sandy chert some 56 mi (94 km) southeast of the 10-8 well (examples in photos 1-3).

The Cadomin Formation   125

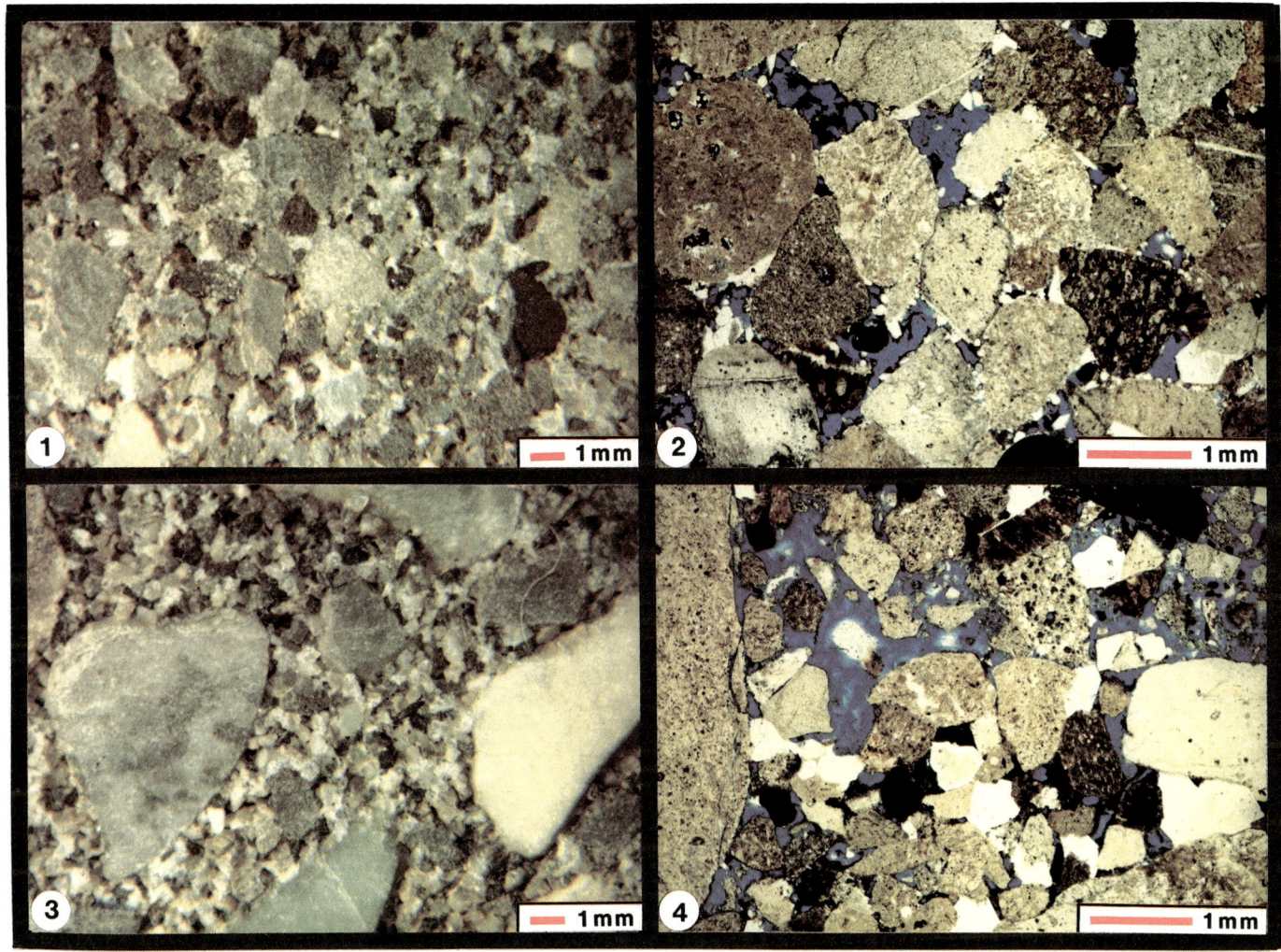

Figure 10. Microscopic views of mixed alluvial plain, Spirit River Cadomin conglomerate facies.

(1) 6-16-68-11W6, Cadomin Formation, 2,631.4m, 6.8×, φ = 10.6%, k = 1,367 md.
The binocular microscope view shown here reveals a well sorted, coarse-grained but dark-colored chert sand matrix.

(2) This thin section view (21×) from the same sample seen in photo 1 illustrates an unusual example of porosity and permeability enhancement due to chert dissolution and pore enlargement by circulating formation waters. At a later stage, druzy quartz developed on pore walls as silica saturation with respect to quartz increased.

(3) 10-21-68-4W6, Cadomin Formation, 2,015.7m, 6.8×, φ = 10.6%, k = 940 md.
Dark chert colors predominate (relative to typical Spirit River facies) but coarse, well sorted chert sands form the conglomerate matrix.

(4) The thin section view (21×) illustrates abundant porosity in the chert sand grain matrix. Quartz grains are not very abundant but where present show evidence of porosity destroying quartz cement overgrowths. Porosity creating leaching, on the other hand, is apparent among the chert sand grains.

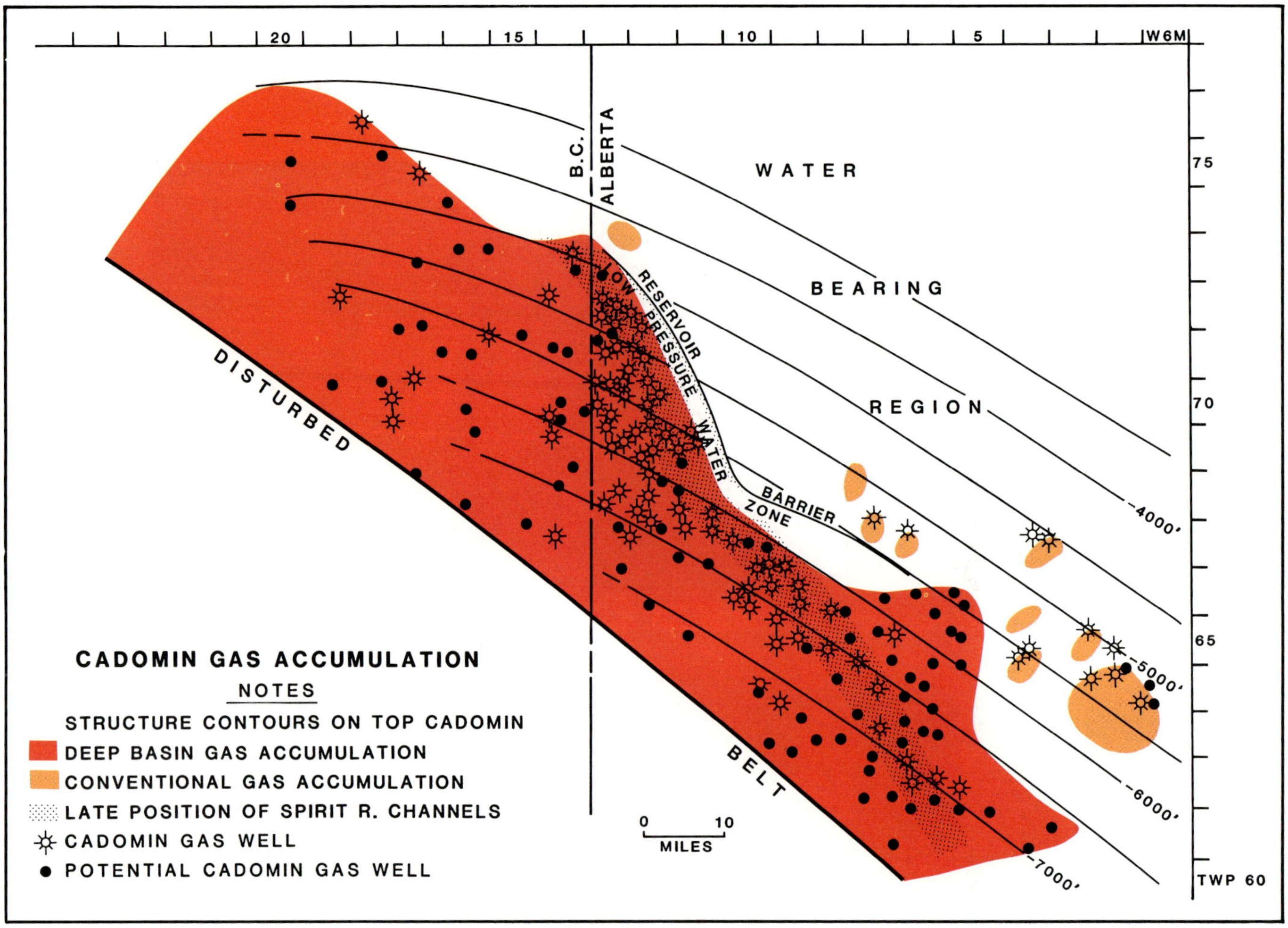

Figure 11. Map showing gas- and water-bearing regions in the Cadomin formation.

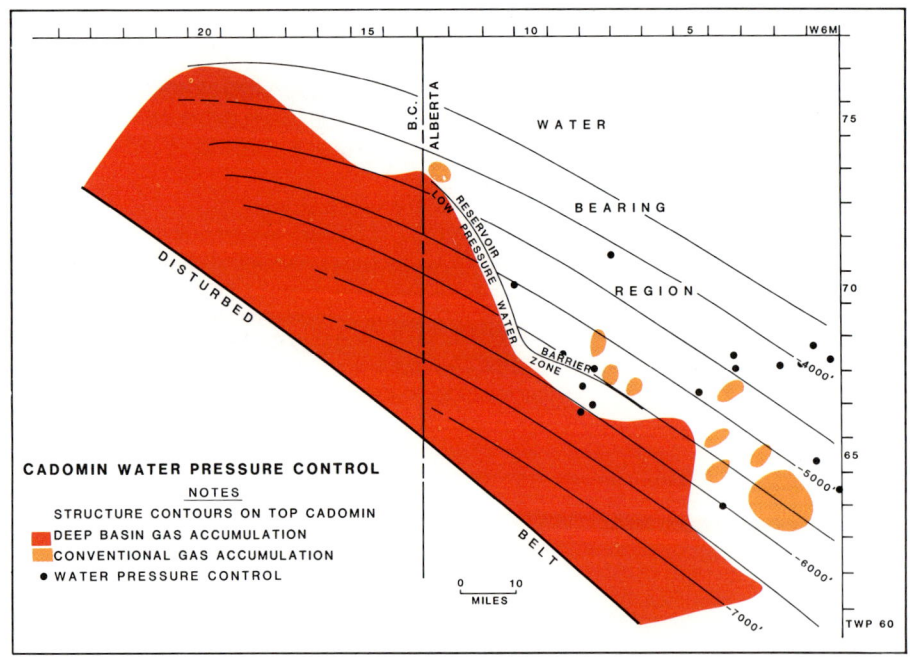

ing. But, there is also an important commingled and interbedded zone of both fluvial types that is 6 mi (9.6 km) or more in width running along the downdip, southwest side of the water/gas contact line. This is represented on Figure 11 by a shaded band which shows the position of the most southwestern extent of the Spirit River channel system.

Reservoir boundaries for this extensive gas field are important from the standpoint of understanding the gas-trapping mechanism. The downdip boundary occurs where the reservoir rock becomes either too tight to be productive, or it is cut off by thrust faulting along the edge of the disturbed belt. There is no evidence for a defined downdip gas/water contact.

Figure 12. Map showing distribution of Cadomin water pressure control points.

Table 1. Typical examples of Cadomin formation water properties.

No.	Well	DST No.	Interval	Recovery	pH	Ion Concentration mg/ℓ						
						Ca	Mg	Na+k	HCO₃	SO₄	Cl	CO₃
1	6-7-67-8W6	6	2,665-2,681.6 m	432 m of salt water - sample from downhole sampler	6.7	3,680	1,115	31,301	810	21	57,400	0
2	11-35-67-9W6	3	2,537-2,550 m	1,185 m of salt water - sample from downhole sampler	5.9	4,709	1,103	35,042	434	6	65,200	0
3	7-1-68-7W6	6	2,306-2,310 m	56 m mud, 224 m salt water - sample taken at top of tool	7.0	1,277	340	26,280	968	16	43,100	0
4	11-35-68-7W6	15	7,324-7,343 ft	1,560 ft salt water - sample taken at bottom of recovery	6.6	1,437	540	30,652	1,051	7	50,761	0
5	11-36-70-11W6	4	2,251-2,262 m	189 m salt water - sample taken at bottom of recovery	7.3	1,020	438	27,801	1,108	27	45,200	0

Notes: Analyses No.'s 1 and 2 are from wells located in the low-pressure water-bearing Cadomin south of the reservoir barrier shown in Figure 10 whereas analyses 3 and 4 are from nearby wells located north of the reservoir barrier and represent typical Cadomin water properties seen in many wells in the regional water-bearing portion of the Cadomin. Note that samples from the low-pressure region south of the barrier show distinctly higher concentrations of calcium, magnesium and chlorine ions. Analysis No. 5 represents Cadomin water properties north of the barrier but 25 mi (42 km) west of analyses 3 and 4.

In the updip region, a water/gas contact is defined from well control at about 4,265 ft (1,300 m) subsea in Township 73 Range 13. This is not believed to be a reservoir discontinuity or barrier, but simply a contact in low-permeability sediments where pressures in the water are in balance with gas phase pressures. It is like a conventional gas/water contact only turned upside-down. The vertical gas column height, defined by these updip and downdip limits, is about 2,600 ft (792 m).

An important reservoir barrier or discontinuity exists over a great distance in the Cadomin and it is closely linked to the gas accumulation because it acts as a lateral boundary. It is defined from well logs, fluid pressures, water analyses and fluid type control. Figure 11 shows the discontinuity oriented in a northwest-southeast direction running oblique to the structural contours. On the northeast side, the Cadomin is, for the most part, permeable but water bearing. But on the west and southwest side, conditions are more complex. We believe that the discontinuity is the result of the northwest Spirit River channel development during late Cadomin times (Fig. 4). The few wells that penetrated the barrier, or discontinuity trend, show that the Cadomin is unusually thin or is absent, with Lower Gething sediments resting directly on the underlying Nikanassin formation. Note that Cadomin water occupies the pore system along both sides of the discontinuity trend, although the water-saturated zone is narrow along the southwest side. Water pressure measurements show that pressures on the northeast side are 160 psi (1,158 kPa) greater than water pressures on the opposite side at common subsea depths. Water analyses data also show a significant change in ion concentrations on opposite sides of the discontinuity trend, but show consistent properties for waters among a number of control points selected along either side (Table 1).

Original reservoir fluid pressures often provide considerable insight into the nature of hydrocarbon traps. The deep basin Cadomin gas accumulation is no exception. Figure 12 is a map of the basin showing well locations where 24 high-quality original formation water pressures were obtained (Table 2). In Figure 13 it is seen that 19 of these values fall along a straight line in a standard pressure-depth plot. The remaining five lower pressure points are from wells located in the water-bearing Cadomin located along the southwestern side of the discontinuity trend. Figure 14 is a map of the same area showing locations of 45 wells where good original Cadomin gas pressure values were obtained (Table 3). In all cases, pressures used were carefully qualified and calculated using standard Horner extrapolation

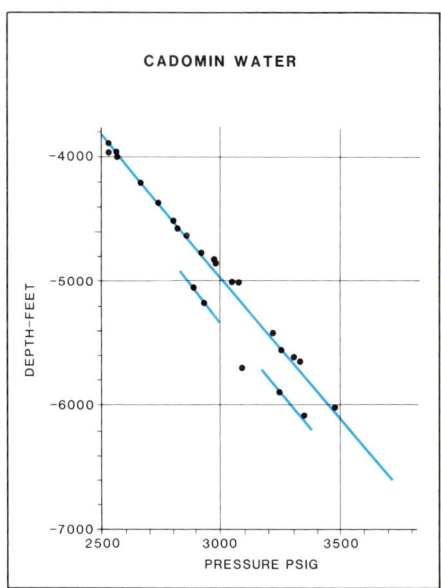

Figure 13. Pressure-depth profile for original Cadomin formation water pressures updip from the deep basin gas accumulation.

methods. These gas pressures were from drill-stem tests supplemented by post-completion pressure build-up tests. The resulting pressure-depth distribution is presented in Figure 15. Note that the gas pressure values appear to define a curved line, although the point spread increases with depth. This turned out to be a sur-

Table 2. Original formation water pressures, Cadomin formation, Elmworth area.

Point	Well	KB Elev. (ft)	DST Inter. Top Base (ft)		REC Depth (ft SS.)	Extr. ISIP (psig)	Extr. FSIP (psig)	Water Recovery (ft)	Remarks
A	7-20-64-1W6	2,845	7,889	7,905	−5,053	3,077	3,077	1,594	
B	11-1-64-5W6	2,936	8,982	9,012	−6,050	3,478		558	
C	7-11-65-2W6	2,843	7,675	7,713	−4,846	2,982	2,982	2,707	
D	10-23-65-5W6	2,586	8,234	8,300	−5,655	3,312		361	
E	10-20-67-5W6	2,342	7,254	7,355	−4,926	2,979	2,903	950	
F	6-7-67-8W6	2,851	8,743	8,798	−5,901	3,247	3,247	1,312	
G	11-35-67-9W6	2,623	8,321	8,364	−5,708	3,094	2,932	3,766	Limited reservoir
H	10-2-67-9W6	2,796	8,888	8,931	−6,128		3,347		Fm. wtr. on comp.
I	6-8-68-1W6	2,408	6,355	6,371	−3,950	2,665	2,552	2,650	
J	11-15-68-2W6	2,273	6,233	6,263	−3,942	2,532	2,539	4,711	
K	6-26-68-2W6	2,214	6,096	6,132	−3,897	2,539	2,533	4,500	
L	11-11-68-3W6	2,343	6,503	6,528	−4,225	2,671	2,612	1,400	
M	11-9-68-4W6	2,220	6,803	6,873	−4,571	2,836	2,830	4,790	
N	10-21-68-4W6	2,196	6,558	6,637	−4,373		2,752	1,000	Pkr. failed
O	6-8-68-8W6	2,750	8,205	8,268	−5,459		3,226	690	
P	6-20-68-9W6	2,552	8,123	8,222	−5,550		3,258	771	8 psi added for depth
Q	11-36-70-11W6	2,326	7,385	7,421	−5,062	3,048	3,048	620	
R	6-25-71-8W6	2,394	6,414	6,450	−4,006		2,574	560	
S	7-6-71-11W6	2,481	7,671	7,736	−5,193	2,987	2,932	184	Limited reservoir
T	10-28-71-11W6	2,647	7,388	7,451	−4,748		2,922	2,038	
U	7-13-71-12W6	2,700	7,779	7,812	−5,060	2,894	2,690	841	
V	7-11-72-12W6	2,679	7,380	7,407	−4,683		2,868	466	
W	10-24-72-12W6	2,588	7,086	7,112	−4,531	2,807	2,804	2,350	
X	11-18-79-16W6	2,683	4,957	5,016	−2,274	1,833	1,833	1,076	Falls on extrap. wtr. pressure gradient
Y	10-31-68-10W6	2,437	8,130	8,153	−5,696	3,335		2,300	

prise because a steeply inclined straight line is usually the case for gas accumulations. Careful re-examination of the pressure data showed nothing apparently wrong and, therefore, the curved gas pressure gradient line represents factual information that must be integrated with other geological data on a physically sound basis.

Figure 16 is a comparison of the pressure-depth relationship for both the Cadomin gas and water. Note that pressures in both fluids are about equal at a subsea depth of about 4,200 ft (1,280 m), which corresponds to the updip water/gas contact position previously described. Gas pressures at all other subsea depths are consistently less than water pressures at corresponding depths throughout the widespread water-bearing region of the Cadomin.

Figure 17 is a comparison between the curved line, representing actual Cadomin gas pressures, and a straight line for the gas-pressure gradient one would expect to see for an accumulation in static equilibrium. The wide deviation between the two is of considerable geological interest and importance. Gas-pressure gradients along the curved line range from about 0.06 psi/ft (1.356 kPa/m) at shallow depths, to 0.65+ psi/ft (14.7 kPa/m) in the downdip area. This compares to the normal static pressure gradient of 0.055 psi/ft (1.243 kPa/m) for the kind of dry gas present in the Cadomin.

The comparison gives rise to the question: What is the explanation for the wide deviation found between the expected straight line, and the actual curved gas-pressure gradient line? There are two possibilities. One is that the pressure points forming the curve are from several pools that each have progressively higher pressures with increasing depth, and, therefore, they only appear to fall along a curve. This would require a straight line with a pressure gradient of about 0.055 psi/ft (1.243 kPa/m) to fit points in each of these individual pools. The second possibility is that there is only a single continuous Cadomin gas accumulation, but it is not in static equilibrium. Rather, it is in a dynamic state of updip gas migration. Although the concept of a dynamic gas accumulation is unusual, it is the preferable interpretation because it fits the geological data without any apparent technical weaknesses. A major difficulty with the first possibility, that is, separate gas pools, is that there is no apparent mechanism for gas to have originally displaced the formation water from the water-wet Cadomin reservoir, and end up in an array of separate gas pools with gas pressures significantly below extrapolated water pressures for the same depths. If separate gas pools existed, then we should expect to find some evidence for lateral seals, or reservoir discontinuities, required to separate the pools. It is difficult to make such a geological case for these, based on the available well control.

The case for a single dynamic gas accumulation fits the geological and reservoir fluid data far better. In order to explain the curved gas gradient line, it is perhaps more easily understood by examining some of the fundamental principles of subsurface fluid hydrodynamics. Figure 18 illustrates the simplest example. A gently dipping porous reservoir is completely saturated with fresh water. Three wells have penetrated the aquifer at points A, B, and C. Water rises to the same position, relative to the horizontal datum, in each well. Note that the height of the water column in each well is a function of pressure in the

Table 3. Original formation gas pressure, Cadomin, Elmworth area.

Well	DST Interval Top	DST Interval Base	REC Depth (ft SS.)	Extr. ISIP (psig)	Extr. FSIP (psig)	Rate mmcf/d	Remarks
7-24-63-3W6	8,655	8,674	−5,652	3,112	3,097		Rec 820 ft oil & 1,800 ft XW
7-8-64-2W6	8,306	8,375	−5,427		3,160	1.74	
10-29-64-3W6			−5,407	3,091		0.74	P.B.U.*
6-2-65-3W6	8,110	8,131	−5,209	3,067	3,055	4.80	
7-32-65-3W6	7,930	7,972	−5,088	2,761	2,747	0.84	
7-7-65-4W6	8,566	8,694	−5,587	3,276	3,276	4.00	
5-12-65-5W6	8,484	8,543	−5,705	3,391		1.20	P.B.U.
6-32-65-7W6			−6,113	3,207		3.95	P.B.U.
5-32-65-9W6	10,446	10,463	−6,626	3,414	3,382	0.04	
10-27-65-10W6			−6,840	3,457		0.82	P.B.U.
7-13-66-9W6	9,144	9,183	−6,239	3,249	3,107	0.27	
10-19-66-9W6			−6,394	3,282		1.00	P.B.U.
12-31-66-9W6			−6,291	3,302		0.68	P.B.U.
9-10-66-10W6	9,527	9,580	−6,606		3,210	0.04	
10-26-66-11W6	9,390	9,432	−6,659	3,483	3,498	1.50	
10-33-67-7W6	7,792	7,884	−5,314	3,150	3,140	2.75	
10-4-67-10W6			−6,421	3,348		0.38	P.B.U.
10-12-67-10W6	9,228	9,255	−6,179	3,363	3,337	1.72	
7-15-67-10W6			−6,269	3,285		0.42	P.B.U.
6-21-67-10W6	9,160	9,199	−6,278	3,398	3,362	5.10	
10-25-67-11W6			−6,292	3,118		1.00	P.B.U.
10-3-68-4W6	6,880	6,915	−4,479		2,855	13.91	
6-11-68-8W6	8,093	8,148	−5,457	3,153	3,096	6.48	
7-4-68-11W6	9,160	9,200	−6,260	3,137		0.10	P.B.U.
6-16-68-11W6			−6,113	3,084		2.60	P.B.U.
7-1-68-12W6			−6,398	3,106		0.21	P.B.U.
11-7-68-12W6	9,261	9,291	−6,456		3,142	0.01	
11-14-68-12W6			−6,212	3,158		0.50	P.B.U.
7-14-68-13W6	9,324	9,363	−6,550		3,202	0.02	Poor Extrap.
6-28-68-13W6			−6,489	3,285		0.44	P.B.U.
10-7-69-12W6	8,707	8,743	−6,042	3,022	3,045	0.10	
6-26-69-12W6	8,284	8,338	−5,802	3,002		0.10	
6-28-69-12W6	8,336	8,372	−5,812		3,102	0.01	
10-34-69-13W6			−5,844	3,003		0.80	P.B.U.
11-1-70-12W6	7,989	8,038	−5,574		2,894	0.06	
11-7-70-12W6			−5,637	2,878		5.07	P.B.U.
11-33-70-12W6			−5,323	2,887		0.46	P.B.U.
11-1-70-13W6	8,392	8,438	−5,740	2,974	2,949	0.07	
10-34-70-13W6	8,202	8,287	−5,351		2,932	0.01	
10-25-70-14W6	8,441	8,497	−5,657		2,892	0.01	
7-5-71-12W6			−5,286	2,793		1.00	P.B.U.
6-18-71-12W6	7,949	7,982	−5,170	2,832	2,827	0.07	
11-4-71-13W6	8,150	8,195	−5,387	2,853	2,851	0.02	
11-11-71-13W6			−5,279	2,831		1.50	P.B.U.
11-19-71-13W6	8,104	8,159	−5,248		2,802	0.06	
10-22-71-13W6	7,841	7,874	−5,105	2,824	2,814	0.83	
13-33-71-13W6	7,815	7,871	−5,072		2,838	1.40	
10-35-71-13W6			−4,999	2,838		1.35	P.B.U.
6-6-72-12W6	7,647	7,664	−4,936	2,794	2,794	0.21	
7-8-72-13W6	7,759	7,803	−4,970	2,789	2,785	0.21	
10-12-72-13W6	7,568	7,615	−4,838	2,767	2,601	0.73	
7-27-72-13W6	7,540	7,558	−4,678		2,754	0.20	
6-5-74-13W6	6,934	7,000	−4,164	2,658	2,617	0.20	
b-82-B/93-P-1	9,268	9,393	−5,965	3,297	3,297	0.01	
a-43-D/93-P-1	9,892	9,934	−6,249		3,624	0.01	
d-68-K/93-P-1	8,543	8,587	−5,308	2,804	2,866	0.04	ISIP not extrap.
b-24-B/93-P-6	9,170	9,258	−5,523	3,362		1.04	
a-57-C/93-P-7			−5,633	3,686		1.60	P.B.U.
a-43-B/93-P-8	7,881	7,906	−4,865		2,705	0.04	
b-46-H/93-P-8			−4,436	2,687		2.50	P.B.U.

*P.B.U. signifies pressure measurement obtained from post-completion pressure build-up analysis.

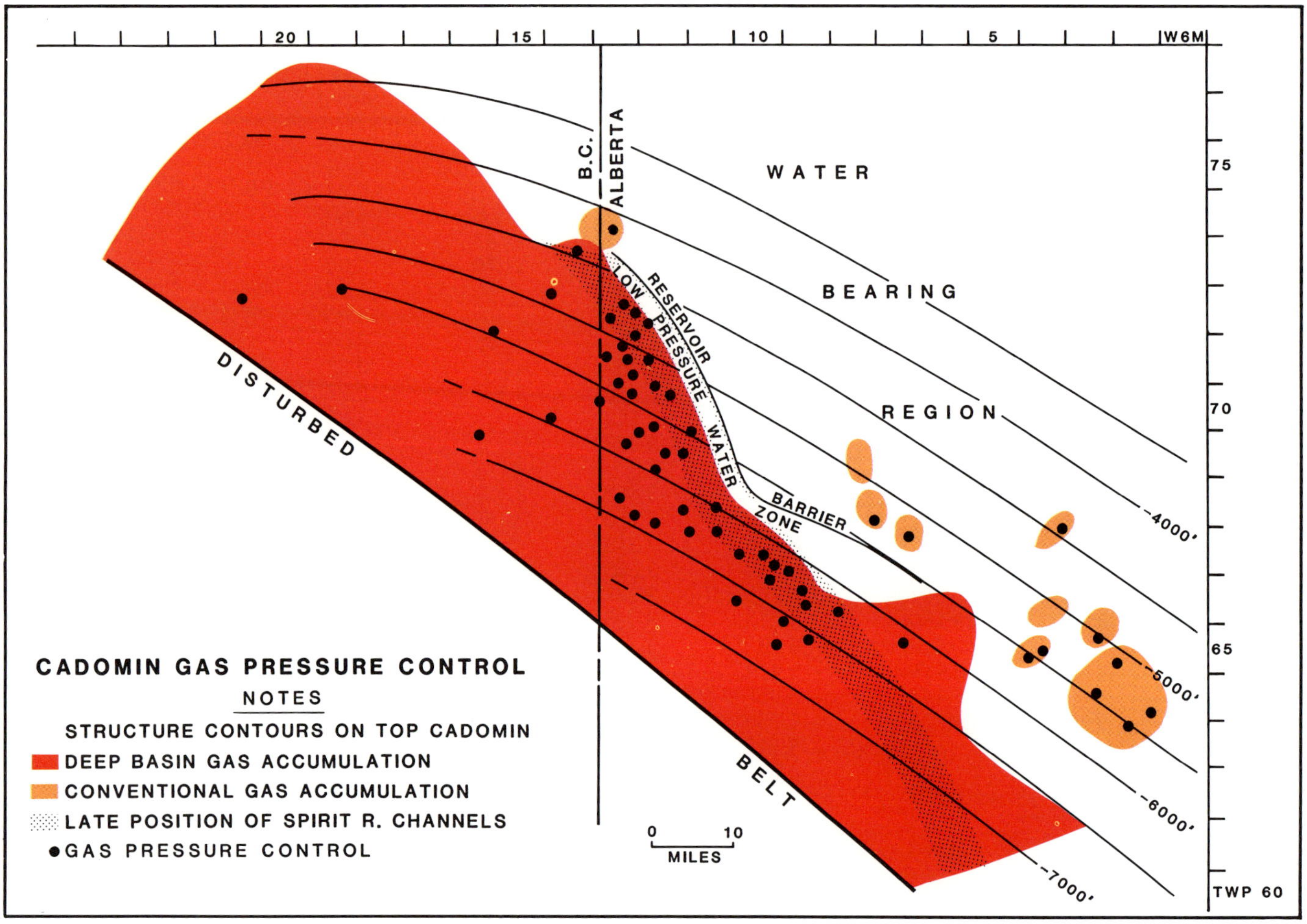

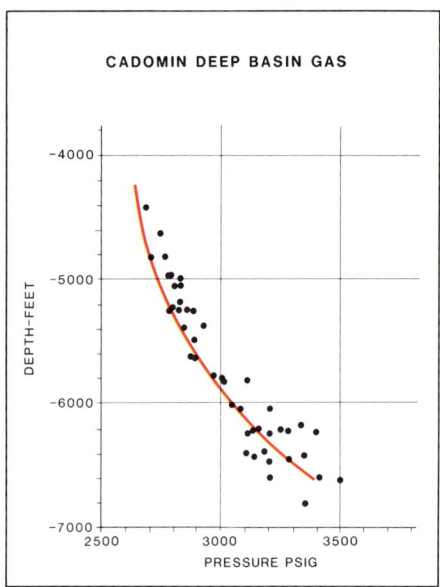

Figure 14. Map showing distribution of Cadomin gas pressure control points.

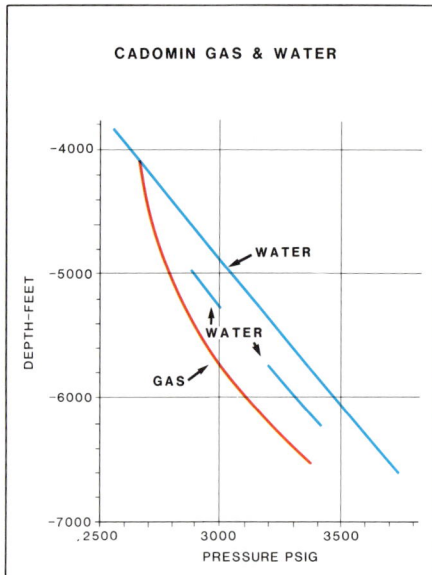

Figure 15. Pressure-depth plot for original Cadomin formation gas pressures throughout the deep basin gas accumulation.

Figure 16. Pressure-depth comparison of Cadomin deep basin gas and regional Cadomin water systems.

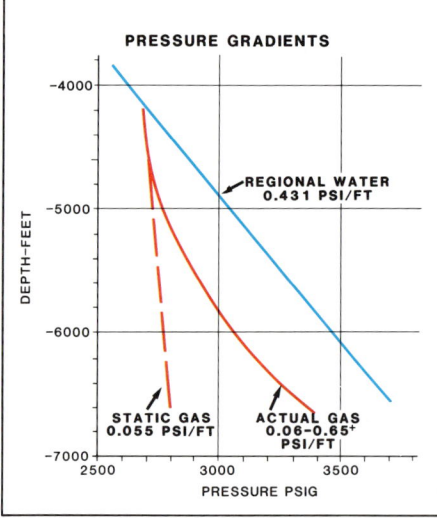

Figure 17. Pressure-depth comparison between actual curved gas pressure gradient line for Cadomin deep basin gas accumulation and the corresponding gradient line for a static gas accumulation.

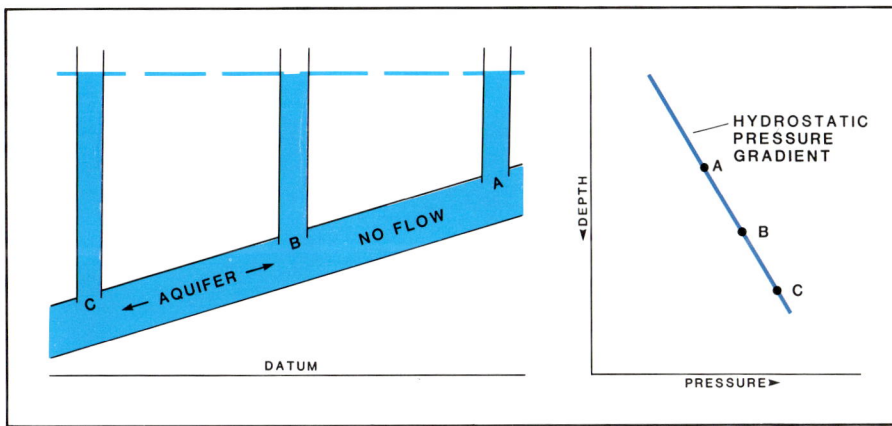

Figure 18. Model of a gently dipping aquifer with three wells located at positions A, B, and C. Hydrostatic conditions prevail in the water phase.

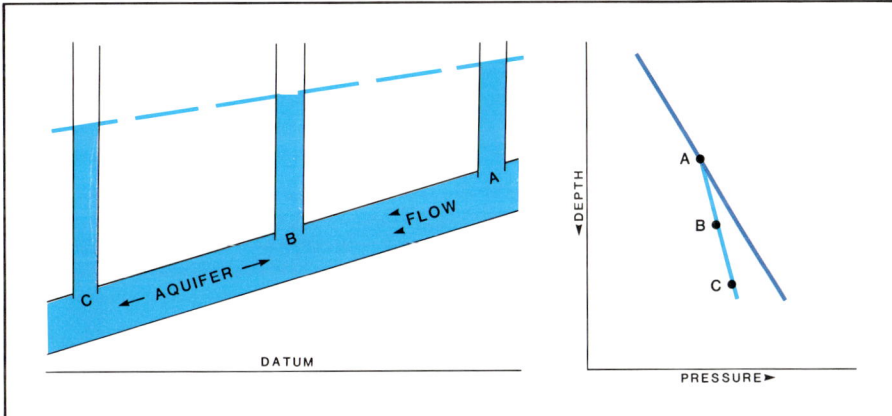

Figure 19. Downdip water flow in aquifer leads to energy loss in dynamic water phase and therefore a steepening of the apparent pressure gradient line.

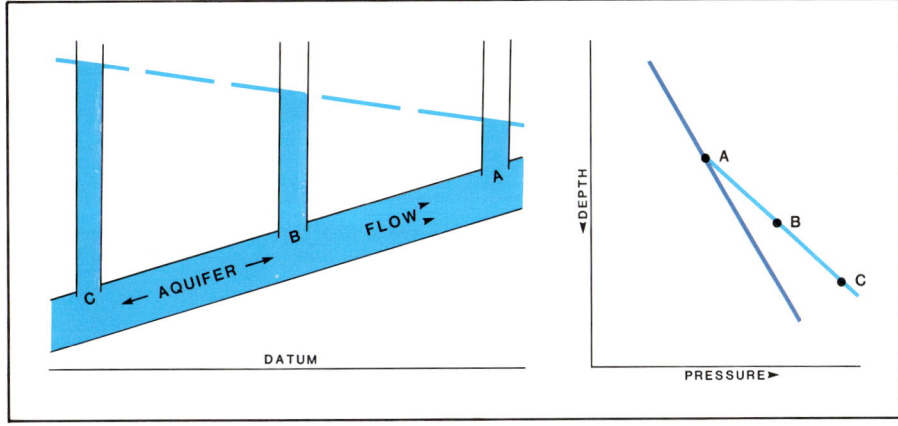

Figure 20. Updip water flow results in energy loss in updip direction and a flattening of the pressure gradient.

aquifer at that point. In this first case, hydrostatic conditions exist. As shown in the adjacent pressure-depth plot, the pressure increases with vertical depth in the water phase at a rate of 0.433 psi/ft (9.79 kPa/m). If, however, the water is in a dynamic state, that is, it is moving slowly through the reservoir, then, from the principles behind Darcy's law for fluid flow through porous media, there will be some energy loss in the water due to frictional resistance to flow through the pore system. The rate of energy loss along the flow path is largely a function of the permeability and rate of water flow. The higher the flow rate and the lower the permeability, the greater is the loss of energy per unit length along the flow path. In the case of downdip water flow, the energy loss, due to frictional resistance to flow, is reflected by a pressure loss in the water at downdip positions. The resultant apparent pressure gradient line, as would be established by making a series of measurements along the water flow path, is steeper than the gradient line for a static system (Fig. 19). Similarly, if water flow was in an updip direction, the energy loss in the moving water is reflected by an increase in the rate of pressure loss in the water phase at updip positions. The resultant pressure gradient line for updip water flow is, therefore, greater (flatter) than the static case (Fig. 20).

The foregoing simplified examples all assumed constant reservoir transmissibility throughout. Figure 21 shows what can happen if, for example, reservoir transmissibility progressively increases updip (perhaps due to increasing permeability, reservoir thickness, or both). Water movement is updip. Note that the water pressure gradient becomes a curve that steepens in the updip direction.

The same principle applies to updip migration of gas in the real world. Slow gas movement through the low-permeability Cadomin reservoir results in an energy loss to the gas phase and this is reflected as a pressure loss updip along the gas migration path. Because the reservoir transmissibility to gas improves updip, the rate of energy loss along the gas migration path decreases, and therefore the apparent pressure gradient curve becomes steeper as it approaches the gradient for gas in static equilibrium (Fig. 17). Corresponding average in situ permeabilities, derived from pressure build-up analyses, show low values generally in the downdip regions in the order of 0.1 md, whereas in the updip areas it is 10 times greater at 1.0 md.

Pressure curves (Fig. 16) indicate that pressures are about equal at the updip position where gas is thought to be in contact with the regional water pressure system. The low gas pressures measured in downdip regions simply reflect the pressure in the gas phase along a gradient that is steeper than that for the regional formation water where both are in balance at some updip position. If the gas was in static equilibrium, then the pressure differences between formation waters northeast of the barrier and the gas southwest of it, would be much greater than they really are, as indicated by Figure 17. Such a high pressure difference would encourage formation water in the northeast to move across the barrier, if possible, and enter the much lower-pressured gas-saturated region.

Water/Gas Contact Phenomena

A question of considerable interest that must be answered is: How is it possible for nature to maintain a gas below water contact in the reservoir, as suggested from the Cadomin situation, without a seal to separate them? The explanation to this very puzzling phenomenon became apparent only after making an exhaustive series of model experiments designed to simulate the field observations.

Figure 22 is a diagram representing one of the models designed to duplicate fluid mechanisms involved. The model is constructed from transparent plexiglass, to permit direct observations of events taking place in the pore system. It consists of a vertical cylinder containing a sand pack having a total length of 28.3 inches (72 cm) and a diameter of 2.24 inches (5.7 cm). The sand pack rests on a porous screen support at its base. Sand permeabilities are high relative to the actual Cadomin reservoir. Permeability of the fine sand (small dot pattern) is 30 darcys while the coarse sand (open circles) is 1,100 darcys for an average sand column permeability of 40 darcys in this case. Sand porosity is 38%. A manometer, tied into the base of the sand column, will be used to measure water pressure. The bypass tube allows short-circuiting of the water from the top of the sand column, through a valve, to the base.

A typical experimental run begins with the total system water saturated to a little above the top of the sand column. The water is colored for better observation of flow. At this point, the valves B and C are open and the water level in the manometer tube is level with the water level just above the sand column. Valves B and C are then closed. An air pump is attached to valve A, which is then opened and air is slowly injected at the base of the sand column. This action simulates generation of gas in deeply buried downdip areas from organic rich source rocks, and entry of gas into adjacent low permeability sands.

As air injection continues, the water level at the top of the sand column is seen to gradually rise (Fig. 23). It does so because the injected air (colored red) slowly displaces the mobile water phase upward and out of the sand column. Soon gas bubbles are seen escaping at the upper water surface. Also, the color of the sand

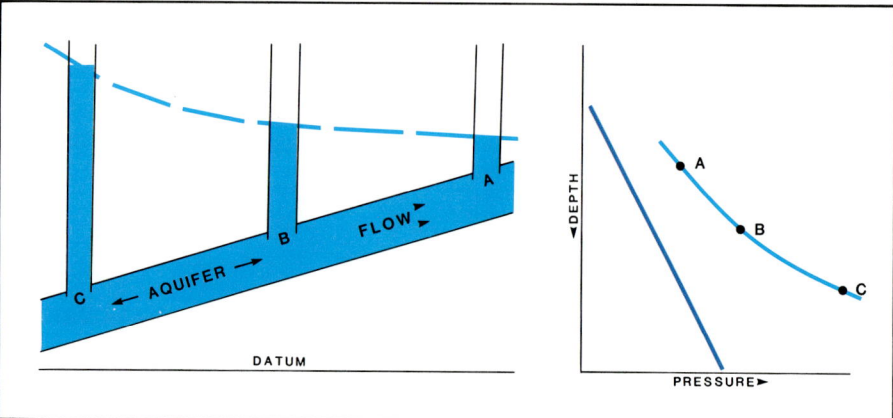

Figure 21. Updip flow in an aquifer with increasing transmissibility updip results in a pressure gradient curve that steepens updip.

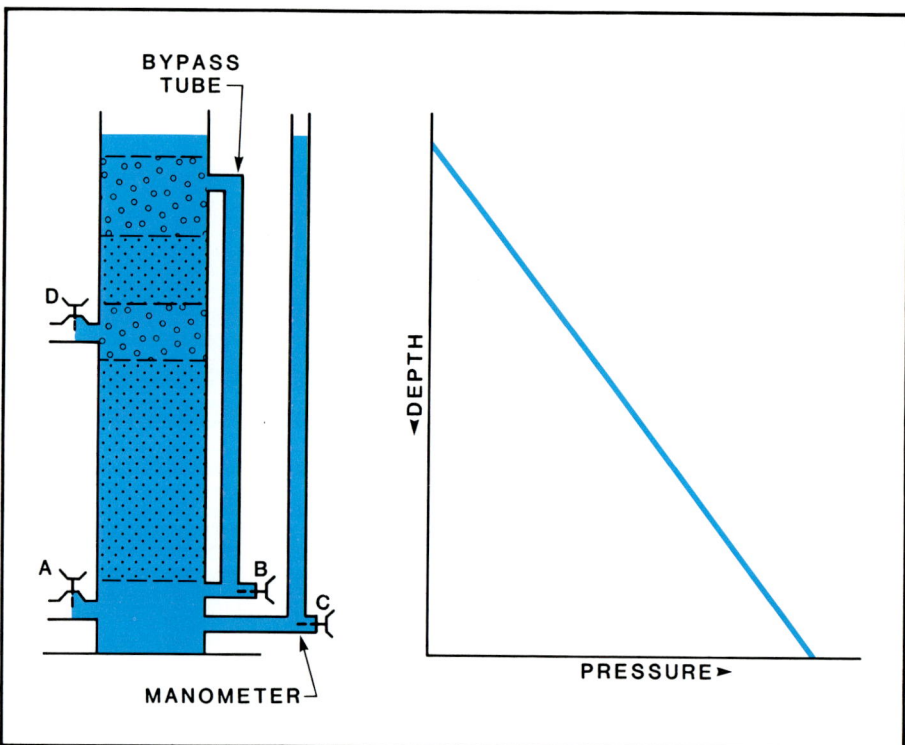

Figure 22. Water-saturated sand column in model designed to investigate gas entrapment principles.

The Cadomin Formation

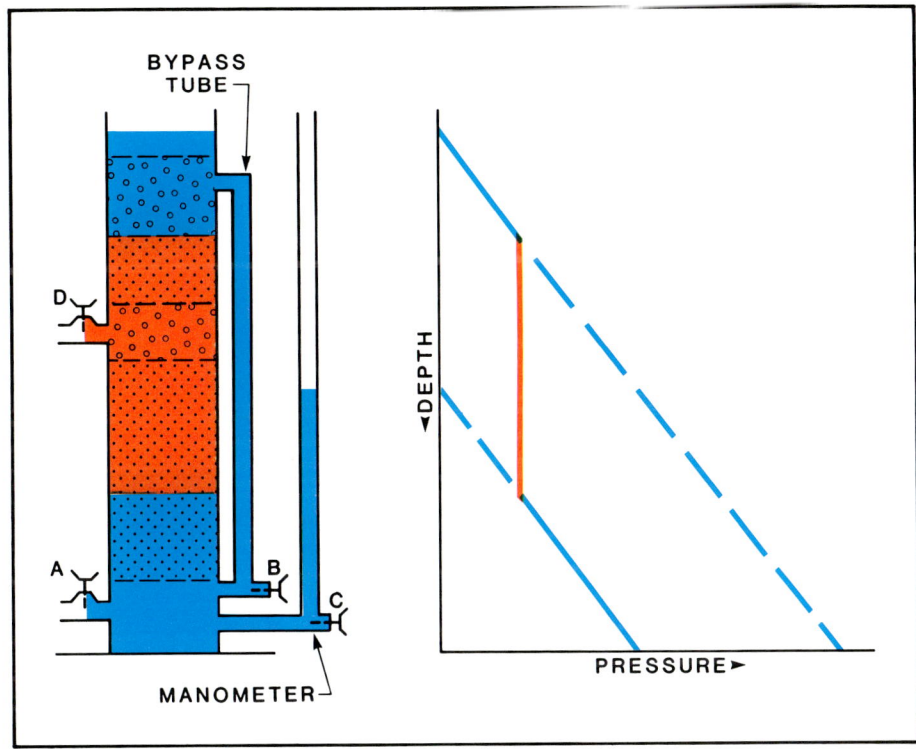

Figure 23. Following downdip gas injection gas accumulates at low pressure beneath water at coarse/fine sand contract.

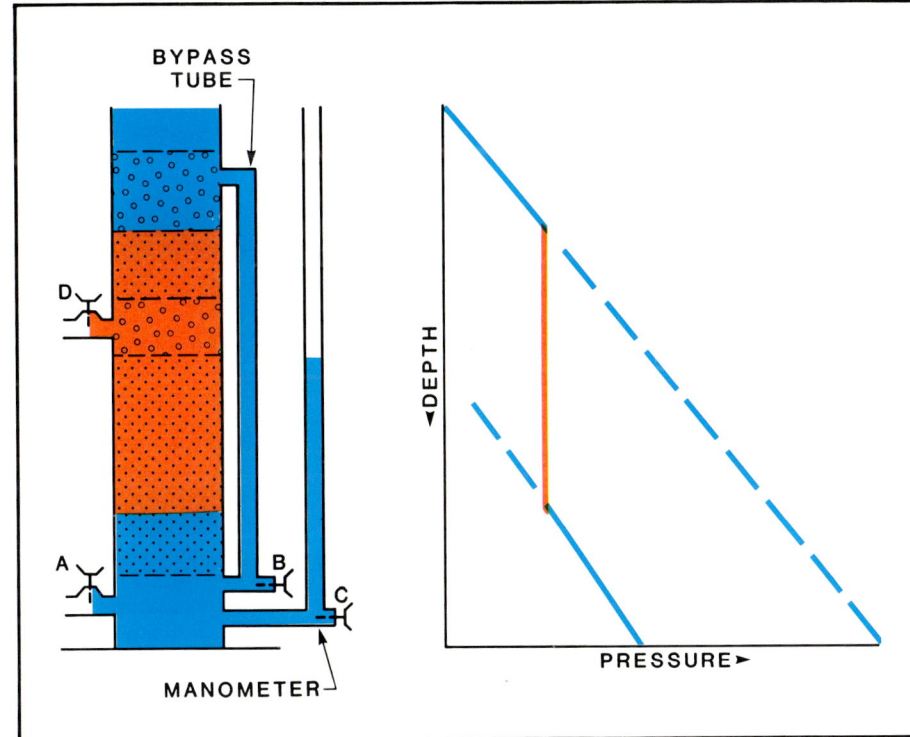

Figure 24. Model conditions following bleed-off of small quantity of gas from coarse sand lens through valve D.

column becomes lighter as the colored water is displaced out of the pore system. Air injection is continued until the water level at the top no longer continues to rise. Valve A is then closed and manometer valve C is then opened.

The water level in the manometer is observed to fall a considerable distance below its original position opposite the top of the sand column. This reaction obviously signifies that the water pressure at the base of the base of the sand column has dropped. (Note, the chamber below the sand column remains filled with water that is in contact with the base of the sand column.) Gas bubbles cease to rise to the water surface and a state of static equilibrium is soon reached in the system wherein the shortened water column height in the manometer remains fixed and stable as long as one wishes to observe it. Meanwhile the water wet sand, as seen through the transparent walls of the plexiglass column, remains coated with a water film and some isolated sand pores below the uppermost coarse/fine sand contact, remain completely water saturated. The top coarse sand segment is also totally water saturated. However, the middle coarse sand segment is water wetted but obviously gas saturated. This segment simulates locally developed permeable areas within the overall low-permeability Cadomin sand.

The model experiments suggest that water is unable to flow downward through the sand column and thereby displace gas out of the pores - even though there is a continuous water-wetting film throughout the gas-bearing sand. This water-wetting film and the isolated water-filled pores that occur throughout the generally gas-filled portion of the column are unable to transmit water flow, and do not transmit water-phase pressures in accordance with the normal static water pressure gradient of 0.433 psi/ft (9.79 kPa/m) from the top water surface to the water-filled chamber linked to the manometer at the base of the column.

If the valve connected to the short, coarse, gas-filled segment is opened momentarily to bleed off gas, and then closed again, the water level in the manometer drops abruptly and slowly rises again to about its previous position. At the same time, the water level above the top of the sand column is observed to fall slightly (Fig. 24).

This little experiment revealed that

with a reduction in gas pressure, some of the water was able to flow downward through the gas column in the fine sands, and accumulate at the base of the column. Initial reduction in gas pressure, due to momentary gas bleed off, reduced water pressure at the base of the column as evidenced by the fall in the manometer water column. Gradual rise again in the manometer indicated repressuring of the gas by water accumulating at the base of the column below the gas accumulation. As the gas pressure rose, however, downward water flow decreased until a state was reached where flow ceased and the manometer column stabilized roughly 12 inches (30.5 cm) below the top of the sand column.

Figure 25 is a simplified diagram that illustrates principles underlying the model reactions described above. This indicates what happens on a microscopic scale. When the non-wetting gas pressure is low, as in the left side of the diagram, a continuous, mobile water phase can exist separating the bound, immobile water film coating pore walls of the water-wet rock minerals, from gas that occupies the center of the pores. Water is able to flow around the gas in this situation. This is what happened when gas pressure in the model experiment was momentarily reduced by bleeding some of it off. As gas pressure increased, due to water accumulation at the base of the gas column, thereby compressing the gas above, the gas deformed into the pore throat displacing and disrupting continuity of the mobile water film. This is depicted in the right half of Figure 25.

The low gas pressure at equilibrium, as evidenced by the manometer water level, duplicates the pattern of field pressure observations. Gas pressures in the gas column below the water-saturated permeable sand, are less than water pressures if projected along a static water pressure gradient tied to the water surface above the column. Even though there is a bottom gas/water contact, there is still no continuously interconnected mobil water phase linking bottom water through the gas column with the upper water-saturated sands. Therefore, there are no buoyancy forces active to displace gas upward through the permeable, fine sand column.

It is important to recognize that when low pressure gas occupied the fine sand column, the relative permeability to water had become zero. Even though a water

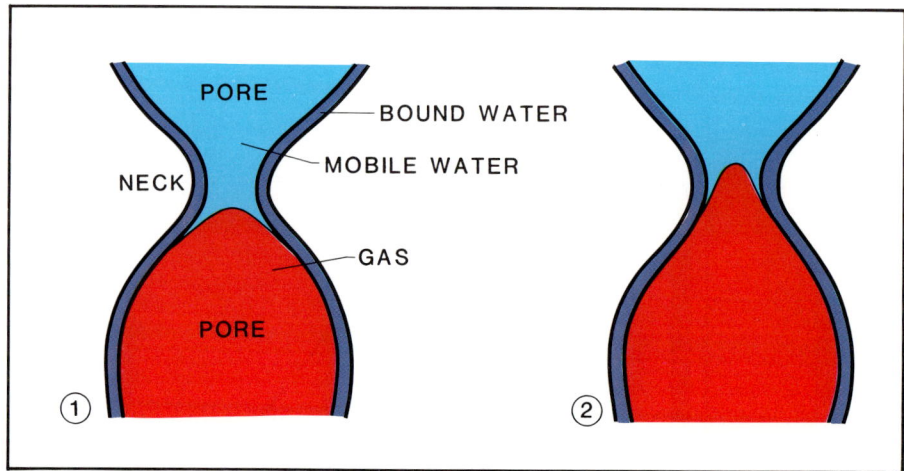

Figure 25. Simplified model of gas and water distribution on a microscopic scale across a pore throat.

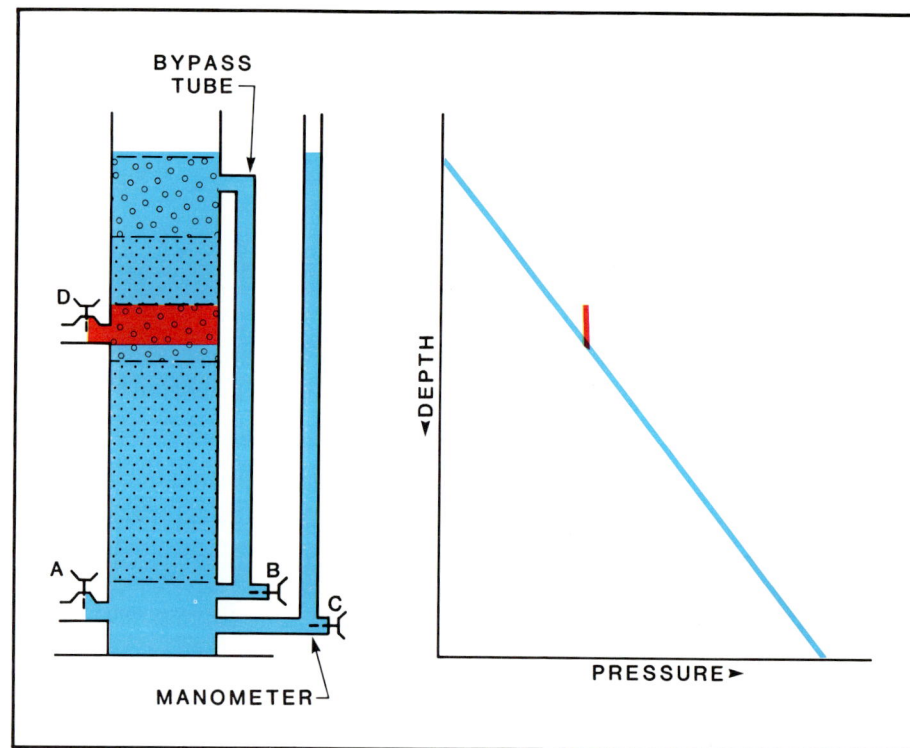

Figure 26. Model conditions after opening bypass control valve B.

film coated all pore walls and many isolated pores were still water filled, this water film was incapable of transmitting water flow or water pressures because water simply cannot flow along the water film unless its thickness reaches a certain minimum value.

A similar situation is believed to be the case of the updip water/gas contact in the Cadomin. At this position, the gas phase pressure in the pores balances and disrupts continuity of the mobile water phase, thereby preventing water in updip sands to flow downdip and displace the gas. Two other conditions may also be developed: (1) the Cadomin may become more permeable immediately updip from the water/gas contact, and (2) the lateral barrier to the northeast may end at that position.

Returning to the model experiments

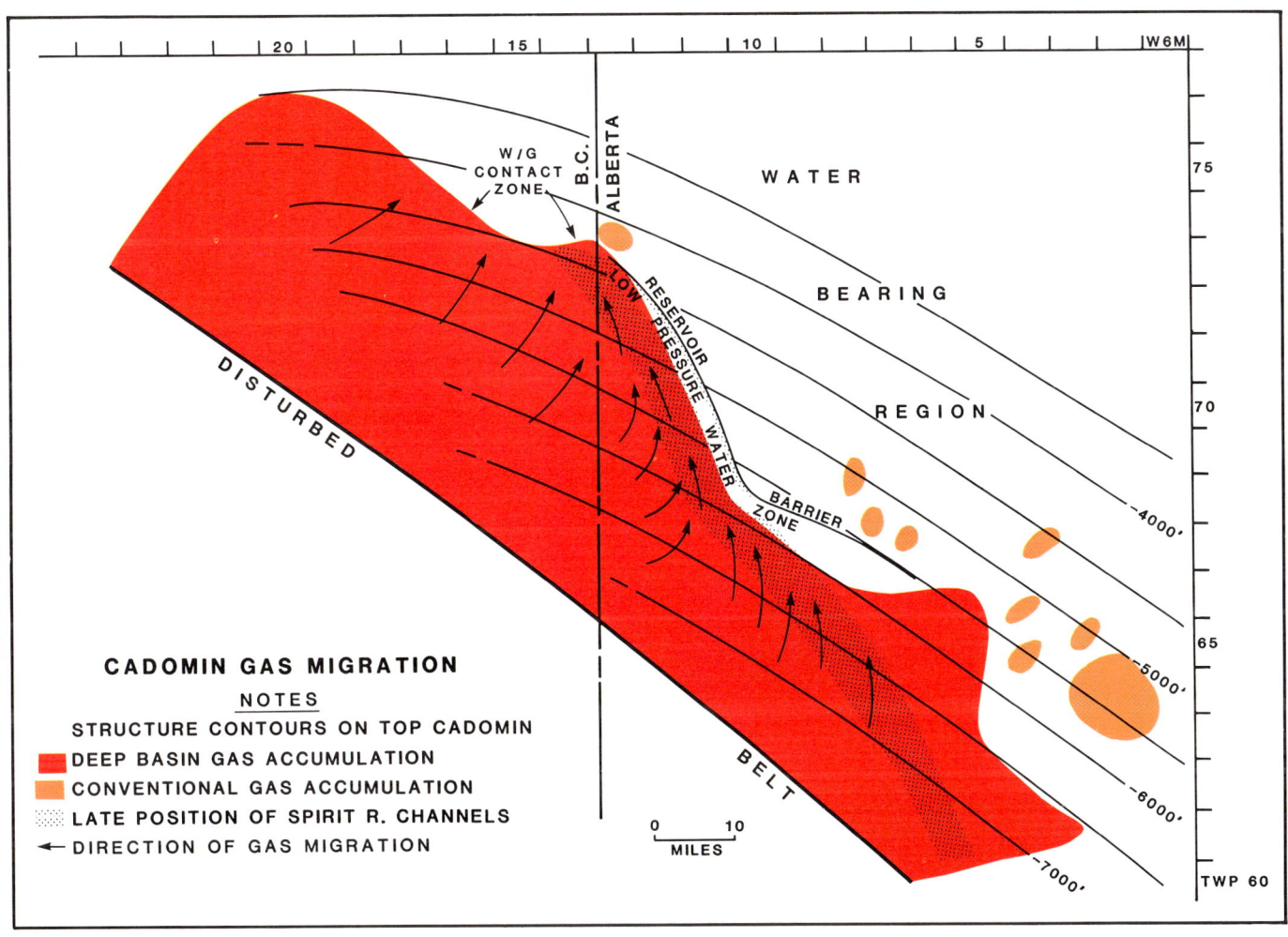

Figure 27. Map showing interpreted updip Cadomin gas migration paths as derived from detailed pressure analysis.

(Fig. 26), if the bypass tube valve B is opened, a continuous path of mobile water is now provided from the top to the base of the trapped gas column. Two things happen very rapidly: (1) the water level in the manometer rises immediately to a position opposite the top water surface in the column, and (2) gas bubbles begin escaping across the top water surface as the gas is gradually displaced by water flowing downward through the bypass tube into the base of the sand column. Eventually a new equilibrium position is reached, with a short gas column remaining trapped in the middle coarse sand interval. If pressures could be measured accurately, it would be observed that the middle gas column pressures are now slightly greater than the water pressures for corresponding depths.

Gas Migration

A question worth considering is: If, as it appears, gas is migrating updip in the Cadomin, why did it not all escape long ago? Based on average reservoir parameters and applying Darcy's Law, it is possible to calculate the rough average rate of gas escape across the updip water/gas contact. The calculated rate is in the order of 100,000 standard cu ft (2,800 cu m) of gas per day across a maximum contact length of 62 mi (100 km). Thus, the entire 15 tcf (425×10^9 cu m) of estimated gas in place could escape in about 400,000 years. The most likely explanation to account for the fact that the accumulation exists despite updip gas leakage, is that gas is still being injected into the reservoir in downdip regions. The most likely source of gas is the organic-rich Cretaceous section in general, and the basal Gething section in particular. This section contains numerous coalbeds throughout the region as well as organic-rich shales and siltstones. The Cretaceous coals in the Elmworth deep basin had enormous gas-generating capability in addition to other organic-rich shale source rocks.

The Cadomin deep basin gas accumulation appears to be a dynamic situation that has persisted for millions of years. Gas, generated in adjacent source rocks, moves vertically into the Cadomin layer and then slowly migrates updip, escaping eventually into the permeable, water-saturated region at the water/gas contact. Updip gas migration is a response to an unbalanced pressure gradient in the accumulation caused by comparatively low pressures at the updip end and higher gas pressures due to gas generation (maturation pressures) in the downdip region. Migration is not likely due to gas buoyancy in water. Low transmissibility throughout most of the gas reservoir tends to retard updip gas flow and thereby allows gas pressure to build up by the process of gas generation downdip, even though the rate of gas generation is

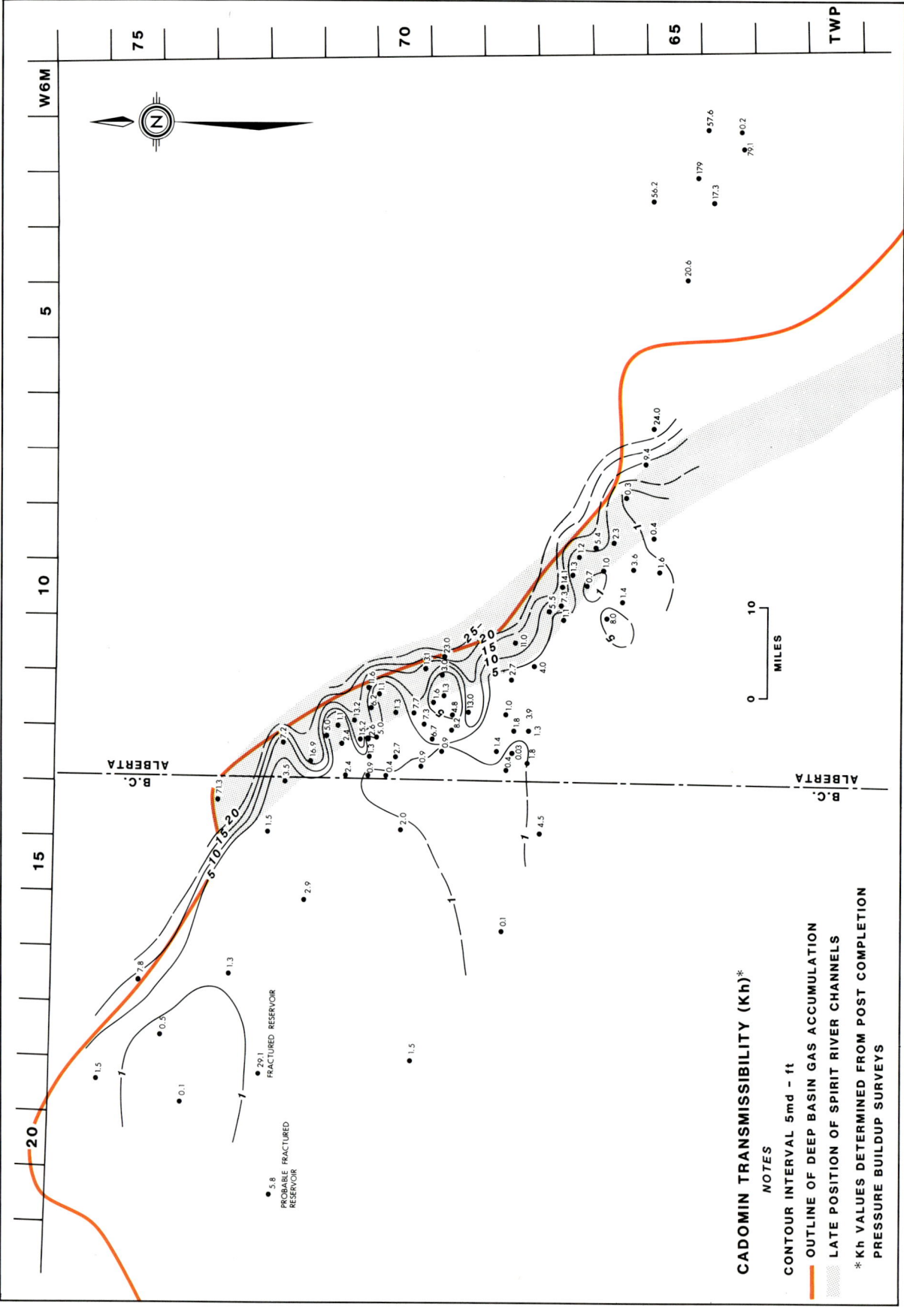

Figure 28. Cadomin reservoir transmissibility (Kh) map showing low Kh values downdip and higher values along the updip edge of the deep basin gas accumulation.

very low. This mechanism probably would not work effectively in relatively higher transmissibility reservoirs because the rate of gas generation could not keep pace with natural gas movement out of the downdip reservoir when water displacement originally took place and therefore buoyancy forces could provide rapid updip migration of generated gas. Because the extent of the low permeability Cadomin is so great in the downdip regions, gas can accumulate from a wide region. Based on the calculated average rate of gas migration across the water/gas contact stated earlier, the average rate of gas migration into the Cadomin from adjacent, active source rocks can be calculated. This rate works out to an average of 30 cu ft of gas (0.85 cu m) per acre (4,047 sq m) per year entering the Cadomin. Constant flushing of gas from downdip entry to an updip escape across the water/gas contact is believed to have resulted in exceptionally low residual water saturations in the Cadomin, as well as in other reservoirs. Under such high gas-saturated conditions, microporosity in the reservoir gains importance since the total gas volume and relative premeability to gas in such pores increase. Recent petrographic research suggests that microporosity is much more abundant than previously thought.

Much work remains to be done at Elmworth to assess indications of possible lower water saturations and higher porosities (due to the microporosity component) than the values in current use. For example, the assigned water saturations in gas-bearing sands usually fall in the 30 to 50+ percent range. Due to the enormous volumes of gas-bearing sandstones at Elmworth, a small reduction in average water saturation translates into a substantial increase in gas-in-place volumes and recoverable gas reserves. We speculate that additional future oil base core control and material balance estimates, based on several years field production history, could lead to reductions in current water saturation values by as much as 50%. Water saturations in the sands may actually lie in the 15 to 25% range.

The significance of microporosity as a contributor to long-term gas production has probably been highly underrated. Recent developments in petrographic techniques permit us, for the first time, to see and more fully appreciate previously unknown characteristics of micropore systems in rocks. This technology is undergoing further testing and improvement. The ability to see micropore networks and to evaluate their interconnection and association with macropore systems (which are necessary for economic recovery rates) should lead to more accurate petrophysically-based predictions of field production performance.

Detailed inspection of the scattered, downdip original gas pressures suggests gas migration paths through the Cadomin reservoir as depicted in Figure 27. In other words, gas would tend to move through the reservoir from locally high pressure areas to lower pressure areas (which, in turn, reflect lower to higher reservoir transmissibilities) assuming lateral reservoir continuity. Thus, gas apparently is migrating, in response to pressure gradients, out of lower permeability, relatively high-pressure alluvial plains sediments, into the comparatively higher permeability, northwestward-oriented Spirit River sediments, and thence updip to the water/gas contact. It is also along this latter trend that gas wells having the highest flow rates have been completed.

Figure 28 is a map showing the Cadomin deep basin gas reservoir transmissibility (Kh) trends. Transmissibility data are derived from analysis of post completion pressure build-up measurements which we consider to be the most reliable source of in situ information. The map clearly depicts improving reservoir transmissibility along the updip edge of the accumulation along the course of the late Spirit River channels.

The narrow, elongate, water-saturated, low pressure zone adjacent to the southwest side of the northwest-oriented barrier (Fig. 27), is indicated to be in pressure balance with the gas accumulation along its length. This is probably water that was bypassed when Cadomin gas originally moved updip into the region. The piston-like displacement of gas, as seen in the model experiments, could have bypassed the water, much in the same manner that injection water during an oil field waterflood will bypass some parts of the pool, due to reservoir continuity and transmissibility configurations that favor such movement.

Gas Accumulation Process

Figure 29 is a simplified diagram representing a gently dipping aquifer comprising low-permeability sands downdip, giving way to higher-permeability sands updip. No permeability barrier exists at the contact to separate the two kinds of sands. In the beginning the sand is totally water saturated as in the previously described model experiment. Note the hydrostatic water-pressure gradient as depicted to the right of the section.

In Figure 30, gas begins to enter the downdip end as a result of large-volume gas generation from adjacent, downdip mature source rocks. Note that the gas pressure at the water/gas contact exceeds the water pressure in order to be able to move through the constricted pore system as a non-wetting fluid phase. Because gas pressures exceed water pressures, water is displaced updip out of the system by the gas.

In Figure 31, continuing gas influx has further displaced much of the original mobile water phase. Gas pressures always equal or exceed water pressures at the advancing water/gas contact front. Note, however, that downdip in the mobile gas saturated reservoir the pressure gradient for the gas now lies below the extrapolated water pressure gradient. As the water/gas contact continues to move updip, the water pressure at the advancing front decreases in accordance with the water-pressure gradient line shown on the graph.

In Figure 32, the moving gas front has now reached the more permeable sands and it cannot progress updip beyond that contact position. The change to a more permeable pore system updip sets the limit for the downdip gas accumulation as in the case for the Cadomin gas. As gas moves into the more permeable sand, the restriction to updip movement is almost eliminated because of much larger pore throats which allow gas to pass readily through them. At this point, buoyancy forces take over and in response, the gas is propelled rapidly updip such that the rate of gas influx across the contact cannot keep up with gas movement updip.

Figure 33 indicates what happens in this situation when the continuous gas supply being injected at the downdip end of the reservoir is shut off. Gas migration ceases and the gas pressure gradient steepens as the accumulation reaches a state of static equilibrium. At this point, the gas is trapped at low pressure relative to the regional water pressure gradient. There are no upward-acting buoyancy forces applied to the gas accumulation because hydraulic continuity of a mobile water phase extending through the reservoir from the updip to the downdip end does not exist.

Exploration Considerations

From the practical point of view of exploration for and exploitation of deep basin type hydrocarbon accumulations, one should bear in mind the following considerations:

The intimate association of reservoir rocks with thick, rich, mature source rocks is a most important factor, and probably is a prerequisite. Such reservoirs are well positioned to capture hydrocarbons as they are expelled from adjacent source rocks. The reservoirs must be capable of retaining their hydrocarbon charge. Elmworth experience suggests that if the reservoir is very permeable, as in the case of a beach conglomerate deposit, then it must be bounded by a very effective shale seal or low permeability sandstone. Proximity to prolific gas source rocks and low reservoir permeability is an important part of the mechanism that allows regionally extensive sands to retain their gas charges despite some continuous gas leakage from the sands updip. Without locally developed reservoir "sweet spots" in these sands however, the prospects for obtaining economic gas productivity are severely diminished. Geologists should consider the various reservoir conditions that could produce locally developed good reservoir rock quality in a generally poor quality reservoir. These would include: conglomerate or sand beaches, densely fractured zones, areas enhanced by diagenetic mineral dissolution processes, local sand mineralogy which inhibits reservoir-destroying diagenetic processes (that is, quartz cementation), late dissolution of calcite cements deposited in sediments soon after burial, marine transgressive (reworked) deposits, etc.

If we recognize that enormous volumes of gas have been generated in the past and that most of it has probably migrated updip out of the deep basin, then one would tend to assume that the adjacent updip region would be a good area to explore for conventional structurally and stratigraphically trapped gas accumulations. In practice, however, this may not be the case in the Elmworth region. Adjacent updip regions of the Mesozoic contain abundant, generally more permeable sands interbedded with potential trap seals such as shales and coals. But the area has a serious disadvantage in that it is more remote from the gas-generating source rocks and therefore there is a greater chance for dispersal and loss of updip migrating gas before it reaches some potential traps. Much of the gas may in fact seek an updip migration route that offers permeable reservoir rock and conti-

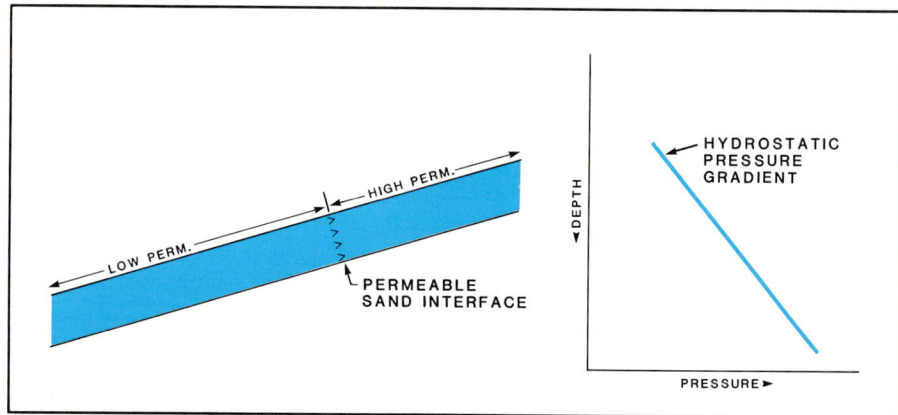

Figure 29. Gas migration model, stage 1, water-filled potential reservoir.

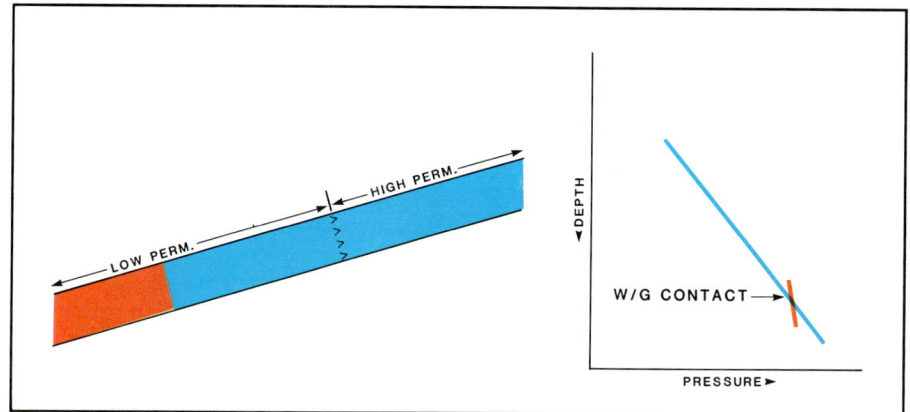

Figure 30. Gas entry begins in downdip region.

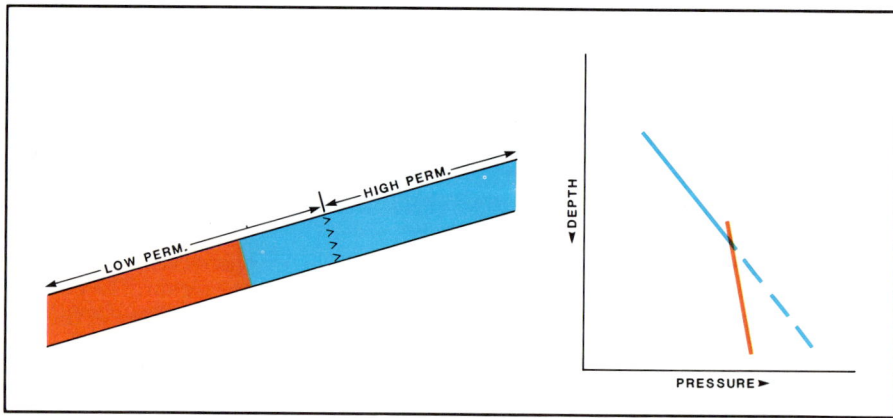

Figure 31. Gas accumulation expands updip as downdip gas influx continues. Note that gas pressure gradient is greater than that for a static gas accumulation, due to updip energy loss in the dynamic gas phase.

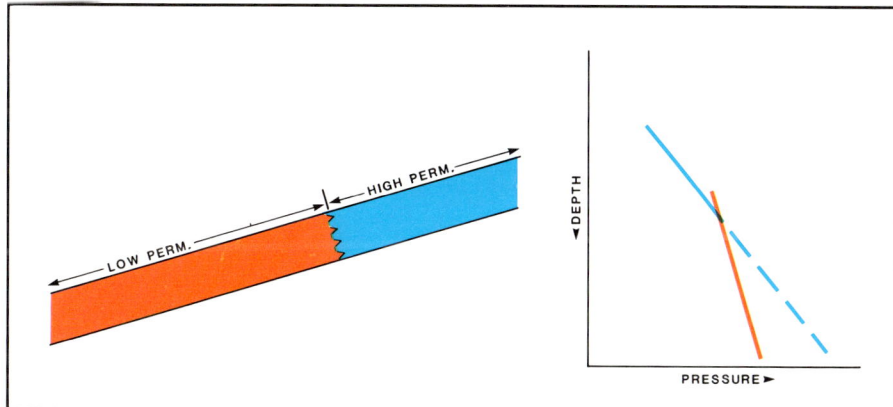

Figure 32. Updip development of the dynamic gas accumulation ends at contact with comparatively high-permeability reservoir.

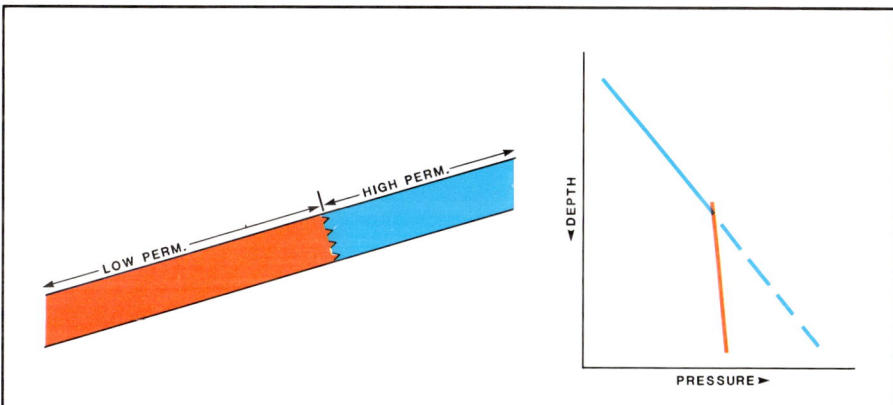

Figure 33. When downdip gas influx ceases gas pressure gradient steepens as accumulation reaches a state of static equilibrium.

nuity more or less directly to the updip outcrop. Also, the high mobility of gas through permeable water-bearing sands can lead to a rapid migration rate to the outcrop. Experience suggests that gas accumulations in adjacent updip sands tend to be small in areal extent and often have a gas/water contact in the interval. Thus, the advantage of high initial gas productivity, because of permeable sands, is offset by bottom water influx. Due to the comparatively limited rate of gas influx to updip traps, the confining seals, unlike in the deep basin, must be much more effective to allow development and retention of large gas accumulations. In this case, reservoir seals containing a continuous micropore system through the seals may not trap gas. Large conventional gas over water traps usually require extremely effective updip and lateral seals having very low permeability. This is because the capillary displacement pressure (that is, gas pressure minus regional water pressure at any given subsea depth) which acts to displace gas into the water-wet pore system, increases very rapidly with increasing gas column height (ref. Fig. 2A). The resistance of fine capillary pore systems in trap seals to gas entry decreases with increasing capillary displacement pressures. Deep basin type gas traps, on the other hand, can exist with comparatively less effective (more permeable) updip seals because the capillary displacement pressure exerted by the accumulation at its updip edge is always low (Fig. 2B). These less rigid seal requirements favor development of very large deep basin type accumulations at Elmworth. The absence of large updip gas accumulations (in consideration of enormous volumes of gas that evolved from deep basin source rocks), suggests there must be effective updip sand continuity which allowed most of the gas to escape through basin edge outcrops.

One last consideration is that the principles of deep basin gas entrapment should be equally valid for oil as well. An example of such a trap may exist at Elmworth in the Cardium formation of Upper Cretaceous age (Fig. 3). The widespread Cardium marine shoreline sandstone virtually blankets the region. It is known to contain high-quality, light-gravity oil throughout an extensive downdip region. The oil probably was derived from underlying organic rich, mature marine shale oil source rock within the Kaskapau. Updip the Cardium becomes water bearing similar to the Cadomin reservoir example. Exploitation of this vast oil resource in the Cardium has not been extensively pursued because the initial well drilling completion attempts experienced only limited economic success. Perhaps greater success would result through highly concentrated efforts to develop the kinds of geological, drilling/completion and supplemental recovery technology most appropriate to the specific reservoir conditions that exist in the Cardium. This would require a dedicated and resourceful management-technical team of experts who could pursue the costly drilling and complex experimental research programs required.

An important analog to the Cardium oil accumulation at Elmworth is the giant Pembina oil field located in the deep basin south of Elmworth and 70 mi (112 km) southwest of Edmonton (Fig. 1). The Pembina pool covers an area of 700,000 acres (283,000 ha). In parts of the field, oil productivity is enhanced by thin, permeable coarse sands and conglomerates at the top of the sandstone reservoir (Kerr, 1980). The oil is a high quality crude with API gravity ranging from 36 to 40°. Recoverable oil reserves, based on detailed study of 4,400 wells and a long production history, are estimated by the Alberta Energy Resources Conservation Board staff at 1.3 to 1.5 billion barrels (207 to 239 million cu m). Two oil reservoir features of particular geological interest and significance are: (1) the absence of a bottom water zone, and (2) unusually low connate water saturations of approximately 10% (based on 24 oil-base cored wells) as described by Purvis and Bober (1979). Both of these features suggest that deep basin trapping processes were active at Pembina.

CONCLUSIONS

Deep basin type gas traps represent a special class of hydrocarbon traps. This Cadomin example of the deep basin gas trapping mechanism points out and explains the unusual conditions that can occur. Other deep basin accumulations at Elmworth show similar relationships but with different features related to different reservoir plumbing developments.

The main points are as follows:

1. There must be a close match between a reservoir's plumbing and its fluid distribution (including fluid pressures);

2. The Cadomin deep basin gas accumulation is indicated to be dynamic where gas losses at the updip water/gas contact are in rough balance with downdip gas influx from adjacent, active source rocks; and

3. The gas accumulation is extensive. It occupies low-permeability rocks out of necessity (although locally high-permeability beds and areas are incorporated which thereby enhance gas well productivity), there is no downdip gas/water contact, and the accumulation is characterized by low pressures relative to projected regional formation water pressures.

REFERENCES

Kerr, W.C., 1980, A geological explanation for the variation in fluid properties across the Pembina Cardium field: Journal of Canadian Petroleum Technology, v. 19, n. 2, p. 76–84.

McLean, J. R. 1977, The Cadomin Formation; stratigraphy, sedimentology, and tectonic implications: Bulletin of Canadian Petroleum Geology, v. 25, n. 4, p. 792–827.

Purvis, R.A., and W.G. Bober, 1979, A reserves review of the Pembina Cardium oil pool: Journal of Canadian Petroleum Technology, v. 18, n. 3, p. 20–34.

Facies Control of Gas Trapping, Lower Cretaceous Falher A Cycle, Elmworth Area, Northwestern Alberta

R.A. Rahmani
Canadian Hunter Exploration Ltd.
Calgary, Alberta

The Falher A Cycle is the youngest progradational, shoreline and nearshore clastic sequence in the Lower Cretaceous Falher Member of northwestern Alberta. In the Elmworth area of the "Deep basin," this unit averages 25 m (82 ft) in thickness and contains vast amounts of gas reserves. The reservoir-quality shoreline sandstone and conglomerate form an east-to-west trending, 25- to 75-km wide (15.5- to 46.6-mi wide) belt. Gas is subnormally pressured and is trapped downdip (westward) from water. The eastern part of the "Deep basin" gas line, which marks the transition from downdip gas to updip water, trends northeast-to-southwest and cuts across the east-west oriented belt of shoreline sandstone and conglomerate. The position of this line is believed to have been controlled by a major distributary channel system, oriented in the same manner, which serves as an eastern updip seal.

The northern part of the "Deep basin" gas line, which has an approximate east-west orientation, occurs at the facies transition between shoreline, reservoir-quality sandstone and conglomerate to the south, and nearshore and offshore, finer-grained and less permeable sandstone, siltstone and shale to the north. This transition acts as a northern updip seal. It is therefore concluded that the trapping mechanism of gas in the Falher A is facies controlled. Gas generated from coal in deeper parts of the basin migrates northeastward (updip) until it reaches the above-mentioned facies transitions where it is presently trapped. This trap is not completely tight, resulting in a slow but continuous eastward leakage of gas and its accumulation updip in conventional traps. This leakage is probably responsible for the subnormal pressure of the downdip gas.

INTRODUCTION

The Falher A Cycle is the youngest progradational clastic shoreline sequence of the Falher Member (Spirit River Formation) in the "Deep basin" of northwestern Alberta and northeastern British Columbia, Canada (Figs. 1 and 2). It is one of the most gas-productive units in the area, and is comprised of sandstone, conglomerate, siltstone, shale and coal of coastal plain, shoreline, nearshore, and offshore environments. The reservoir-quality sandstone and conglomerate form a belt 25 to 75 km (15.5 to 46.6 mi) wide, oriented west to east. The average thickness of the Falher A Cycle is about 25 m (82 ft). The gas is trapped in low-permeability (below 0.5 md), tight sandstone and is producible from associated higher porosity and permeability conglomerates. As in other units within this basin (Davis, 1984; Gies, 1984) and in a few other basins (for example, the San Juan basin; Hunter, 1979; Cumella, 1981; Davis, 1984), gas is found trapped downdip from water and is subnormally pressured. Extensive development drilling indicates no gas/water contact downdip from the subnormally-pressured gas. In almost all of these cases, it was found that the downdip gas occurs in tight, low-permeability rock, and the updip water in a considerably more porous and permeable rock. Such clear and almost abrupt reservoir quality change is related to markedly different and environmentally-controlled diagenetic processes (Hunter, 1979; Cumella, 1981; Cant, 1983). Trapping mechanism suggestions ranged from downdip water flow holding the gas back, water blocks in narrow pore throats (Masters, 1979), a lack of buoyant forces on the gas (Gies, 1981; 1984), faulting (Jones, 1980), or combined stratigraphic-diagenetic trapping (Cant, 1983).

The objective of this paper is to describe fluid distribution in the Falher A Cycle and attempt to relate such distribution to rock characteristics, and to explain the trapping mechanism.

FLUID AND PRESSURE DATA

Figure 3 shows structure contours, fluid and pressure control points of the Falher A, and the "Deep basin" gas line that separates the water from the gas-charged zone. On Figure 4 the pressures measured from these locations were plotted against their respective subsea depths. This plot clearly shows the two markedly different pressure environments for the water and gas. As discussed by Davis (1984), a normal pressure environment exists, in the updip water section, until downdip one reaches 2,700 ft, or 823 m (subsea), then a subnormal pressure environment prevails. The subnormally-pressured environment occurs downdip in the gas-saturated section. The gas accumulation has at least 2,000 ft (610 m) of gas column, limited

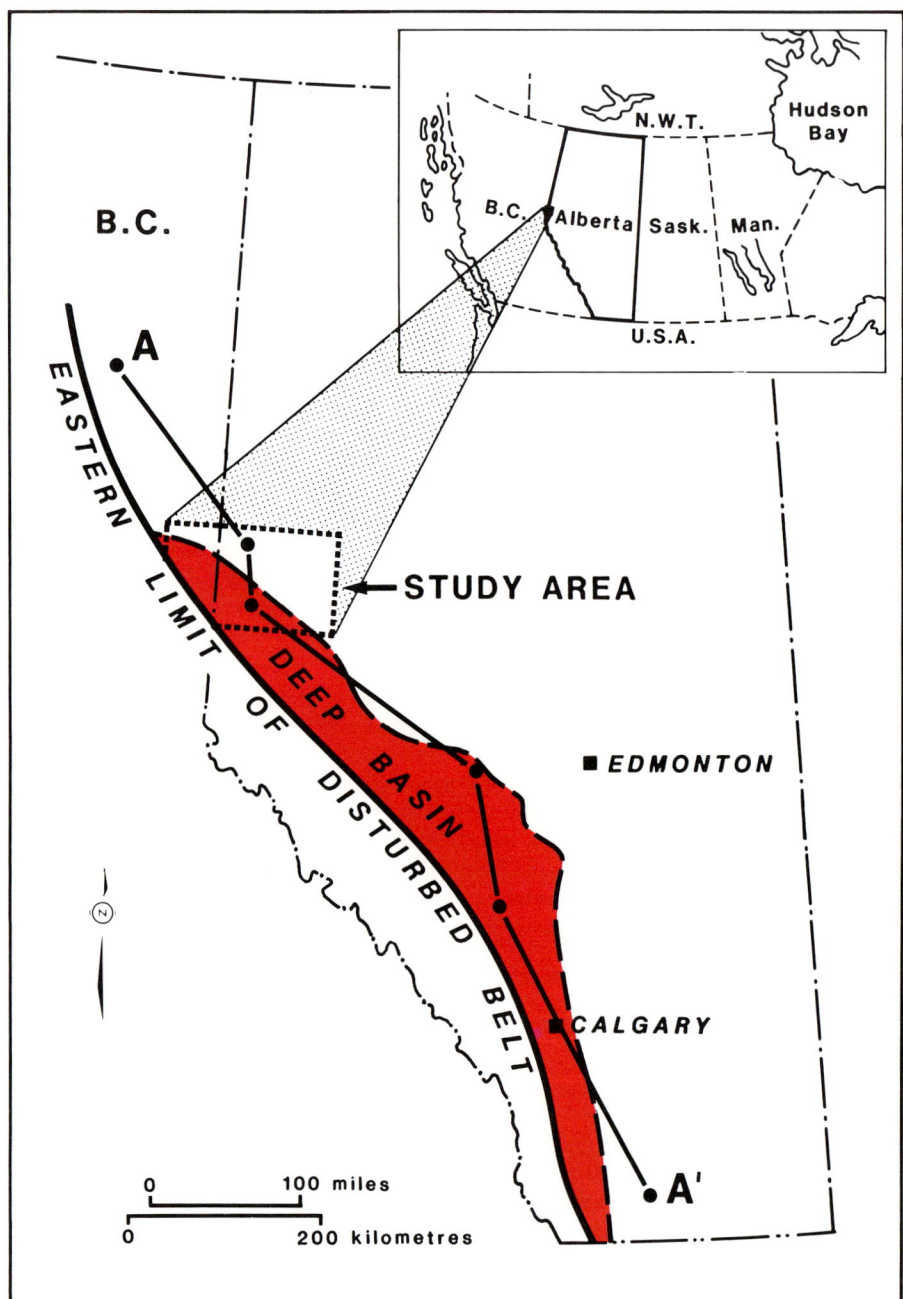

Figure 1. Location map showing study area relative to the Deep basin of Alberta, and location of regional cross-section A-A' (Fig. 2).

only by the absence of commercial reservoir rock on the downdip side.

RELATIONSHIPS OF FLUID DISTRIBUTION TO FACIES

It is evident from examining Figure 3 that a relationship exists between the location of the "Deep basin" gas line and the facies distribution. The eastern, northeast-southwest oriented part of the gas line coincides with the area where the zone of conglomerate and clean sand narrows considerably (townships 67 to 70 and ranges 5 to 9, W6M). It was also found that this part of the line follows the course of either one major trunk channel or a swarm of distributary channels (Fig. 5). This observation was further confirmed when the Falher A capping coal was isopached (Fig. 5). The coal distribution shows a 10-km wide (6.2-mi wide) belt of no coal or poor coal development (<0.5 m) coinciding with the location of the eastern part of the Deep basin gas line. This map (Fig. 5) also shows another such belt to the west (almost parallel to range 12). These belts of poor coal development strongly suggest the occurrence of distributary channels contemporaneous with the surrounding coal-forming swamps. Figure 6 shows the approximate location and trend of the channel or channel system that parallels the eastern updip limit of "Deep basin" gas, as well as two gamma log cross sections along and across the channel. The east-west section, perpendicular to the "Deep basin" gas line, demonstrates how the channel at 11-30-68-8W6 separates the updip wet reservoir quality rocks (regressive beach facies in 11-3-69-8W6 and 11-26-68-8W6) from the downdip gas-saturated rocks (also regressive beach facies in 7-26-68-9W6 and 7-5-69-9W6). Section SW-NE of Figure 6 is oriented along the channel trend. This section clearly shows the considerable thinning of the conventional, blocky-looking, regressive beach, reservoir rock along this trend. The channels eroded the upper and most permeable parts of the regressive sequence. There is only a small part of one of the channels' wells that is cored (11-30-68-8W6). This core consists of shale, siltstone, and very fine-grained sandstone. Examination of digitized logs of four of the seven channel wells indicates that the fill of these channels consists of fine-grained, low-porosity rocks. One of those seven wells was tested and yielded no recovery (6-27-67-9W6).

Seismic detection of areas with extremely thin coal or with no coal at all, for example along the course of and adjacent to the channel in Figure 5, depends on many geological variables and is by no means a simple procedure. The Spirit River Formation, which contains Falher A, is comprised of up to seven northerly-thinning clastic wedges capped by coal

Figure 2. Regional cross-section A-A' (location on Fig. 1) depicting Lower Cretaceous stratigraphic units and facies along the axis of the Western Canada Sedimentary Basin. Falher A Cycle is indicated by large solid arrow in 11-28-70-11 W6 (after Smith, 1984).

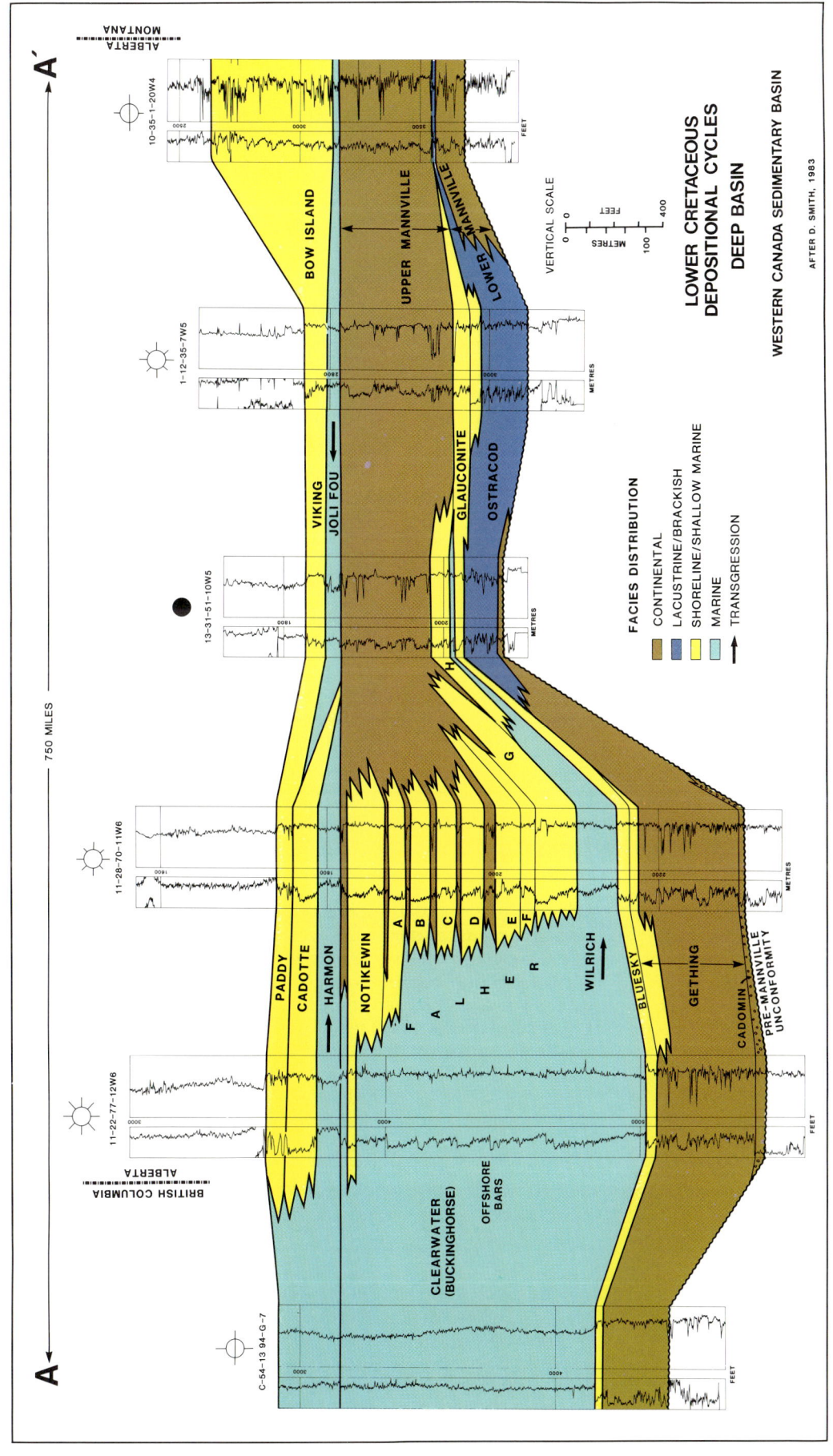

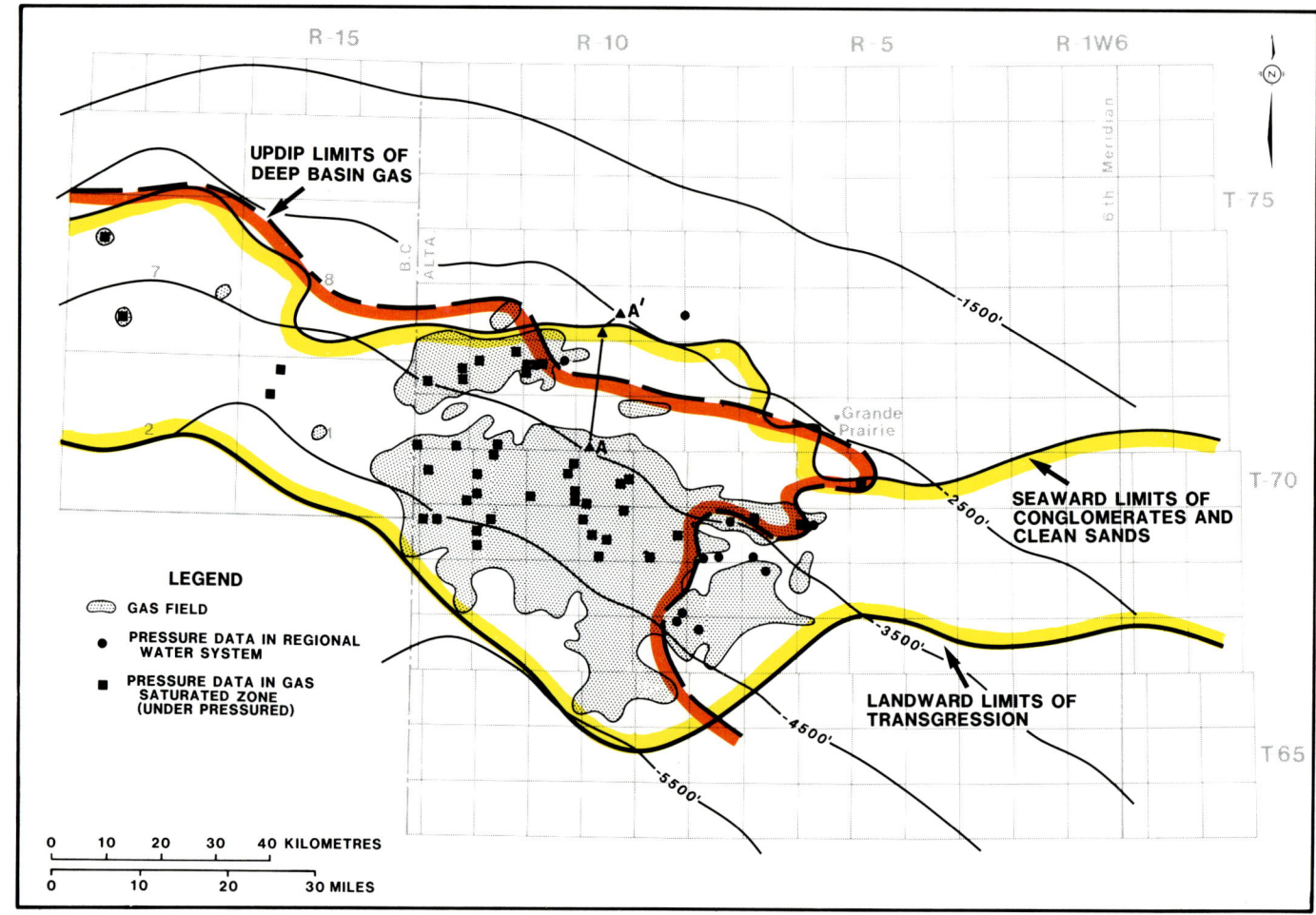

Figure 3. Falher A major facies trends and locations of DST and pressure data from wet and gas-saturated wells. Structure contours drawn on top of Falher A. Line A-A′ locates section shown in Figure 7.

seams similar to Falher A. These multiple coals render seismic resolution rather difficult, especially in detecting thickness changes in only one coal seam that lies within a relatively thin sequence that contains several seams. Furthermore, true coals are seismically indistinguishable from carbonaceous shale, as both lithologies exhibit a strong low velocity contrast with the surrounding silts and sands. The channel fills in the Falher A beach zone are similar in lithology to the enclosing rocks. Thus they cannot be distinguished directly with seismic.

The northern updip limit of the gas line, oriented east-southeast to west-northwest, also shows strong correspondence with facies boundaries (Figs. 3 and 5). The gas line seems to closely follow the line that separates the seaward limit of conglomerate and clean sand to the south (shoreline deposits) from the sands and finer-grained sediments to the north (nearshore and offshore deposits). Available DST data, as well as log analysis, show that rocks on the north side of the Deep basin gas line are wet, and those to the south of the line are gas-saturated. In the area of townships 71 and 72 and ranges 8, 9, 10, and 11, on both sides of the "Deep basin" gas line, Falher A is not producing because of a lack of coarse-grained and/or permeable reservoir rocks. A cross section, based on digitized logs (Fig. 7), across the gas line shows the northward trend from gas-saturated to mixed gas and water, to water-saturated tight rocks.

RELATIONSHIPS OF FLUID DISTRIBUTION TO RESERVOIR ROCK QUALITY

The general facies map (Fig. 3) shows that conglomerates and clean sands occur in a belt oriented west-northwest to east-southeast and suggests that, in general, reservoir characteristics deteriorate north of this belt. From core and sample control it was found that conglomerate and/or pebbly sandstone extend as far east as Range 25 W5M. Figure 8 is a map showing the average grain size distribution in a large area within and considerably beyond the Elmworth field. It confirms a predictable trend of grain size decrease to the northeast and east, suggesting a sediment source from the west and southwest. The average grain size in the Elmworth gas field is coarse-grained sandstone (ranges between very fine-grained sandstone and conglomerate), certainly coarser-grained, and consequently more permeable rock than that in the wet zone to the north. There is no indication that the wet rocks are more porous or permeable than the gas-charged rocks. This is contrary to

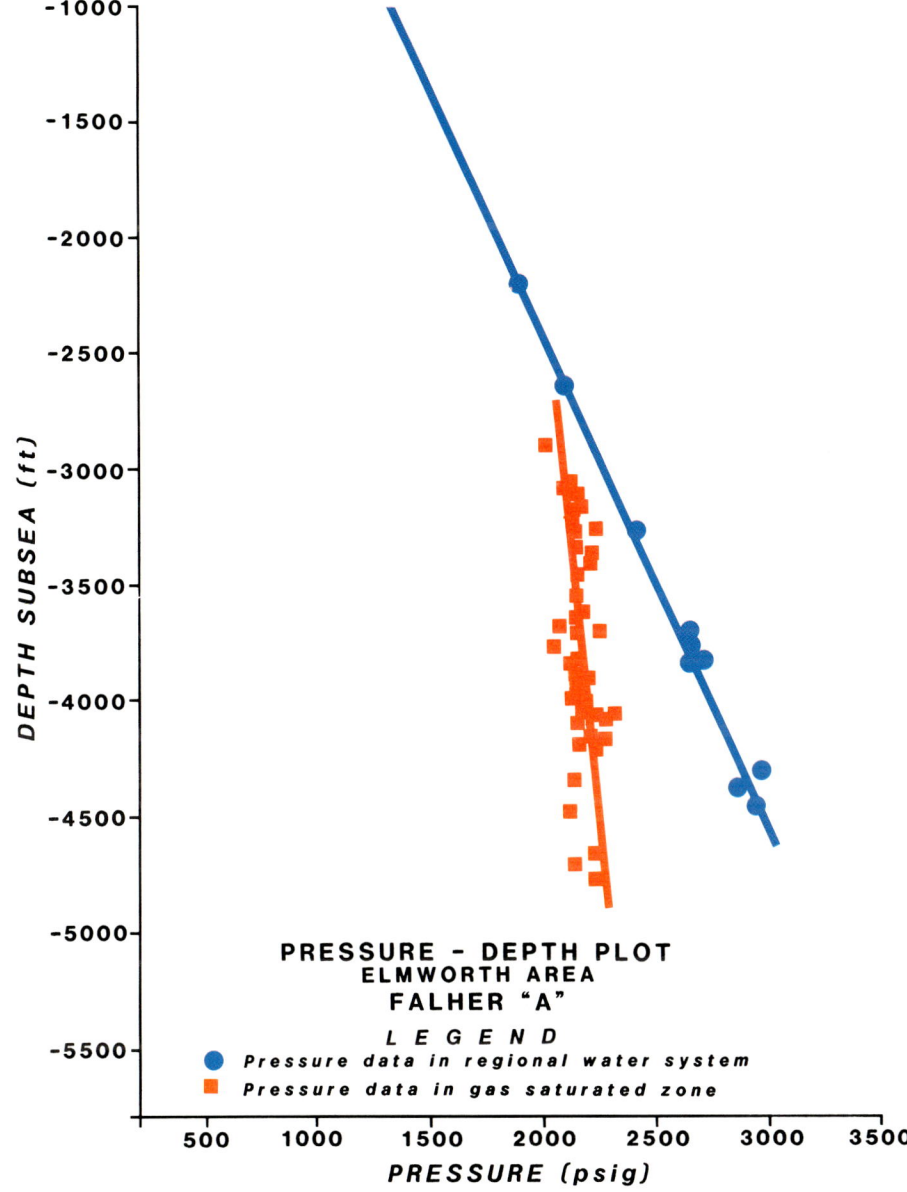

Figure 4. Pressure-depth plot, Falher A (after Davis, 1984).

easterly decreasing grain size and deteriorating rock quality trends.

DISCUSSION

All the available data discussed above appears to point to the conclusion that the updip "Deep basin" gas line in the Falher A is controlled by the initial characteristics of the rocks (facies changes). This relationship is most clearly seen in the northeast-to-southwest trending part of the gas line (Figs. 3, 5, and 6). Here, shoreline conglomerate and clean sand extend on both sides of the gas line in an east-to-west direction. Conglomerates and sandstones on the immediate east and west side of the line appear to have had identical reservoir properties prior to the arrival of the gas. As was mentioned earlier, the location of this part of the gas line is most probably controlled by channels (Figs. 5 and 6). Although there is little core control in wells along this channel, digitized logs and one DST strongly indicate that the channel fill consists of rocks that are considerably tighter than the surrounding reservoir rocks. This sudden decrease in width of pore throats appears to have been sufficient cause for downdip trapping of the gas. The sealing effect of the channel fill is further corroborated by the pressure data (Fig. 4). Pressure differences between gas-saturated rocks on one side of the channel and wet rocks on the other side show pressure difference in the order of 500 psig at similar subsea depths (at approximately 3,700 ft, or 1,128 m [subsea] and 4,300 ft, or 1,311 m [subsea] on Figs. 3 and 4) indicating very little communication across the channel in that area (Township 67, 68, and 69; Range 7, 8, and 9W6). At shallower depths (between 2,700 and 2,900 ft, or 823 and 884 m [subsea], on Figs. 3 and 4), however, pressure differences between wet and gas-saturated rocks are in the order of 50 to 100 psig, perhaps suggesting a less tight seal in these locations (Township 70 and 71; Range 5 and 6 W6). This deterioration of the effectiveness of the sealing capacity happens near the seaward (northward) end of the channel fill, and points to a likely location through which updip gas leakage could have been taking place.

The conventionally trapped gas, (that is, gas over water) immediately to the east of the gas line is testimony to the contention that gas generated deeper in the basin has

previous findings from other rock stratigraphic units (Hunter, 1979, for the San Juan Basin; and Cant, 1983, for the entire Spirit River Formation of the Elmworth field). However, it was found that the water-bearing rocks immediately to the east of the northeast- to southwest-trending part of the "Deep basin" gas line are slightly more porous and permeable than the gas-charged rocks west of the same line. In Figure 9 the average rock permeability is portrayed and it also confirms a previously predicted trend of northeastwardly and eastwardly deteriorating reservoir characteristics. The permeability categories (I, II, and III) portrayed on Figure 9 are those used by Sneider et al (1981) in the detection and characterization of reservoir rocks in the "Deep basin." In areas of good well control, it appears that there exists a band of fairly tight Falher A rock, on both sides of the "Deep basin" gas line, that separates the gas-saturated rocks to the south from the wet rocks to the north (that is, townships 71 and 72, ranges 9 and 10, W6M).

Figure 10 shows a selected number of gamma-log traces of Falher A from wells throughout the study area. These gamma traces clearly show the easterly and north-

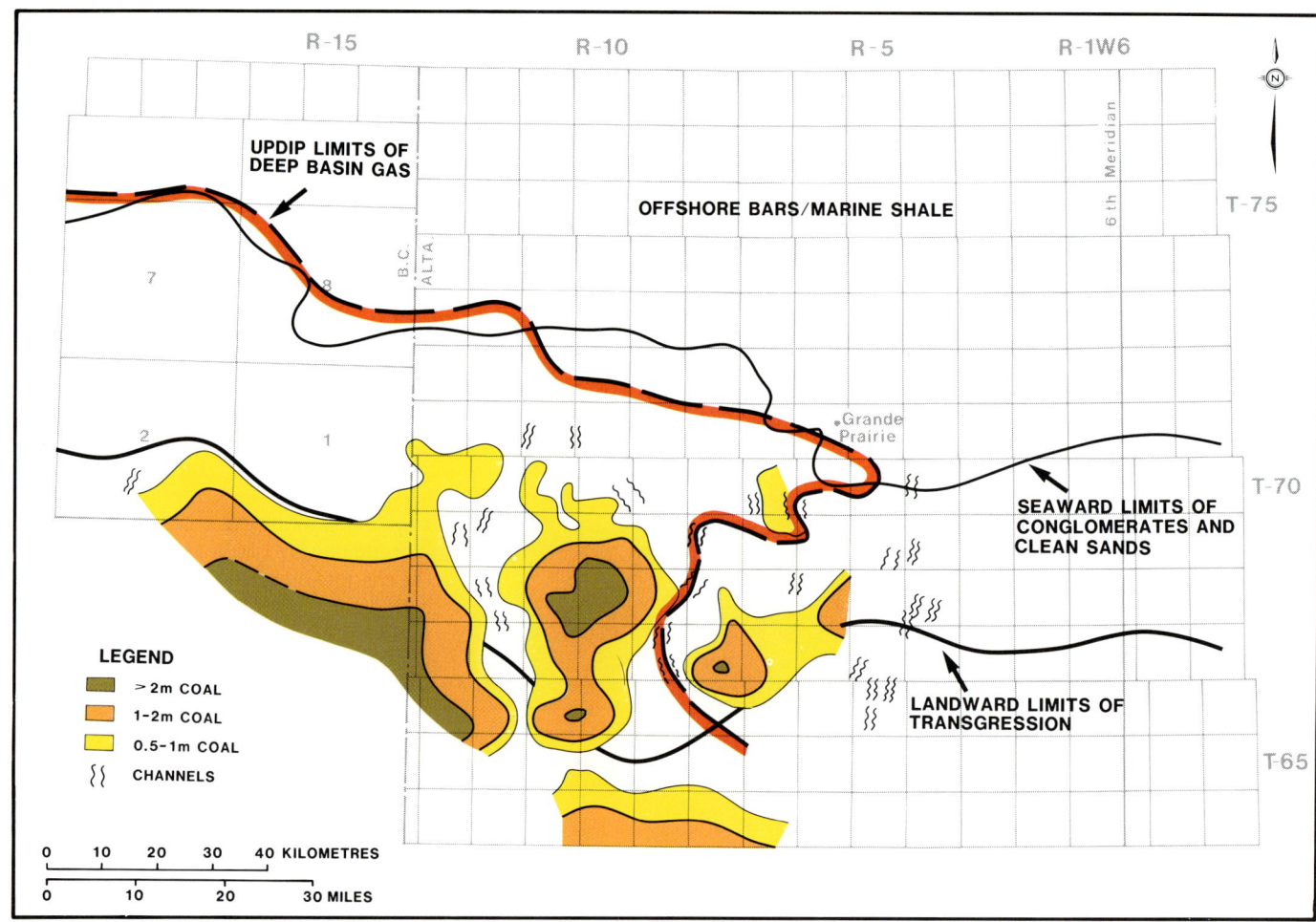

Figure 5. Isopach map of Falher A capping coal, superimposed on facies trends and related to fluvial channels. Coal thickness is based on data from 408 wells.

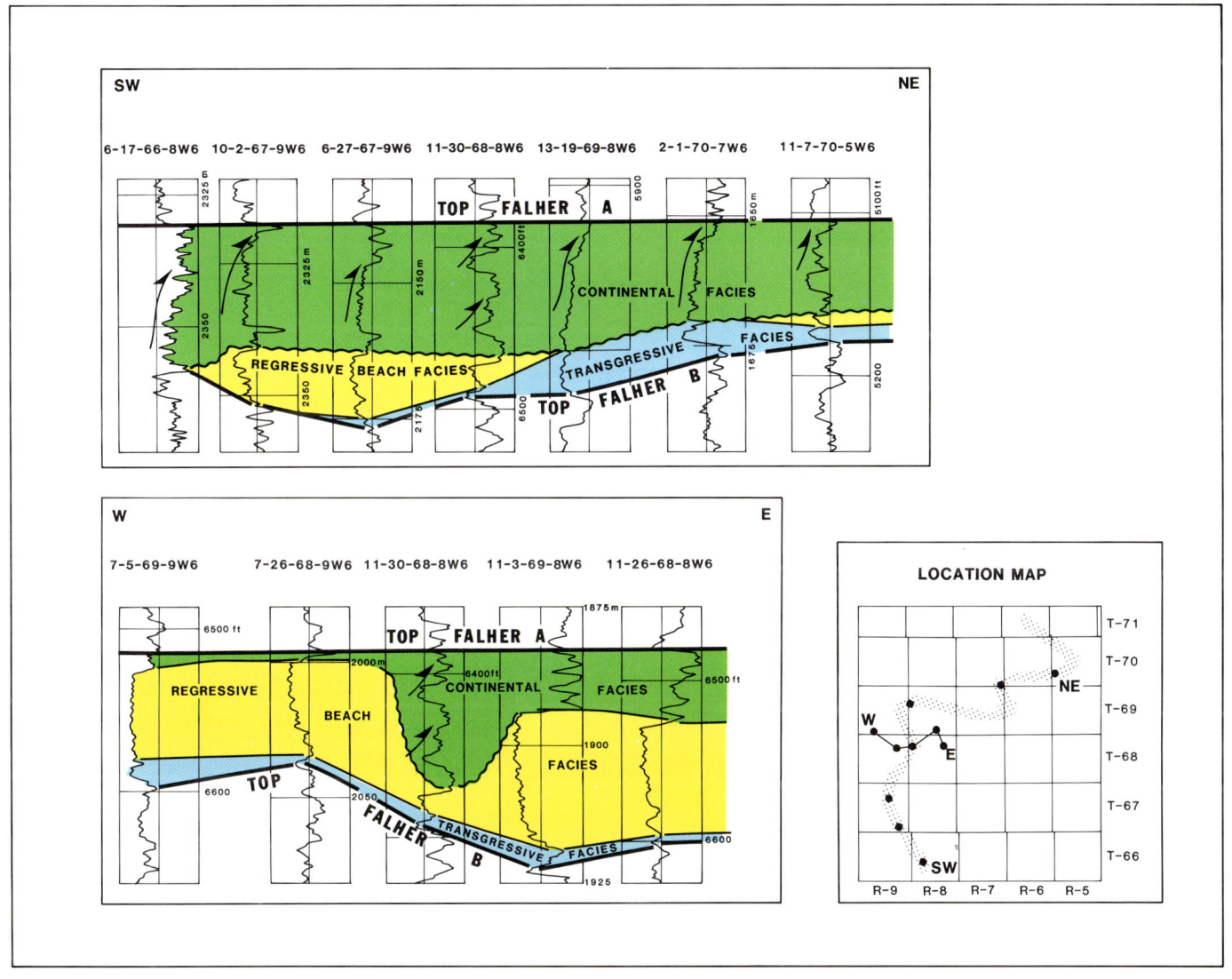

Figure 6. Gamma log sections across and along the major channel system shown in Figures 3 and 4. This channel system coincides with the eastern updip limit of gas, and probably served as a seal against further major updip migration.

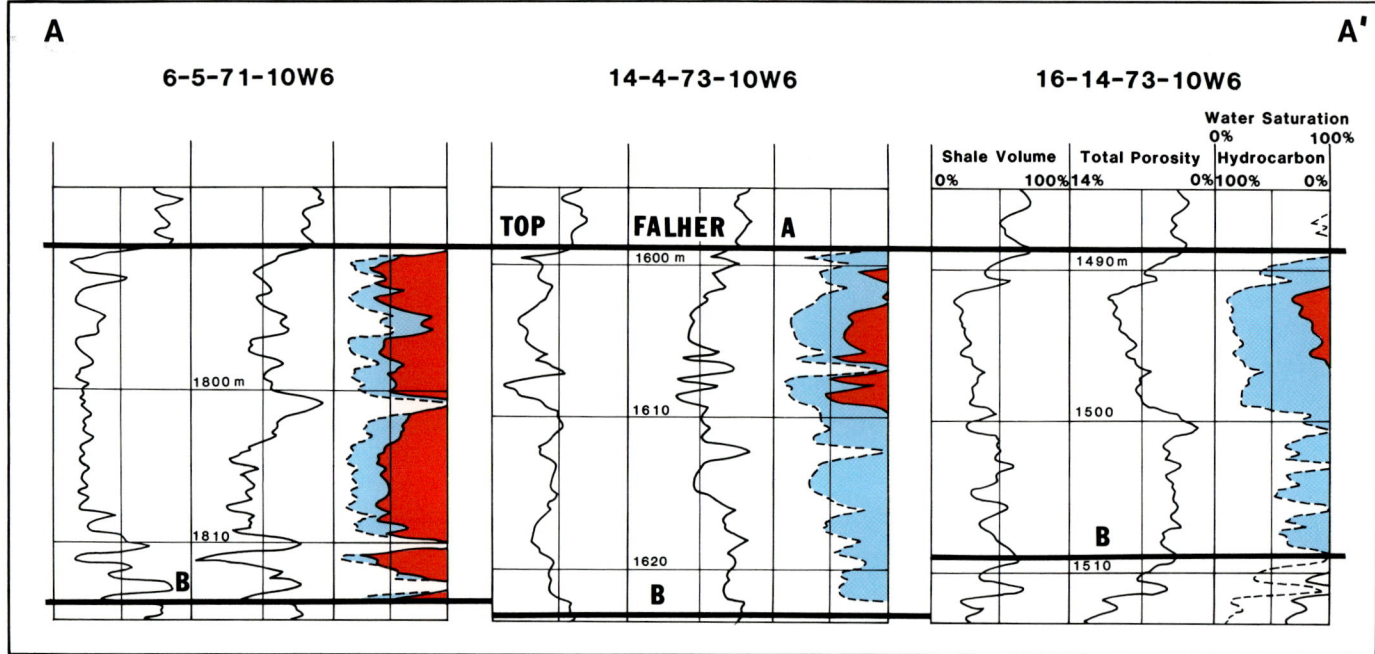

Figure 7. A cross section, constructed from digitized logs, showing shale volume, porosity, and fluid content of three wells through tight Falher A rock. This section runs across the "Deep basin" gas line. See Figure 3 for location.

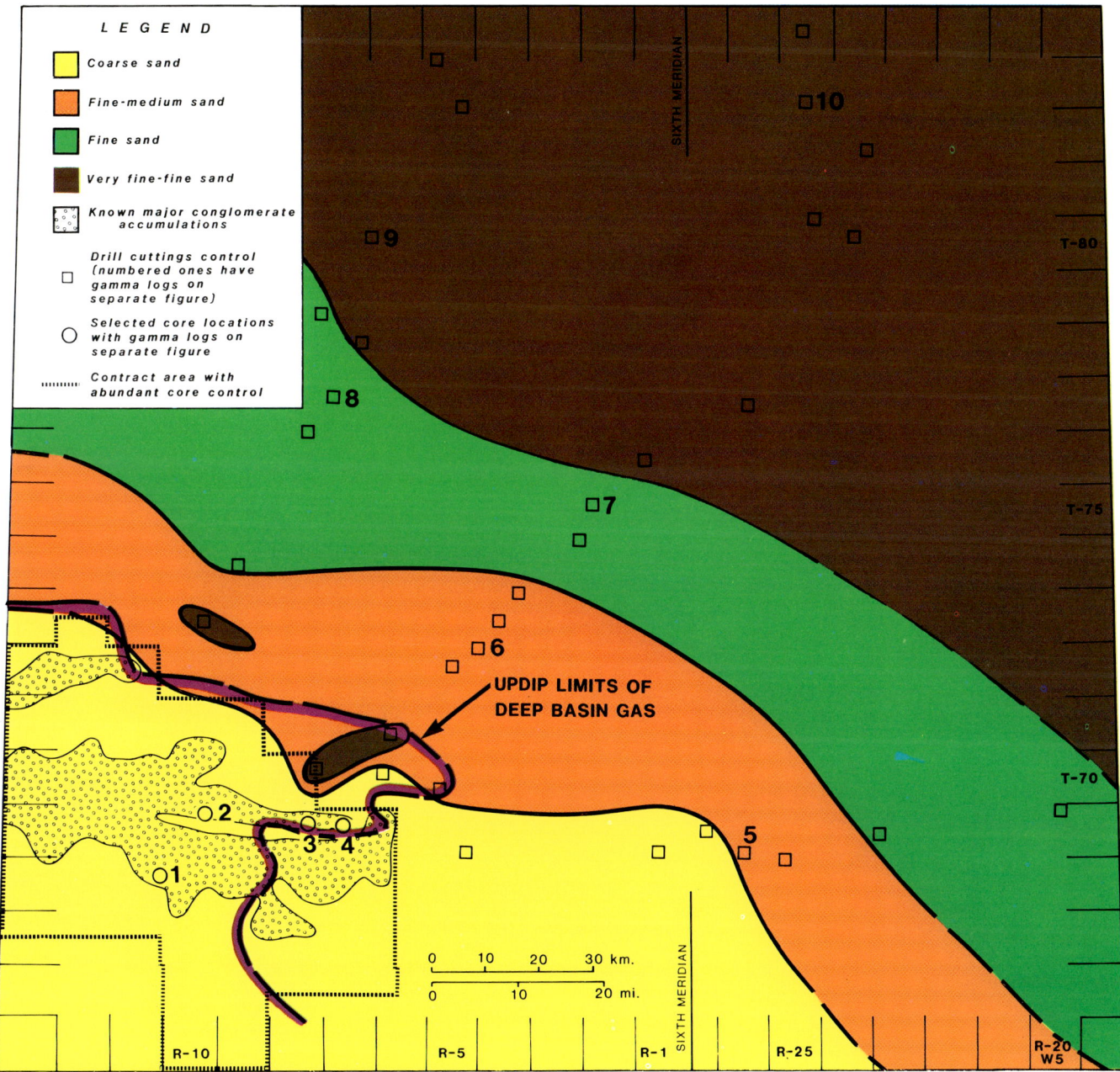

Figure 8. Average grain size of sandstone and conglomerates of Falher A Cycle of northwestern Alberta.

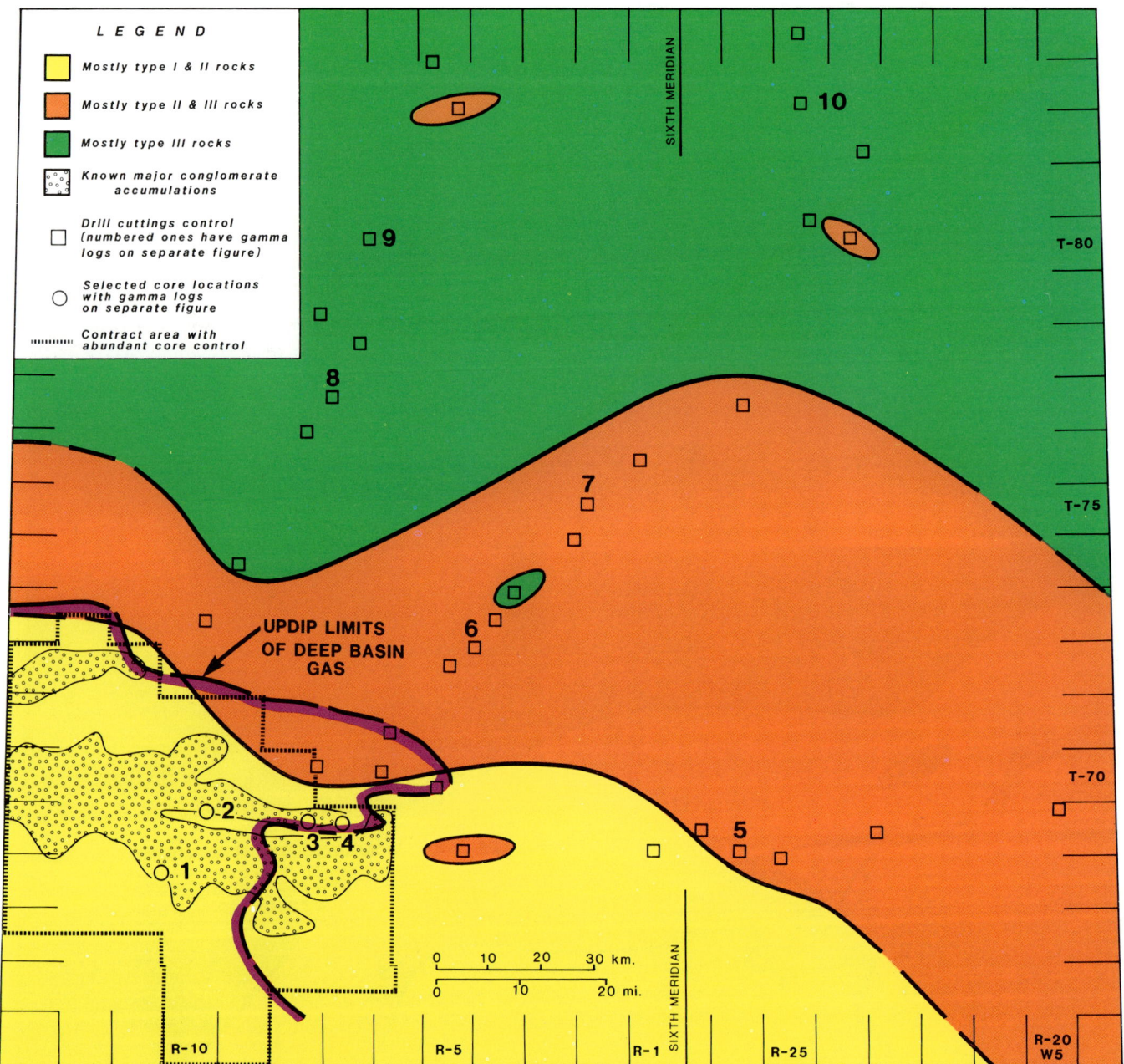

Figure 9. Average permeability (rock types I, II and III) of Falher A Cycle of northwestern Alberta. Type I, >0.5 md; Type II, 0.5 to 0.07 md; and Type III, <0.07 md. For permeability types see Sneider et al (1981).

been slowly but continuously leaking updip. This process of updip leakage would also explain the fact that downdip gas is subnormally pressured, and assumes that the channel does not constitute a complete non-leaking seal. It is also speculated that updip leaking gas travelled long distances eastward, to eastern parts of Alberta and appeared in the form of numerous Lower Cretaceous gas traps as well as gas seeps on Athabasca River in the Fort McMurray area (Gies, personal communication, 1983). This is logical since in the eastward direction coarse-grained rocks are considerably more continuous and permeable than those to the north and northeast.

The northern west-northwest to east-southeast trending part of the gas line also appears to be controlled by facies change from shoreline to nearshore and offshore sediments. However, the situation here is a little more complex than it initially appears. It was explained above that there is a belt of nonproducing (when unstimulated) tight Falher A rocks on both sides of the gas line in townships 71 and 72 and ranges 8, 9, 10, and 11. Digitized logs across this belt (Fig. 7) indicate these tight rocks are wet on the north side and gas-saturated on the south side of the gas line. It is this belt of tight rocks, separating the northern wet province from the southern

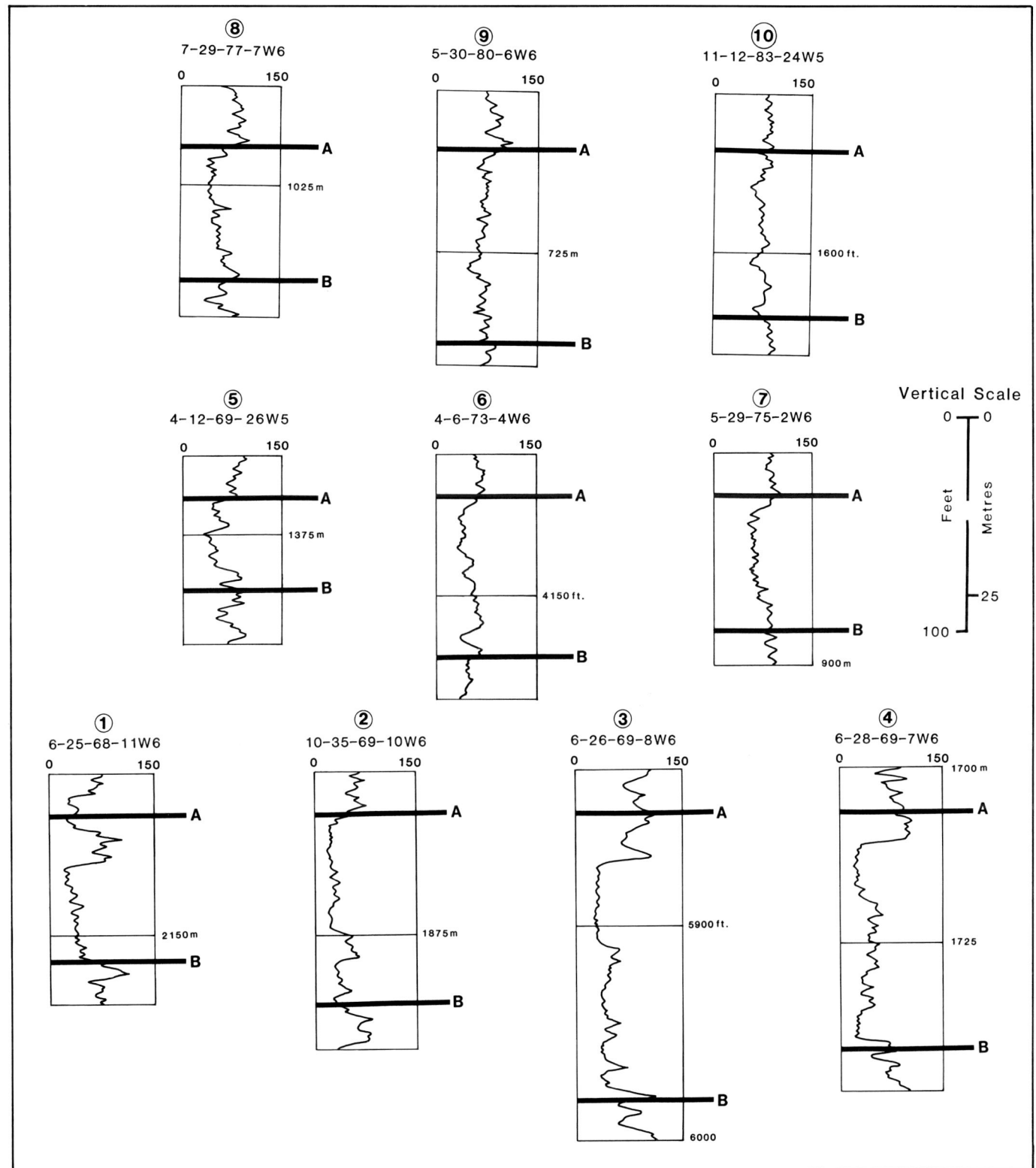

Figure 10. Typical gamma log patterns from Falher A Cycle. See Figures 8 and 9 for location and relationship of gamma logs to grain size and permeability trends.

gas-saturated province, that needs to be further investigated (detailed petrographic examination of cores and samples), along with a detailed and thorough study of all geophysical log and rock data available on either side of this part of the "Deep basin" gas line. Until then, only speculative conclusions can be made.

CONCLUSIONS

Based on the available data it can be concluded that trapping of subnormally-pressured gas in the Falher A Cycle is facies controlled. The eastern updip limit of the gas appears to be controlled by the sealing effect of a major channel system. The northern and northeastern updip limit closely follows a major facies change from coarser and more permeable shoreline to finer and less permeable nearshore and offshore deposits. Those seals, as concluded by several previous authors, have not been and are not now completely tight, allowing continuous but slow updip leakage of gas. The most likely direction of this leakage is eastward, since rocks are more permeable in that direction. This path is also suggested by the fact that gas exists in conventional traps immediately eastward of the updip limit of the "Deep basin" gas line. However, based on previous works, it is generally accepted that gas is continuously being generated downdip to nearly balance the updip leakage.

It should be stressed, however, that I do not wish to imply that the mechanism for Falher A gas trapping is of universal application to explain trapping in the entire "Deep basin" (see for example Gies, 1984). Some of the questions that remain unanswered ask: how an enormous volume of gas in a dozen other reservoirs is still contained downdip from water, and why the gas-water line in the "Deep basin" occurs in approximately the same position for a vertical stratigraphic interval of about 4,000 ft (1,219 m)? These and other questions deserve thorough investigation through detailed studies of the stratigraphy, sedimentology, and fluid distribution of the rock units of interest. Justification for such studies are obvious. Models derived from the investigations can be applied for more effective exploration and development of these intriguing gas occurrences that form some of the largest gas fields in North America.

ACKNOWLEDGMENTS

The author would like to acknowledge J. A. Masters, R. E. Wyman, R. M. Gies, G. Staples, and P. Jackson for critically reading the manuscript and making several constructive comments. A. Mercer typed the manuscript, and G. Jones and G. Cone drafted the figures.

REFERENCES CITED

Cant, D. J., 1983, Spirit River Formation - a stratigraphic-diagenetic gas trap in the "deep basin" of Alberta: AAPG Bulletin, v. 67, p. 577-587.

Cummella, S. P., 1981, Sedimentary history and diagenesis of the Pictured Cliffs Sandstone, San Juan Basin, New Mexico and Colorado: University of Texas at Austin, Master's thesis, 219 p.

Davis, T. B., 1984, Subsurface pressure profiles in gas saturated basins, in J. A. Masters, ed., Deep basin gas: AAPG Memoir 38, this volume.

Gies, R. M., 1981, Lateral trapping mechanisms in "deep basin" gas trap, western Canada (abs.): AAPG Bulletin, v. 65, p. 930.

——, 1984, Case history for a major Alberta "deep basin" gas trap; the Cadomin Formation, in J. A. Masters, ed., Deep basin gas: AAPG Memoir 38, this volume.

Hunter, B. E., 1979, Regional analysis of the Point Lookout Sandstone, Upper Cretaceous, San Juan Basin, New Mexico–Colorado: Texas Tech University, Ph.D. thesis, 118 p.

Jones, R. M. P., 1980, Basinal isostatic adjustment faults and their petroleum significance: Canadian Society of Petroleum Geologists Bulletin, v. 28, p. 211-251.

Masters, J. A., 1979, "Deep basin" gas trap: AAPG Bulletin, v. 63, p. 152-181.

Smith, D. G., 1984, Paleogeography of the Lower Cretaceous in and adjacent to the "deep basin" of the Elmworth Area, western Canada and northeastern British Columbia, in, J. A. Masters, ed., Deep basin gas: AAPG Memoir 38, this volume.

Sneider, R. M., et al, 1981, Methods for characterization of reservoir rock, deep basin gas area, Western Canada: San Antonio, Society of Petroleum Engineers 56th Annual Fall Meeting, SPE Paper 10072.

Gas Reserves and Production Performance of the Elmworth/Wapiti Area of the Deep Basin

Richard D. Smith
Canadian Hunter Exploration Ltd.
Calgary, Alberta

Detailed reservoir mapping of conventional permeability sands within four gas contract areas in the Elmworth field yields 7.5 tcf of contractible gas and another 9.6 tcf of remaining probable gas with total natural gas liquids of 1 billion barrels. Still unresolved is the amount of commercial production which can be obtained from an immense amount of gas-in-place in very low permeability sands. Production in the field areas has increased from 20 mcf/d to current peaks of close to 1.0 bcf/d in just four years. This pace has made essential a close integration of exploration and development geology, petrophysics, and engineering.

INTRODUCTION

Large quantities of gas have been discovered along the western side of Alberta in what is now referred to as the Deep basin (Masters, 1979). A thick wedge of gas-saturated Mesozoic clastics, approximately 50 mi (80 km) wide and 600 mi (965 km) long, stretches from Dawson Creek, British Columbia, to the Canada/United States border (Fig. 1). The northern part of the Deep basin has been referred to as the Elmworth/Wapiti area, and it is in this area that this paper will concentrate. Gas-saturated rocks reach thicknesses of 6,000 ft (1,835 m) along the western side of the Western Canada basin and thin to zero approximately 50 mi (80 km) eastward. Every conglomerate, sandstone, siltstone, coal, and shale is charged with gas. The total gas resource that can be calculated in the clastic rocks in the Elmworth/Wapiti area is approximately 800 tcf (Masters, 1984). Development to date has focused on the high porosity and permeability facies in this area. Net pay mapping has demonstrated contractible reserves of 7.5 tcf and potential contractible reserves of 9.6 tcf, hence, the ultimate contractible reserves in the mapped area could reach as high as 17.1 tcf.

In the Elmworth/Wapiti area, approximately 1,125 wells have been drilled and production casing run on 875 of these since its discovery in 1976. Drilling density varies from one well per two and one-half sections in the most developed part of Elmworth, to less than one well per township in the lesser developed area. Ten gas plants have been built with a combined processing capacity of greater than 1 bcf/day. Although much of the gas supply in this area remains a resource at this time, improved economics would have an immediate impact on converting this resource to deliverable reserves.

GAS RESERVES

The gas reserve described in this paper will be that which has been identified and mapped within the contract areas shown on Figure 2. A substantial part of the gas-saturated area lies outside the contract areas and has not been the subject of detailed reserve mapping. Masters (1984) has calculated a total resource gas volume within the northern area of the Deep basin of 800 tcf. This number represents the total volume of gas that can be placed in a total clastic isopach of the gas-saturated Mesozoic section. The reserves described in this paper have been conventionally mapped and represent a presently economically recoverable gas reserve. Due to the unconventional nature of this area, arguments and controversy will continue on just how much of the resource will become economically developable.

Categories of Gas Reservoirs

Gas reserve estimates have been broken down into several specific categories for the purpose of this discussion. It is necessary to make sure each category is properly understood.

The following definitions describe reserve category:

Proven Reserves—are those reserves delineated by drilling and testing that are considered to a high degree of certainty to be recoverable at commercial rates.

Probable Reserves—are those reserves that can be reasonably counted on. They lie within a stratigraphic or structural trap generally delineated by geological data in conjunction with fluid interfaces determined from offset well control and are indicated present by log evaluation of wells which are untested.

Note: The combination of these two categories of reserves represent the gas reserve calculation this paper will describe. Established, Incremental Probable, and Ultimate Possible reserves are various combinations of Proven and Probable that will be used.

Established Reserves (Marketable or Contractible)—Established reserves are those reserves that are attained by adding proven reserves to that component of probable reserves that can be shown to be geologically continuous and supported by pressure data that places them in association with proven reserves.

Incremental Probable Reserves—Incremental probable reserves are those reserves that represent that component of probable reserves that can be shown to be geologically continuous but that are not supported by pressure data that places them in association with proven reserves.

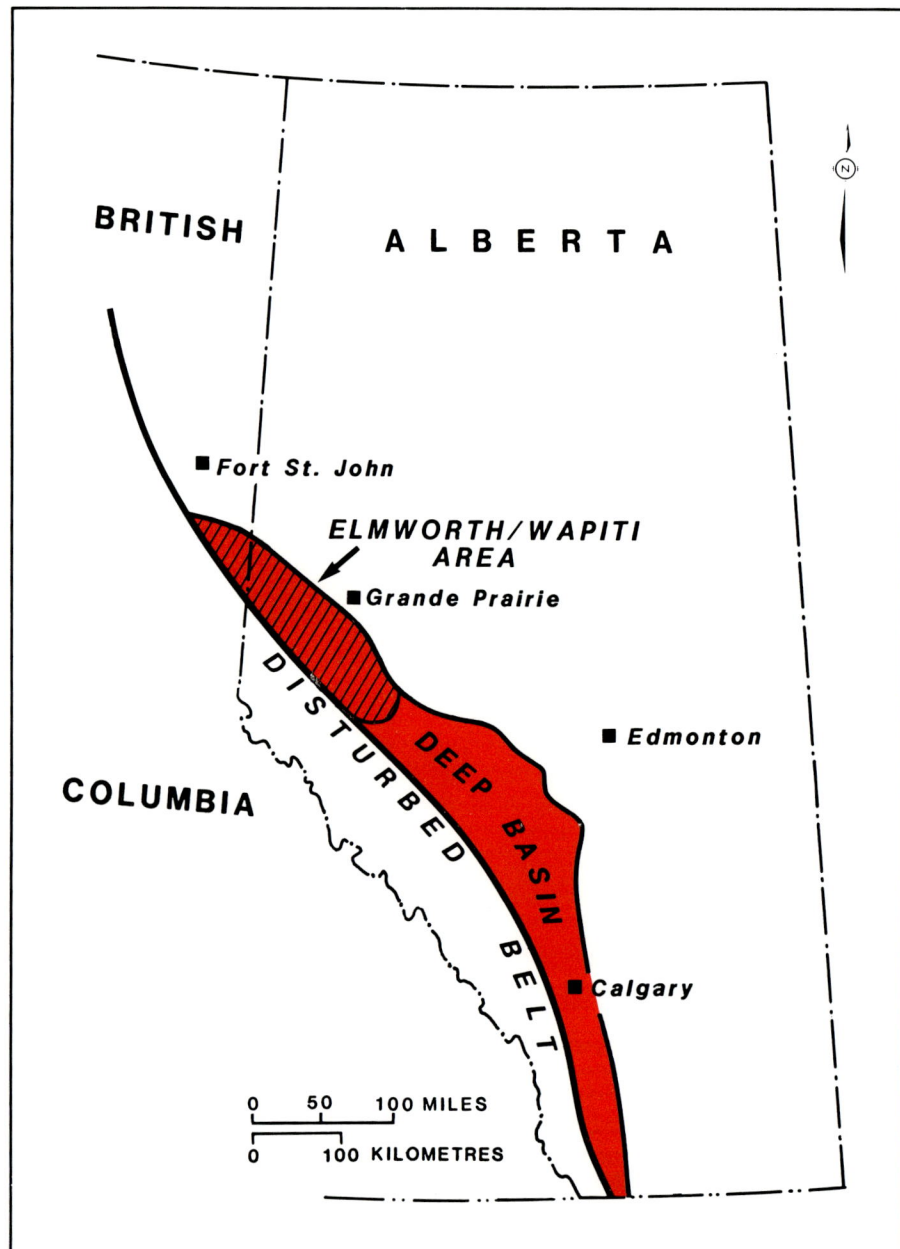

Figure 1. Location map showing the position of the Elmworth/Wapiti area relative to the Deep basin of Alberta.

Ultimate Possible Reserves (Proven plus Probable)—Ultimate possible reserves represent the total reserve attained by adding the established reserves and incremental probable reserves.

Net pay maps for the major gas-contributing zones in Elmworth/Wapiti indicate the areas which represent established and incremental probable gas reserves. Red areas on the net pay maps (Figs. 5 to 13), represent those on which established gas reserves were calculated and non-red areas represent those on which incremental probable gas reserves were calculated. The ultimate possible reserve is the total gas volume volumetrically calculated inside the zero isopach on the net pay maps.

Elmworth/Wapiti Gas Reserves

Table 1 gives a zone-by-zone reserve summary of reserves in the Elmworth/Wapiti area of the Deep basin. This reserve summary represents a volumetrically calculated gas reserve based on net pay mapping and one section acreage assignment for zones that could not be mapped with the present well control.

The average reservoir parameters and the average gas properties used in this reserve summary are found on Tables 2 and 3 respectively.

GAS BEARING RESERVOIRS

Regional geological mapping in the Elmworth/Wapiti area led to the interpretation of the depositional framework of the gas saturated section and played a major role in projecting drilling programs and development of the area.

Initial mapping of reservoirs throughout this area began with isopachs of genetically related sand and conglomerate bodies.

Net pay maps were constructed by using net pay thickness determined from logs, cores, and samples, and gas productivity from drill-stem tests and completion. Pool continuity was confirm by ensuring reservoir pressures were the same in each pool.

Geological Mapping

The 6,000 ft (1,830 m) of Cretaceous sandstone along the west side of the Western Canada basin has been described (Smith, Zorn, and Sneider, 1984). These sediments were deposited during a series of major depositional cycles, each characterized by a basinward progradation of thick and extensive fluvio-deltaic complexes which were periodically inundated by a rapid marine transgression. Figure 3 is a stratigraphic column showing the gross lithology and depositional environments of the gas-bearing zones in the Elmworth/Wapiti area. Figure 4 illustrates this series of transgressive/regressive episodes of sedimentation in the Elmworth/Wapiti area. The massive thickness of potential reservoir rock resulted from the stacking of these continental marine transitions as the west side of the basin subsided. The high gas delivery reservoirs of the Elmworth/Wapiti area are the rocks deposited in the high-energy, depositional environments. In most cases therefore, net pay maps are representative of prograding high-energy beaches and beach bars.

Logging Evaluation and Cutoffs

Routinely a comprehensive suite of open hole logs is run on all Deep basin wells. This program consists of: (1) dual induction and/or dual laterolog; (2) borehole compensated sonic logs; (3) compensated neutron/formation density log; and (4) proximity or microlaterolog.

Rock-log calibration and productivity results from drill-stem testing and completions have established the shale volume, porosity, and water saturation cutoffs in the Elmworth/Wapiti area to be those shown on Table 4.

Productivity Evaluation

All zones that may have the capability of commercial production are tested by Canadian Hunter in Deep basin wells. Details of the testing techniques follow:

Drill-Stem Testing—Testing results in the Elmworth/Wapiti area have indicated that severe damage occurs to low-permeability gas zones when drilling the well bore. Consequently, it has become customary to drill-stem test all potential gas zones as soon as possible after penetration. Due to the limited flow usually encountered in

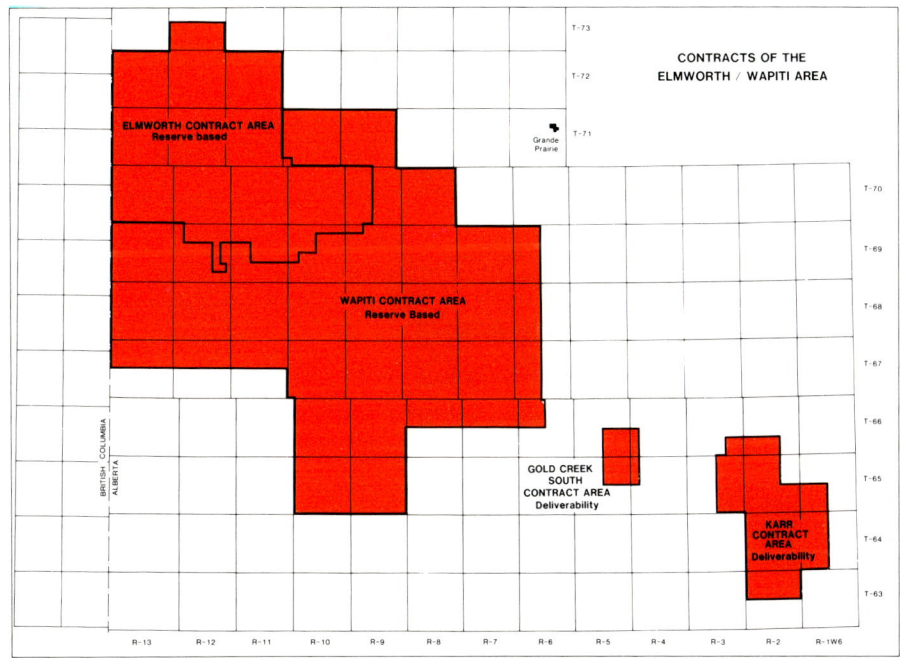

Figure 2. Map showing the position of gas contract areas held by Canadian Hunter Exploration Ltd. in the Elmworth/Wapiti area.

Table 1. Elmworth/Wapiti area reserve summary (bcf @ 14.4 psia and 60°F).

Zone		Established Reserves	Incremental Probable Reserves	Ultimate Possible Reserves
Chinook		4.2*	—	4.2
Cardium		4.2*	—	4.2
Doe Creek		2.1*	—	2.1
Dunvegan		15.3*	—	15.3
Paddy		216.3	49.3	265.7
Cadotte		473.0	1,324.2	1,797.2
Notikewin	– Middle	137.5	1,110.6	1,248.1
	– Basal	29.2	200.7	229.8
Falher A	– Conglomerate	958.7	341.8	1,300.5
	– Sand	1,243.0	1,297.9	2,540.9
	– Basal	191.1	176.0	367.1
Falher B	– Conglomerate	691.5	43.6	735.0
	– Sand	587.3	860.6	1,447.8
Falher C	– Conglomerate	127.8	19.8	147.6
	– Sand	114.2	340.4	454.6
	– Basal	22.9	238.1	261.0
Falher D	– Conglomerate	256.0	154.7	410.7
	– Sand	247.8	903.8	1,151.6
	– Basal	125.5	266.5	392.0
Falher F	– Conglomerate	52.0	156.1	208.1
	– Sand	91.2	330.0	421.3
Bluesky-Gething		146.7	149.9	296.6
Lower Gething		121.9*	—	121.9
Cadomin		1,175.8	1,616.3	2,792.1
Nikanassin		297.1*	—	297.1
Charlie Lake		3.9*	—	3.9
Doig		65.0*	—	65.0
Halfway		27.0*	—	27.0
Belloy		24.3*	—	24.3
TOTALS		7,452.5	9,580.3	17,032.7

*Reserves based on one section acreage assignments on proven well.

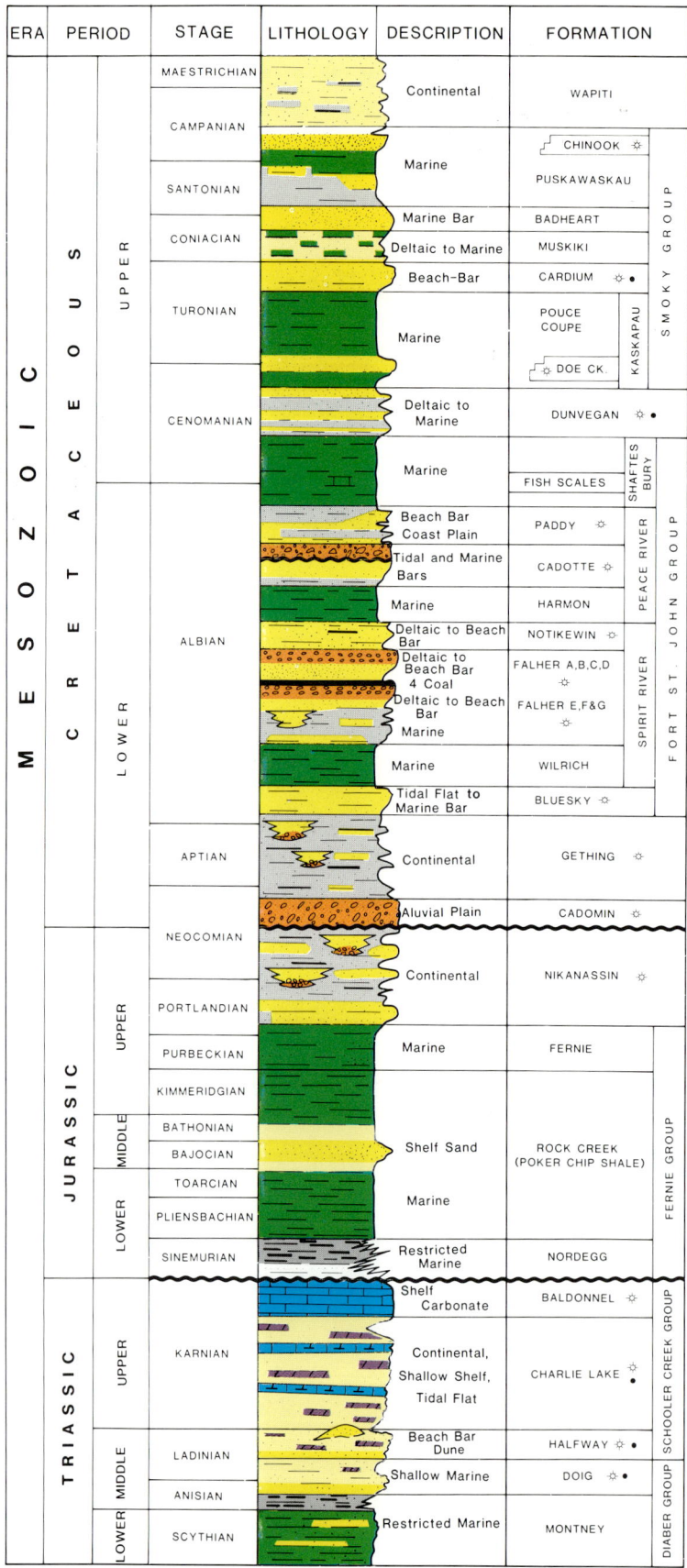

these reservoirs, the closed-chamber technique has been used to get an estimated flow rate for complete data analysis. In more conventional quality reservoirs, a Horner analysis can be used to estimate Kh, damage ratio, and estimate flow volumes if the damage was removed.

Well Completions—Wells in the Elmworth/Wapiti area almost always contain low-pressure, low-productivity gas zones that damage from drilling fluids. For this reason, well completions are both expensive and demanding.

The completion technique for wells in the Deep basin area has become a standardized procedure. Pressure transient data is used extensively to evaluate the various stages of completion, predict zone performance, and justify additional stimulation.

Net Pay Mapping

The extensive areal distribution, as well as the multi-layered nature of the Elmworth/Wapiti gas reservoirs has resulted in the net pay mapping being a complex and challenging task. Total volumetric reserves were mapped in 234 pools on 23 zone maps. The integration of genetic geology mapping, log net pay thicknesses, productivity from drillstem tests or completions, and pool continuity from recorded pressures on drill-stem tests or pressure buildup on completion, have been combined to give the accompanying series of maps. Figures 5 to 13 represent a sample of the net pay maps and acreage assignments that were used when calculating the volumetric reserves that are summarized on Table 1. As previously stated, established reserves are those shaded red, and incremental probable reserves are those not shaded red but within the zero net-pay contour. The total of both of these represents the ultimate possible reserves in this area.

RESERVOIR TYPES

Reservoirs of the Elmworth/Wapiti area can be divided into conventional and tight categories. Conventional reservoirs are usually coarse sandstones or conglom-

---

Figure 3. Stratigraphic column showing the gross lithology, depositional environment and gas/oil producing zones of the Elmworth/Wapiti area.

erates with porosities in the 8 to 12% range and permeabilities in the 0.5 millidarcy to 5 darcy range. Productivity from these reservoirs (unstimulated) varies from 1 mmcf/d to 28 mmcf/d and when stimulated, these rates can increase to as high as 100 mmcf/d. Tight reservoirs are usually medium to fine grained sandstones with porosities ranging from 4 to 7% and permeabilities from 1 microdarcy to 0.5 millidarcys. Productivity from these reservoirs is normally 20 to 100 mcf/d unstimulated, and from 500 mcf/d to 1 mmcf/d stimulated.

High-Productivity Reservoirs

The high-productivity or conventional reservoirs of the Elmworth/Wapiti area were deposited in high-energy depositional environments where finer-grained rock was winnowed out of the coarser grained rock. This took place on the beaches, in tidal channels of the shoreline, and in the channel bottoms and point bars of the fluvial environments. Wells completed in these zones currently supply 95% of the gas being produced through the gas plants of this area. As large volumes of gas are withdrawn from the conventional reservoirs, material balance calculations will give a more accurate assessment of reserves and new pay mapping will take place.

Well performance and sustained production has aided in our understanding of pool continuity. An interference test on several wells in one Falher B Conglomerate pool has recently indicated that this pool is made up of two pools instead of one. This multi-well test was conducted using four observation wells and one producing well. Material balance calculations show the 11-4-71-13W6 well is effectively draining a 33 bcf reservoir. This volume represents about 46% of the 71 bcf volumetrically mapped. Production from the 11-4 well has resulted in pressure decline on the 12-2-71-13W6 and the 11-6-71-13W6 wells. No pressure response has been noted on the 7-5-71-12W6 or the 16-3-71-12W6. The results of the interference test have led to the remapping of the Falher B Conglomerate #804 pool into two separate pools (Fig. 15).

Low-Productivity Reservoirs

The low-productivity, tight reservoirs of the Elmworth/Wapiti area contain large volumes of gas, but productivity from these zones individually is very often subeconomic. Reserves in the Elmworth/Wapiti area have only been assigned to low-productivity reservoirs that: (1) have tested gas at greater than 100 mcf/d either from drillstem test or post-completion, and (2) are overlain by a high-productivity conventional reservoir through which the low-productivity gas will drain.

To date, a variety of projects and studies have been undertaken to assess the productivity of tight sand reservoirs.

Tight Sand/Conglomerate Simulation Studies—A series of numerical model studies have been carried out based upon a hypothetical reservoir representing the "typical" Falher package of sand with overlying conglomerate. The purpose of this work was to determine what level of gas recovery from the sands would be

Table 2. Elmworth/Wapiti area average reservoir parameters (bcf @ 14.4 psia and 60°F).

Formation	Porosity %	Water Saturation %	Oil Saturation %	Reservoir Pressure (psia)	Reservoir Temperature (°R)	Z
Chinook	12.40	49.0	0.00	1,456.0	568.0	0.796
Cardium	10.85	41.5	0.00	1,038.5	546.5	0.838
Doe Creek	12.5	54.7	0.00	980.3	564.3	0.851
Dunvegan	12.02	46.8	0.00	1,175.8	569.0	0.831
Paddy	11.7	28.0	0.00	1,857.0	596.0	0.823
Cadotte	7.8	39.0	0.00	2,271.0	620.0	0.871
Middle Notikewin	8.4	45.0	0.00	2,574.0	612.0	0.843
Basal Notikewin	9.0	36.0	0.00	2,907.0	626.0	0.855
Falher A Conglomerate	7.5	37.0	0.00	2,343.0	607.0	0.836
Falher A Sand	8.6	45.0	0.00	2,240.0	603.0	0.830
Basal Falher A Congl.	8.7	44.0	0.00	2,259.0	608.0	0.859
Falher B Conglomerate	7.4	34.0	0.00	2,119.0	614.0	0.850
Falher B Sand	8.8	33.0	0.00	2,935.0	634.0	0.890
Falher C Conglomerate	9.1	37.0	0.00	2,865.0	627.0	0.857
Falher C Sand	8.0	42.0	0.00	2,822.0	631.0	0.862
Basal Falher C Congl.	8.7	32.0	0.00	2,204.0	623.0	0.847
Falher D Conglomerate	6.2	28.0	0.00	2,305.0	618.0	0.817
Falher D Sand	7.3	33.0	0.00	2,500.0	626.0	0.846
Basal Falher D	7.8	30.0	0.00	2,650.0	628.0	0.886
Falher F Conglomerate	8.4	33.0	0.00	3,602.0	629.0	0.917
Falher F Sand	7.9	30.0	0.00	3,602.0	629.0	0.917
Bluesky-Gething Sand	7.4	33.0	0.00	3,048.0	628.0	0.893
Lower Gething Sand	6.9	35.0	0.00	3,191.0	642.0	0.876
Cadomin	6.1	29.0	0.00	3,123.0	636.0	0.888
Nikanassin Sand (Upper, Middle & Lower)	9.3	28.0	0.00	3,657.0	647.0	0.910
Charlie Lake	13.1	9.0	0.00	4,150.0	662.0	0.847
Doig Sand	7.7	31.0	0.00	4,306.0	657.0	0.980
Halfway Sand	5.9	47.0	0.00	4,800.0	653.0	0.977
Belloy Sand	11.3	47.5	0.00	4,327.0	659.0	0.970

possible for different levels of sand permeability and for varying sand thickness if only the overlying conglomerate was completed and produced. These model studies indicated that 20-year recoveries in the order of 80% from the sand were possible if vertical sand permeability exceeded 1 microdarcy (Fig. 16). As sand permeability decreases, the 20-year recovery also decreases. But, 20-year recoveries in excess of 60% are still expected for those sands displaying vertical *in situ* gas permeability down to 0.05 microdarcy (0.00005 millidarcys).

Tight Sand Permeability Studies—Laboratory studies were undertaken to determine *in situ* gas permeabilities in the Falher sandstones below the high-deliverability conglomerates. To date, 89 Falher sandstone samples that ranged in porosity from 3.5 to 11.5% have been analyzed for gas permeability at simulated *in situ* conditions of stress and water saturation. Without exception, these studies indicate that provided *in situ* water saturations are less than 50%, gas permeabilities will substantially exceed 0.05 microdarcies. In other words, for all samples tested, the vertical permeability to gas was found to be sufficient at average predicted brine saturations for vertical transmission of the gas in the sandstones, up into the conglomerates and out to the well bore (Fig. 17).

Long Term Pressure Buildup—A four-point production test on the Falher B Conglomerate in the 10-11-71-11W6 well indicated that the reservoir intersected a boundary less than 1,000 ft (305 m) away from the well bore. With large withdrawals of gas from this zone, deliverability declined dramatically during the first three months and material balance calculations indicated the pool to be limited to approximately 16 bcf. The well was shut-in and a pressure buildup was monitored over the next 13 months. During this time, pressure buildup in the Falher B conglomerate reserve gave evidence that gas was migrating vertically out of the underlying sandstone and recharging the reserves of the pool. The 10-11 well has been returned to production and the drawdown performance is being monitored in an effort to obtain more evidence of vertical gas migration out of the underlying tight sands and into the overlying conglomerate.

Commingled Production—All zones have been completed to permit segregation of flow in the well bore either through single or dual completion. Commingling of production from various Falher members in some wells is operationally imposed when fracture stimulation was not contained in-zone and crossflow results between reservoirs.

The potential exists to combine flows from low-productivity zones with high- or low-productivity zones in a commingled well bore. Thus zones which are individually non-commercial may be combined with others to obtain commercial production rates. Although there are wells suitable for this type of completion, none have been attempted to date.

At this early stage of development in Elmworth/Wapiti, we have endeavored to bring onstream reservoirs of conventional quality as a simple consequence of cost. Future development will ultimately include some component of commingled

Table 3. Elmworth/Wapiti area typical gas composition (bcf @ 14.4 psia and 60°F).

Formation	Specific Gravity	C_1-C_4 (%)	C_{5+} (%)	H_2S (%)	CO_2 (%)	N_2 (%)
Chinook	0.691	96.97	0.66	0.00	0.38	1.99
Cardium	0.668	94.88	1.225	0.00	0.135	3.76
Doe Creek	0.684	97.47	0.70	0.00	1.05	0.78
Dunvegan	0.687	97.23	0.91	0.00	1.02	0.84
Paddy	0.656	98.54	0.43	0.00	0.64	0.39
Cadotte	0.626	97.65	0.42	0.00	1.28	0.65
Middle Notikewin	0.668	97.55	0.95	0.00	1.19	0.31
Basal Notikewin	0.661	98.05	0.37	0.00	0.74	0.84
Falher A Conglomerate	0.655	97.91	0.67	0.00	0.85	0.57
Falher A Sand	0.647	98.53	0.43	0.00	0.61	0.43
Basal Falher A Congl.	0.616	98.07	0.19	0.00	1.55	0.19
Falher B Conglomerate	0.640	98.22	0.60	0.00	0.76	0.42
Falher B Sand	0.630	98.04	0.50	0.00	0.87	0.59
Falher C Conglomerate	0.670	97.58	0.83	0.00	1.24	0.35
Falher C Sand	0.671	97.38	0.94	0.00	1.40	0.28
Basal Falher C Congl.	0.678	96.59	1.39	0.00	1.29	0.73
Falher D Conglomerate	0.701	97.85	1.14	0.00	0.62	0.39
Falher D Sand	0.670	97.95	0.83	0.00	0.98	0.24
Basal Falher D	0.612	97.93	0.20	0.00	1.14	0.73
Falher F Conglomerate	0.629	97.43	0.43	0.00	1.71	0.43
Falher F Sand	0.629	97.43	0.43	0.00	1.71	0.43
Bluesky-Gething Sand	0.624	97.41	0.24	0.00	1.96	0.39
Lower Gething Sand	0.628	96.26	0.23	0.00	2.60	0.91
Cadomin	0.656	95.60	0.59	0.00	3.07	0.74
Nikanassin Sand (Upper, Middle & Lower)	0.675	94.79	0.72	0.00	3.68	0.81
Charlie Lake	1.090	78.63	9.49	9.86	0.22	1.80
Doig Sand	0.602	95.74	0.29	0.74	2.85	0.38
Halfway Sand	0.651	85.81	1.14	9.37	2.60	1.08
Belloy Sand	0.642	93.29	0.16	2.56	2.91	1.08

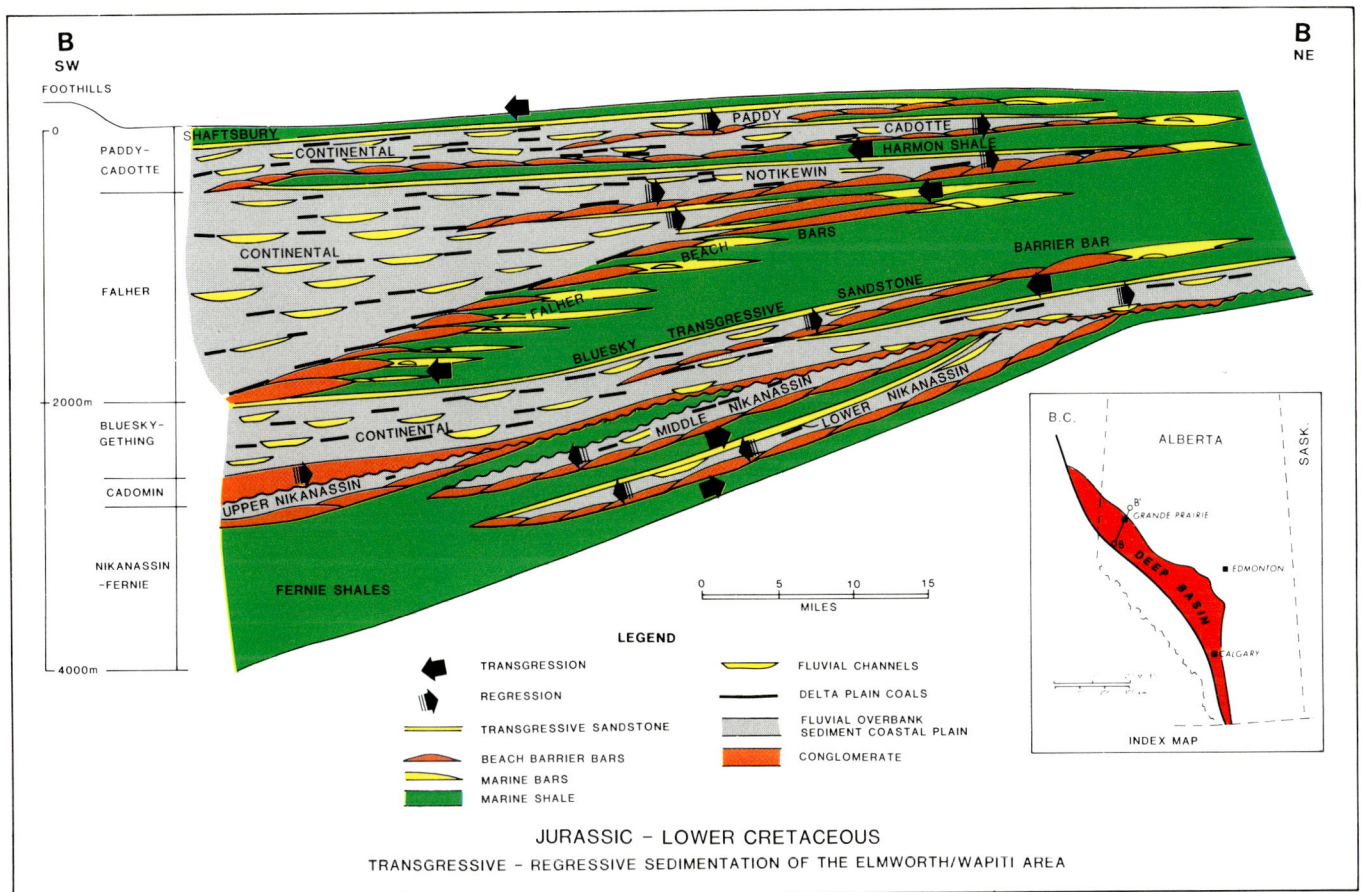

Figure 4. Illustrative cross section showing the transgressive-regressive sedimentation of the Jurassic-Lower Cretaceous rock of the Elmworth/Wapiti area.

Table 4. Shale volume, water saturation, and porosity cutoff used for reserve assessment in the Elmworth/Wapiti area.

Formation	Volume Shale	Water Saturation	Porosity Sand	Porosity Conglomerate	Sand I/C* Conglomerate
Chinook	0.3	0.7	0.07	0	0
Cardium	0.3	0.7	0.07	0	0
Doe Creek	0.3	0.7	0.07	0	0
Dunvegan	0.3	0.7	0.07	0	0
Paddy	0.3	0.7	0.08	0.03	0.04
Cadotte	0.3	0.7	0.07	0.03	0.04
Notikewin					
– Middle	0.3	0.7	0.07	0	0.04
– Basal	0.3	0.7	0.07	0	0.04
Falher A	0.3	0.7	0.07	0	0.04
Falher B	0.3	0.7	0.07	0	0.04
Falher C	0.3	0.7	0.07	0	0.04
Falher D	0.3	0.7	0.07	0	0.04
Falher F	0.3	0.7	0.07	0	0.04
Bluesky/Gething	0.3	0.7	0.05	0	0
Cadomin	0.3	0.7	0	0.03	0
Nikanassin	0.3	0.7	0.07	0	0
Charlie Lake	0.3	0.7	0.03	0	0
Doig	0.3	0.7	0.05	0	0
Halfway	0.3	0.7	0.05	0	0
Belloy	0.3	0.7	0.07	0	0

*Sand in contact with conglomerate. Net pay if: 1) shale content is greater than 30% or 2) Sw is greater than 70% or 3) ∅ is less than amount shown.

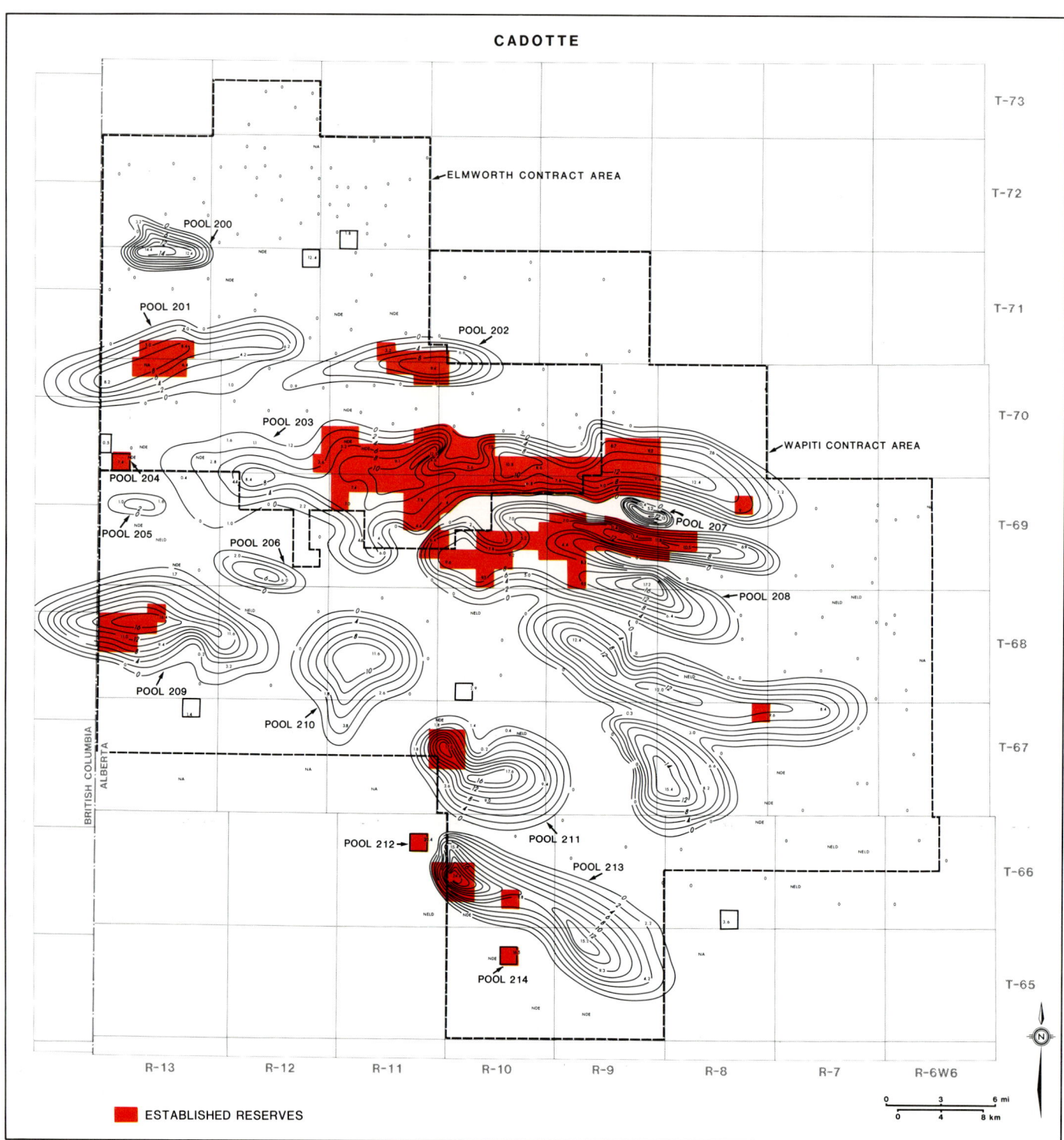

Figure 5. Cadotte net pay isopach and acreage assignments.

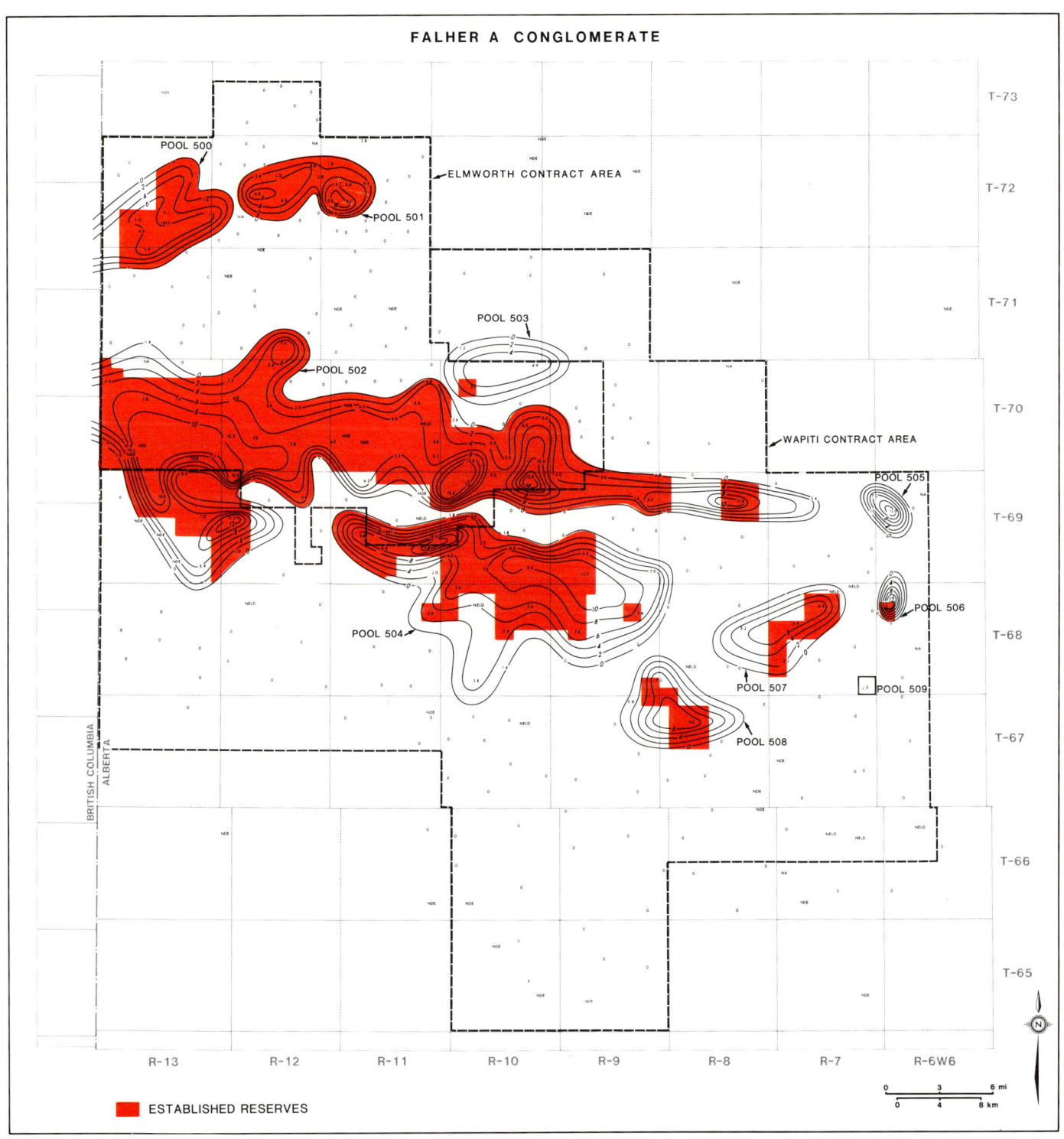

Figure 6. Falher A Conglomerate net pay isopach and acreage assignments.

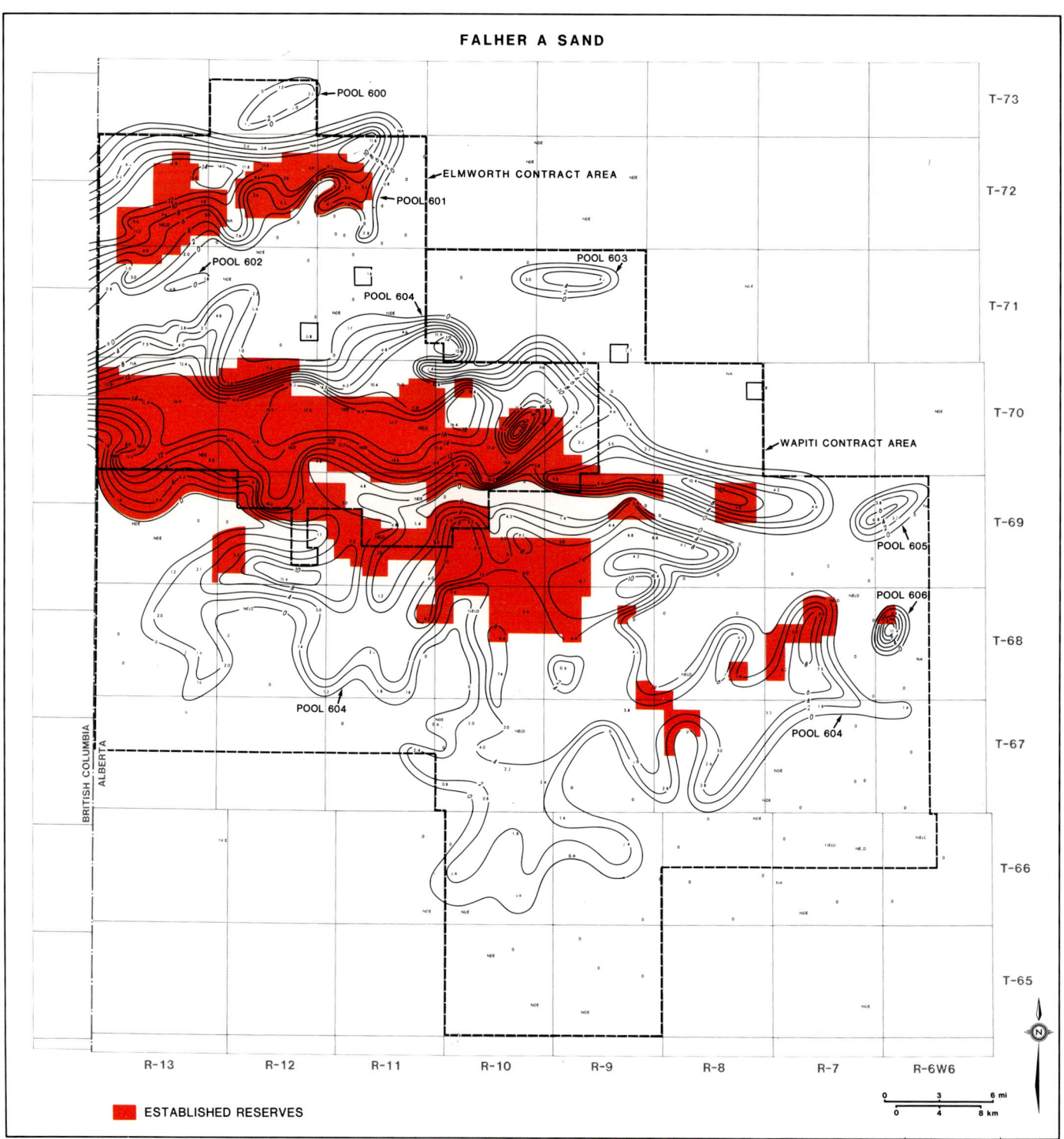

Figure 7. Falher A Sandstone net pay isopach and acreage assignments.

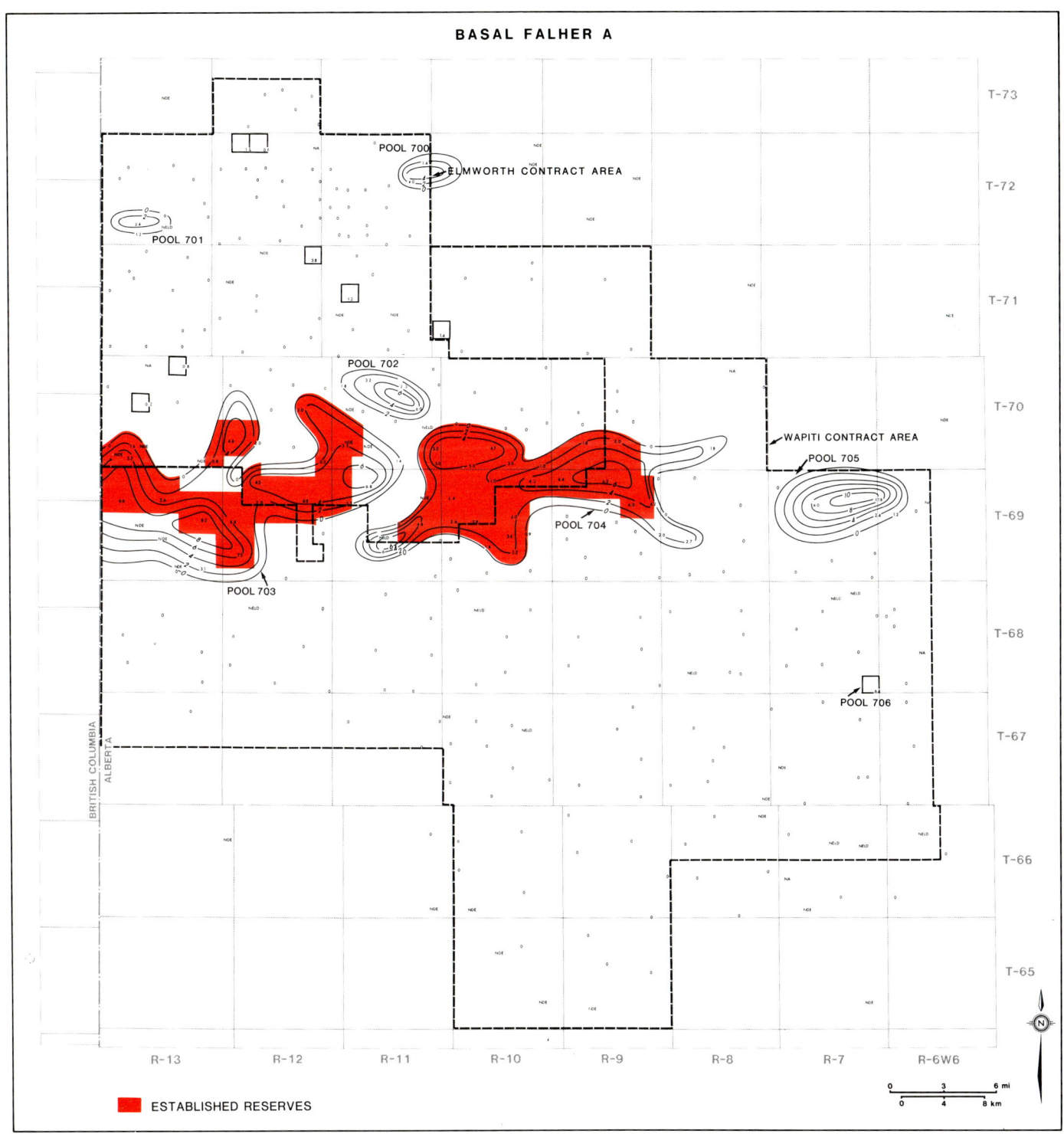

Figure 8. Basal Falher A net pay isopach and acreage assignments.

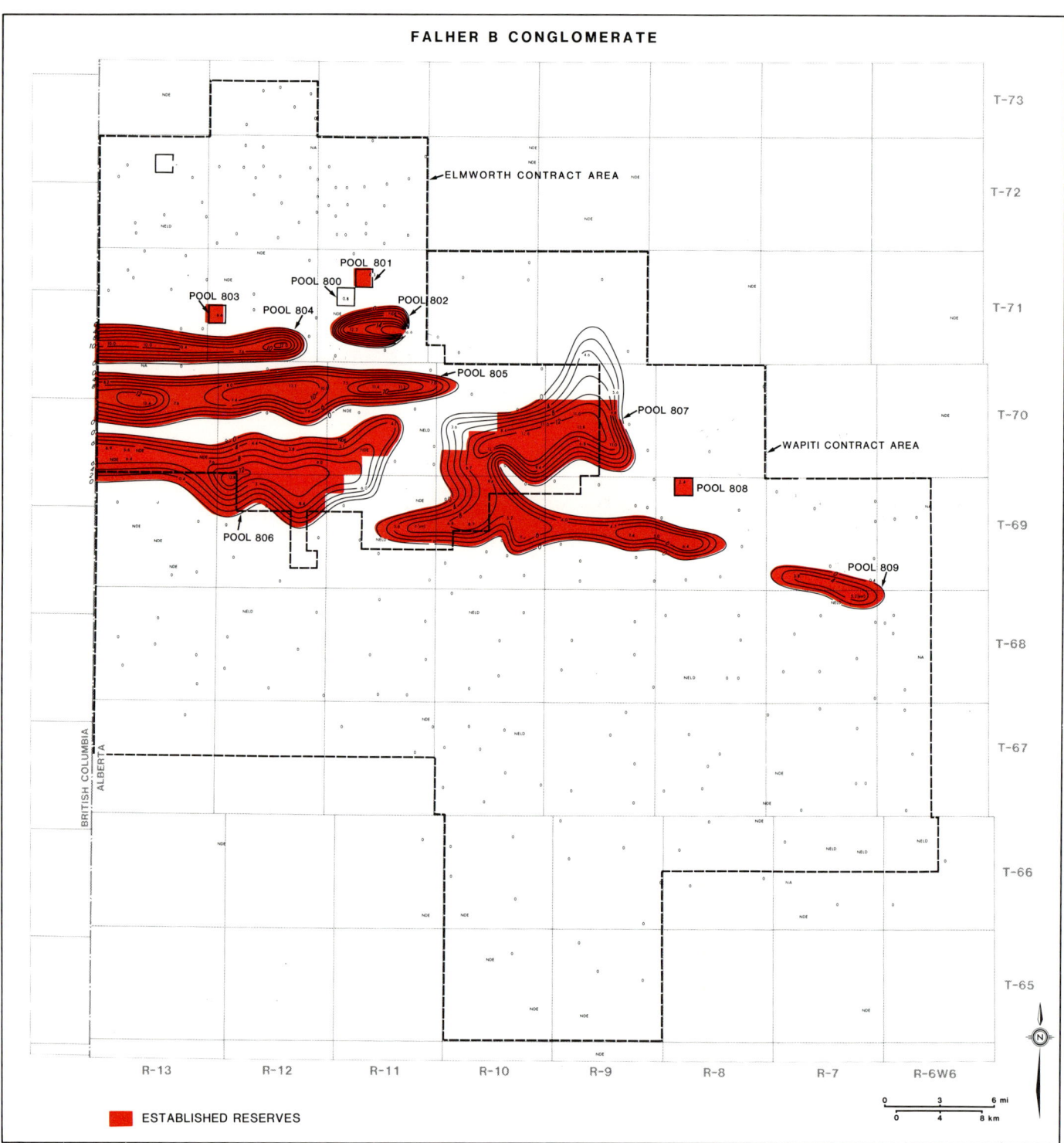

Figure 9. Falher B Conglomerate net pay isopach and acreage assignments.

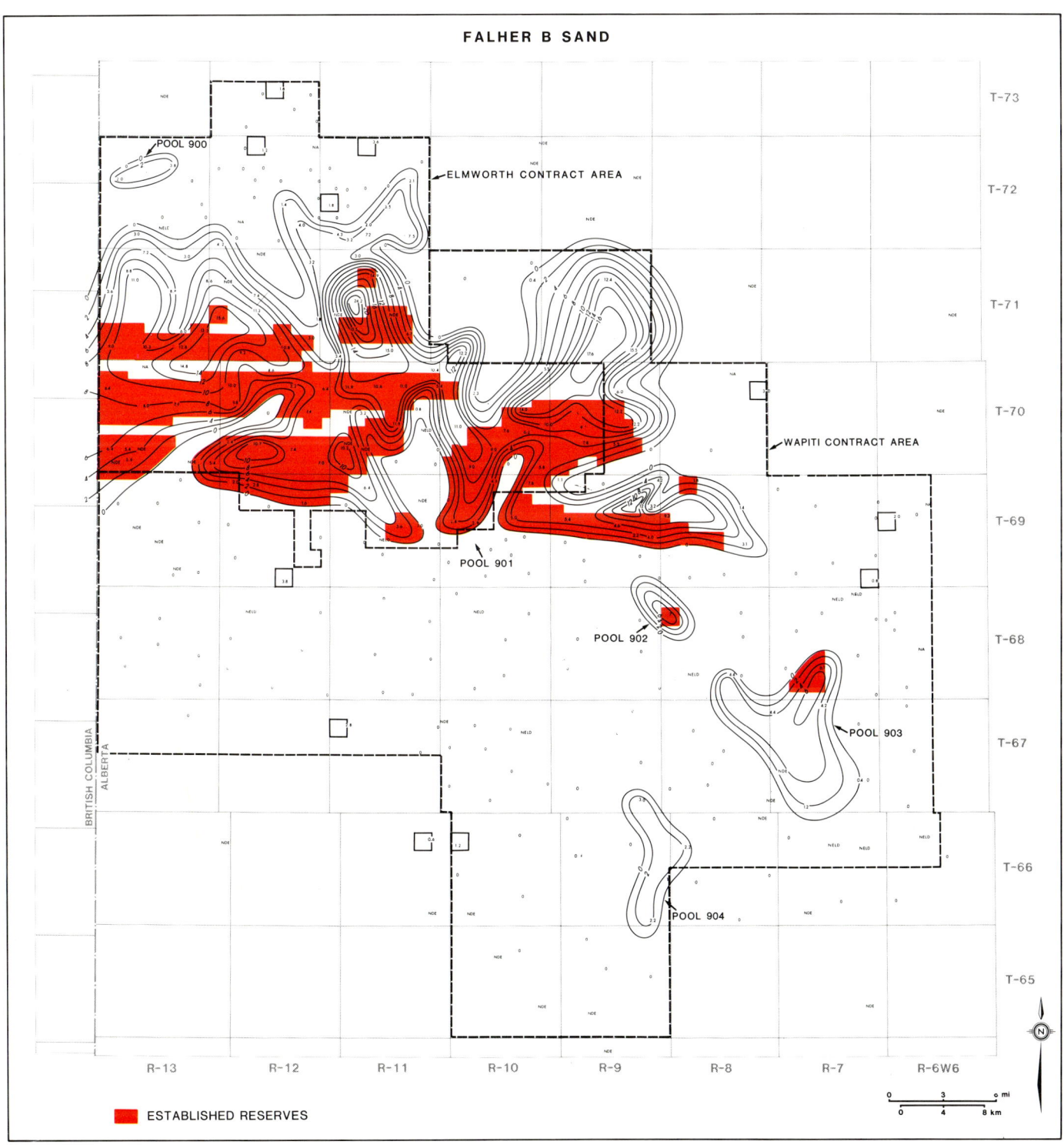

Figure 10. Falher B Sandstone net pay isopach and acreage assignments.

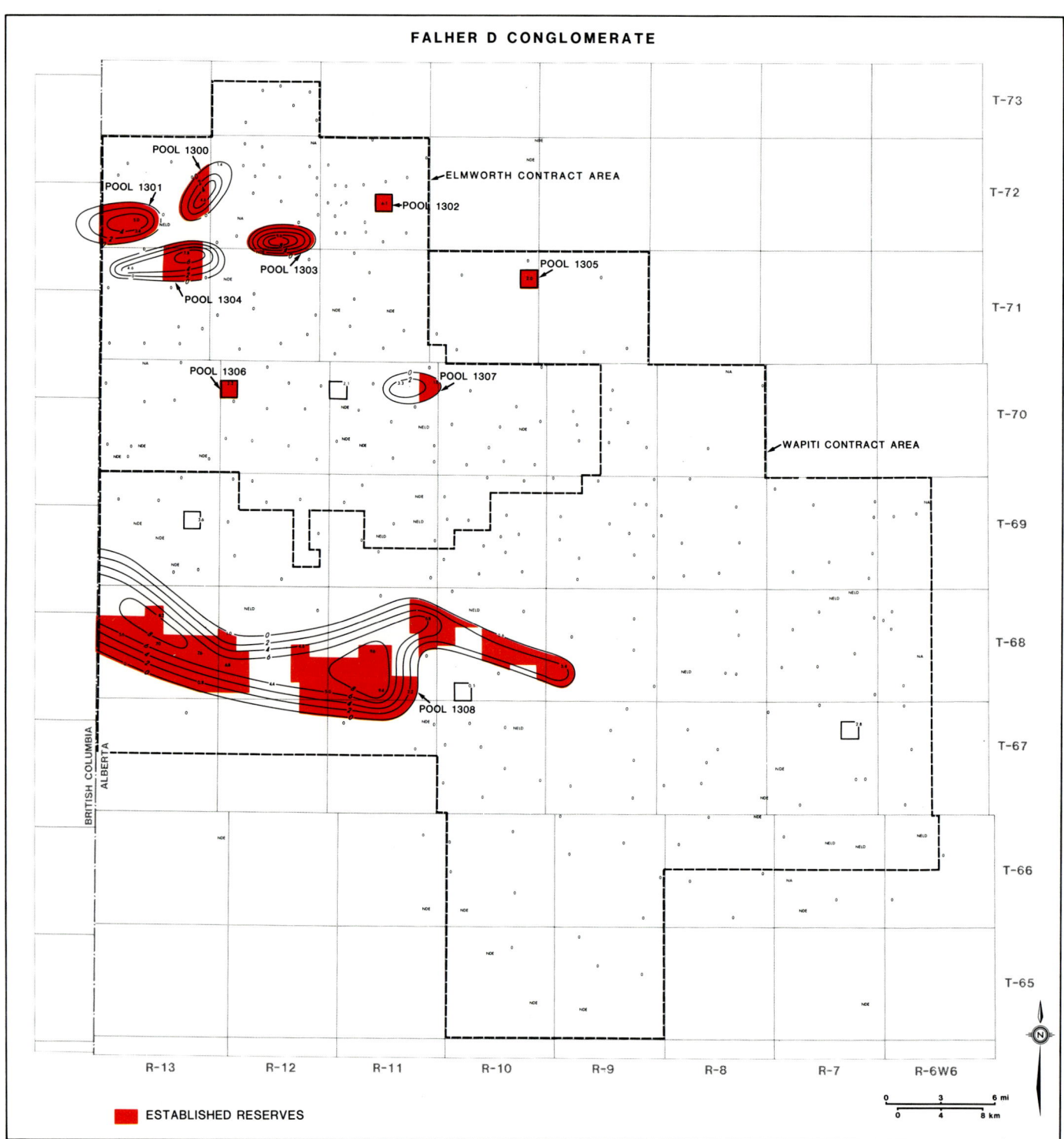

Figure 11. Falher D Conglomerate net pay isopach and acreage assignments.

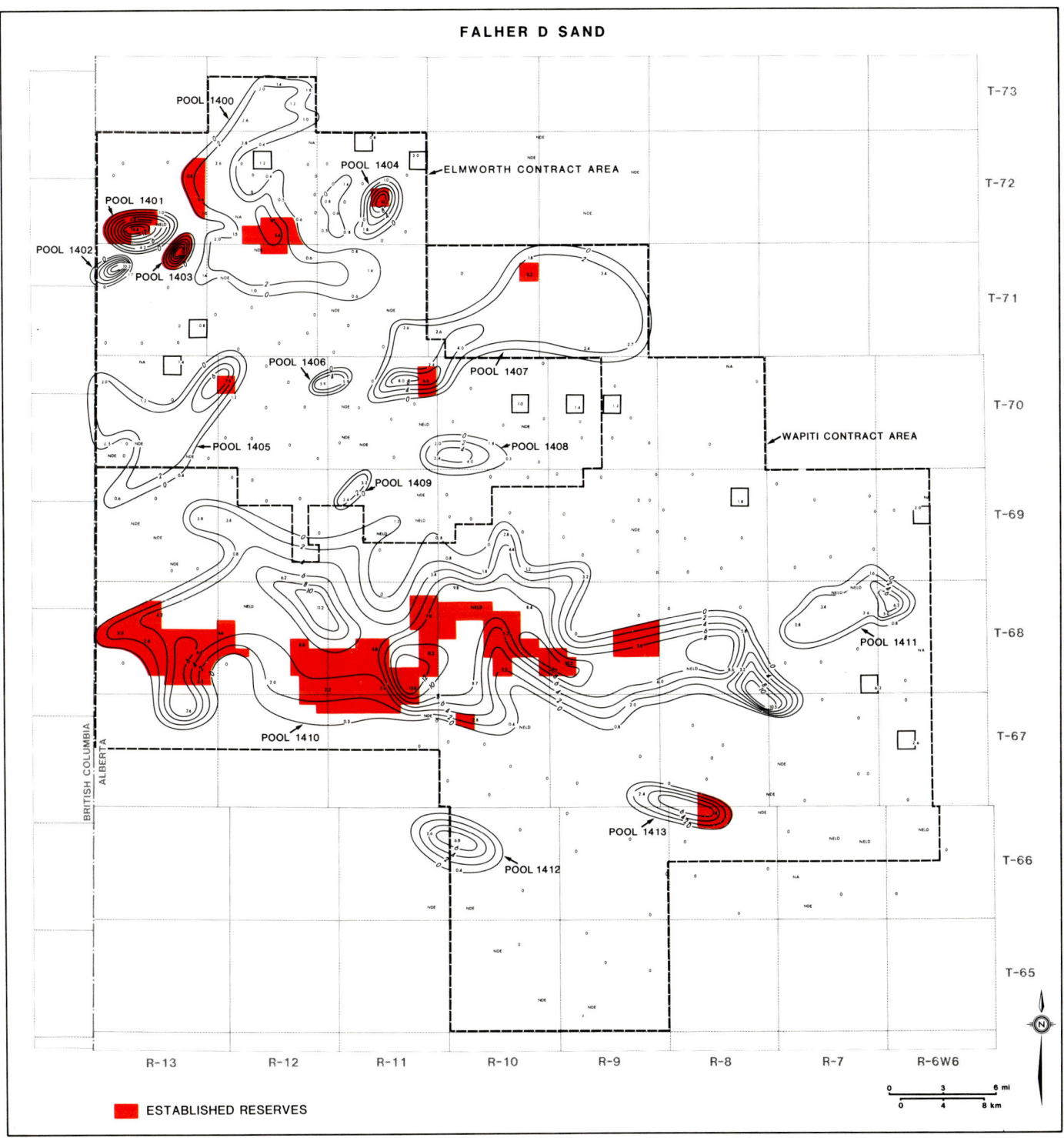

Figure 12. Falher D Sandstone net pay isopach and acreage assignments.

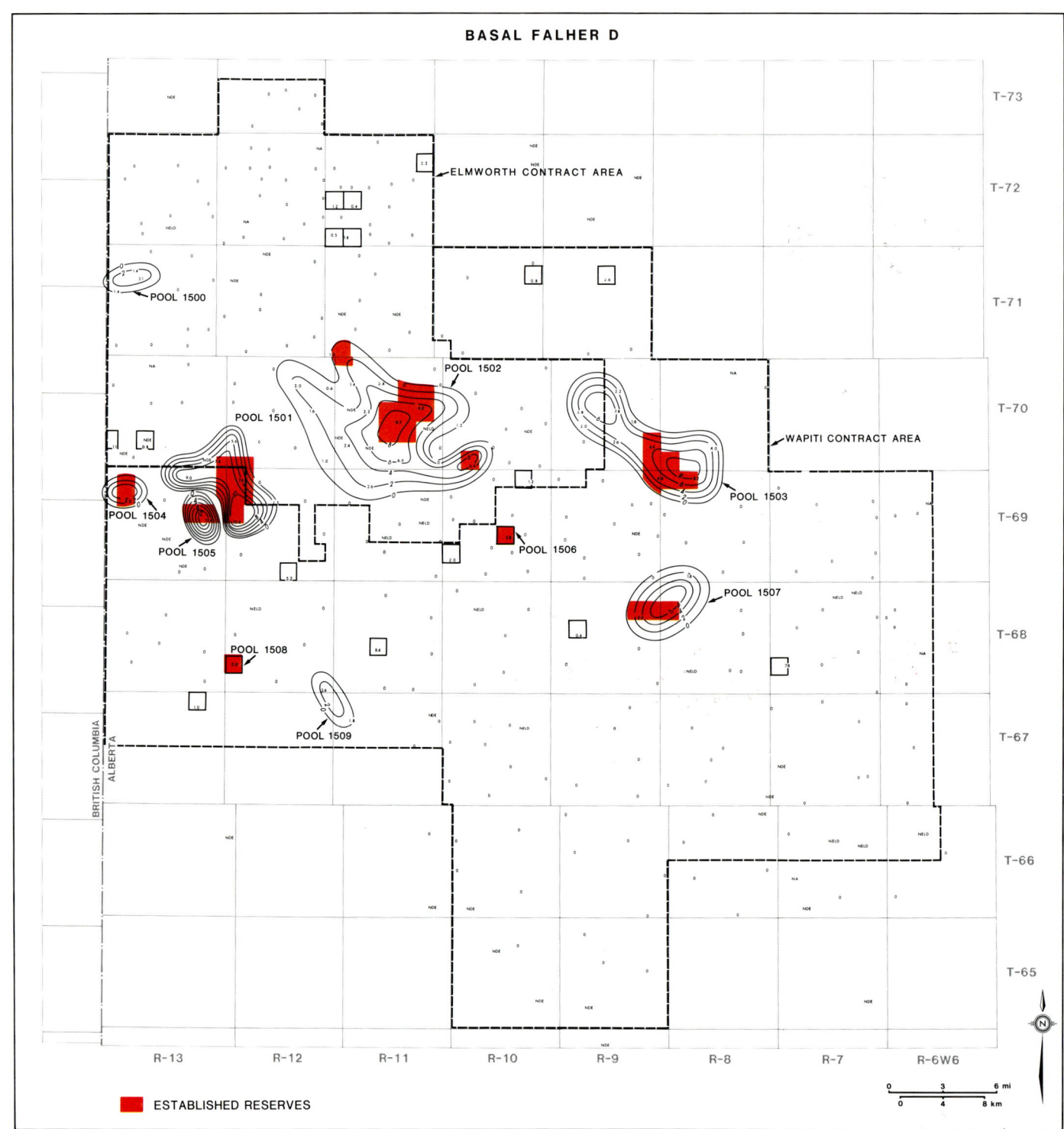

Figure 13. Basal Falher D net pay isopach and acreage assignments.

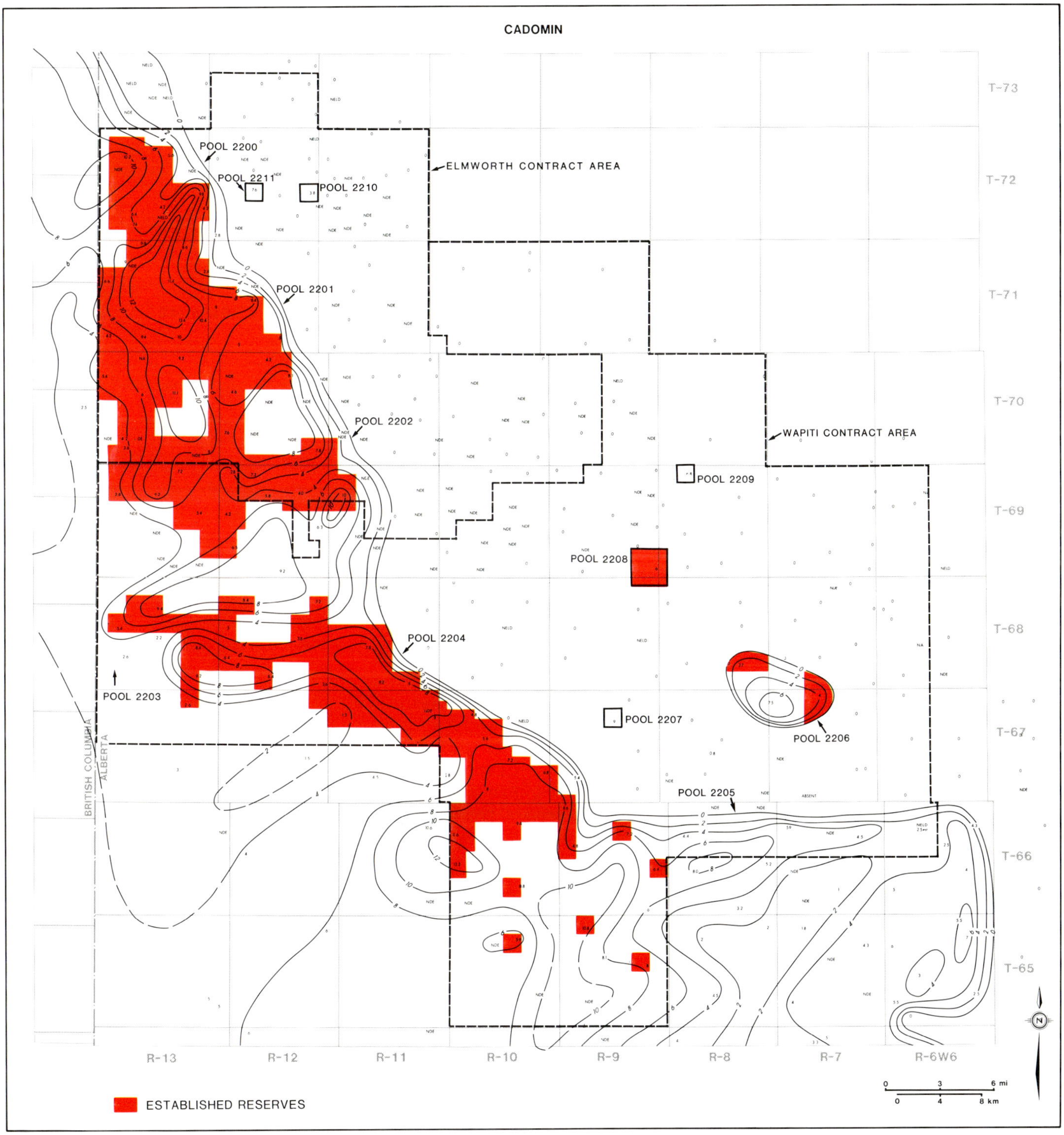

Figure 14. Cadomin net pay isopach and acreage assignments.

Table 5. Elmworth/Wapiti statistics.

Number of wells drilled	1,125
Number of wells cased	875
Cumulative production (1983 12/31 est.)	450 bcf
Number of gas plants	10
Total plant capacity	1,248 mcf/d

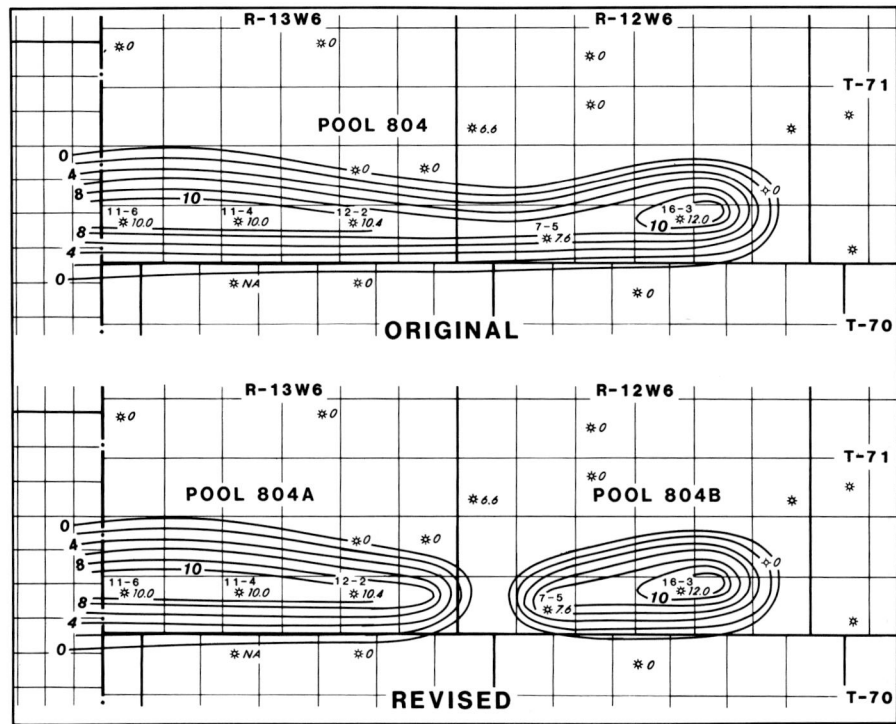

Figure 15. Two net pay isopachs of the Falher B Conglomerate #804 pool. First map reflects original mapping, second map shows influence of material balance after production.

completions to permit gas recovery from poorer quality reservoirs.

PRODUCTION AND PRODUCTION FACILITIES

Well Capacity

Well performance in the Deep basin covers a broad spectrum. High-productivity, conventional wells can be capable of producing as much as 100 mmcf/d from a high-permeability conglomerate sand, while at the other end, low-permeability tight sands, although gas charged, are incapable of producing at commercial rates into conventional gathering systems. High-productivity wells have permeabilities often in excess of 200 millidarcys. The Falher conglomerates produce sweet, fairly dry gas to 950 psig plant inlet.

Production of the tighter sands will depend on several factors. First, additional completion experience will be necessary to ensure minimum damage and effective stimulation. Second, it will probably be necessary to have further economic incentives to provide the producer with adequate return to pursue these tighter sands. And third, these wells will probably have to be produced into a lower pressure gathering system, such as the system currently in place for six wells in the Deep basin area.

The current low-pressure gathering system permits production from these wells at 90 psig wellhead pressure. The wells have been on production for approximately two years and the performance suggests that with the appropriate system sustained production can be assured from the tighter formations. The nine completions in five wells typically have permeabilities in the range of 0.1 to 2 millidarcys. Production rates vary from 120 mcf/d to 2,100 mcf/d per completion.

The very large volumes of tight gas will probably be produced through two primary mechanisms. First, horizontal flow directly to vertical well bores. This conventional method depends on our ability to stimulate the tight sands, and ensure unimpeded flow to the well bore. The second method, which may have equivalent volumes of gas available, is related to tighter sands vertically adjacent to the high permeability Falher conglomerates. Production from these sands will depend on pressure drawdown within the conglomerates to allow the gas to seep vertically into the high-permeability conglomerate "pipelines." Reservoir engineering studies have demonstrated that this phenomenon will occur and in fact result in high ultimate recoveries from the sands vertically adjacent to the conglomerates. To date insufficient field performance has been experienced to verify this phenomenon. However, additional cumulative production and special pressure surveys should improve our understanding.

To access the very large resource of gas available in the Deep basin, ongoing geological and engineering studies will continue. Based on experience to date and using the categorization of reserves shown earlier, only 1% of the resource is now included in the proven category.

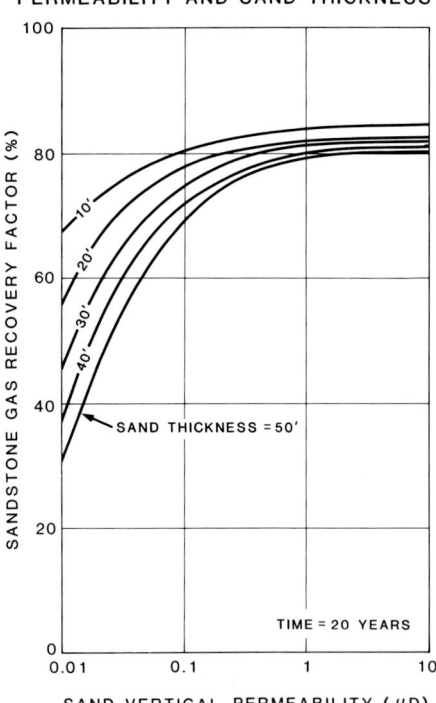

Figure 16. Graph showing sandstone gas recovery factors as a function of vertical permeability and sand thickness.

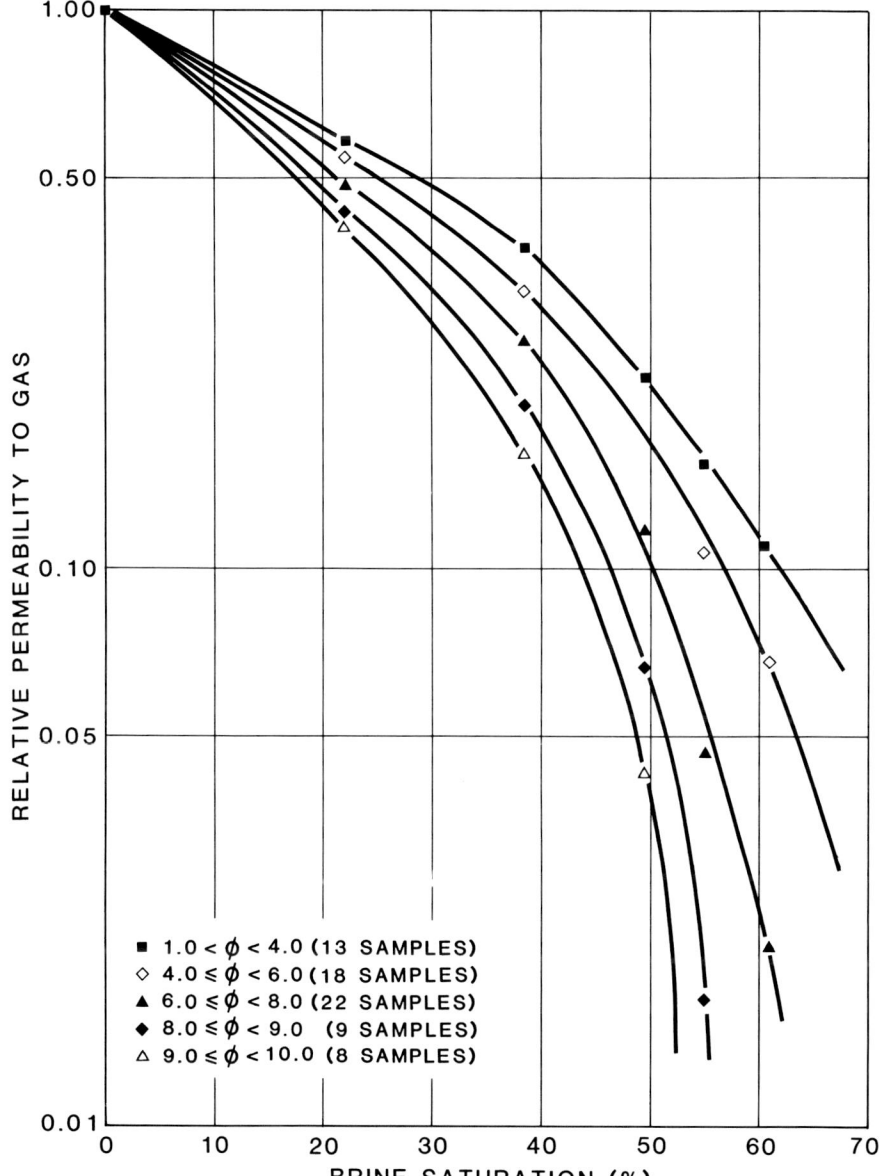

Figure 17. Graph showing relative permeability as a function of brine saturation and porosity in the Falher A sandstone.

Plant and Gathering System

The Deep Basin System—Approximately 1,125 wells have been drilled to date in the total field outline. Many of these wells have been tested in an average of more than two zones. Some 200 gas wells have now been tied-in and are capable of production. Approximately 650 drilled, cased, and completed wells are waiting on the need for additional deliverability.

Total plant capacity in the Deep basin is approximately 1.25 bcf/d from ten plants.

The ten plants vary in size from 10 mmcf/d to the largest which is the Elmworth plant with a capacity of 450 mmcf/d.

Elmworth Contract Area—Canadian Hunter is the operator and prime interest holder in the Elmworth Plant. The plant has a capacity of 450 mmcf/d through three trains. The plant employs a dessicant process that removes water and processes the gas to pipeline specs.

Approvals are now completed for the next phase of the Elmworth plant development. A deep cut facility will be installed to extract natural gas liquids from the raw inlet gas. Liquid yield will be some 55 barrels from each million cubic feet of processed gas. Based on the ultimate possible reserves of 17.1 tcf, the volume of liquids could be a high as 1 billion barrels. This product will be sold through normal liquid marketing outlets.

Field Performance—The field performance to date has been as expected. Approximately 450 bcf of gas has been produced to date. This represents only 6% of the established reserves. Detailed pressure surveys are taken on a regular basis on producing wells and many of the surplus, not currently tied-in wells. This data will allow us to simulate the performance of the field, better estimate the reserves, and optimize the timing of additional compression and drilling.

The pressure data confirms good geological and pressure continuity between most of the conglomerate wells.

Figure 18 shows the gas production performance of wells within the Deep basin. First significant production commenced in 1979 and then expanded with the development of plants and reserves. It should be noted that peak annual production is significantly higher than annual average.

Peak daily production from the Deep basin occurred in December 1983 at close to 1 bcf/d, responding to the very high demand caused by severe winter weather.

SUMMARY

Gas reserves and production found in the Elmworth/Wapiti area of the Deep basin have been extremely impressive to date. From 1976 to 1983, this area has made a significant contribution to the energy supply of North America. Under today's gas economics, 7.5 tcf of established, contractible gas has been developed. Further drilling and testing within net pay pools is expected to convert a large portion of 9.6 tcf from potential contractible to established contractible reserves. Additional exploratory drilling will add reserves outside the presently mapped pools.

Technology and facilities have already tapped this large resource for deliveries of

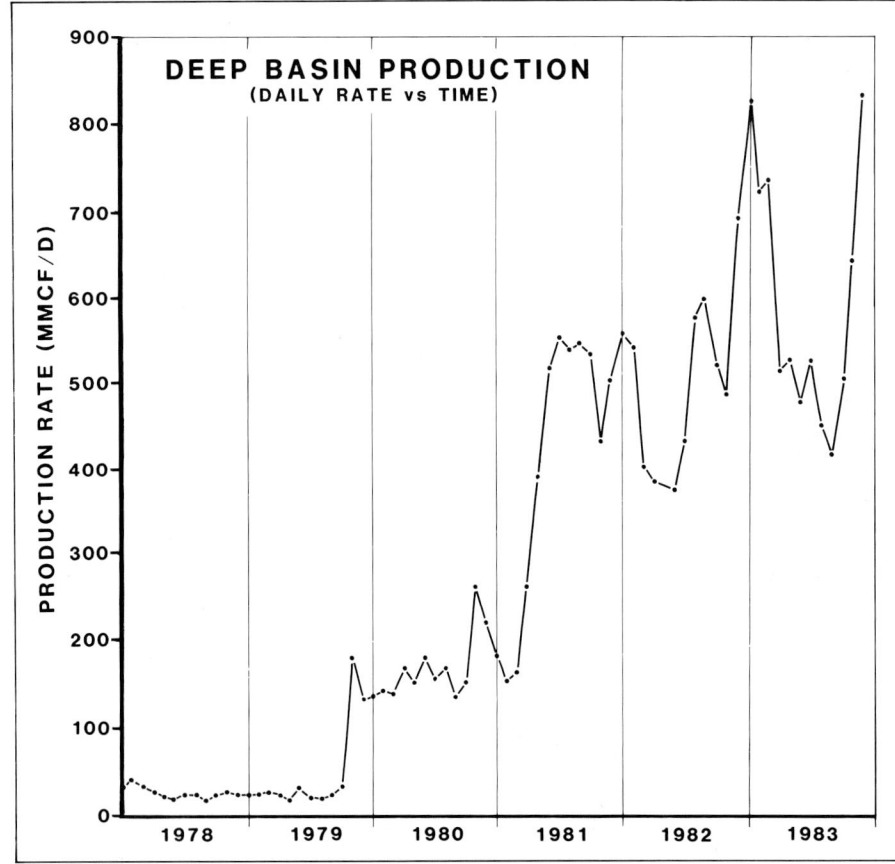

Figure 18. Graph showing Deep basin gas production, 1978-83.

gas to a maximum of 1 bcf/d, and the only obstacle in the way of increasing this number is economic gas sales opportunity. Just how much of the 800 tcf resource will be converted to deliverable reserves will take many years to determine. The potential of this area unquestionably establishes it as a gas giant.

REFERENCES CITED

Masters, J. A., 1979, Deep basin gas trap, Western Canada: AAPG Bulletin, v. 63, p. 152–181.

———, 1984, Lower Cretaceous oil and gas in Western Canada, in J. A. Masters, ed., Deep basin gas: AAPG Memoir 38, this volume.

Smith, D. G., C. E. Zorn, and R. M. Sneider, 1984, The paleogeography of the Lower Cretaceous of western Alberta and northeastern British Columbia in and adjacent to the Deep basin of the Elmworth area, in J. A. Masters, ed., Deep basin gas: AAPG Memoir 38, this volume.

Gas Resources in Elmworth Coal Seams

R.E. Wyman
Canadian Hunter Exploration Ltd.
Calgary, Alberta

Abundant coal seams occur in the Lower Cretaceous section of the Elmworth area. Gas desorbed from pressurized cores of coal indicates there are about 500 cu ft of methane per ton of coal. In addition to being a significant source for gas in the Deep basin, the coalbeds themselves contain about 50 tcf of gas in place. It is probable that some of this gas can be recovered through processes of diffusion from the matrix and Darcy flow in natural fractures. Where coal is adjacent to producible sands or conglomerates, mathematical modeling shows that at least half of the gas contained in the adjacent coal can be recovered. Additional gas may be recovered from isolated coal seams; further field testing will determine this potential.

INTRODUCTION

During Early Cretaceous time, the "Deep basin" of northwestern Alberta underwent periodic regression and transgression of the epicontinental sea. This resulted in the accumulation and preservation of several, widely spread peat beds. With time, heat, and pressure these peat beds eventually formed the coal seams which are found in abundance in the Lower Cretaceous interval between the Paddy and Nikanassin formations. Most of these coal seams cap upward-coarsening nearshore-to-shoreline beaches and deltaic sequences. The seams vary in thickness from less than 1 to about 8 m (3.3 to about 26.3 ft).

During the early development of the Elmworth field, most of the attention was given to the gas-productive potential of the sandstones and conglomerates. However, it was also recognized that the coals held large potential for gas. Studies were undertaken to quantify this potential and identify commercial opportunities within the Elmworth area.

GENERAL CHARACTERISTICS OF METHANE AND COAL

Methane may be derived from coal by one of three basic processes:

1. Synthetic Natural Gas (SNG) from gasification plants. Coal may be mined and subsequently gasified to synthetic natural gas in plants located at the surface.
2. In situ gasification. Coal may be gasified by injecting air or oxygen in one hole to sustain a burn front. Gas is recovered from a second nearby hole. With injection of air, this gas will have a low BTU content.
3. Demethanization of coal seams. This is the process that is believed to be most feasible for the Elmworth area and will be discussed further.

Methane resides in the coal seams as free gas in larger pores and fractures, and as gas physically adsorbed onto the coal surfaces (Curl, 1978). Most of the gas will be adsorbed on the large surface area in coal. A pound of coal may have about 500,000 sq ft of surface area which is available to adsorb methane. The sorption isotherm shown in Figure 1 indicates how gas will desorb as pressure is decreased. Gas in fractures and large pores makes up the balance of the total. The amount of gas that is adsorbed in a particular coal seam will depend upon the depth and pressure, the rank, the moisture content and the temperature. Variations of methane adsorption with coal type are shown in Figure 2. Increased temperatures will reduce the amount of gas that can be adsorbed in a particular type of coal (Fig. 3).

Increased moisture content will decrease the adsorptive capacity of coal to a certain point. In the case of the Pittsburgh coal shown in Figure 4, increasing the moisture content from 0 to 2% caused the volume of adsorbed methane to decrease by one-third. Above 2%, the moisture content did not seem to have a significant effect. Of course, if the moisture is high enough to have free water in the fracture system this will be another impediment to production.

The nature of the coal matrix is such that the flow of methane may be limited by the ability of the methane molecules to pass from one micropore to another. Studies of the internal structure of coal indicate that pores of about 40 Å or less are connected by passages of 5 to 8 Å. Since methane molecules may be only 4 Å in diameter this would be just barely large enough to allow diffusion from one micropore to another. The rate at which the methane molecules will migrate through the coal is described by the diffusion coefficient which may vary from 1.0×10^{-13} sq cm/sec to 5.0×10^{-5} sq cm/sec. The diffusion process is described by Fick's law or other diffusion models (Smith and Williams, 1982). The amount of time needed for 50% of the gas to diffuse out of a one-half centimeter piece of coal will vary from a few days to thousands of years depending upon the diffusion coefficient. It then becomes obvious that in addition to the diffusion coefficient it is important to estimate the size of coal particles between natural fractures. The nature and frequency of fractures will control the transmissibility of natural gas in coal. When methane migrates from the matrix to a larger pore or fracture it will then flow according to Darcy's law. A mathematical simulation of the flow processes will be discussed later.

ANALYSIS OF ELMWORTH COAL

A core taken with a pressurized core barrel was recovered from the "Fourth Coal" in Elmworth 15-16-68-13 W6 (Fig. 5). After the core reached the surface (still under in situ pressure), it was bled down under controlled conditions; the gas bleed-off was carefully measured. The coal was then transferred immediately to sealed containers; gas was allowed to evolve and was measured for the next 200 days. Finally the coal was pulverized to measure any residual gas. Assistance for the analysis of the coal was given by the Colorado School of Mines Research Institute (Larsen and Knowles, 1982).

Gas Content

These results indicated that the Fourth Coal is rich in methane with an indicated content for the clean coal of over 500 cu ft per ton (Table 1).

The shaly coal and coaly shale were found to have a much lower retained gas content. For instance, the ash content, which reflects the shaliness, is four to five times higher for sample 26 which had only 347 cu ft per ton compared with samples 24 and 25 which had over 500 cu ft per ton (Tables 1 and 2). Discounting nitrogen, some of which may have come from the core barrel or cannister, most of the gas desorbed from the coal was methane (96%) with 1 to 2% ethane (Table 3). It is interesting to note that with increased time, traces of heavier hydrocarbons start to appear in the analysis. This data indicates that the appearance of the heavier molecules is slightly retarded with time, giving a chromatographic effect during the desorption process.

The rate of gas desorption from the Fourth Coal sample is shown in Figure 6. After about 90 days the rate of effluent had decreased to a very small amount. In order to see if a freshly exposed surface from the center of the coal would result in additional gas emission the core was cut in half and replaced in the sealed cannister. As seen in Figure 6 there was a small increase of about 3% in gas emissions, an insignificant volume compared to the amount of gas that evolved in the first month. Therefore it would appear that under these conditions (non-stressed) diffusion would not be a limiting factor to gas production.

Diffusion Coefficient

Laboratory measurements of diffusion coefficient indicated a value for the Fourth Coal between 1.0×10^{-8} and 1.0×10^{-11} sq cm/sec; this falls within the range

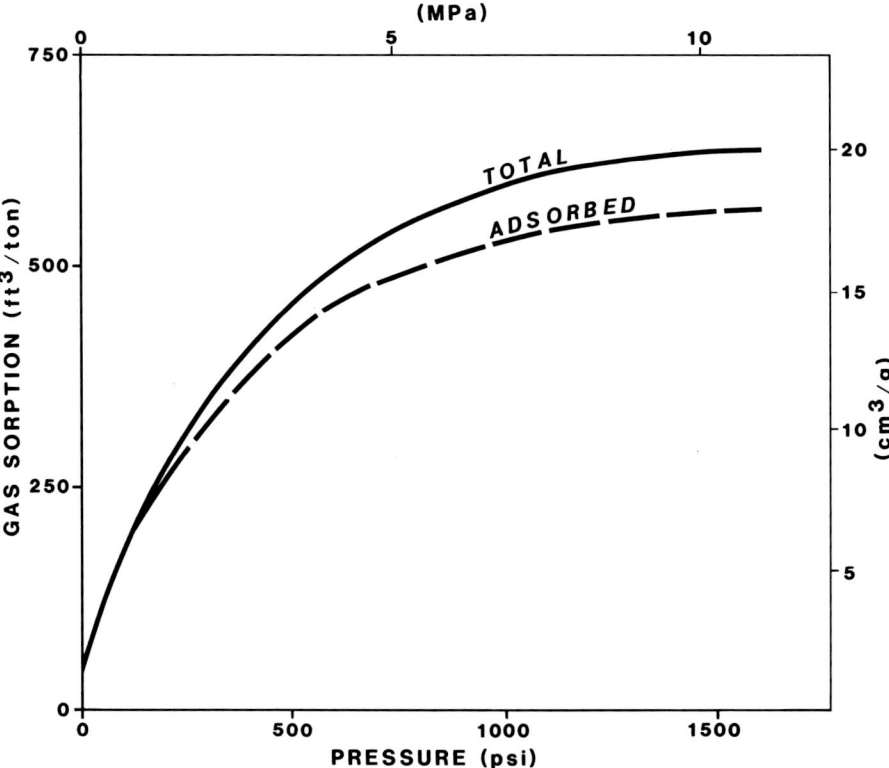

Figure 1. Sorption isotherm for typical coal.

Table 1. Gas content of Fourth Coal, Elmworth 15-16-68-13W6.

		Sample No. (cu ft/ton)		
		24	25	26
1.	Lost in transfer of core to container	11	13	19
2.	Desorption in 200 days	418	450	262
3.	Residual (estimated)	85	90	66
	TOTAL:	514	553	347

Table 2. Analysis of coal samples, Elmworth 15-16-68-13W6.

Sample #	Sample No. (cu ft/ton)		
	24	25	26
Core depth (meters)	2608.7	2608.9	2609.4
Permeability (μd) [at in situ cond.]	0.2	0.1	0.1
Moisture content (%)			0.92
Pressure (psi) [estimated only]	3500.0	3500.0	3500.0
Vitrinite reflectance (%)	1.63	1.65	1.43
Density (gm/cu cm)	1.3	1.3	1.6
Ash (wt. %)	3.7	4.2	17.4

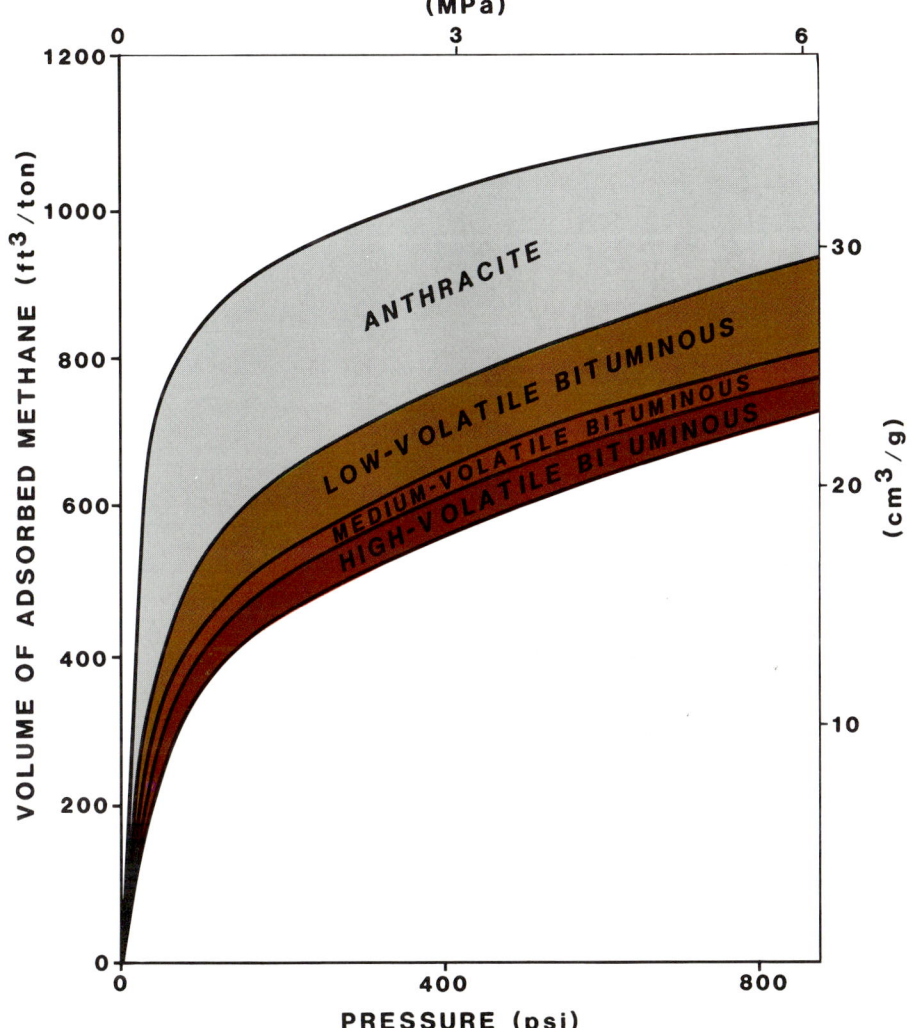

Figure 2. Adsorption of coal types as a function of pressure.

Table 3. Gas composition (Vol. %), Elmworth 15-16-68-13W6.

Time	15 Days	34 Days
Methane	96.78	96.11
Ethane	1.50	2.04
Propane	0.021	0.031
Iso-Butane	0	0.00082
N-Butane	0	0.00057
Iso-Pentane	0	0.000013
N-Pentane	0	0
CO_2	1.64	1.51
H_2	0.0016	0.0013
SO_x	0.053	0.016

of reported values for methane in other coals. An interesting observation of the tests done on the Fourth Coal was that at higher pore pressures the diffusion coefficient appeared to increase (Doherty, 1982). The highest pore pressure measured in the laboratory was 1,500 psi. This infers that at in situ conditions in Elmworth the diffusion coefficient might approach 1.0×10^{-5} sq cm/sec. Later in this paper mathematical simulations that cover this possible range of diffusion coefficients will be presented.

Permeability

In addition to the diffusion coefficient and particle size, the transmissibility or permeability of the coal to gas is very important to recovery of gas. As with sandstone or carbonate rocks, reasonable permeability is essential to commercial recovery. Laboratory measurements of permeability are very difficult to obtain because of the tendency of coal to chip or disintegrate while preparing the sample. After selecting a sample that will allow a permeability plug to be cut one must be concerned whether it really represents the typical in situ condition. It is possible that the laboratory measurements may represent the low end of the permeability distribution since plugs that fall apart (and therefore may be more fractured) are not measured. At least one hopes this argument will prevail after seeing the results of several coal samples from the Fourth Coal. Under simulated in situ stress conditions the permeabilities ranged from 0.01 to 0.2 microdarcys. While this would be adequate permeability to drain gas from coal into adjacent permeable sands or conglomerate formations it would be too low to be of commercial interest in isolated coal seams.

Moisture Content

As seen in Table 2 the moisture content is less than 1%. Figure 4 indicates that this would be a relatively dry coal and could contain about 30% more gas than would be found in a coal with 2% or greater moisture content. This may imply that the fracture system is dry, but this has yet to be documented.

GEOCHEMISTRY OF ELMWORTH COAL

Measurements of thermal maturity indicate this coal to be very mature to severely altered. Vitrinite reflectance of the coals from the 15-16 well indicate values ranging from 1.25 to 1.91, a low-volatile bituminous coal. As can be seen in Figures 2 and 3, the relative volume of adsorbed methane should be high for such a thermally mature coal. This is in agreement with measurements of thermal maturity for coals of this age over a broad area in Elmworth. As can be seen in Figure 7 the maturity as indicated by the vitrinite reflectance generally increases in a southwesterly direction (Weiss, 1983). The main generation of gas from coals will come from those with a vitrinite reflectance over 1.0.

It is well known that coals are powerful generators of methane (Tissot and Welte, 1978) because the quantity of organic carbon in coal may reach 95%; a rich source shale will rarely exceed 5%. Over the lifetime of coal, it may generate over

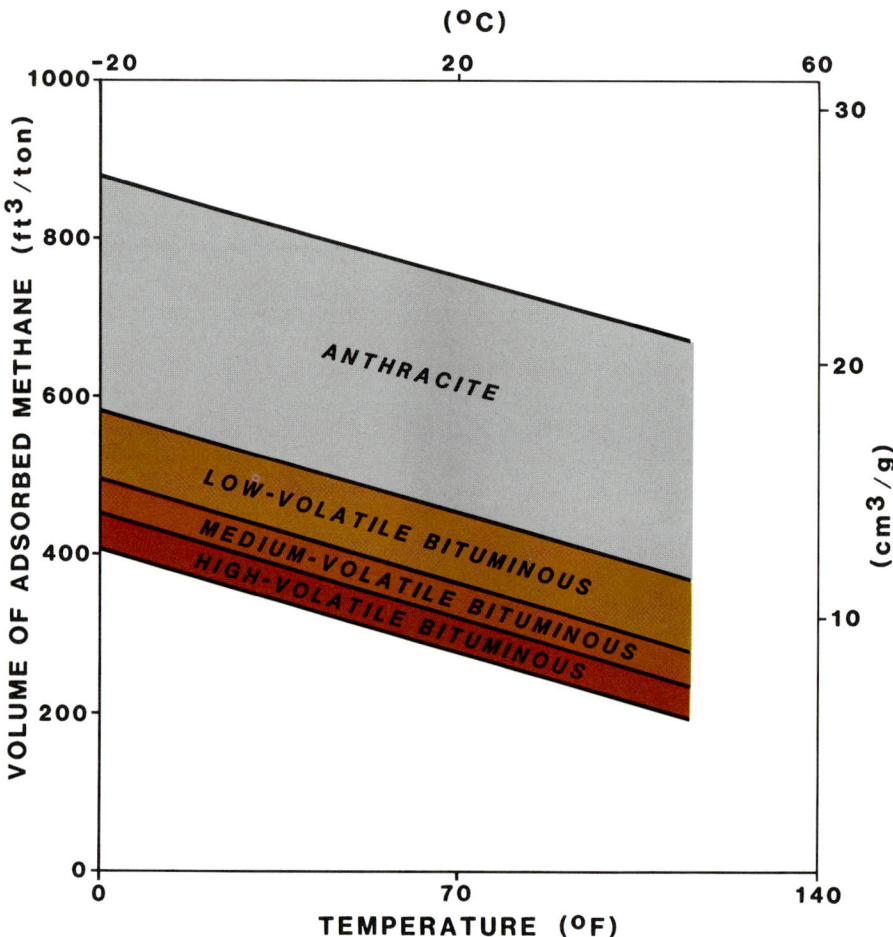

Figure 3. Adsorption of coal types as a function of temperature.

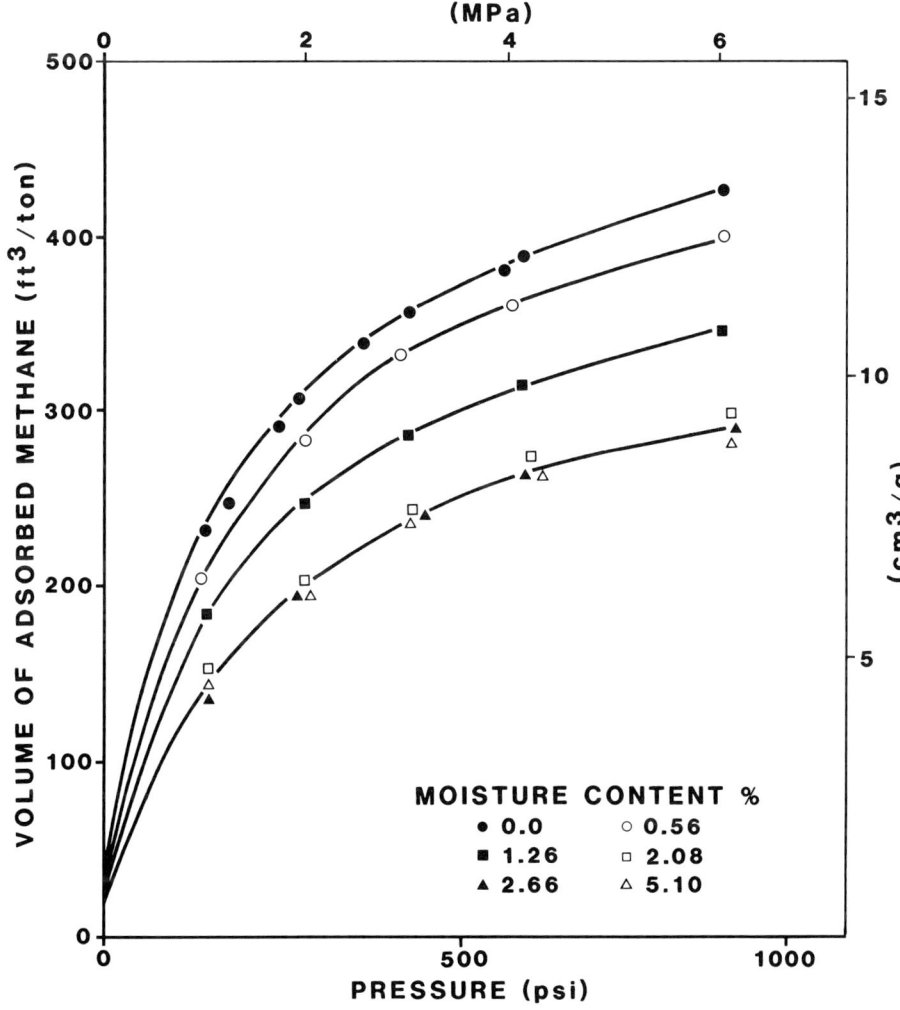

Figure 4. Effect of moisture upon adsorption for a Pittsburgh Coal.

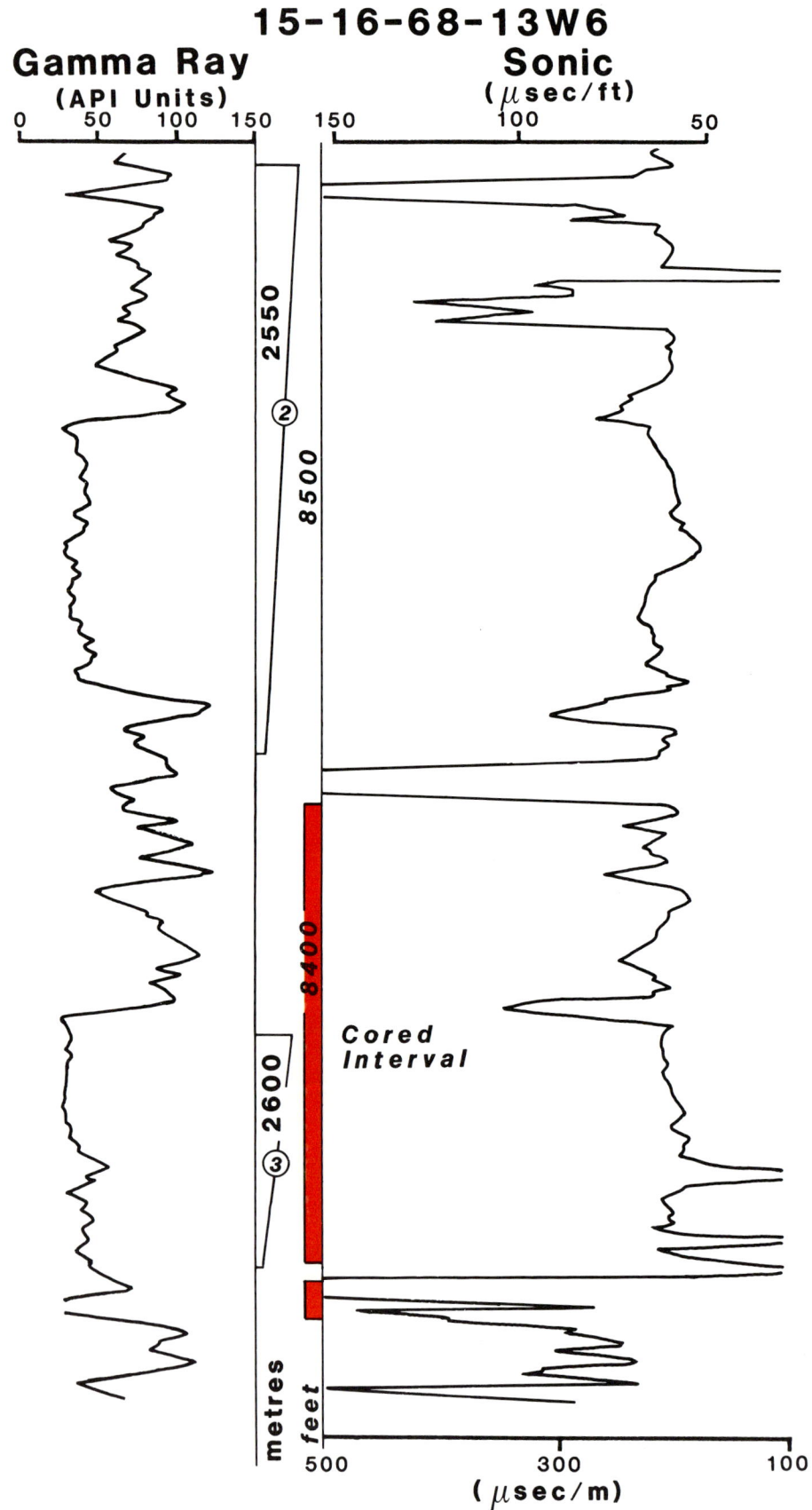

Figure 5. Log response of coal recovered from Elmworth well 15-16-68-13W6.

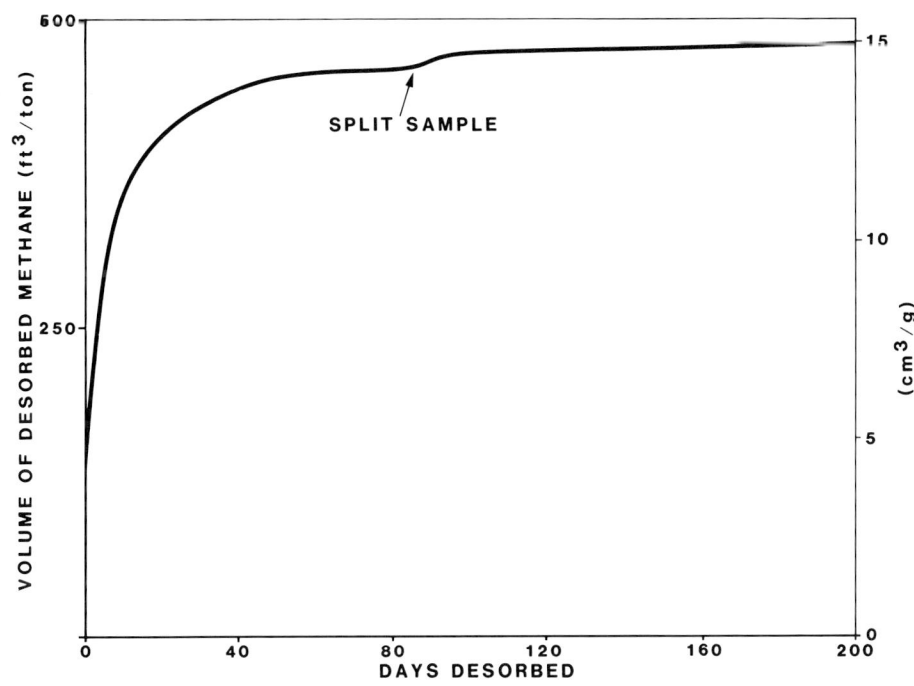

Figure 6. Desorption of gas from pressurized coal.

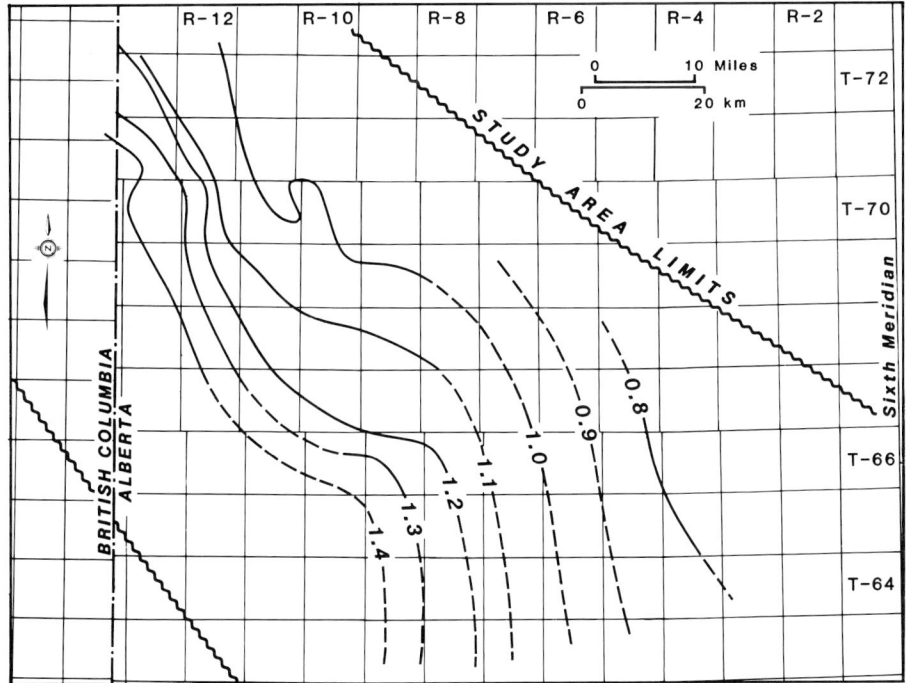

Figure 7. Vitrinite reflectance of "Fourth Coal."

3,000 cu ft per ton. However, at any one time only about 10 to 20% will be found in the coal depending upon the variables discussed previously. It is reasonable to assume that the Elmworth coals along with the organic shales have expelled enormous amounts of gas over geological time (Masters, 1984).

PETROGRAPHIC ANALYSIS

Insight into the storage capacity and transmissibility of coal can be achieved by means of the Scanning Electron Microscope (SEM). The gas storage capacity of coal is a function of three basic pore types: matrix porosity, phyteral porosity, and fracture (cleat) porosity.

Matrix Porosity

Most of the gas in coal is stored in the matrix system or as adsorbed molecules. This system is too small to be observed with the standard SEM. The laws of diffusion control the movement of gas through the matrix.

Phyteral Porosity

Phyterals are coalified plant fossils and are easily observed under both binocular and scanning electron microscope magnification. An example of phyteral porosity under non-stressed conditions is shown in Figures 8A and 8B. Phyterals can have 50% pore space. However, this type of pore probably will be significantly reduced under in situ stress. Although phyterals may contribute significantly to porosity it is doubtful they help the overall permeability.

Fracture Porosity

The fracture system is the most important for the transmissibility of gas in coal. Usually three sets of cleat orientation can be observed. The spacing between cleats and the width of fractures that are permeable are the most important factors controlling gas production from coals. Fracture surfaces are frequently seen with mineralization. An example of mineralization that appears to be porous and probably permeable is seen on the horizontal plane of Figures 9A and 9B. In this case the natural fracture appears to have been propped open, a microversion of a sand-propped hydraulic fracture. Some fracture surfaces have no mineralization and may represent fractures caused by coring or handling. A

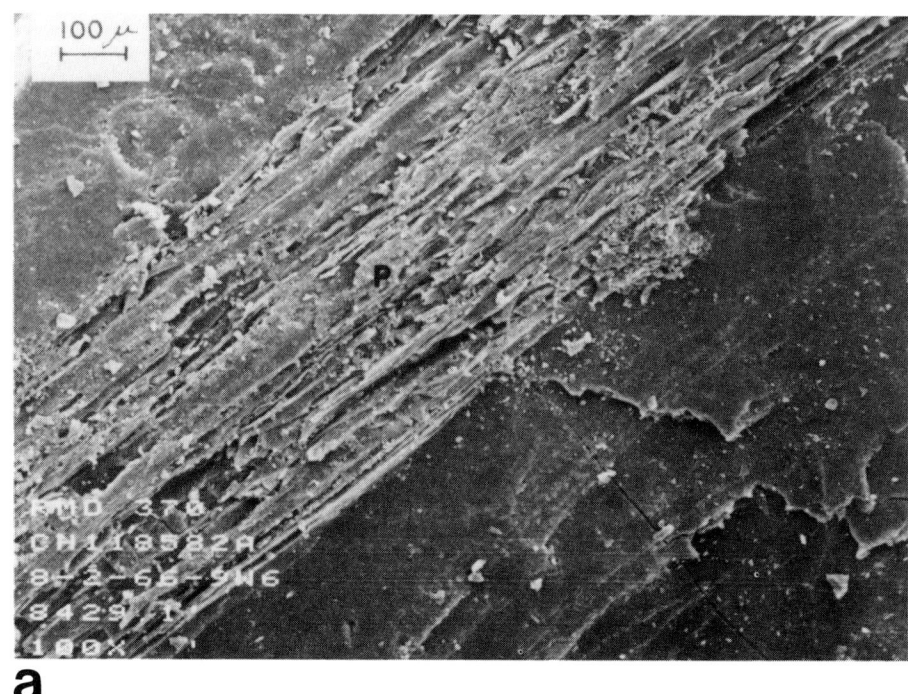

a

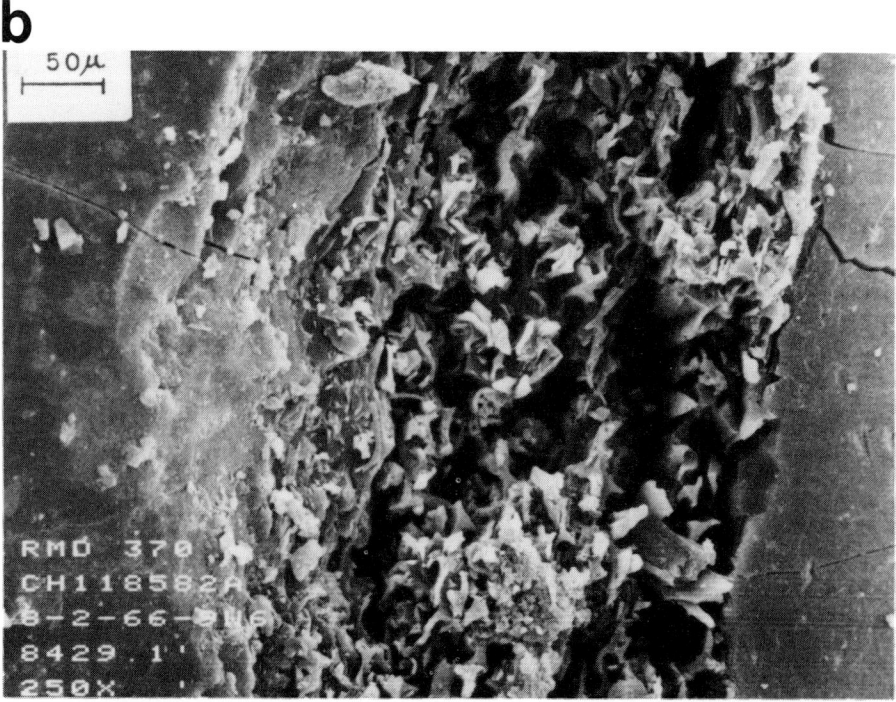

b

Figure 8. Photomicrographs of coal showing phyteral porosity.

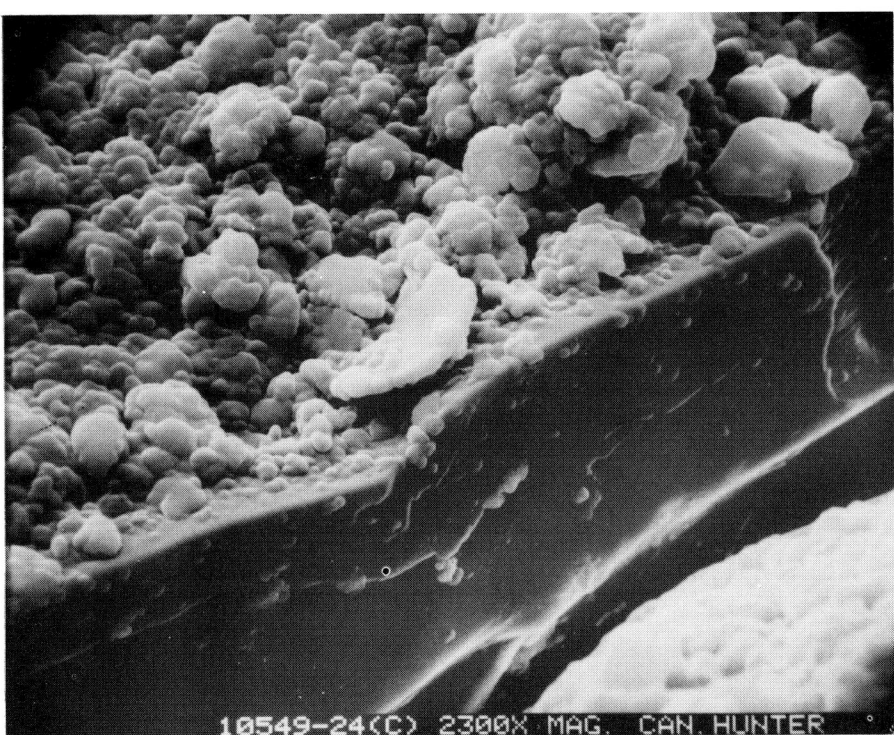

Figure 9. Photomicrographs of coal fractures.

clean surface that may have resulted from handling is seen perpendicular to the mineralized surface in Figure 9B.

In the Falher samples three basic types of cleat systems are observed; they are face, butt, and horizontal systems. The face cleat is the most continuous and usually the best developed; the horizontal system, although weakly developed along bedding planes, may also be reasonably continuous. In the Falher samples fractures vary in width from less than 0.1 micron to over 20 microns. The fracture spacing ranges from 3 to 20 mm (0.12 to 0.79 in). A thin section showing fracture patterns is illustrated in Figure 10.

The fracture width and spacing are important input parameters to the mathematical models for describing gas production. A discussion of these follows.

SIMULATION OF GAS PRODUCTION FROM COAL

Two basic configurations for demethanization of the Falher coals were studied. One was production from an isolated coal seam and the other was transfer of gas from a coal seam to an adjacent permeable formation. To study this, a three-dimensional, two-phase gas/water model with adsorbed gas was used (Intercomp, 1982). This model assumed gas diffusion from the interior of a coal particle to a fracture (or cleat) and then two-phase (gas/water) Darcy flow through the fracture system.

Production from an Isolated Coal Seam

There are many isolated coal seams in the Spirit River Formation that vary from less than 1 to about 8 m (3.3 to about 26.3 ft). One of the thickest occurs beneath the Falher D sequence and is often referred to as the Fourth Coal (Fig. 11); this isolated coal seam served as the basis for our model studies.

Sensitivity studies were run with the model for effective permeabilities of 0.1 to 10 md. Microfracture spacing of 0.3 to 1.0 cm (0.12 to 0.39 in) were assumed to exist in the coal matrix. A hydraulic fracture length of 500 ft (152.4 m) was modeled with a well spacing of 640 acres. With a 20 ft (6.1 m) coal seam, no water in the system, and a diffusion coefficient of 1×10^{-10} sq cm/sec, the model indicates great sensitivity to the assumed permeability (Fig. 12). Increasing the diffusion coeffi-

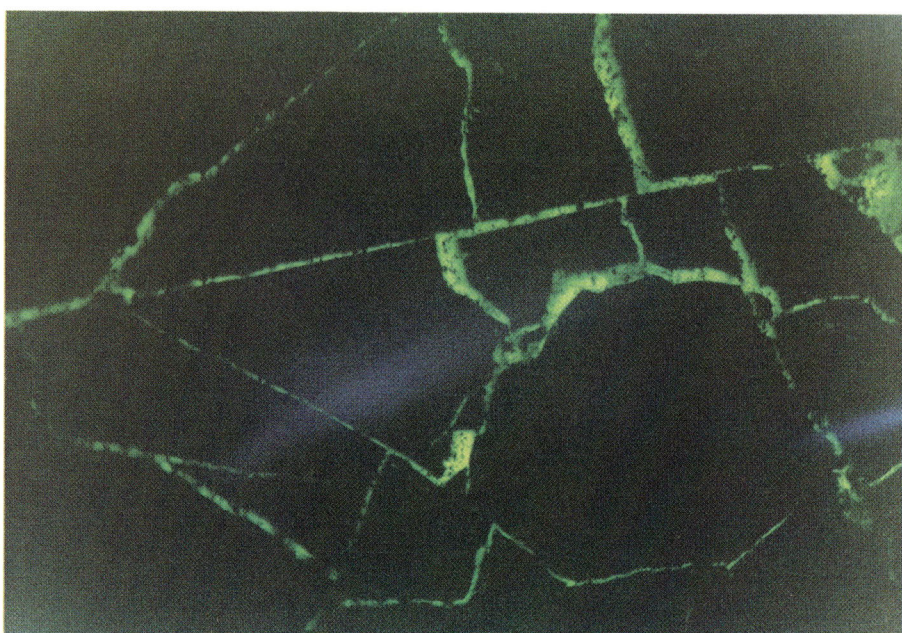

Figure 10. Thin section showing fracture patterns.

Figure 11. North-to-south cross section showing coal seams in study area.

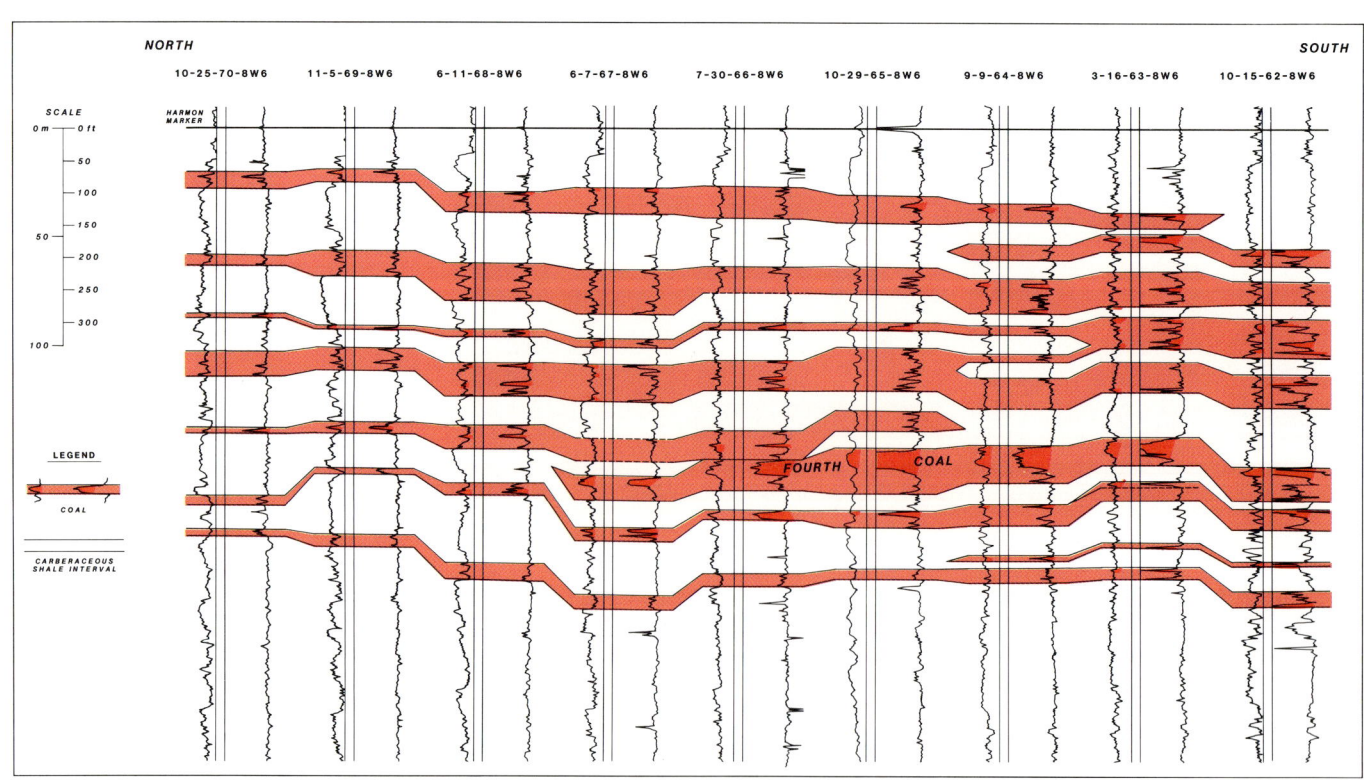

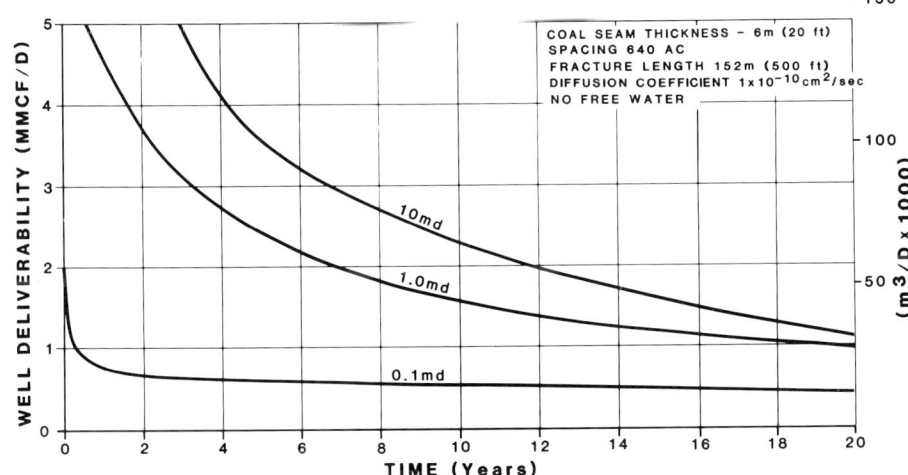

Figure 12. Effect of permeability on production from an isolated coal bed.

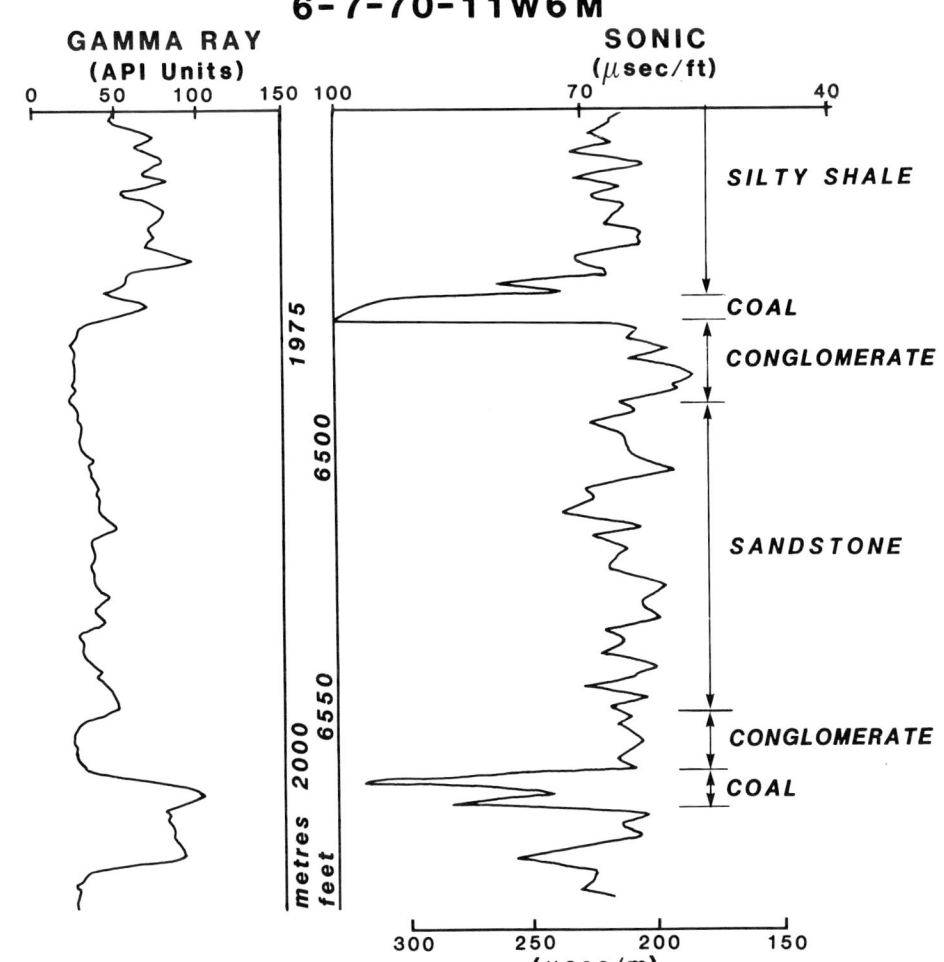

Figure 13. Log of coal adjacent to conglomerate.

cient to 1.0×10^{-5} sq cm/sec does not significantly change these results. However, if free water is introduced into the model, production rates are decreased by an order of magnitude during early times. The best guess is that the Falher coals are comparatively dry. Unfortunately it is difficult to have a very high degree of confidence in the magnitude of permeability; the only measured values that exist are considerably less than the lowest values used in this analysis. However, as mentioned earlier these may not be representative of in situ values. Effective permeability levels in producing U.S. coalbeds, based on history matching, range from 0.1 to 250 md (Intercomp, 1977). These are at shallower depths than the Falher Member coals.

These observations indicate a wide range of possible production rates from an isolated coal seam. Key variables that must be quantified are in situ permeability and free water saturation. Productivity can not be pinpointed until it is carefully field tested. Even then, an understanding of the lateral distribution of fractures or permeabilities would require several tests.

Production from a Coal Seam Adjacent to a Permeable Reservoir

A more encouraging picture of coal demethanization (that is, fewer uncertainties) arises when the coal seam is adjacent to a permeable conglomerate. This configuration often occurs with the Falher Member conglomerates (Fig. 13). A mathematical model was constructed assuming a 5 ft (1.5 m) coal seam directly adjacent to a 20 ft (6.1 m) conglomerate (Fig. 14).

The estimated cumulative gas production from 640 acres is shown in Figure 15. After 20 years the conglomerate alone will yield 4.4 bcf. With the coalbed contribution another 1.9 bcf of gas is produced. This adds over 40% to the conglomerate reserves. It is interesting to note that the recovery is relatively insensitive to the coal permeability. Even at 0.01 microdarcys, which is considerably less than the average permeability measured, the total reserves are reduced only 5%. Figure 16 shows that under these conditions about one half of the gas-in-place in the coal seam will be recovered within 10 years. This assumes a bottom-hole pressure of 200 psi. Significantly more gas will be recovered if this pressure can be lowered below 200 psi. Figure 17 shows that this

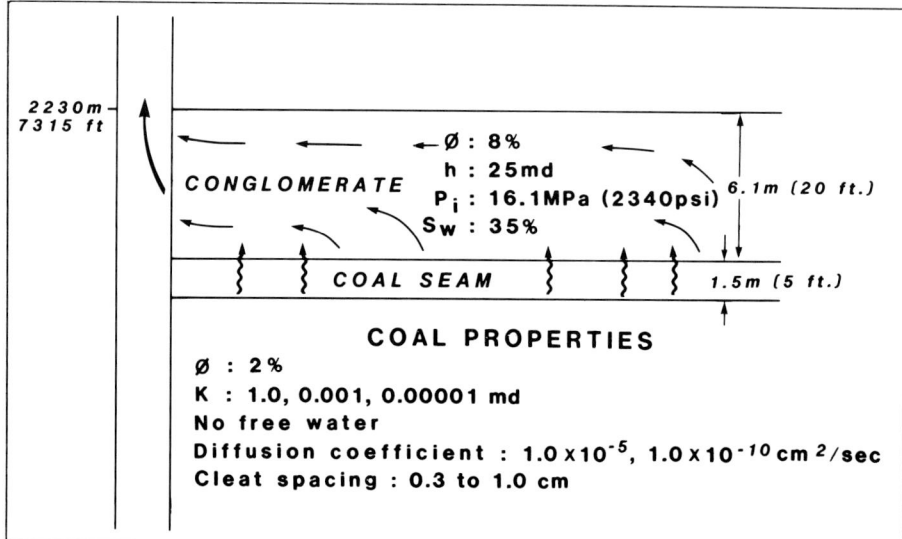

Figure 14. Model of coal adjacent to a conglomerate bed.

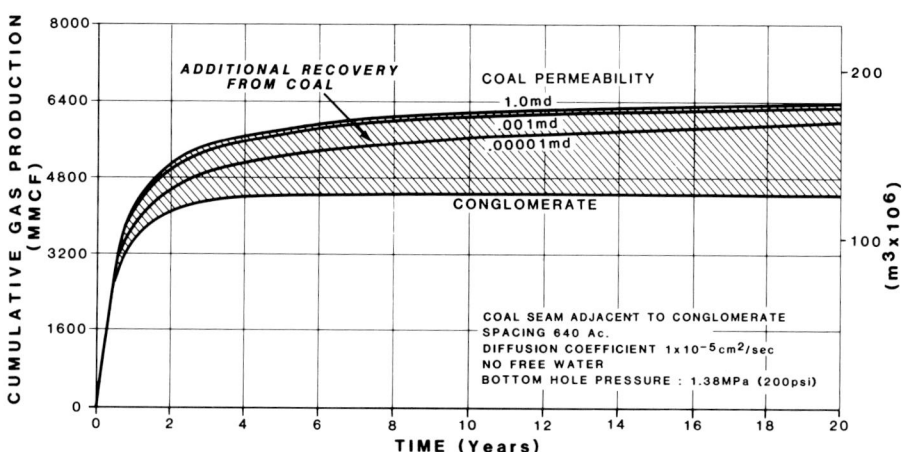

Figure 15. Effect of coal permeability on gas recovery from coal adjacent to conglomerate.

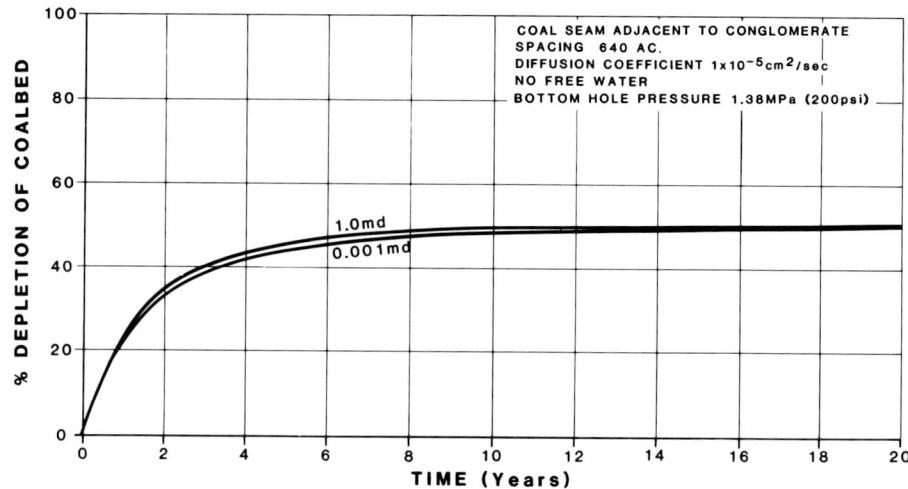

Figure 16. Depletion of coal bed adjacent to permeable bed.

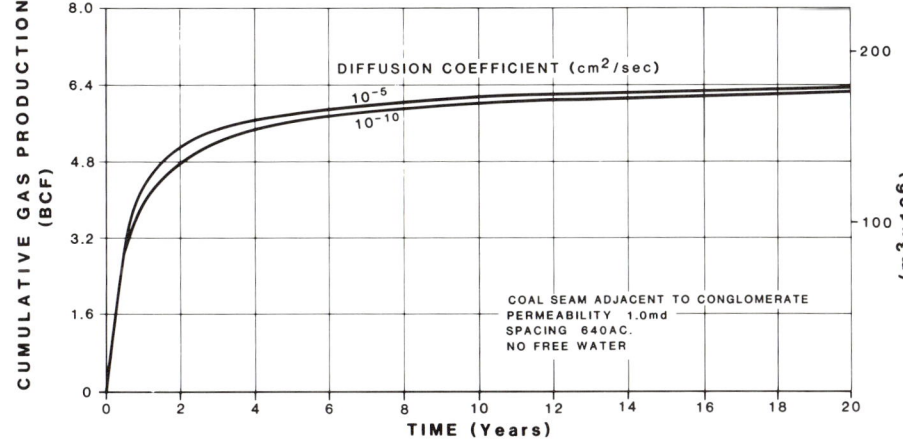

Figure 17. Effect of diffusion on cumulative gas production.

model is practically insensitive to the diffusion coefficient.

The extent of coal adjacent to the Falher A conglomerate in Elmworth is shown in Figure 18. These coal seams which are generally 1 to 2 m (3.3 to 6.6 ft) wide contain about 400 bcf of gas. The total gas-in-place in these and other coal seams adjacent to the B, C, and D sands and conglomerates add up to over 1 tcf; at least half of this should be recoverable.

This gas is above and beyond any reserves, or even speculative potential, now attributed to the producing sands and conglomerates with their adjacent tight sands.

IDENTIFICATION OF COALS FROM LOGS

Normally, coals are easy to identify on logs because of the relatively high interval

Figure 18. Coal directly adjacent to producible Falher Cycle.

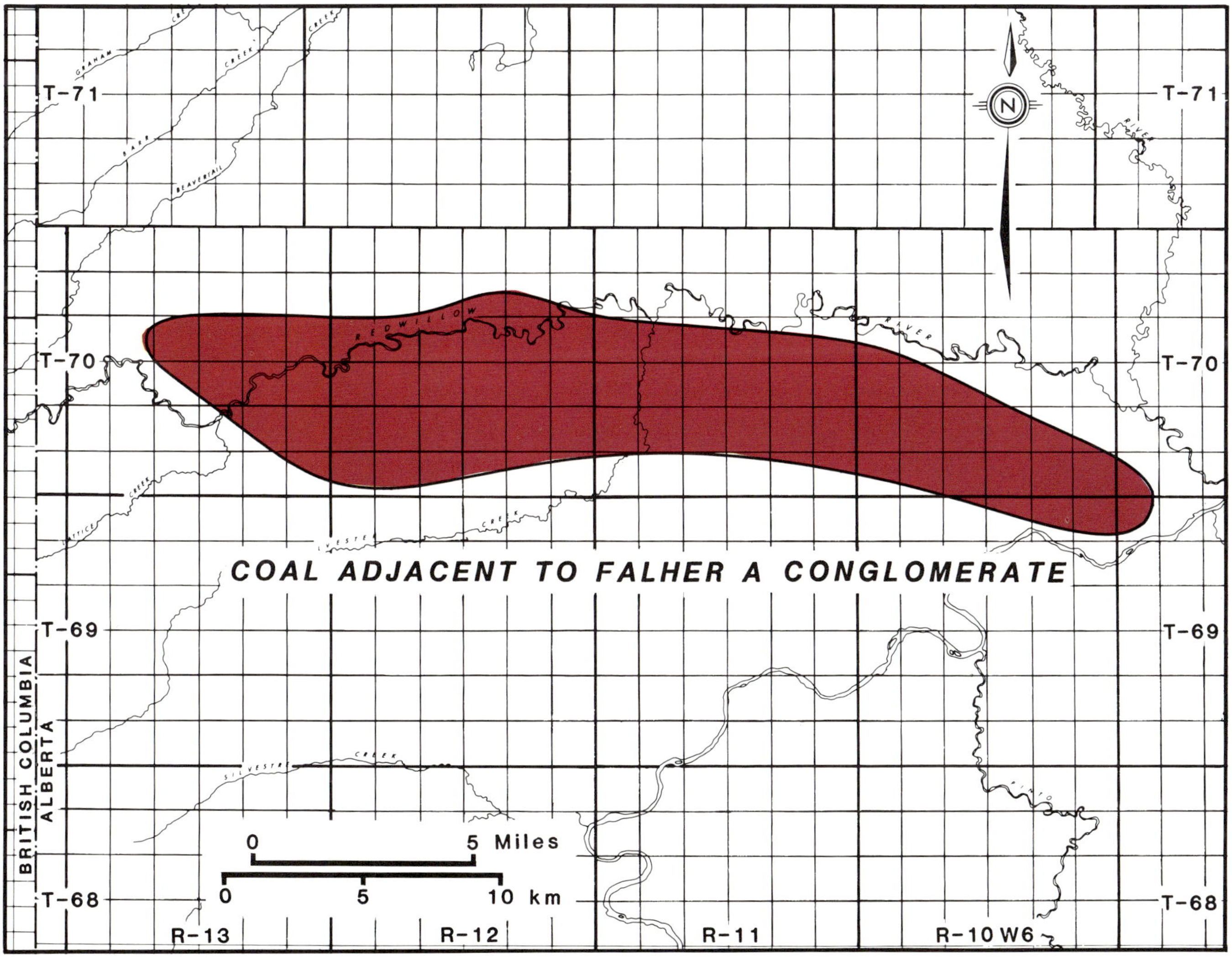

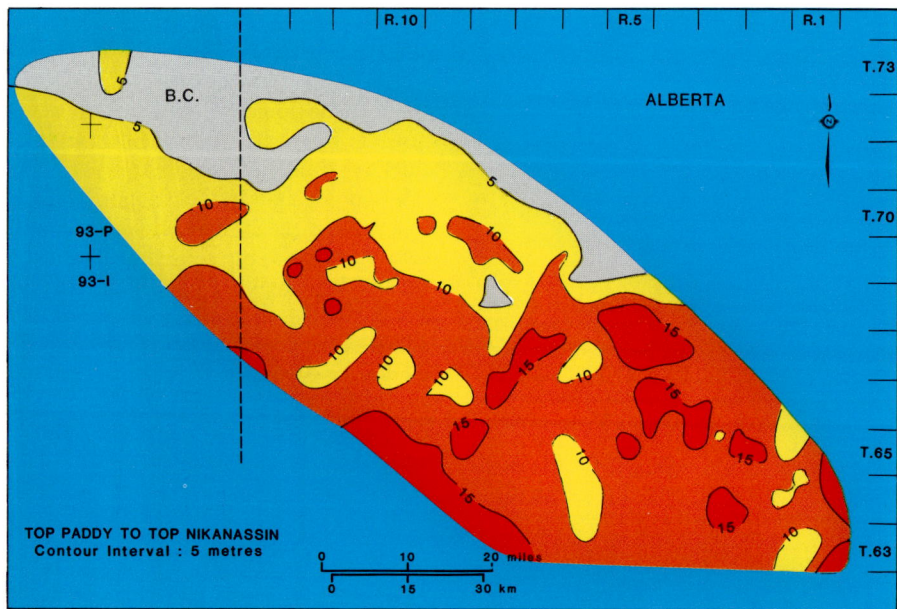

Figure 19. Coal isopach, measured from the top of the Paddy to the top of the Nikanassin, study area.

GAS RESOURCES IN COAL

Isopach maps have been made of the various qualities of coal as indicated by the shaliness. The total thickness of coals with less than 50% shale in the Lower Cretaceous is shown in Figure 19. This represents 164 billion tons of coal at depths ranging from 1,550 to 3,350 m (5,085 to 10,991 ft).

The quantity of these coals as determined by quality or shaliness shows a familiar triangular distribution where the least-shaly coals with the highest concentration of gas occupy the top part of the triangle (Fig. 20). The total gas-in-place in coals within the study area represented by Figure 19 is 50 tcf.

As discussed previously some of these seams are directly adjacent to sands or conglomerates of producing formations and contain about 1 tcf of gas; however, the bulk of the coal is isolated within shales or silty shales (Fig. 11). The thick Fourth Coal alone is estimated to contain 10 tcf of gas-in-place (Fig. 21).

By considering the quality of coal, the gas contained in the various qualities and the net thickness, the distribution of gas within coals is summarized on a "bcf per section" map shown in Figure 22. On the Alberta side of this study area there is an average of 15 bcf per section in the coal where the sum total of all seams was over 5 m (16.4 ft). Some areas have over 25 bcf per section. None of this gas has yet been recognized as proven reserves or even speculative potential supply.

CONCLUSION

Within the greater Elmworth area of the Deep basin in British Columbia and Alberta, there are about 164 billion tons of coal within the Lower Cretaceous formations at depths ranging from 1,550 to 3,350 m (5,085 to 10,991 ft).

The coal seams in this area vary from less than 1 to about 8 m (3.3 to 26.2 ft). The thickest single coal seam, one known as the Fourth Coal, near the bottom of the Falher Member, is about 8 m (26.2 ft) wide in some places; it contains some 10 tcf of gas and is a prime candidate for further

Table 4. Logging criteria used to identify coal.

Logging Curve	Coal assumed for all values that are:
Interval transit time	Equal to or greater than 300 μs/m
Density correction	Density correction is less than 0.1 gm/cu cm
Density	Equal to or less than 2.1 gm/cu cm
Gamma Ray	Normalized gamma ray reading indicates shale volume equal to or less than 50%
Resistivity	Equal to or greater than 60 ohm-metres

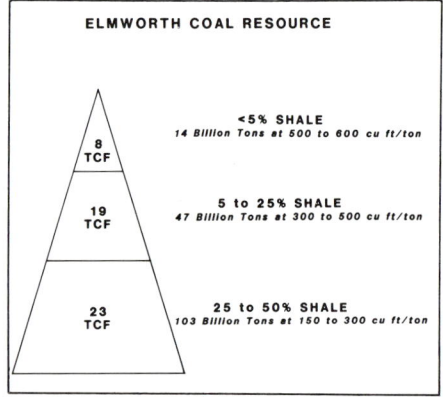

Figure 20. Resource triangle of coal.

Table 5. Gas content and density of coals related to shaliness

Volume of Shale	Gas Content cu ft/ton	Average Density of Coal gm/cu cm
<5%	550	1.250
5-25%	400	1.385
25-50%	225	1.630

transit time seen on the acoustic log and the low-density values. However, one must discriminate relatively clean coal seams from low-density shales, shaly coal or false readings due to washed-out zones. In Elmworth, a combination of interval transit times, density correction, caliper, gamma ray response, and resistivities can be used to identify coalbeds. To quantify the quantity and gas content of coals in the Elmworth area the cutoff values shown in Table 4 were used.

From analysis of the pressure core the amount of gas in the Elmworth coals appears to be closely related to the shaliness or ash content. Since the volume of shale can be calculated from the logs, a good indication of gas volume and coal density can be made directly from the digitized logs (Table 5). These relationships can then be used to compute the total amount of coal and its contained gas.

experiments in completion. If successful, multiple completion techniques can easily be tried to enhance the commercial recovery of gas from coal.

The methane contained in coal of this area is estimated to be 50 tcf. Although not all of this gas could be recovered it has some significant implications for the Elmworth area:

1. At least 500 bcf of this gas should be recovered indirectly through adjacent permeable sands and conglomerates;
2. A potentially larger amount may be recovered directly from the Fourth Coal and other isolated coal seams;
3. Geochemical analysis indicates that the coals and carbonaceous shales in the Elmworth area were the most prolific source beds for gas; and,
4. Gas is still being generated today from the coals.

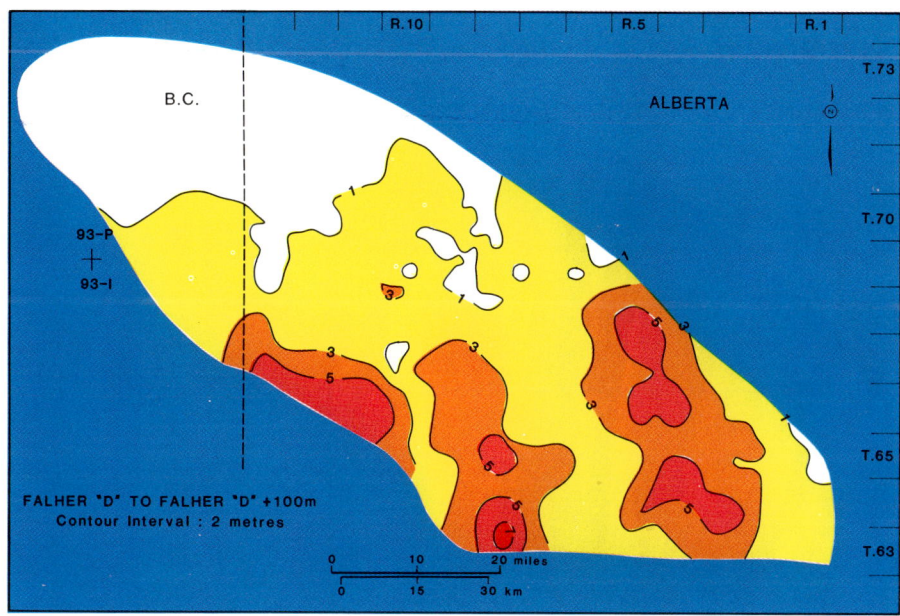

Figure 21. Isopach of "Fourth Coal."

REFERENCES CITED

Curl, S. J., 1978, Methane prediction in coal mines: London, International Energy Agency Coal Research, Report No. ICTIS/TR 04.

Doherty, M. G., 1982, Unpublished report on privately funded research done by The Institute of Gas Technology.

Intercomp, 1977, Feasibility of methane production from coalbeds — theory and application of coalbed degasification: unpublished report on privately funded project.

Larsen, V., and J. H. Knowles, 1982, Evaluation of the methane content of the Fourth Falher coal bed, Canadian Hunter 15-16-68-13W6 well, Elmworth gas field, Alberta, Canada: Internal report prepared for Canadian Hunter Exploration, Ltd. by Colorado School of Mines Research Institute.

Masters, J. A., 1984, Lower Cretaceous oil and gas in western Canada, in J. A. Masters, ed., Deep basin gas: AAPG Memoir 38, this volume.

Smith, D. M., and F. L. Williams, 1982, Diffusional effects in the recovery of methane from coalbeds: Pittsburg, Paper presented to The Unconventional Gas Recovery Symposium of The Society of Petroleum Engineers.

Tissot, B. P., and D. H. Welte, 1978, Petroleum formation and occurrence: New York, Springer-Verlag.

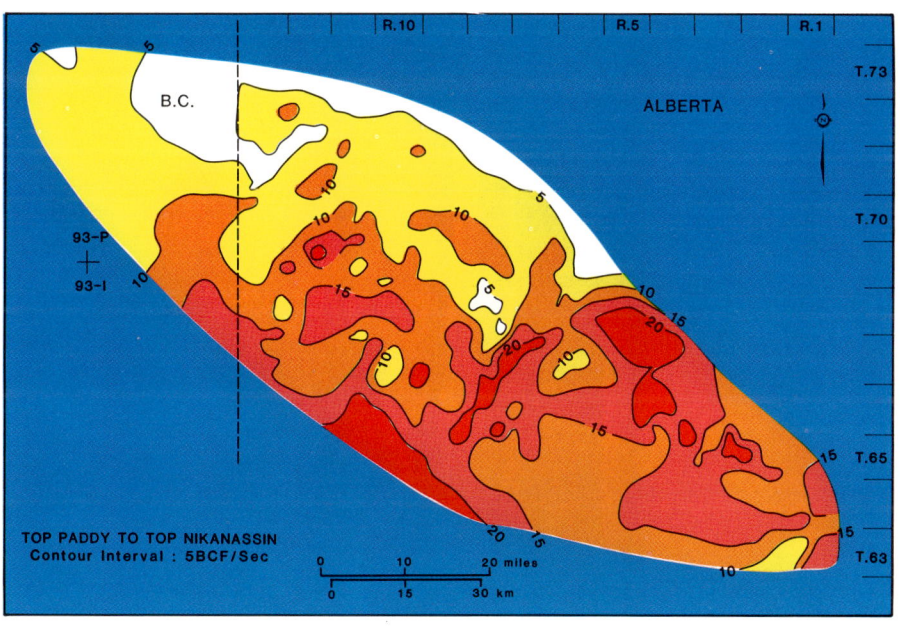

Figure 22. Map of potential gas in coals, mapped by bcf per section (interval is 5 bcf/sec).

Weiss, H. M., 1983, Organic facies of the "Fourth Coal" (Falher Member, Spirit River Formation, Lower Cretaceous) in the Elmworth area, Alberta, Canada: RFA Julich, Federal Republic of Germany Institute for Petroleum and Organic Geochemistry (ICH-5), Thesis in progress.

Subsurface Pressure Profiles in Gas-Saturated Basins

T. B. Davis
Canadian Hunter Exploration Ltd.
Calgary, Alberta

Basins which have gas-saturated sections exhibit a characteristic pressure profile. They may be subnormally or supernormally pressured, but never normally pressured. The pressure profile of numerous gas-saturated sections in North America has been documented by constructing pressure-depth plots. The approach here is descriptive. Interpretation of the data has not yet yielded a unique solution.

Examples of gas saturation in subnormally pressured Upper and Lower Cretaceous rocks of Alberta, Canada, and Lower Silurian rocks of eastern Ohio, U.S.A., and supernormally pressured Tertiary and Upper Cretaceous rocks of Wyoming, U.S.A., are presented. These pressure-depth plots should help identify similar gas-saturated sections in other petroleum provinces.

INTRODUCTION

Many sedimentary basins in North America contain widespread, gas-pervasive sections. These basins exhibit characteristic pressure profiles over the gas-saturated section. The profile may be subnormally or supernormally pressured, but never normally pressured.

The "Deep basin" area in the Western Canada basin of Alberta (Fig. 1, Area A) was discovered in March, 1976, and is still being developed today. Formation pressures collected from drill stem tests and production tests in Cretaceous rocks indicate that the productive formations in the gas saturated part of the basin are subnormal (underpressured) for their depth.

Two other documented areas in North America have gas-saturated formations and subnormal pressure responses. These areas are Medicine Hat, Alberta in the Upper Cretaceous, and eastern Ohio in the Lower Silurian (Fig. 1, Areas B and C).

The Green River and Red Desert basins in Wyoming (Fig. 1, Areas D and E) have a supernormal pressure profile (overpressured for their depth). This occurs in gas-saturated rocks of Tertiary and Upper Cretaceous age.

ELMWORTH DEEP BASIN AREA, ALBERTA, CANADA

Within the Elmworth Deep basin area there are seven formations from Cretaceous rocks that are gas productive. A majority of this production comes from an area that is considered gas pervasive; that is, a thick section of rock is gas saturated; every stringer of porosity, even the shales. Characteristically, free water is updip of the gas accumulation (Masters, 1979). Two typical gas-pervasive formations, the Cadotte and the Falher A, are presented.

The Cadotte Formation is of Lower Cretaceous age (Fig. 2). In the study area it consists of continental and marine sands of western source. The marine and beach sands trend in a west to east direction.

Figure 3 is a fluid map from the Cadotte Formation. Gas and water tests are shown as squares and dots respectively. The water-saturated area is blue and the gas-saturated area is pink. Commercial gas fields are red. Structure contours indicate the Cadotte is dipping to the southwest at approximately 50 to 80 ft per mi (9 to 16 m per km).

Formation water has not been found in the shaded pink area, and the commercial gas pools in this gas-saturated section are dark red. The blue area, is the normally-pressured system with Cadotte gas pools which produce from structural and/or stratigraphic (conventional) traps. These pools have water present down-dip of the accumulation.

The pressure-depth plot of the Cadotte Formation (Fig. 4) indicates two major systems. A normally-pressured water system (blue), which has conventional gas accumulations (gas trapped above water) associated with it, and a subnormally-pressured gas column (red). The pressure-depth plot indicates that the start of the gas leg is at 3,000 ft, or 914 m (subsea) and the bottom is at 4,200 ft, or 1,280 m (subsea). There is 1,200 ft (366 m) of gas column present with no associated down-dip water. The subnormal pressure response seen on the pressure-depth plot falls in the area, where only gas is present (Fig. 3, shaded pink area).

Cross section A-B (Fig. 5) illustrates the fluid relationship in the Cadotte Formation in the Elmworth area. Water occurs in the updip region, and gas saturation occurs deeper in the basin.

A similar subnormal pressure response is also found in the Falher A cycle (uppermost Falher Member) in the Elmworth "Deep basin" area. This rock unit, of early Lower Cretaceous age (Fig. 2), consists of continental to coastal to marine deposits of western provenance and grades to a marine shale further to the northeast.

Figure 6 is a map indicating fluid and pressure control for the Falher A. Squares indicate gas recoveries and dots are water recoveries. The red outlines are gas fields in the Falher A cycle. As with the Cadotte the gas fields in the blue area are accumulations which have associated water and are considered to be local structural and/or stratigraphic (conventional) traps. The

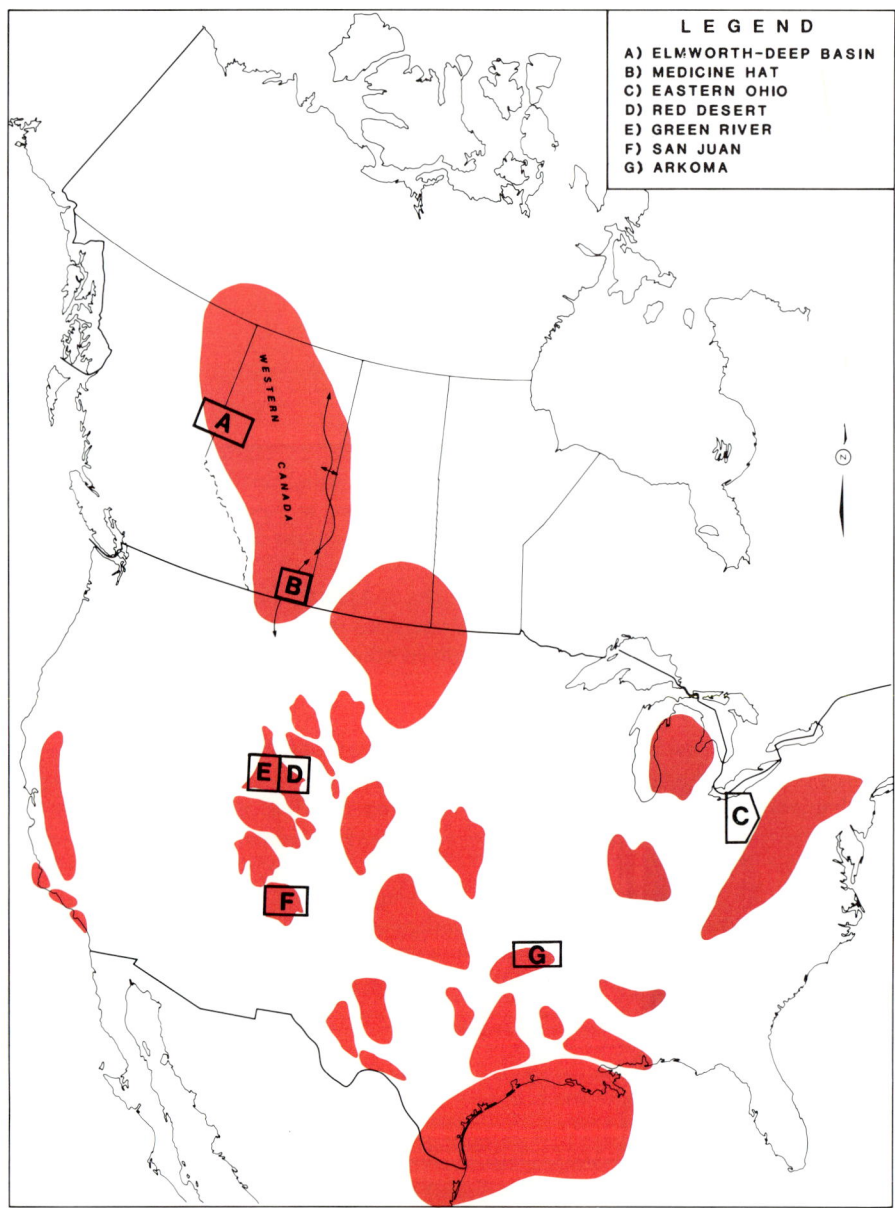

Figure 1. Map of North America sedimentary basins and index map.

Figure 2. Stratigraphic correlation chart for Cretaceous rocks in the Western Canadian basin, Alberta (E.R.C.B., 1979).

charged throughout a large area.

MEDICINE HAT AREA, ALBERTA

The Milk River Formation (Fig. 2), of Late Cretaceous age, consists of nearshore/shoreface sediments in the southwest, prograding into lower shoreface, offshore sediments and marine shales to the northeast.

Figure 8 is a map showing fluid tests from the Milk River Formation. The dots indicate water and the squares are original pressures of gas fields producing from the Milk River. The red outline is the gas producing limit of the Milk River.

A pressure-depth plot of the Milk River (Fig. 9) indicates a regional water system (blue) extending from the outcrop to a depth of approximately 1,300 ft, or 579 m (above sea). At a point of 1,900 ft, 579 m (above sea), a subnormally pressured gas profile (red) is present. This subnormal pressure response extends to 700 ft, or 213 m (above sea). There is approximately 1,200 ft (366 m) of gas column present with no indication of downdip water. A diagrammatic north to south cross section (Fig. 10) illustrates this fluid relationship.

The subnormal pressure system coincides with the producing area for the Milk River Formation.

fields in the pink area are commercial gas accumulations in the gas-saturated section of the Falher A. The pink area indicates where the Falher A is gas saturated, and exhibits generally lower permeability. Formation water has not been tested in this area. The rocks dip to the southwest at 50 to 80 ft/mi (9 to 16 m/km).

The pressure-depth plot (Fig. 7) illustrates the pressure profile that is present as one goes downdip from the water system into the gas saturated Falher A. A normal pressure system (blue) exists downdip until 2,700 ft, or 823 m (subsea), then a subnormal pressure profile is present. This subnormal pressure profile occurs when the gas-saturated Falher section is reached.

Figure 7 indicates that there is approximately 2,000 ft (610 m) of gas column present with no associated downdip water. The fluid relationship is illustrated in cross section A-B (Fig. 5).

The Cadomin Formation is also subnormally pressured in this area, and has been discussed in detail by R. M. Gies in this volume.

These examples from the Elmworth area, also known as the Deep basin, indicate that subnormal pressure responses are associated with formations that are gas-

Subsurface Pressure Profiles in Gas-Saturated Basins 191

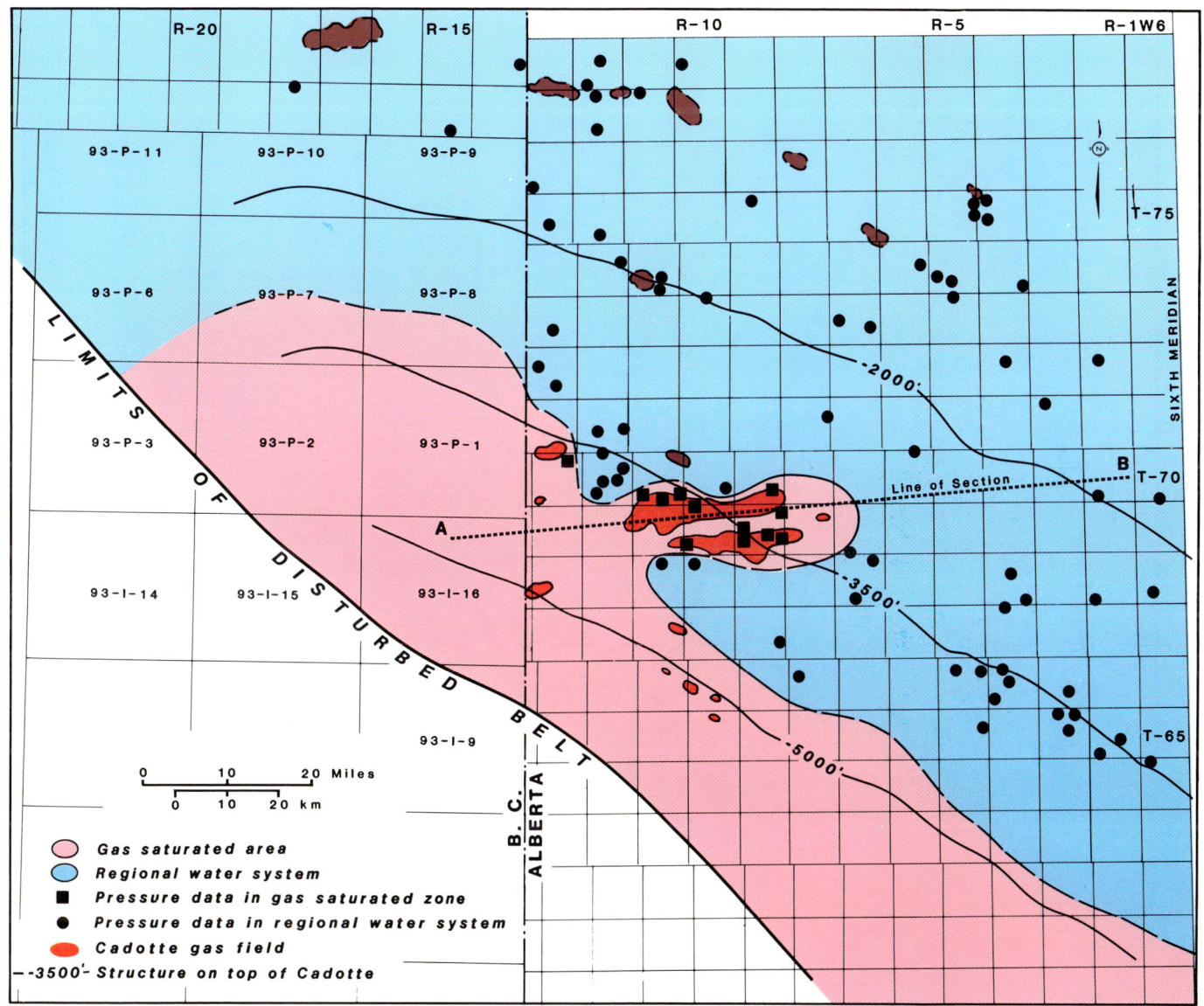

Figure 3. Fluid and data map of the Cadotte formation, Elmworth, Deep basin, Alberta.

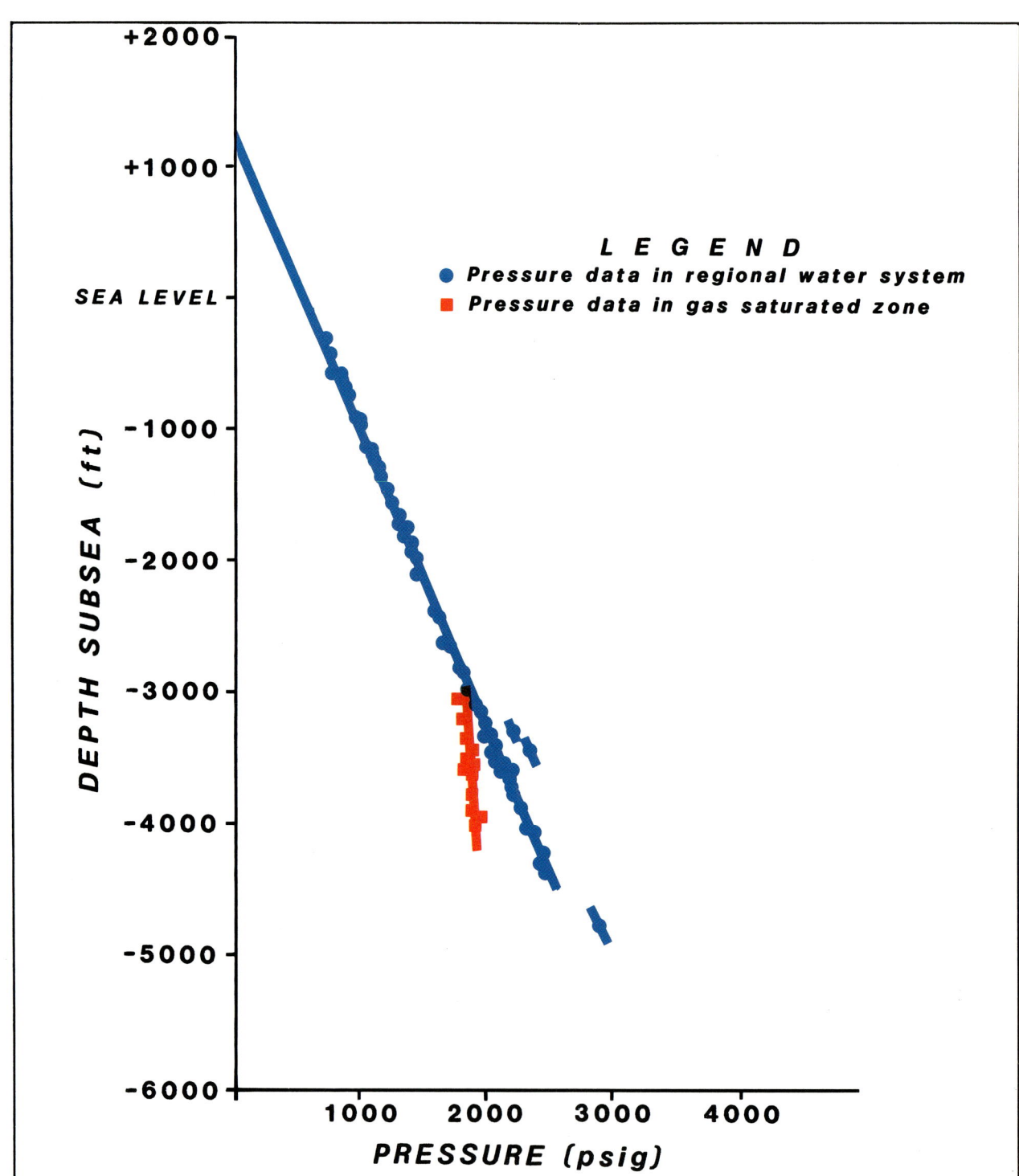

Figure 4. Pressure – depth plot, Elmworth area, Cadotte Formation.

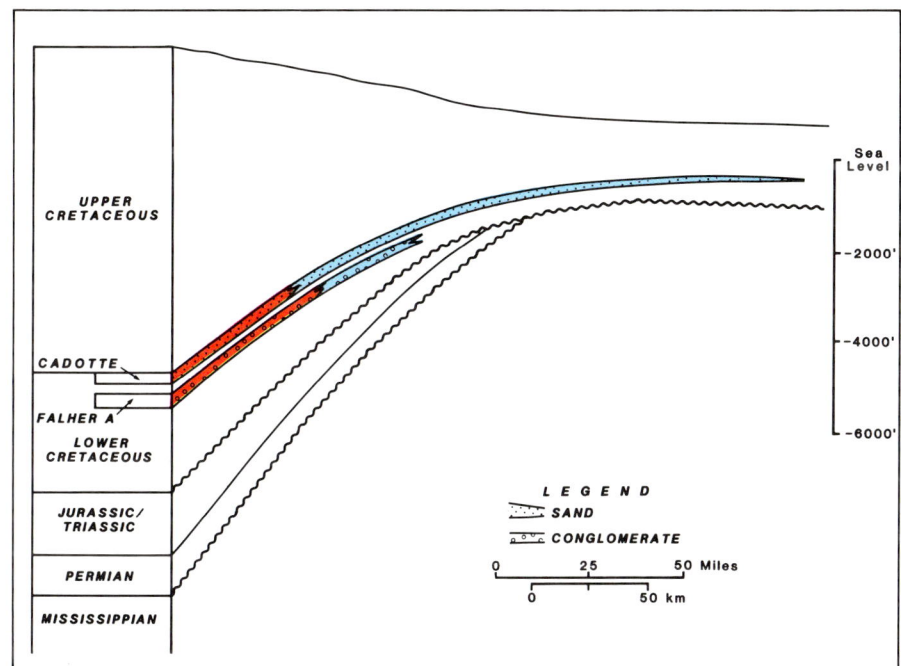

Figure 5. Generalized fluid cross section, Northern Alberta.

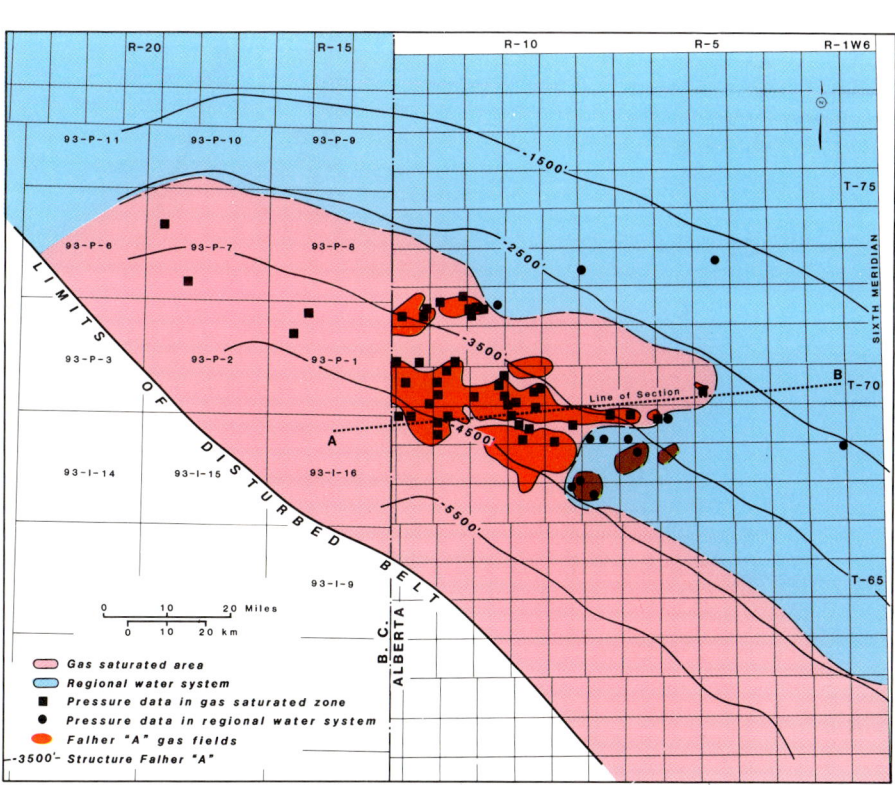

Figure 6. Fluid and data map of the Falher formation, Elmworth, Deep basin, Alberta.

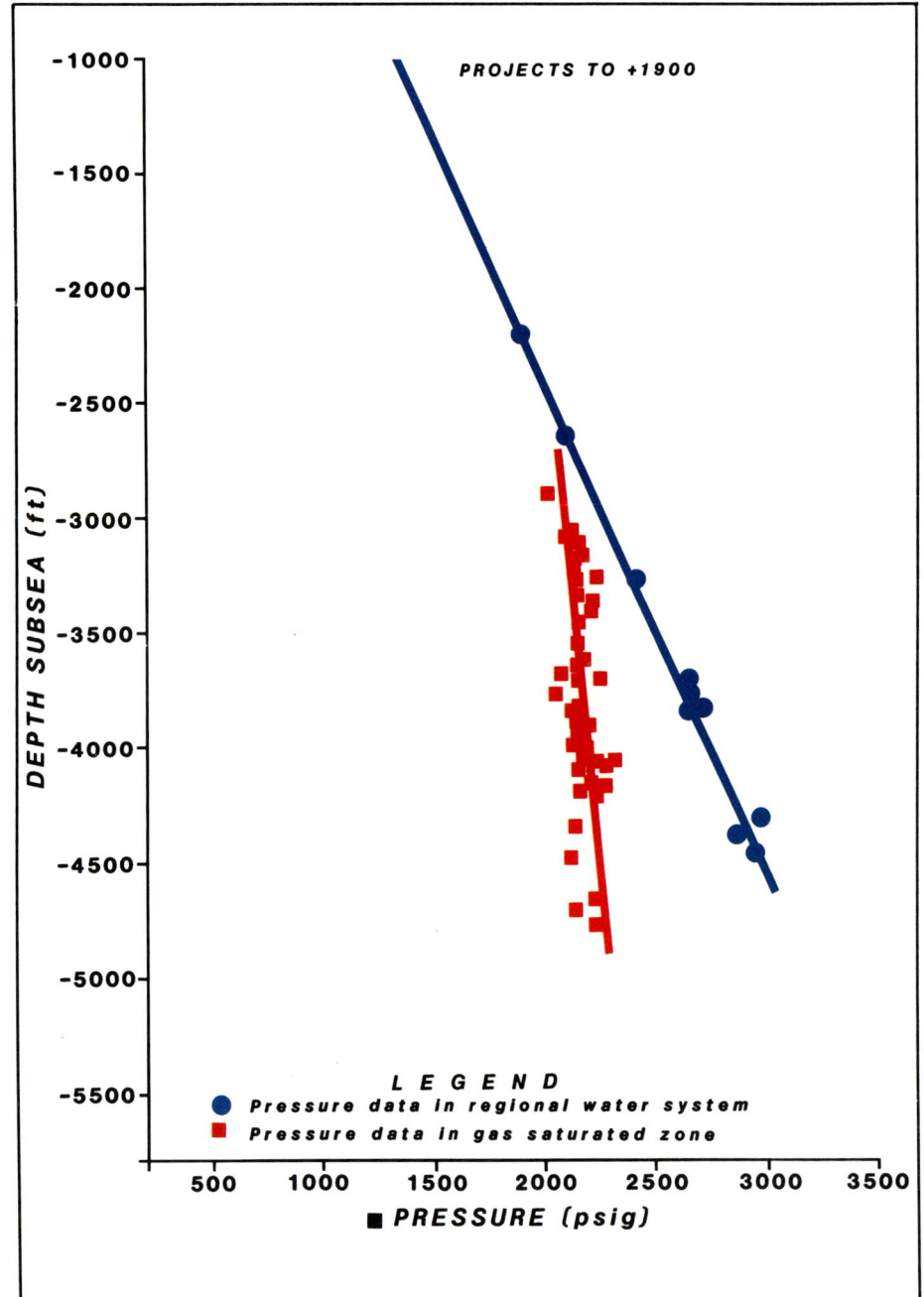

Figure 7. Pressure – depth plot, Elmworth area, Falher "A."

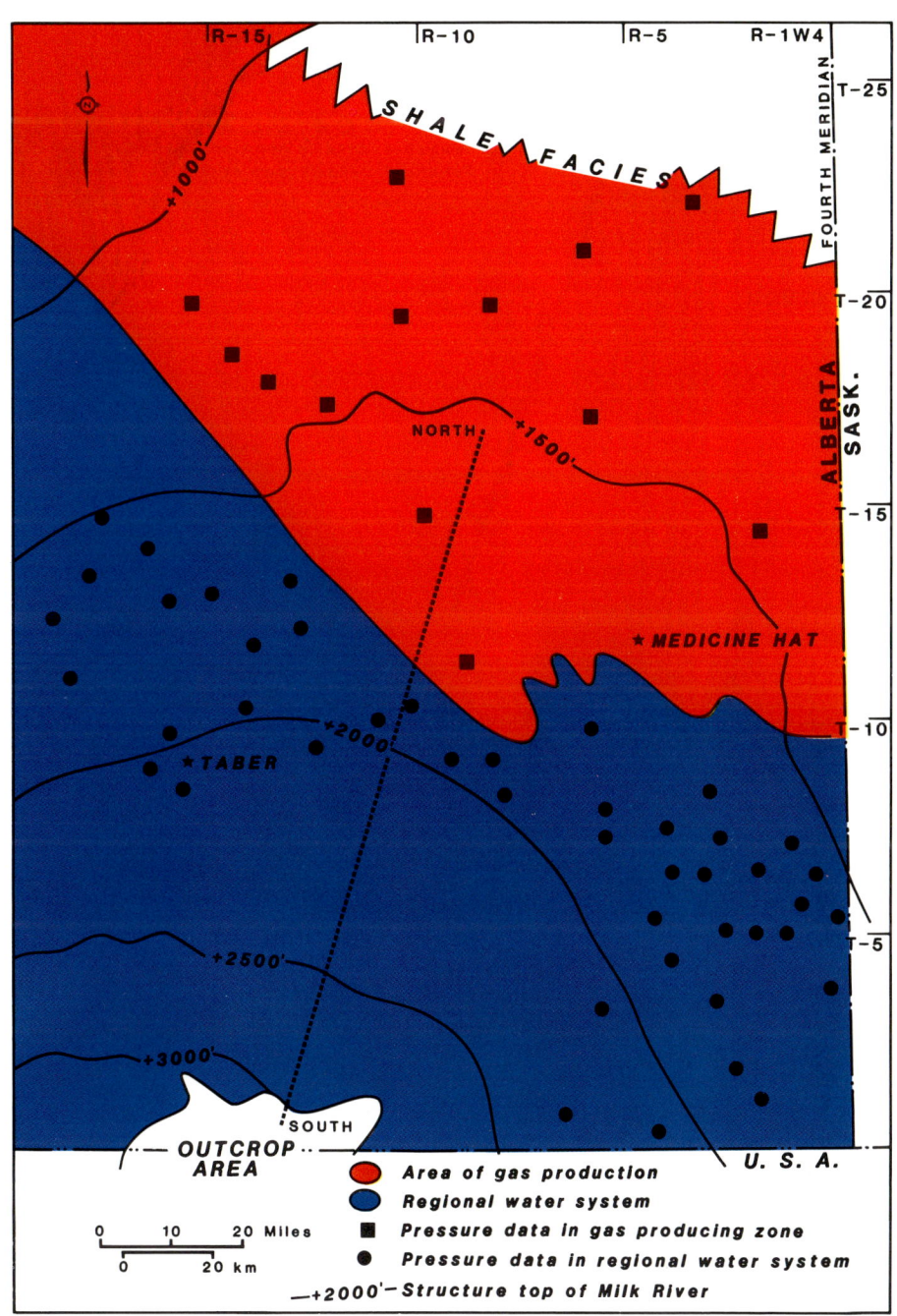

Figure 8. Fluid and data map of the Milk River Formation, Medicine Hat area, Alberta.

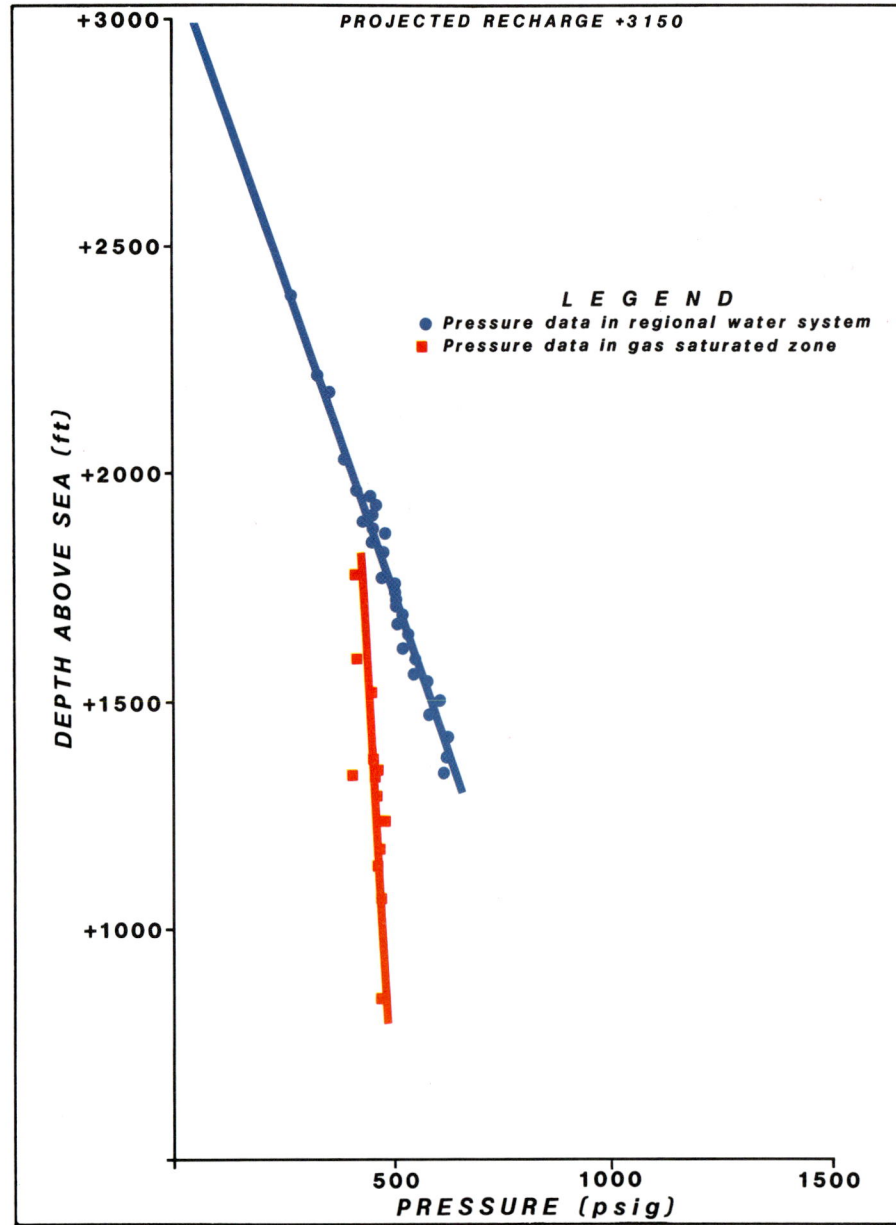

Figure 9. Pressure – depth plot, Southern Alberta, Milk River Formation.

Figure 10. Generalized fluid cross section, Southeastern Alberta.

Figure 11. Stratigraphic correlation chart for Ohio and Northwest Virginia (after Smonsa and Patcher, 1978).

EASTERN OHIO

The Clinton Formation sandstone is of Early Silurian age (Fig. 11). The depositional environments have been interpreted as an eastern source of sediments forming distributary channels and bars associated with shallow marine deltaic environments, prograding to marine shales to the west.

Figure 12 is a fluid map of the Clinton sandstone. The dots and squares are pressure and fluid data. The green and red outlines are Clinton oil and gas fields. An area of gas saturation with no formation water is shaded in pink. The formation dips to the east at 45 ft/mi (8.6 m/km).

Oil and gas fields in the blue area are on the normal pressure system. These oil and gas accumulations are trapped along the regional pinchout of the sand on the east side of the Cincinnati Arch. These fields have associated water downdip.

A pressure-depth plot (Fig. 13) indicates a normally-pressured water system (blue) and two subnormally-pressured gas systems (red). At approximately 3,200 ft, or 975 m (subsea), a subnormally-pressured gas profile is seen on the pressure-depth plot and can be followed to 5,200 ft, or 1,585 m (subsea; red). A smaller subnormal pressure profile occurs at 4,600 ft, or 1,402 m (subsea) and ends at 4,800 ft, or 1,463 m (subsea). The two different red systems are probably due to poor reservoir continuity in the gas-saturated section. These data points fall within the pink shaded area on Figure 12.

There is approximately 2,000 ft (610 m) of gas column with no downdip water associated with it. This fluid relationship is shown on the diagrammatic west to east cross section (Fig. 14).

A subnormal pressure zone is present over a large area of the Clinton Formation. This zone appears to coincide with a gas-saturated section in the Clinton sand.

RED DESERT BASIN, WYOMING

The Mesaverde Group of Wyoming is of Late Cretaceous age (Fig. 15). The depositional environment varies vertically. The Erickson Formation is comprised of fluvial, continental and deltaic sediments. The Almond Formation is continental to deltaic with a transgressive marine sand at the top capped by the Lewis Formation. The Lewis Formation is marine shale with occasional isolated sandstone beds. The Fox Hills Formation consists of sands deposited in beach to shallow marine environments. The Lance Formation and the overlying Tertiary sediments are of fluvial and lacustrine origin.

Figure 16 is a fluid map of the Upper Cretaceous and Tertiary sandstones in the Red Desert basin. Dots indicate water tests and the squares indicate gas tests. The green and red areas are commercial oil and gas fields and the pink shade indicates an area of gas saturation with no formation water. The sediments generally dip from the west and east into the bottom of the basin as shown on Figure 16.

A pressure-depth plot (Fig. 17) indicates a normally-pressured water profile (blue), and a supernormally-pressured gas profile (red). The oil and gas fields in the blue area (Fig. 16) fall on the normally-pressured system and have downdip-associated water. The gas fields in the pink area fall in the supernormally-pressured system. Formation water has not been found in this pressure profile. Several gas columns are seen in the supernormal pressure system. This is due to poor reservoir continuity in the gas-saturated area.

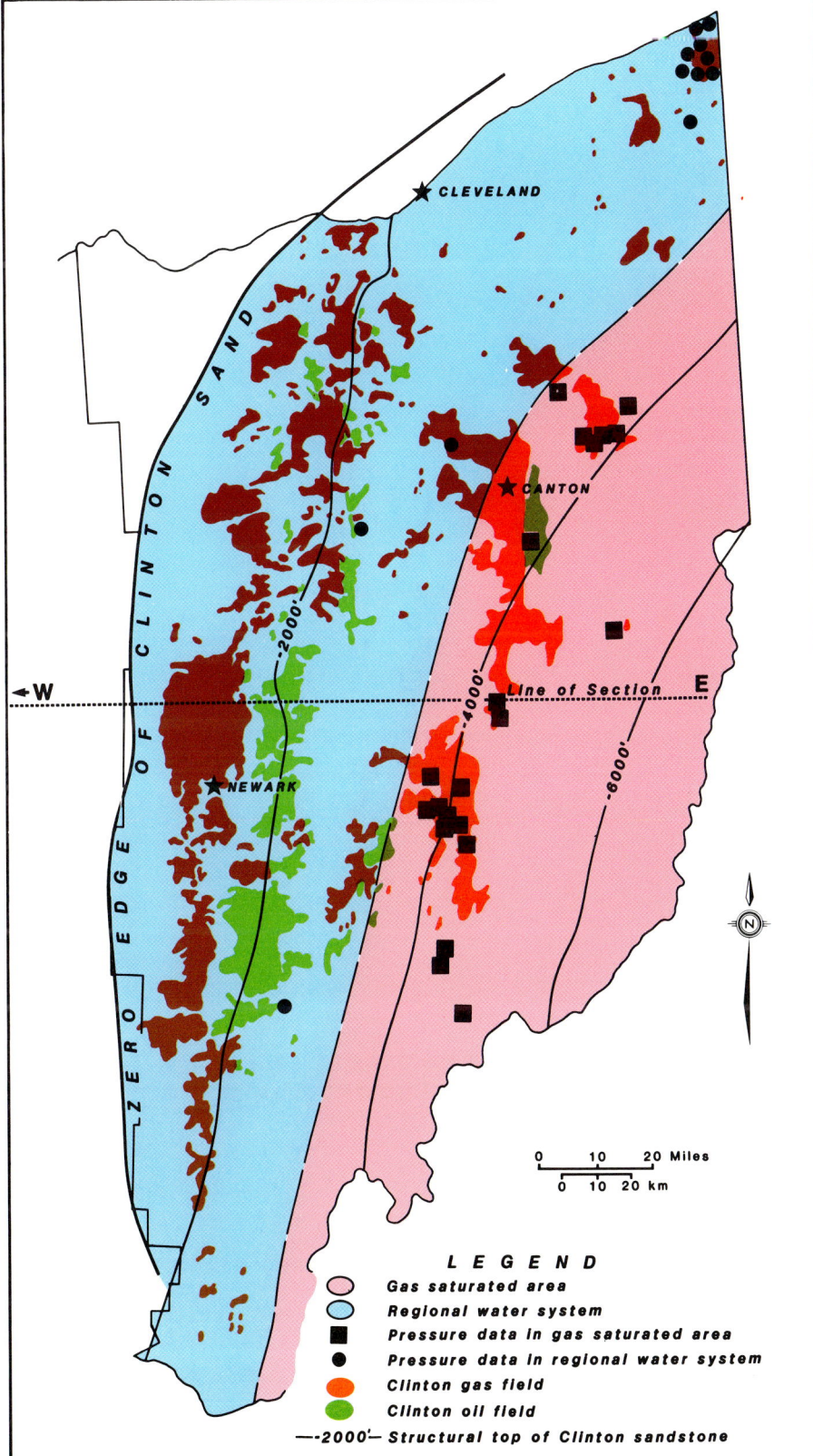

Figure 12. Fluid and data map of the Clinton Sand, Eastern Ohio.

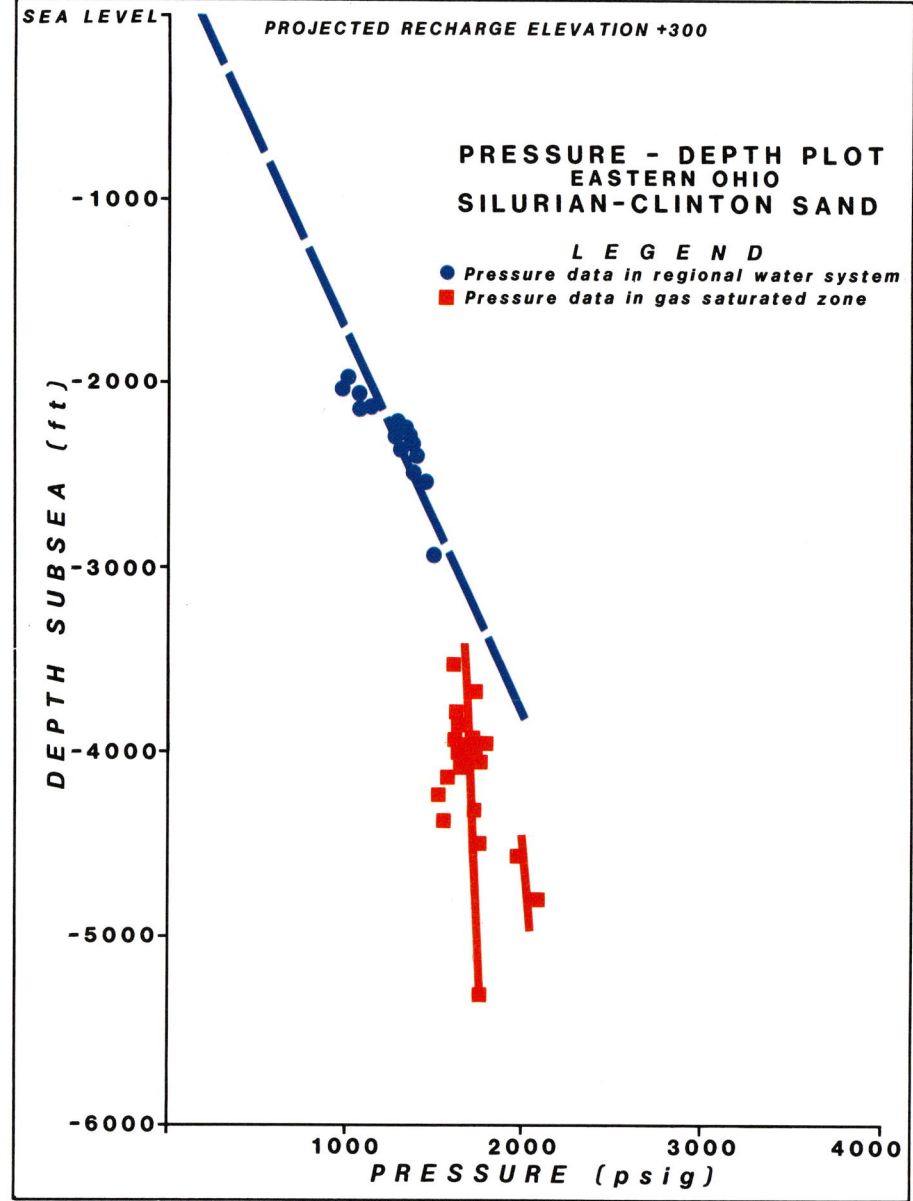

Figure 13. Pressure – depth plot, Silurian-Clinton sand, Eastern Ohio.

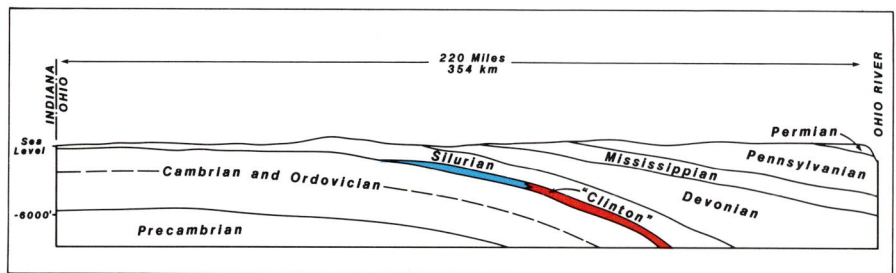

Figure 14. Generalized fluid cross section, Eastern Ohio.

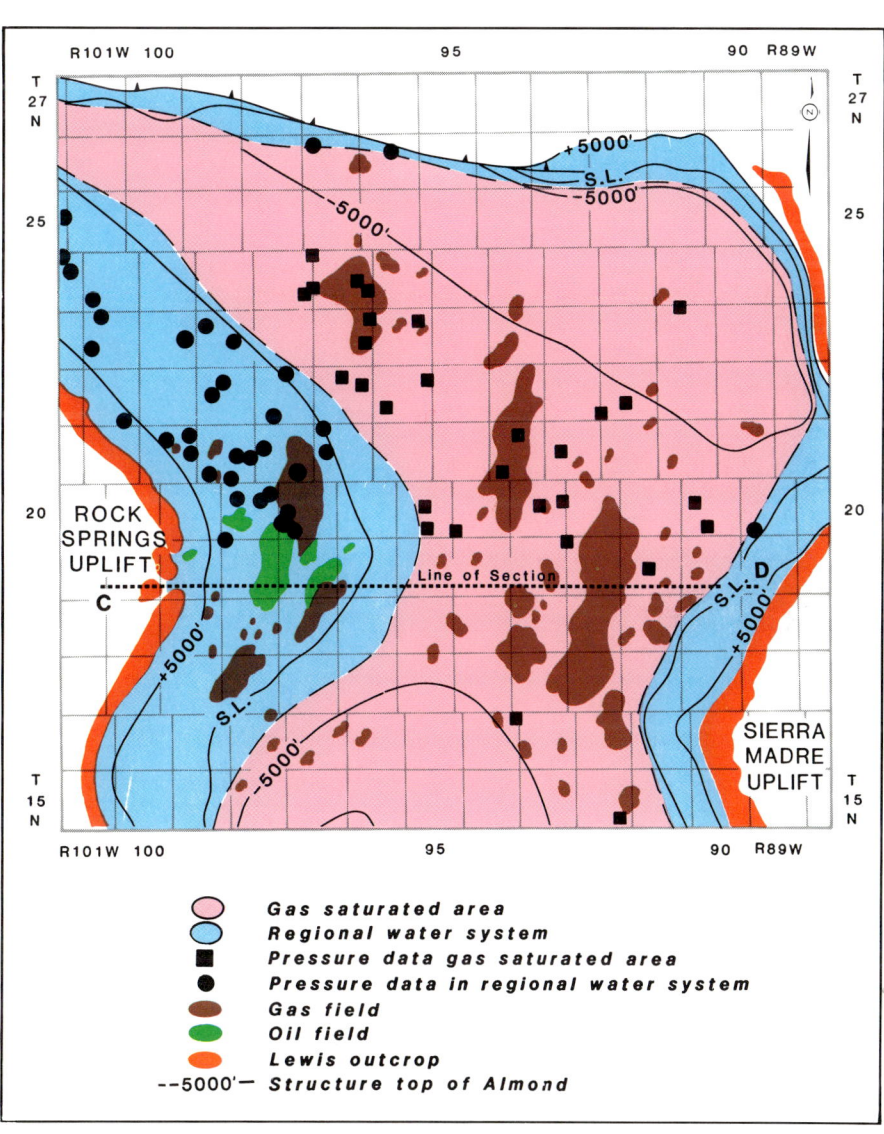

Figure 15. Stratigraphic correlation chart for Tertiary and Cretaceous rocks in Southwestern Wyoming (after Newman, 1981).

Figure 16. Upper Cretaceous fluid map, Red Desert basin, Wyoming.

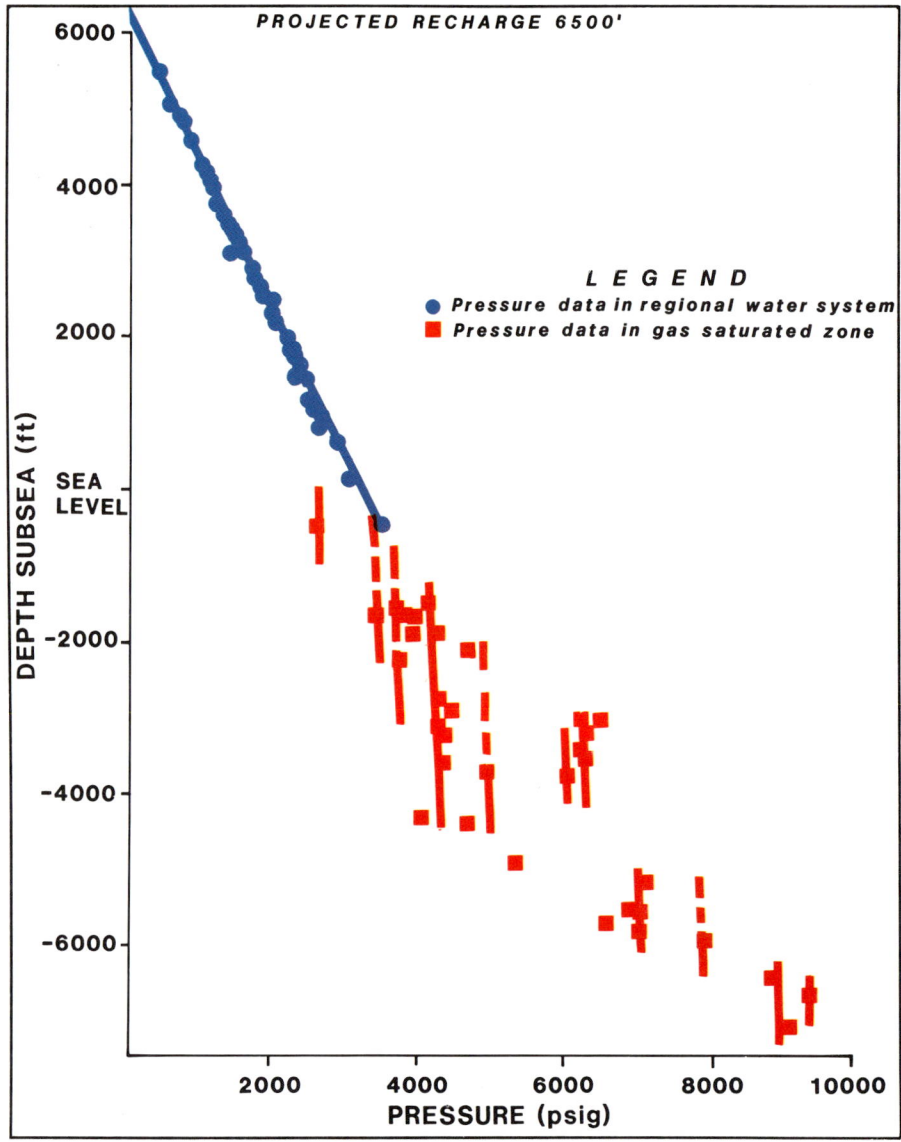

Figure 17. Pressure – depth plot, Cretaceous Sands, Red Desert basin, Wyoming.

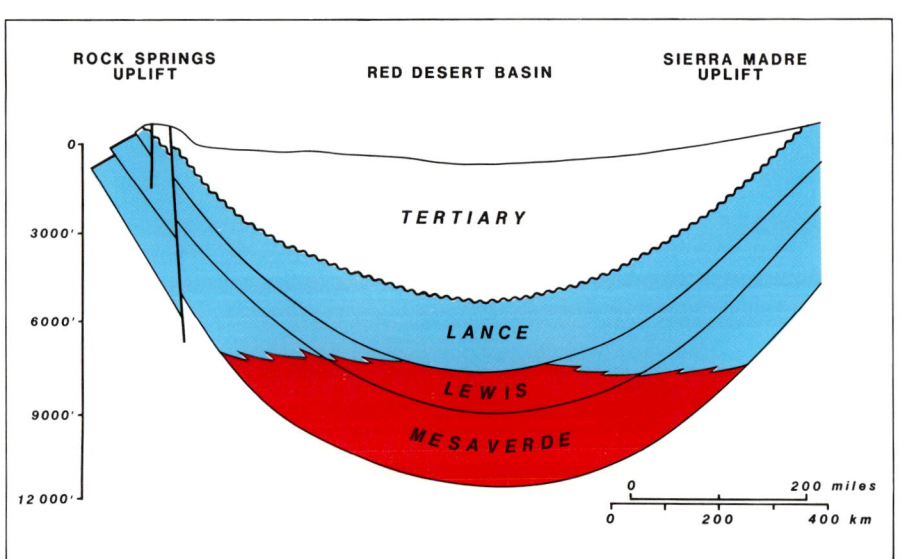

Figure 18. Generalized fluid cross section, Red Desert basin, Wyoming

The fluid distribution of the supernormally- and normally-pressured sections of the Red Desert basin is illustrated in cross section C-D (Fig. 18). The gas-charged area (shaded pink, Fig. 16) appears to coincide with the supernormally-pressured zone.

GREEN RIVER BASIN, WYOMING

The Wasatch and Fort Union formations (Tertiary) and the Mesaverde Group (Upper Cretaceous) of the Green River basin, Wyoming, consist mainly of fluvial, deltaic and shallow marine sandstone and shale.

Figure 19 is a fluid map of the Cretaceous and Tertiary rocks in the Green River basin. The dots are water and the squares are gas tests. The green and red outlines are oil and gas fields from Cretaceous and Tertiary rocks. A structure contour on top of the Mesaverde Group indicates that these sediments dip to the east at approximately 200 ft/mi (38 m/km).

The pressure-depth plot (Fig. 20) shows two basic pressure regimes. A normally-pressured water system in blue and a supernormally-pressured gas system below that (red). On the normally-pressured system, oil and gas are trapped in structural and/or stratigraphic accumulations. Below approximately 500 ft, or 152.4 m

Figure 19. Tertiary and Upper Cretaceous fluid map, Green River basin, Wyoming.

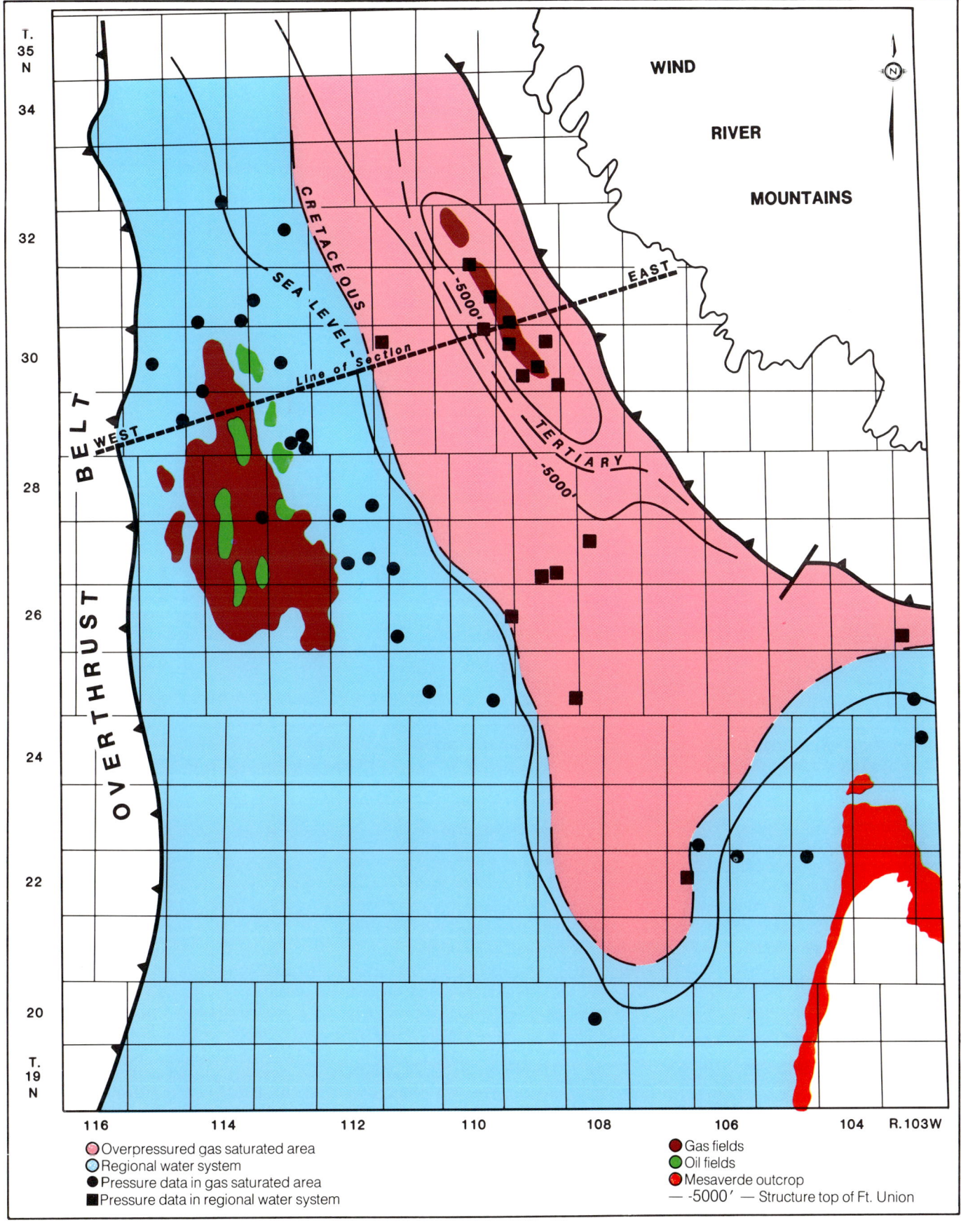

(subsea), a supernormally-pressured profile is observed. There has been no formation water tested below this depth.

Oil and gas fields that are on the normal pressure system occur in the blue area (Fig. 19). These fields have associated downdip water. The gas fields in the pink area are in the supernormally-pressured system and have no downdip water associated with them. The different gas columns that occur in the supernormal pressure profile suggest very poor reservoir continuity in the gas area, which is probably due to the depositional environment of these sands. The pink shaded area on Figure 19 is the outline of where the supernormally-pressured system starts in these rocks.

The fluid relationship in the Green River basin is illustrated by a west to east diagrammatic cross section (Fig. 21).

The supernormally-pressured zone in the Green River basin appears to coincide with the gas-saturated section in the lenticular sands.

OTHER GAS-SATURATED SECTIONS

Although not documented in this paper, other areas in North America have similar types of fluid distribution.

Barry (1959) described the gas trapping in the San Juan basin (Fig. 1, Area F) as a pressure sink. Limited original pressure data in the gas-saturated part of this basin makes it difficult to construct pressure-depth plots to see if a subnormal profile is present, but potentiometric surface maps made by Barry strongly suggest a subnormally-pressured profile in the gas-saturated section. There is no doubt that there is a large pervasive gas-saturated section downdip of water.

The Basal Atoka sand in the Arkoma basin (Fig. 1, Area G) appears to be another subnormally-pressured area, where very little or no water is produced. Lack of good quality pressure data makes it very difficult to construct pressure-depth plots.

CONCLUSIONS

1. There are many gas-saturated basins that are characterized by a large areal extent of pervasive gas saturation with no downdip water.
2. These pervasive gas sections can be described on a pressure-depth profile

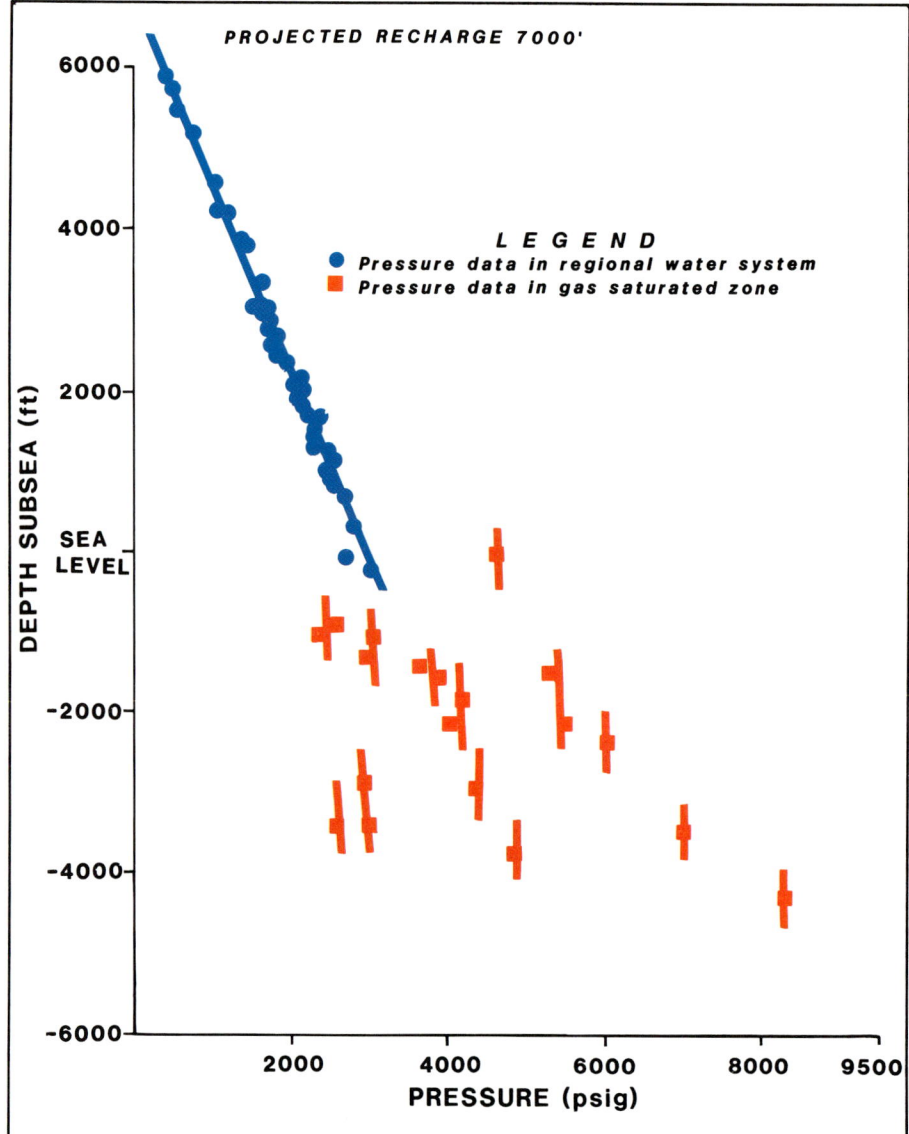

Figure 20. Pressure – depth plot, Tertiary and Upper Cretaceous sands, Green River basin, Wyoming.

and may be subnormally- or supernormally-pressured in relationship to the regional water profile.
3. Such accumulations are always associated with low-permeability formations although high-permeability areas may occur within the gas-saturated section.
4. The use of pressure-depth plots may help in identifying similar hydrocarbon occurrences in other petroleum provinces.

REFERENCES

Barry, F. A. F., 1959, Hydrodynamics and geochemistry of the Jurassic and Cretaceous systems in the San Juan basin, northwestern New Mexico and southwestern Colorado: Stanford University, Ph.D. thesis.

Brenan, C. B., Jr., 1968, Natural gas in Arkoma basin of Oklahoma and Arkansas, in B. W. Beebe, ed., Natural gases of North America, pt.3; natural gases in rocks of Paleozoic age: AAPG Memoir 9, v. 2, p. 1616-1635.

Gies, R. M., 1982, Origin, migration, and entrapment of natural gas in the Alberta Deep basin, part II: Paper presented at the AAPG annual meeting (Calgary, June).

Masters, J. A., 1982, Deep basin gas trap,

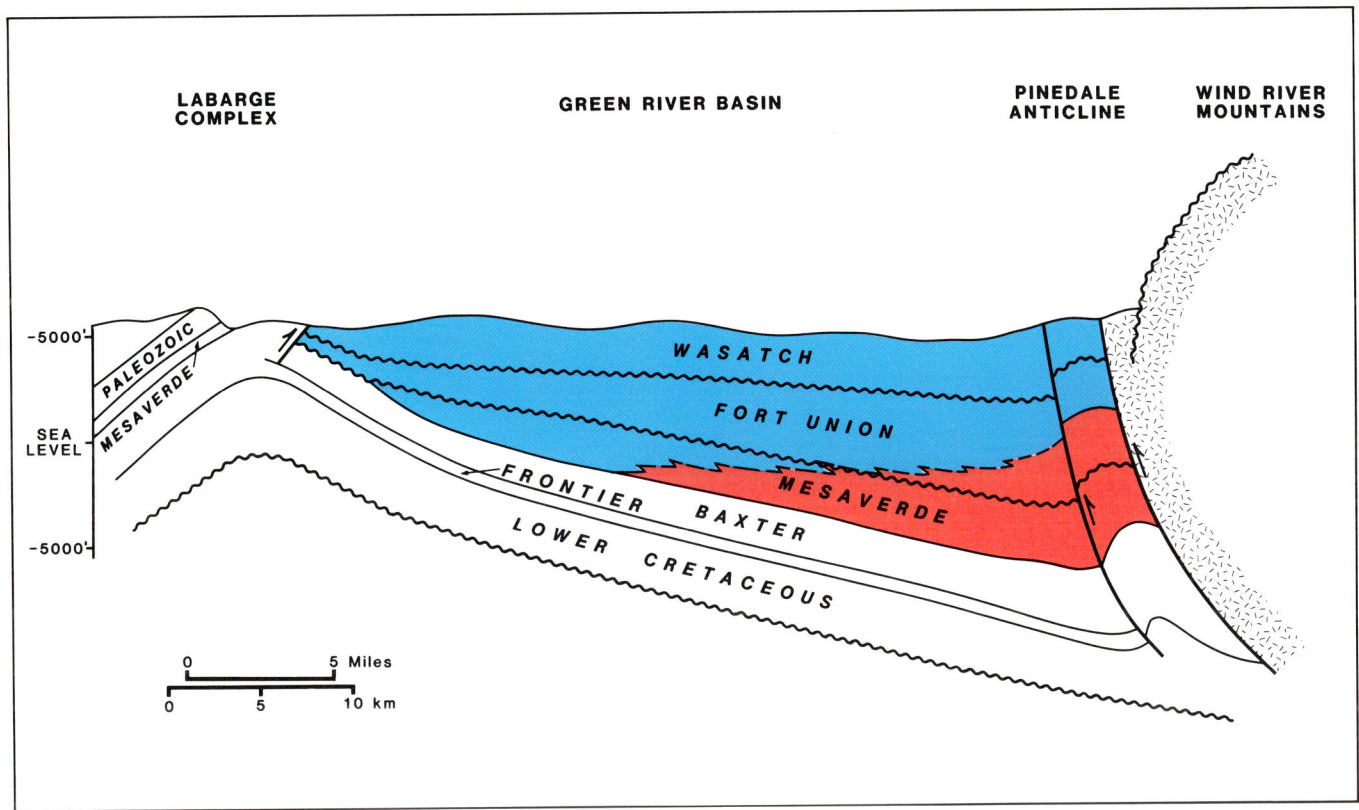

Figure 21. Generalized fluid cross section, Green River basin, Wyoming.

Western Canada: AAPG Bulletin, v. 63, p. 152-181.

McPeek, L. A., 1981, Eastern Green River basin; a developing giant gas supply from deep, overpressured Upper Cretaceous sandstone: AAPG Bulletin, v. 65, p. 1078-1098.

Myhr, D. W., and N. C. Meijer-Drees, 1976, Geology of the southeastern Alberta Milk River gas pool, in M.M. Lerand, ed., The sedimentology of selected clastic oil and gas reservoirs in Alberta: Canadian Society Petroleum Geologists, Memoir, p. 96-125.

Newman, H. E., III, 1981, Greater Green River basin stratigraphy as it relates to natural gas potential: Paper presented at the Society of Petroleum Engineers/ Department of Energy Low Permeability Symposium (Denver, May).

Pendergast, R., 1969, Correlating cardium stratigraphy using static bottom hole pressures: CWLS Journal, v. 2, p. 19.

Randolf, P. L., 1979, Massive hydraulic fractures tests huge Wyoming gas basins: The Oil and Gas Journal, (Nov. 1), p. 47–54.

Russell, W. L., 1972, Pressure – depth relations Appalachian region: AAPG Bulletin, v. 59, p. 528–536.

Smith, R. D., 1982, Deep basin gas, "The sleeping giant": Paper presented at the AAPG Annual meeting, Calgary, Alberta, June 27–29.

Smonsa, R., and D. G. Patcher, 1978, Silurian evolution of central Appalachian basins: AAPG Bulletin, v. 62, p. 2,308–2,328.

Integrated Rock-Log Calibration in the Elmworth Field - Alberta, Canada

Integrated geological-petrophysical-reservoir engineering studies conducted within the Alberta basin, Canada, by Canadian Hunter Exploration, Ltd., resulted in the discovery of a number of gas fields, including the giant Elmworth field (Masters, 1979; and Sneider et al, 1983). These studies utilized rock-fluid data from cuttings, cores, well logs, and drill-stem and production tests to determine reservoir-rock potential and hydrocarbon saturation of thick, multiple Cretaceous sandstone and conglomerate intervals. These rocks typically range in porosity from 3 to 15% and in permeability from a few microdarcys to several darcys. Key elements in the exploration search and field exploitation are the evaluation of reservoir-rock quality, determination of the depositional facies containing the best reservoir rocks, and integrated log evaluation that identifies and quantifies lithology, porosity, and fluid type and distribution.

This paper focuses on the rock-log part of the integrated studies and discusses the methodology used to establish accurate evaluations of pore space and hydrocarbons of all drilled wells. The paper is presented in three parts. The first part covers the evaluation and characterization of reservoir-rock properties primarily from well cuttings. The second part addresses rock-log calibration methods and describes "quick-scan" and digital log analysis techniques for accurate porosity and hydrocarbon saturation determination. The third part presents some examples that illustrate the methodology and results of rock-log calibration.

Part I

Reservoir Rock Detection and Characterization[1]

R. M. Sneider
Robert M. Sneider Exploration, Inc.
Houston, Texas

H. R. King
Canadian Hunter Exploration, Ltd.
Calgary, Alberta

INTRODUCTION

This paper describes the concepts and methods used to establish and characterize the reservoir-rock potential of the thick, multiple Cretaceous sandstone and conglomerate intervals in the Deep basin. The detailed rock studies were made as part of a comprehensive and integrated study of well logs, DST, production tests and regional facies analyses, which are described elsewhere in this memoir. We have emphasized in this paper the methodology that might be useful to geologists and engineers who need to make rapid, accurate estimates of reservoir-rock quality in clastic intervals.

Approach and Considerations for Reservoir-Rock Evaluation

Prior to discovery of the Deep basin gas reserves in 1975, some 95 wells had been drilled completely or partially through the objective Cretaceous section in an area covering about 190 townships. From these wells, several tens of thousands of feet of well cuttings and a few hundred feet of core were examined with a binocular microscope to assess rock quality. This examination was supplemented with petrographic and scanning electron microscope studies and X-ray diffraction analyses. Analysis of DSTs, production tests, and well logs were compared with rock-pore type descriptions to assess which rock-pore type could or might be capable of hydrocarbon production with or without artificial stimulation. Comparisons were made with fields in the United States that were known to be productive from similar rock types. For example, when we first studied the Cretaceous Falher sandstones and conglomerates, we used the Pennsylvanian sandstones and conglomerates from the Anadarko basin, Oklahoma (Sneider et al, 1977), for calibration until cores were cut in the Falher. Once new wells were drilled in the Deep basin, comparisons between rock-pore types, log analyses, drill-stem and production tests, and production, helped establish the characteristics and potential of the various rock types. To establish our criteria for reservoir potential, over 10,000 ft (3,045 m) of conventional cores were analyzed in addition to several tens of thousand of feet of cuttings.

Petrologic studies of the Mesozoic gas-bearing section show that the potential reservoirs have varied and complex lithology and pore systems. Most of the rocks are cemented, compacted, and contain varying amounts of the clay minerals kaolinite, illite, chlorite, and smectite. Because of the clays, some reservoir rock is highly sensitive to drilling and completion fluids. In addition, the pore system of most reservoir rock is complex and variable because of altered primary macro pores, micro pores in clays or chert, and secondary pores developed through leaching. Commercial production is from rocks that range in porosity from about 5 to 15% and in permeability from a few microdarcys to more than 10 darcys.

Comparison of the geological and petrophysical properties including porosity-permeability and pore-size distribution show that most petrophysical properties of sandstones and conglomerates can be related to: (1) grain size and sorting; (2)

---

[1] Portions of this paper were published in the September 1983 Journal of Petroleum Technology (Sneider et al, 1983). Permission has been granted by the JPT to reproduce many of the illustrations.

degree of rock consolidation; (3) volume percent of clays, cements and other pore-filling materials; and (4) sizes of pores and pore interconnections.

We have found that there are a finite number of rock types and corresponding pore geometries (pore types) when considering the pore-space parameters that control fluid flow. We recognize the complexities of size, shape, orientation, and continuity of pores in sandstones; however, our experience shows that we can usually make a good estimate of the important petrophysical properties, especially porosity and permeability of clastic rocks from the parameters listed above.

Many of the low-permeability rocks have complex pore geometries more typical of carbonate rocks (Archie 1950; 1952).

PORE-TYPE CLASSIFICATION FOR CLASTIC ROCKS

The simple pore-type classification (Table 1) we use is designed primarily for a wellsite geologist or engineer who looks at rocks in cores and cuttings with a binocular microscope. Porous rocks are subdivided into three classes (I, II and III) based on the type, amount, and distribution of visible and small (pinpoint) pores observed on *freshly-broken, dry rock surfaces* with a binocular microscope at *20-power magnification*, and the way the rock breaks.

Comparison of pore types with production tests and log analyses shows that:

- Type I are capable of gas production without natural and/or artificial fracturing;
- Type II are capable of gas production when interbedded with Type I rocks or with natural and/or artificial fracturing; and,
- Type III rocks are too tight to produce at commercial rates even with natural or artificial fracturing.

In deciding on the pore type(s), we pay particular attention to estimating: (1) particle size and sorting; (2) consolidation, the way the rock breaks apart (see Table 2); (3) the amount (volume) of cements and pore-filling materials; (4) an estimate of visible pores; and (5) the size and distribution of pore throats.

We make the estimates of grain size and sorting using standard size-sorting comparators and we estimate volume percentages using the Terry-Chilingar charts made for volume estimates. Consolidation is an arbitrary, but practical scheme (Table 2). The above-mentioned observations are recorded on a data sheet like Figure 1.

Figures 2 and 3 show SEM photographs of typical pore types I, II, and III. We find that in many reservoirs the pore types grade into one another or that more than one pore type is present in a sample. We record the relative percentage of each pore type in the samples described.

As will be demonstrated in the next section, we can usually make reasonably good guesses of permeability of most rocks

Table 1. Pore-type classification and characteristics, Deep basin, Canada.

Type	Characteristics of Dry, Freshly Broken Rock Surfaces at 20X Magnification	Remarks

I	• Visible φ – very abundant to common
	• Pinpoint φ – very abundant to common
	• Pore interconnection is visible on many pores
	• Neddle probe can easily dislodge some grains from rock surface and reveal pore
		• Reservoir quality rock for gas without natural and/or artificial fracturing (if thick enough)
		                            Permeability, md
		    IA > 100
		    IB 10-100
		    IC 1-10
		    ID ± 0.5-1

II	• Visible φ – scattered • Pinpoint φ – abundant to common • Needle probe can only occasionally dislodge a grain from rock surface	• Capable of gas production if interlayered with Type I rock, or has natural, open fractures and/or is artificially fractured and is thick enough • Permeability is > ±0.07 to 0.5-1.0 md (Depending on particle size, sorting, and clay mineral content)
III	• Visible φ – none to very isolated • Pinpoint φ – none to few, scattered pores • Usually very well consolidated and/or pore filled with clays or other pore filling material	• Usually too tight to produce at commercial rate with natural or artificial fractures or when interlayed with Type I rock

Table 2. Consolidation classification for silicate-rich clastic rocks.

Descriptive Term	Sample Description
Unconsolidated	Sample disaggregates into individual particles before or after hydrocarbons are removed.
Slightly Consolidated	Sample easily disaggregates or crumbles into individual particles when rubbed between fingers.
Moderately Consolidated	Sample disaggregates only after rubbed vigorously between fingers.
Moderately-Well Consolidated	Sample will not disaggregate when rubbed vigorously between fingers. Forceps or steel probe will disaggregate this sample into individual particles and smaller pieces containing several particles.
Well Consolidated	Sample disaggregates with great difficulty, using forceps or steel probe, into smaller pieces containing several particles.
Very Well Consolidated	Sample will not disaggregate with forceps or probe. A hammer disaggregates the sample into small pieces; pieces break across particles.

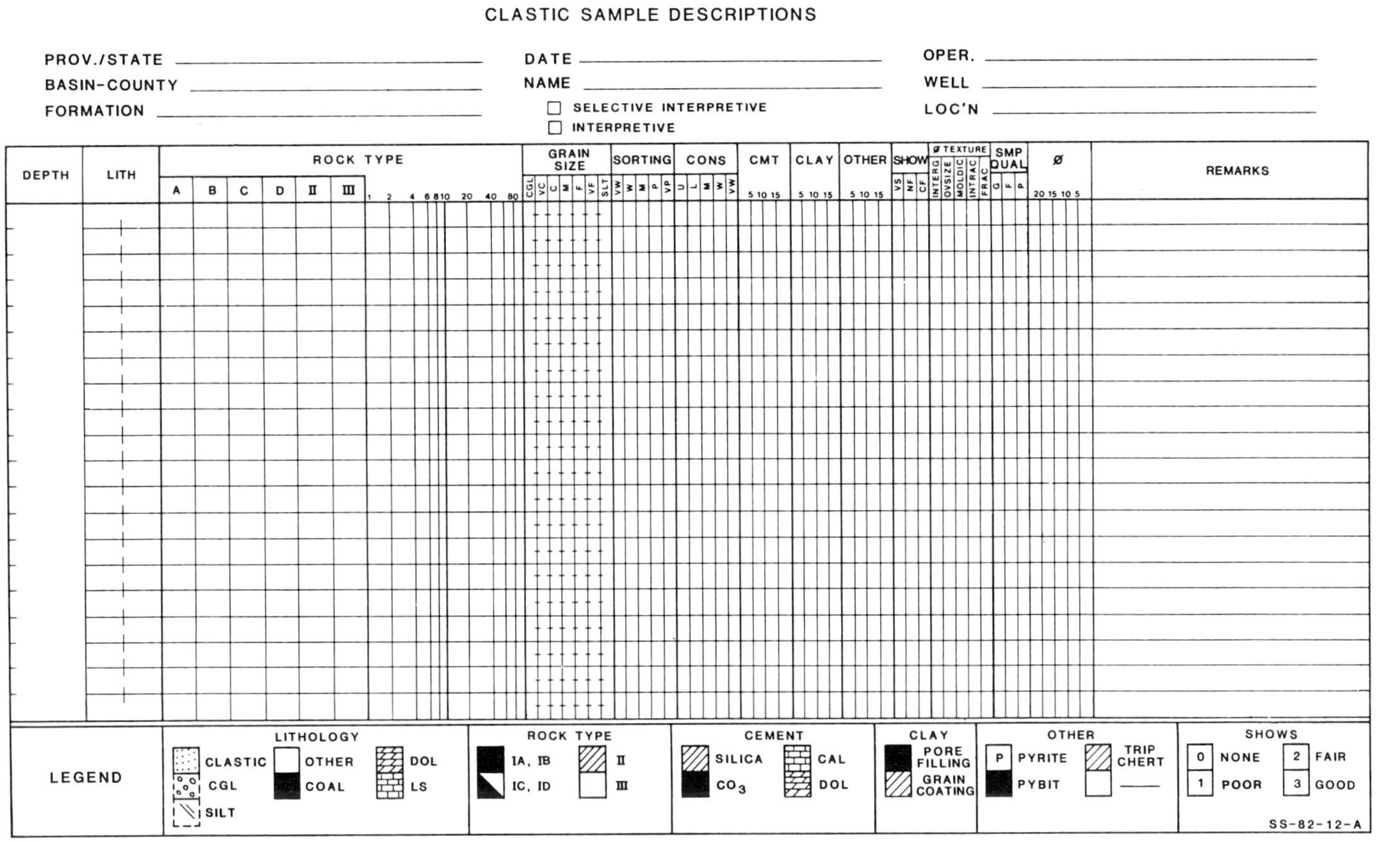

Figure 1. Sample description sheet for cuttings and cores.

and therefore have subdivided Type I rocks into four subgroups with the following air permeability ranges:

- IA, greater than 100 md;
- IB, 10 to 100 md;
- IC, 1 to 10 md; and,
- ID, ± 0.5 to 1 md.

We cannot overemphasize that all our work is based on observing freshly-broken, dry rock surfaces (cores, cuttings, outcrop samples) at 20X magnification. Most of the key rock-pore properties we use are not observable at 10 to 12X, the usual magnification of many binocular microscopes. In practice, the observations we make at 20X are usually confirmed with a "quick view" of the rock at 40 to 60X magnification.

Methods and aids useful to recognize and classify pore types and estimate permeability are discussed in the next section.

AIDS TO RECOGNIZE PORE TYPE AND ESTIMATE PERMEABILITY

The key to our ability to recognize and classify pore types and estimate permeability is detailed studies of a variety of different reservoir rocks using thin sections, the scanning electron microscope, and X-ray diffraction. These rocks are from conventional core. We compare the rock-pore description derived from these studies with detailed petrophysical studies and with our binocular microscope description (at 20 to 60X). Figure 4 shows some of the tests made on whole and plug-size core samples and the steps in preparing and calibrating the pore-type comparators.

The primary observations made with the binocular microscope (at 20X) on the core samples are grain size, sorting, consolidation, and volume and distribution of cements, pore-filling material, visible pores, and pore throats. We compare these with observations from thin sections, photomicrographs of thin sections, SEM photographs of the rock itself (at magnifications up to 5,000 times), and plastic replicas (pore casts) of the pore space. Next, we compare all these rock-pore observations with the various petrophysical analyses, especially porosity and permeability (measured unstressed and under stress). Calibrating rock-pore and petrophysical parameters on the same samples is of fundamental importance in learning how to estimate pore types and permeability from chips or plugs of rock.

We have established a reference collection of rock-pore-petrophysical properties for typical reservoir rocks for the Deep basin. In new basins or formations where we have no core, that is, where there are only cuttings, we examine the cuttings, decide what rock and pore types may be present and pick reference rocks from basins which we think have similar rock-pore types. As mentioned previously, when cuttings of the Falher sandstones and conglomerates were first studied, we used the Pennsylvanian sandstones and conglomerates from the Anadarko basin of Oklahoma for calibration until Falher cores were cut. Figure 5 shows a cross plot of porosity versus permeability (corrected for Klinkenberg effect) for these Pennsylvanian sandstones and conglomerates. In this Oklahoma oil field, as in the Falher, we found that porosity, permeability, and

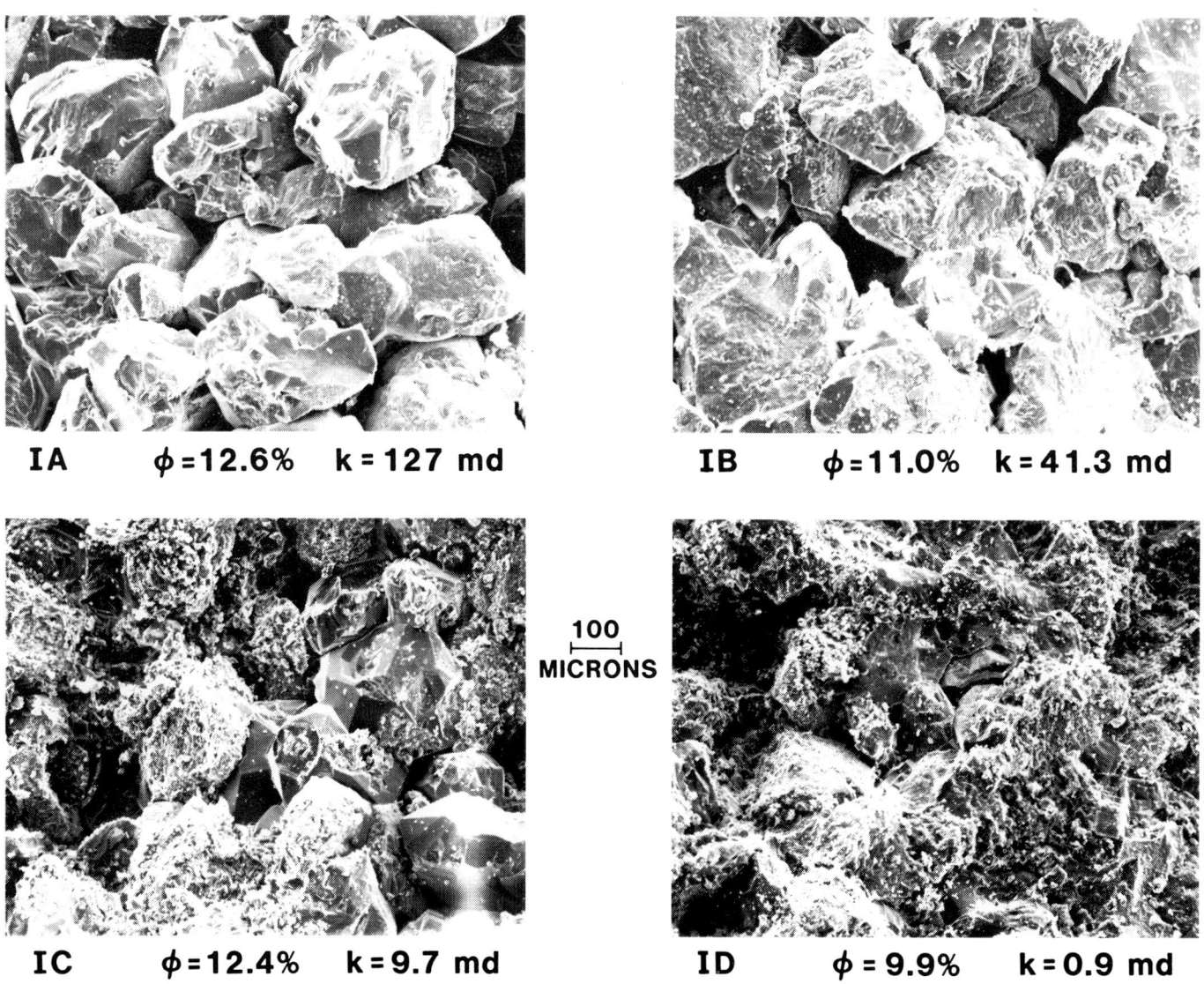

Figure 2. Scanning electron microscope photographs of pore types IA, IB, IC, and ID.

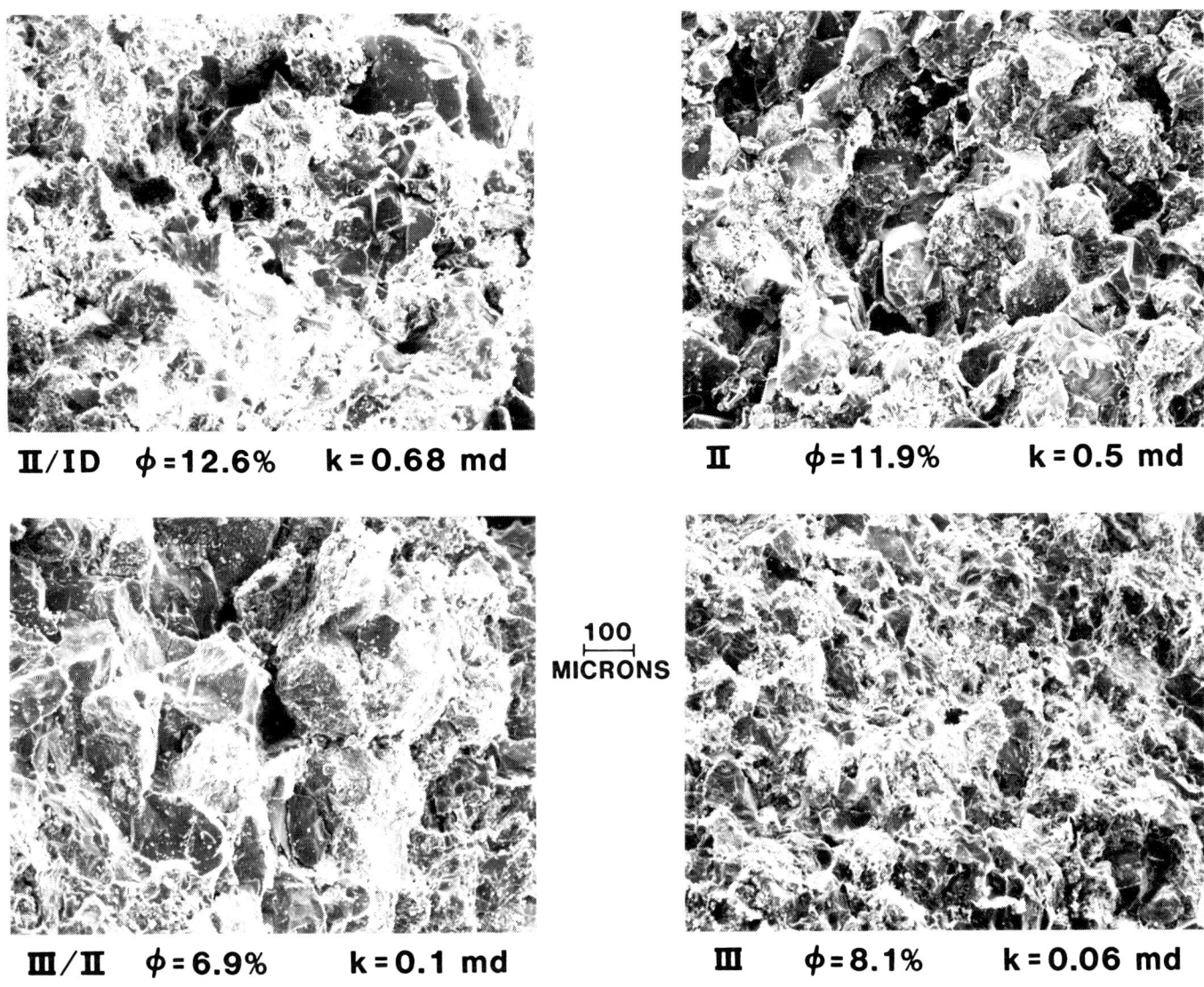

Figure 3. Scanning electron microscope photographs of pore types II/ID, II, III/II, and III.

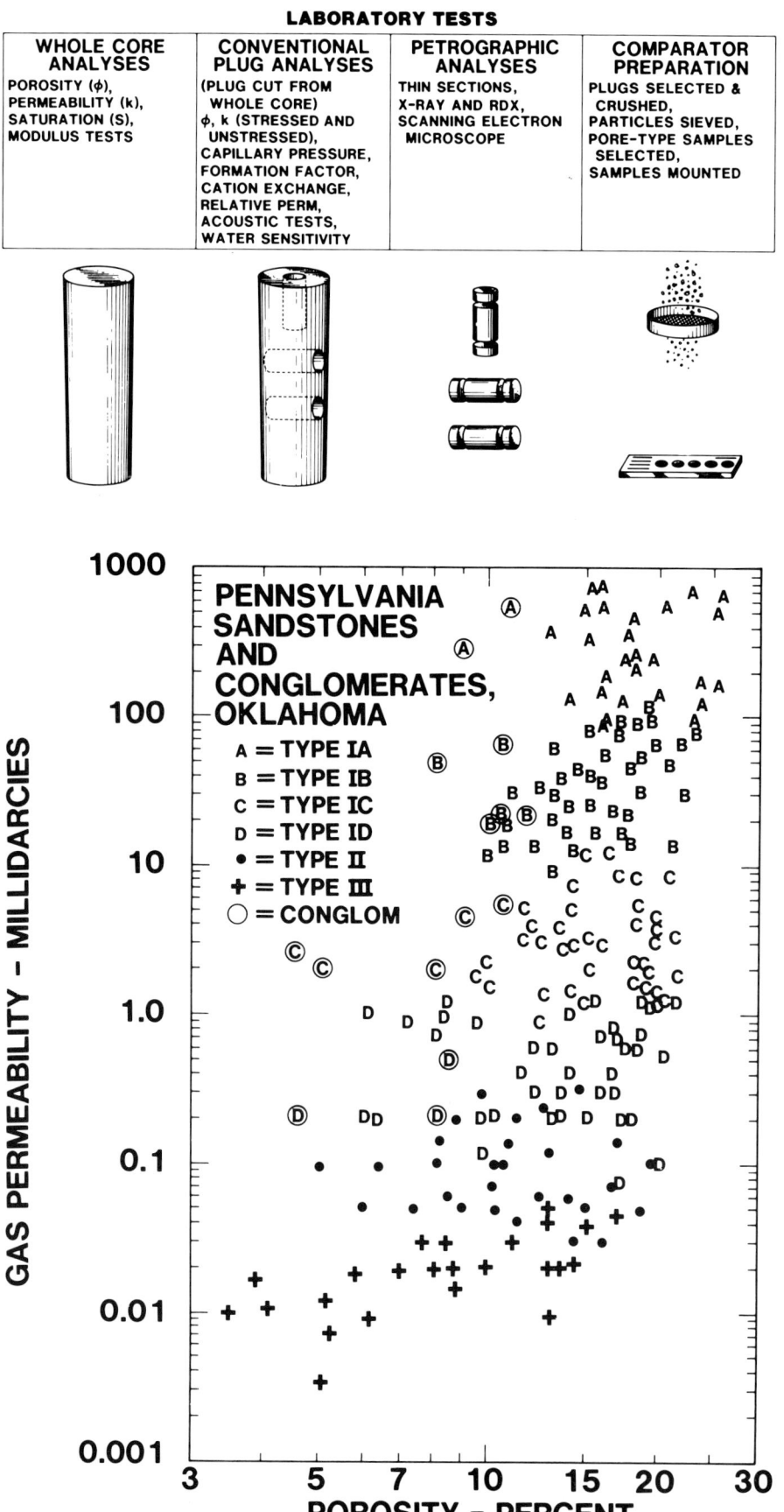

Figure 4. Laboratory tests, measurements and procedures for preparing and calibrating pore-type comparators.

pore geometry are related to grain size and sorting, cementation and compaction, consolidation and the amount of pore-filling clays (Sneider et al, 1977).

To set up our rock-pore type calibration for the 15 major Cretaceous reservoir units in the Alberta basin, over 10,000 ft (3,045 m) of conventional core in over 200 wells were examined. Over half the core was specifically taken by Canadian Hunter to obtain new petrophysical parameters, as well as to provide stratigraphic and depositional/environmental information to define plays and prospects, and for development drilling. To determine the characteristics of the pore types of intervals that will or will not produce, the critical step is to compare (1) the rock-pore properties observed using a binocular microscope (at 20X) aided with petrographic and scanning electron microscope studies, with (2) petrophysical properties established from well logs and cores which in turn are compared, with (3) drill-stem and production tests and production of the intervals of interest.

The following visual aids are helpful to train geologists and engineers to make rapid and accurate estimates of porosity, permeability and a rock's reservoir potential: (1) plastic trays of cutting-size rock chips crushed from conventional cores of known rock-pore types, porosity, and permeabilities (Figs. 4 and 6); (2) calibration sets of rock slabs and simulated cuttings (Fig. 7); and (3) photographs of rock-pore types (Fig. 8) taken of a freshly-broken rock surface at 20X, thin sections at 32X and 125X, a rock surface with the SEM at 100 to 5,000X and a replica of the pores (pore casts). Figure 8 is a sample page for the rock-pore type comparator set shown in Figure 7 and pore type IB of the topmost comparator in Figure 6.

Figure 5. Porosity-permeability cross plot showing pore types of Pennsylvanian sandstones and conglomerates, Elk City field, Oklahoma.

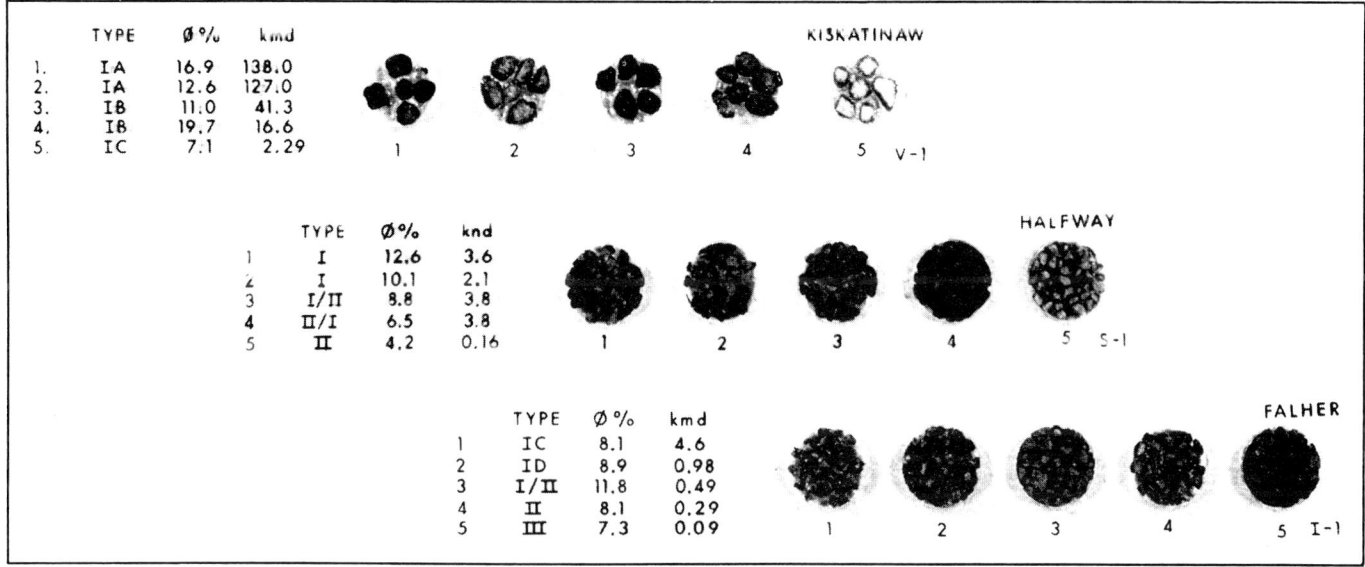

Figure 6. Pore-type comparators for three Alberta Basin producing formations.

SPECIAL ROCK-PORE EVALUATION PROBLEMS

When working with cuttings, three rock-pore types present difficult evaluation problems. These are conglomerates, very shaly or clay-mineral rich sandstones, and rocks with abundant microporosity within clay-mineral aggregates or microporous chert.

Fragments of gravel-size particles develop as the drill bit cuts through a conglomerate. The cuttings that result rarely show any pores. Careful attention must be paid to those fragments which show part of the original particle surface. Small crystals of cement on these surfaces commonly are the only clue that an open

Figure 7. Pore-rock type calibration sample, Kiskatinaw formation, Alberta.

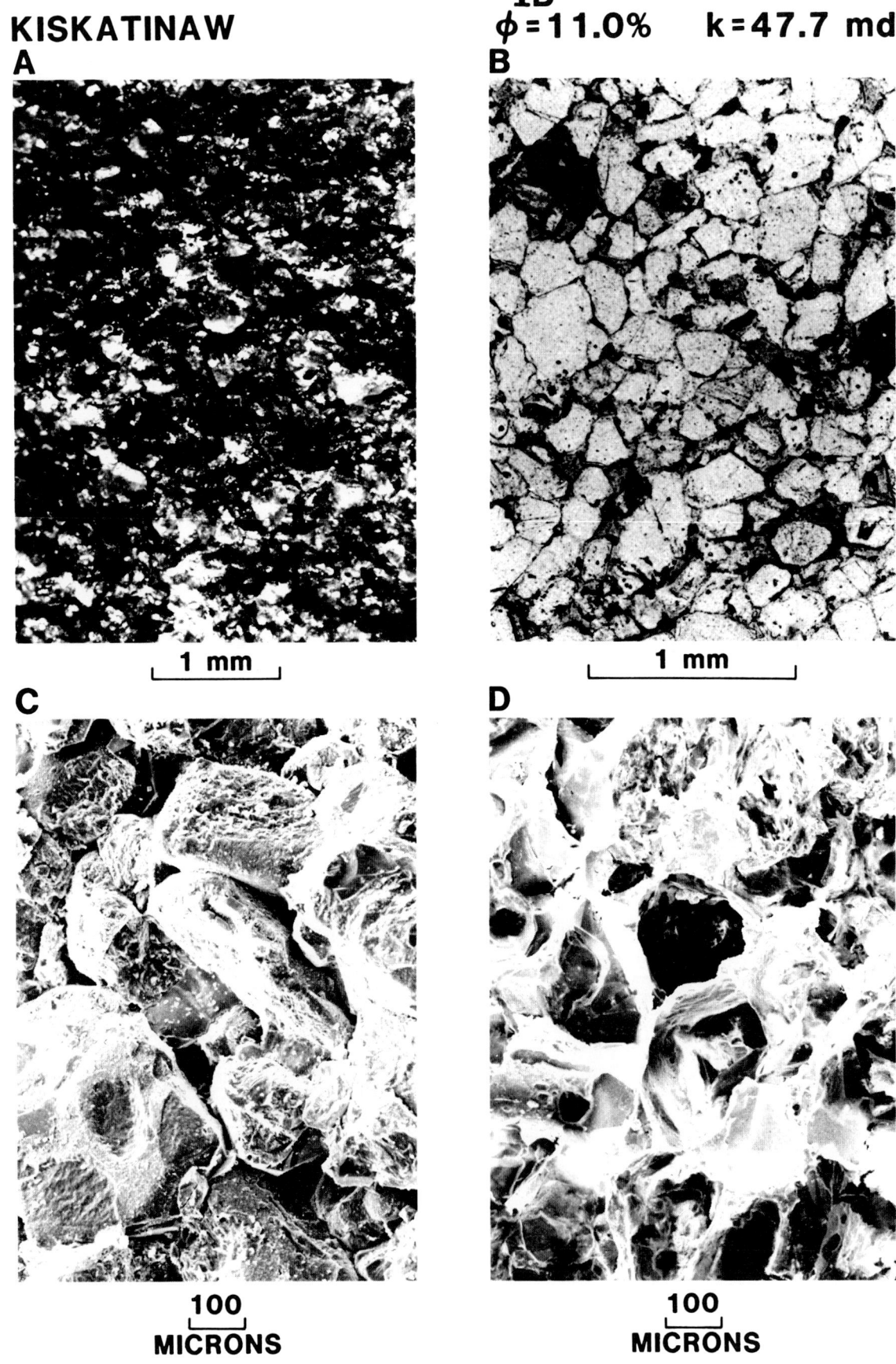

Figure 8. Photographs illustrating pore structure of a Kiskatinaw IB pore type. (A) freshly broken surface at 20x, (B) thin section, (C) SEM view, and (D) SEM view of pore cast.

pore was there before fragmentation into a chip. Grain surfaces that have parts of large equant carbonate or quartz crystals on them generally indicate cemented rock in many reservoirs. Our experience suggests it is very difficult to accurately pore-type most conglomerate rock. Porous and permeable conglomerates generally damage easily, especially by plugging with whole mud. Evaluation of a conglomerate interval by drill-stem testing can be inconclusive or indicate an apparent "tight" rock. We have found many examples of low-porosity conglomerates (less than 7 to 8%) that flow substantial amounts of hydrocarbons because of excellent permeability.

Evaluation of shaly or clay-rich rocks are a problem encountered by all who work with them (Keighin, 1979; Keighin and Sampath, 1980; Neasham, 1977a; Pittman and Thomas, 1978; and Wilson and Pittman, 1977). Our calibration rock-pore sets of these rocks require special handling. The permeabilities used in the comparators must be measured under stress and corrected for Klinkenberg effects (Klinkenberg, 1941) to obtain a reasonable value of permeability. Water-sensitivity tests of core samples are very helpful to evaluate the effects of swelling clays. With the aid of SEM studies, especially of pore casts, we believe we can evaluate the pore type in these difficult rocks with careful study. Putting fresh water on cuttings and seeing how the clays respond helps to evaluate the influence of clays on the pore system and thus one can have some basis for assigning a pore type. Most clay-rich, fine-grained sandstones usually are Type II rocks.

Microporosity of cherts and clay aggregates (Fig. 9) are another evaluation problem (Kieke and Hartmann, 1973; Neasham, 1977b; Pittman and Thomas, 1978). We believe that using 20X magnification, we can "see" a lot of particles that have secondary porosity. Using binocular microscopes that can view rocks at 40 to 60X magnification, we believe we can make reasonable estimates of the volume of secondary pores. Study of pore casts (Fig. 9) are helping us to better recognize and evaluate microporosity.

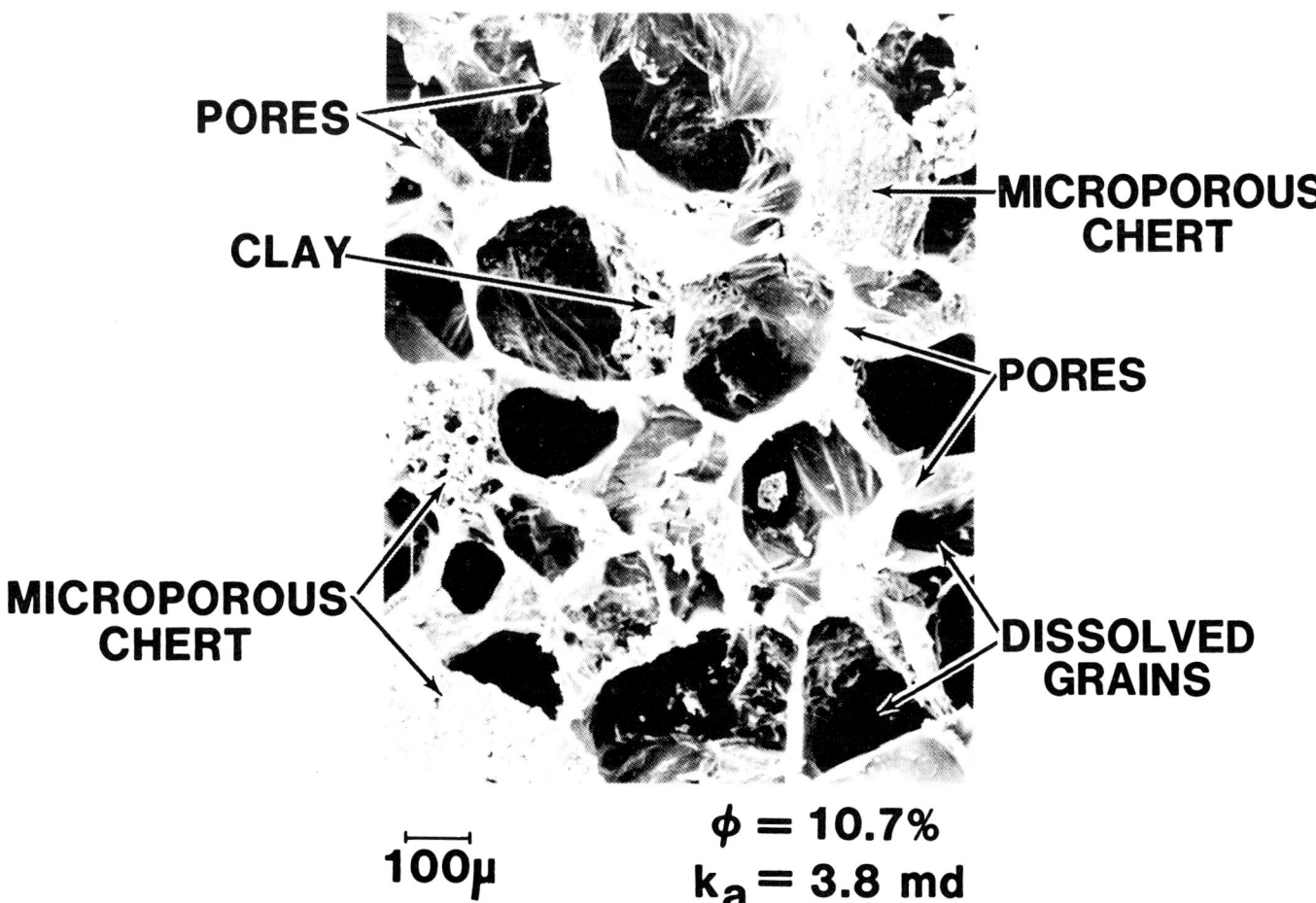

Figure 9. A scanning electron microscope photograph of pore cast showing micro pores in a Kiskatinaw sandstone, Elmworth field, Alberta.

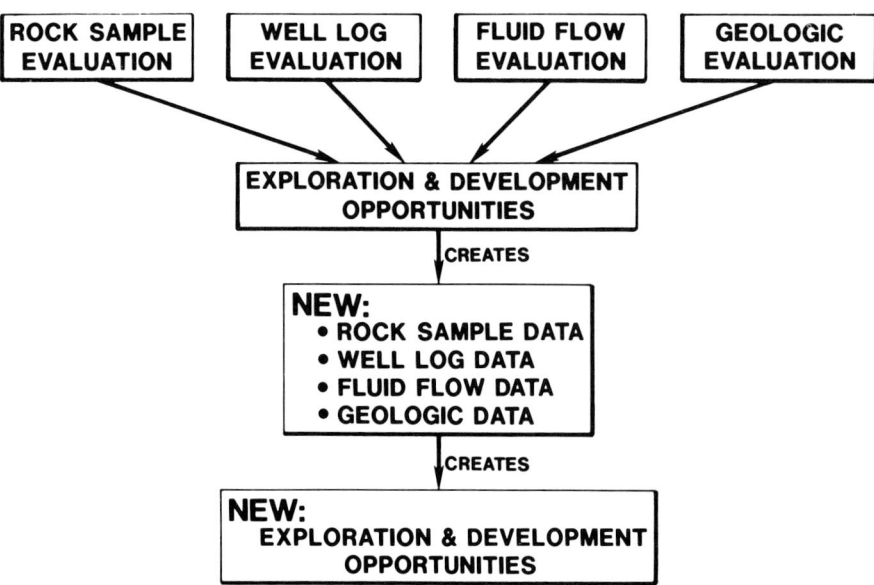

Figure 10. Value of integrated geological-petrophysical-petroleum engineering studies.

In cutting samples composed mostly of loose grains one is faced with the decision of whether or not the loose grains represent unconsolidated or loosely-consolidated sands or sandstones (Type I) or rocks that have abundant clays which fall apart when exposed to drilling fluids or rocks that "explode" apart when brought to the surface. We know of no easy way to consistently tell why a sample contains loose grains. Examination of grain surfaces sometimes is a clue to the origin of loose "grain" cuttings.

CONCLUSIONS

We found it very useful to classify sandstones and conglomerates into pore types for the purpose of establishing reservoir-rock potential. The system we use, together with the variety of visual aids illustrated in this paper, makes possible rapid detection and characterization of reservoir rocks in drill cuttings at the well site, using a binocular microscope equipped with at least 20-power magnification. Comparators are useful to estimate permeability.

Integration of rock-pore type data with laboratory petrophysical data, log-derived data, and geology has resulted in the discovery of major new reserves. We believe that the concepts and methods we use make for more efficient and effective delineation of exploration opportunities, identification of bypassed pay zones, well-log interpretation, and identification and evaluation of completion intervals. Figure 10 expresses our view of the value of continued integration of geologic and petroleum engineering data in exploration and development. We found that continued incorporation of petroleum-engineering data and ideas into exploration is an essential ingredient to successful ventures in many of today's reservoirs.

REFERENCES

Archie, G. E., 1950, Introduction to petrophysics of reservoir rocks: AAPG Bulletin, v. 34, p. 943–961.
——— , 1952, Classification of carbonate reservoir rocks and petrophysical considerations: AAPG Bulletin, v. 36, p. 278–298.
Keighin, C. W., 1979, Influence of diagenetic reactions on reservoir properties of the Nelsen, Farrer and Tuscher formations, Uinta Basin, Utah: Denver, Society of Petrology Engineers Symposium, SPE-7919.
——— , and K. Sampath, 1980, Evaluation of pore geometry of some low-permeability sandstone, Uinta Basin: Journal of Petroleum Technology, v. 34, p. 65–70.
Kieke, E. M., and D. J. Hartman, 1974, Detecting microporosity to improve formation evaluation: Journal of Petrology Technology, v. 26, p. 1080–1086.
Klingkenberg, L. J., 1941, The permeability of porous media to liquids and gases: Drilling and Production Practice, American Petroleum Institute, p. 200–213.
Masters, J. A., 1979, Deep Basin gas trap, Western Canada: AAPG Bulletin, v. 63, p. 152–181.
Neasham, J. W., 1977a, The morphology of dispersed clay in sandstone reservoirs and its effect on sandstone shaliness, pore space and fluid flow properties: Annual Meeting, Society of Petroleum Engineers, SPE-6858.
——— , 1977b, Applications of scanning electron microscopy to the characterization of hydrocarbon-bearing rocks: Scanning Electron Microscopy, v. 1, p. 101–108.
Pittman, E. D., and J. B. Thomas, 1978, Some applications of scanning electron microscopy to the study of reservoir rock: Journal of Petroleum Technology, v. 31, p. 1375–1380.
Sneider, R. M., et al, 1977, Predicting rock geometry and continuity in Pennsylvanian reservoirs, Elk City Field, Oklahoma: Journal of Petroleum Technology, v. 29, p. 851–866.
——— et al, 1983, Methods for detection and characterization of reservoir rock, Deep Basin gas area, Western Canada: Journal of Petroleum Technology, v. 35, p. 1725–1734.
Wilson, M. D., and E. D. Pittman, 1977, Authigenic clays in sandstones; recognition and influence in reservoir properties and paleoenvironmental analysis: Journal of Sedimentary Petrology, v. 47, p. 3–31.

Part II

Well Log Analysis Methods and Techniques

R. W. Hietala
E. T. Connolly

*Canadian Hunter Exploration, Inc.
Calgary, Alberta*

INTRODUCTION

During the exploration and exploitation phases of a major hydrocarbon-bearing basin (Masters, 1982; Smith, 1984), Canadian Hunter Exploration developed and applied useful techniques to integrate subsurface well-log data with rock information. Integration of rock-log data is essential for determining answers to three basic questions: (1) Where do producible hydrocarbons occur in a basin? (2) What is the lithology and pore volume of the reservoir units? and (3) What are the hydrocarbon volumes and flow characteristics?

Resolution of the above points have useful application as an analog to future fields.

The Elmworth field in the Deep basin of western Alberta is the area of study in this paper. Figure 1 illustrates the geographical location of this major hydrocarbon-producing area. A Cretaceous correlation chart (Fig. 2) indicates the formations that are the subject of discussion for the rock-log comparisons. Principal gas-bearing zones in this vertically-stacked sequence have been highlighted.

The application of rock-log calibrations using pore-type classifications and Canadian Hunter-developed digital log analysis techniques has resulted in an accurate porosity and hydrocarbon analysis method. Successful delineation of hydrocarbons in the Deep basin requires the understanding of subtle changes in rock characteristics and well-log responses.

In the early stages of basin exploration large amounts of well-log data were visually scanned for log anomalies. Every

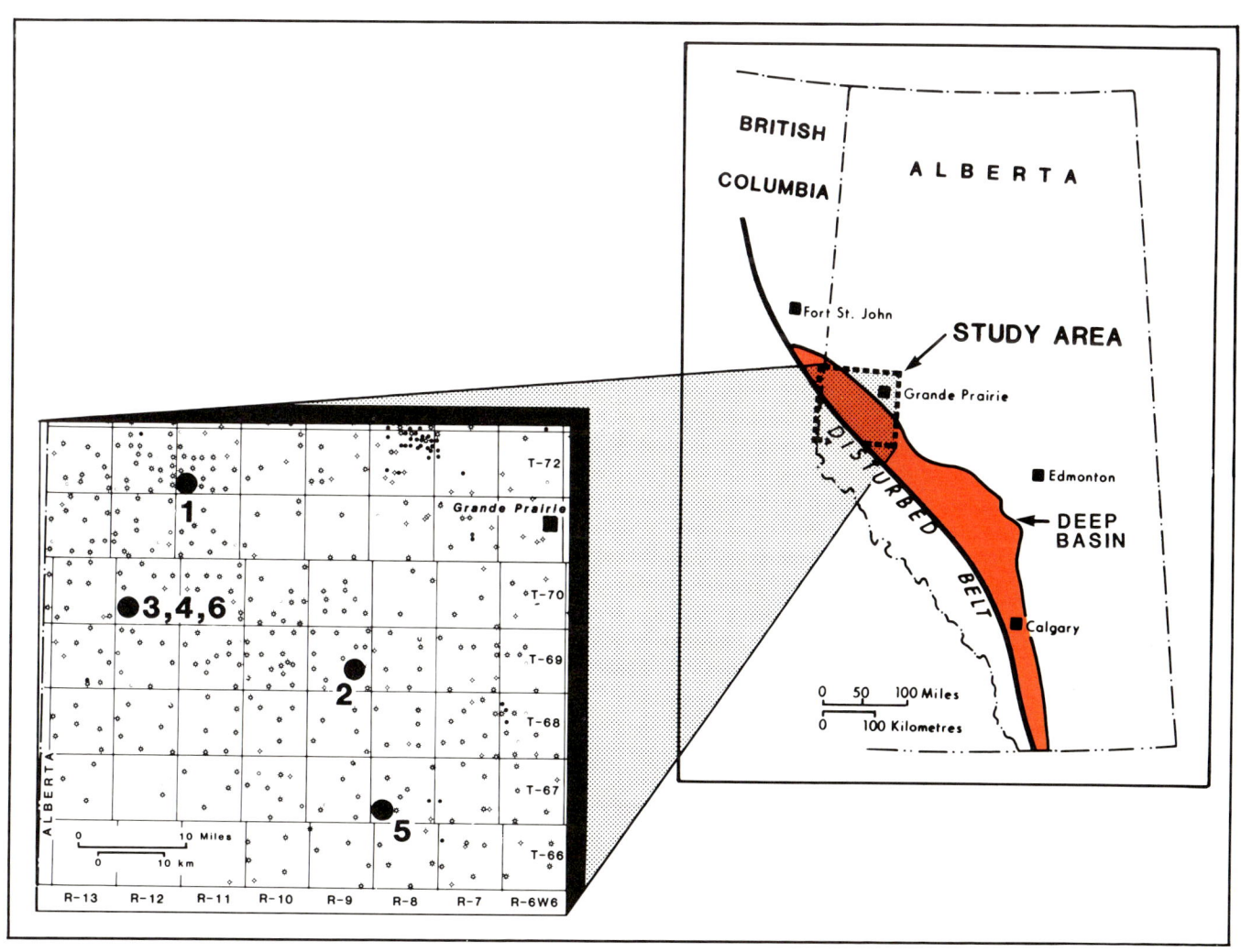

Figure 1. Location map of study area.

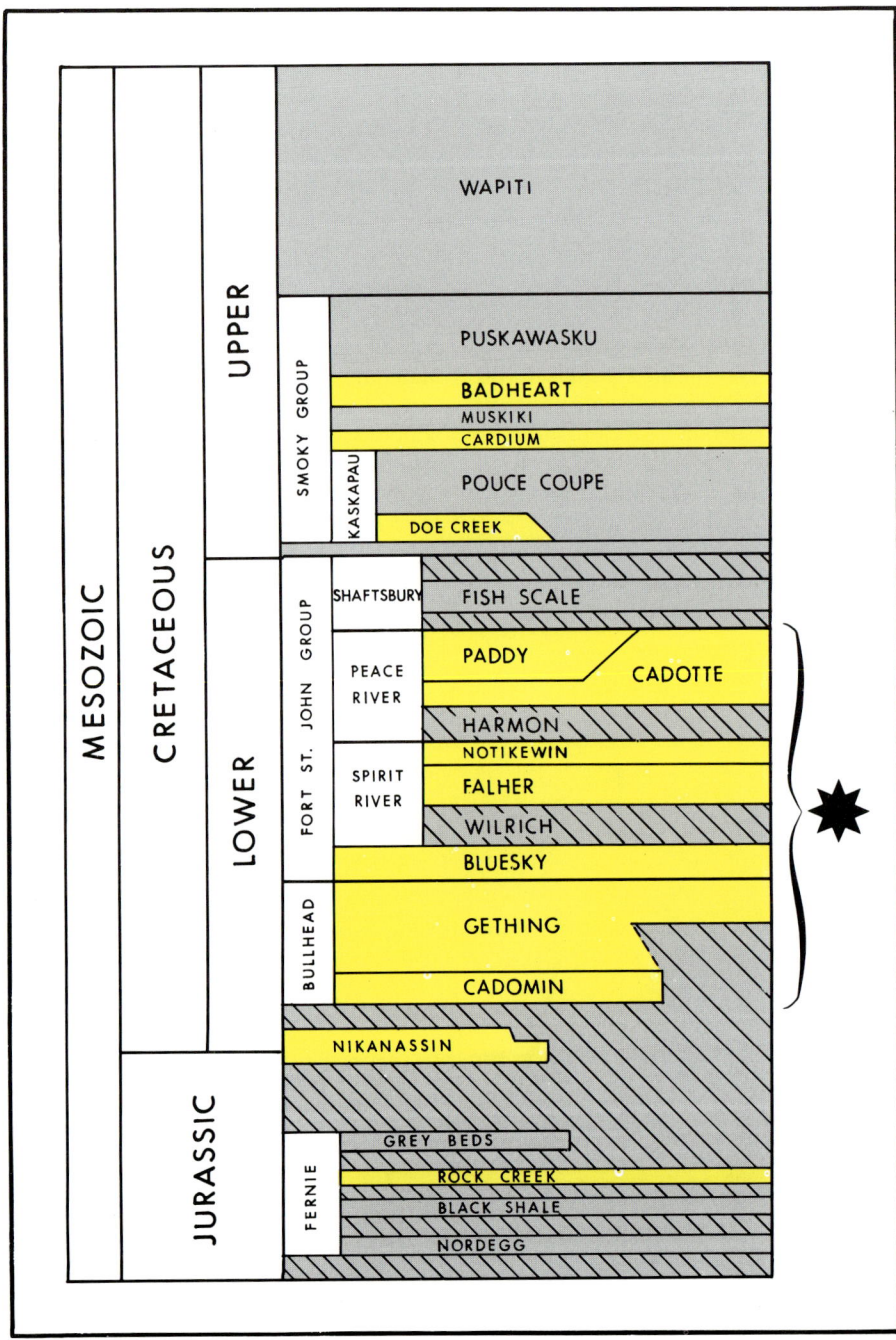

Figure 2. Stratigraphic correlation chart, Jurassic-Cretaceous.

available well log was examined for potential shows and the shows were mapped for trend-setting patterns. Prospective wells were first analyzed by hand computation techniques and curve overlay methods. It is important to note that all porosity within a well bore was evaluated and recorded regardless of zone. This was necessary because input constants and cutoff values usually change with increased control, and having the basic data available for re-calculation results in earlier area interpretations.

QUICK-SCAN ANALYSIS

"Quick-scan" analysis enables one to gather a large amount of data on an area in a short time. The procedure can be summarized as follows: (1) Visual comparisons are made for resistivity-porosity anomalies to determine areas of prime potential; (2) Porosity and water saturations are calculated for porous zones in all wells; (3) Input data are keyed into a computer for basic log analysis and information storage; and, (4) Porosity versus water-saturation crossplots are made for recognition of hydrocarbon limits.

Composite porosity versus water-saturation crossplots for a given rock type were made on a formation-by-formation basis. Figure 3 (Connolly et al, 1983) illustrates both productivity and rock-type classification on a porosity versus water-saturation crossplot from initial basic "Archie type" log analyses (Archie, 1942). Flow tests for each series of data points were plotted, and porosity-water saturation product lines to be used as cutoff or screening parameters were determined. It is important that this cutoff be related to rock type when comparisons are made. Intervals meeting individual zone cutoff values were used to calculate hydrocarbons in place and recoverable reserves in terms of billions of cubic feet (bcf) per section. Porosity, water saturation, and depth were used to make empirical estimates of gas-flow rate, and these rates were calibrated against drill-stem test and production data. Individual formation maps of bcf/section and computed flow-rate values (XOF) were constructed to highlight prospective areas of acreage acquisition.

While the reconnaissance or "quick-scan" is being continued a more complete computerized analysis is made of local areas that have been highlighted as having potential. To maintain flexibility and handle large amounts of data an in-house minicomputer system dedicated to log analysis is used.

The use of both "quick-scan" and digital analysis techniques provides a rock-log calibration method that satisfies large-area basin studies. For the determination of rock-log calibrations a multi-discipline approach, taking expertise from fields of geology, petrophysics, and hydrodynamics, was found to be the most effective.

DEVELOPMENT OF EVALUATION TECHNIQUES

Historical Need for Rock-Log Calibrations

Before discovery of the Falher sand in the Elmworth field of Alberta nearly 100

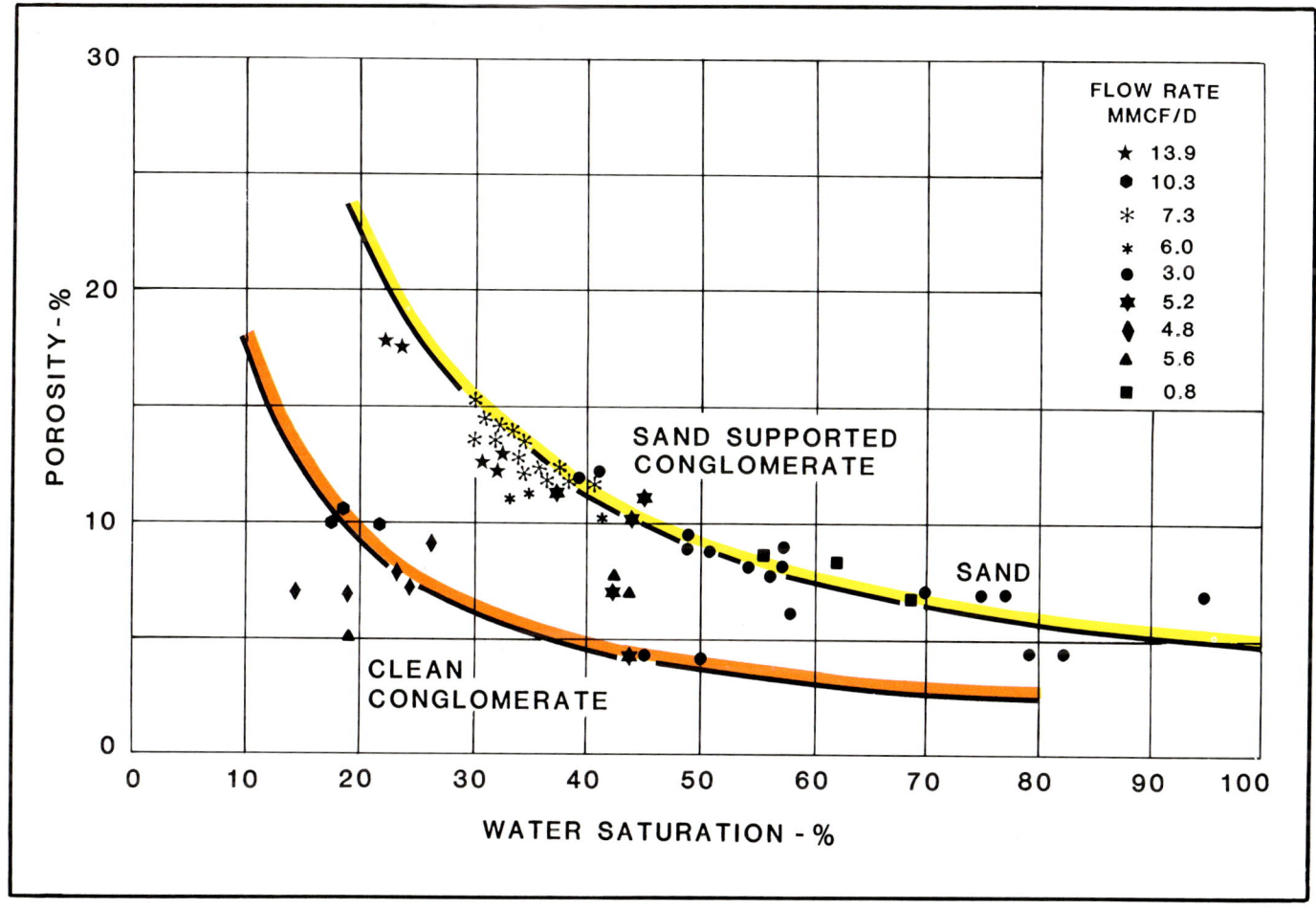

Figure 3. Composite porosity - water saturation cross-plot.

wells had been drilled in a 7,000 sq mi (18,130 sq km) area. Most of these wells were deep Paleozoic tests or shallow Cretaceous oil plays. The major hydrocarbon-producing zones in the Cretaceous of the Deep basin were by-passed because of a combination of factors listed below:

(1) Major companies were drilling for deeper Paleozoic carbonate oil production;
(2) Non-carbonate lithologies were not closely examined;
(3) At the time of initial drilling, shallower low-productivity gas was not economic;
(4) Formation damage in the Cretaceous sands was not understood;
(5) Hydraulic fracturing had not been applied to stimulate formation-damaged wells;
(6) Clastic facies interpretations were not fully understood when the basin was first explored;
(7) Porosity and permeability relationships of the main hydrocarbon zones were not clearly known when drilling first started;
(8) Log relationships in the shallow zones were different from the standard clastic assumptions; and,
(9) Integrated rock-log calibrations were not applied.

The purpose and aims of a fully integrated analysis to evaluate the Deep basin can be outlined as follows: (1) lithology identification (sandstone, conglomerate, coal, and shale); (2) facies interpretation, (channels, beaches, bay, lagoon); (3) porosity identification, (capacity of a rock volume to contain fluids); (4) determination of fluid distribution, (relative proportion of each pore fluid); (5) estimation of flow potential (ability of fluids to flow through the rock being analyzed); and (6) parameter identification (provision of calibrated parameters to replace extensive coring).

A continuous recording of rock- and fluid-calibrated well-log curves is the goal sought in petrophysical calibrations. Drilling and coring costs indicate this to be the most practical approach, as cores are not always available. In a vertically-stacked pay section such as that shown in the correlation chart (Fig. 2) a well-calibrated data set is needed to optimize completion intervals and procedures. In the Deep basin the identification of lithology from well logs calibrated to core data has proved to be a valuable tool in the integrated, formation-evaluation approach.

Until recent years production of hydrocarbons in commercial quantities from clastic rocks with porosities less than 10% was felt to be uneconomic. This perception left a "gray zone" open for opportunity, but close integration of rock and log information is needed to make this opportunity an economic venture.

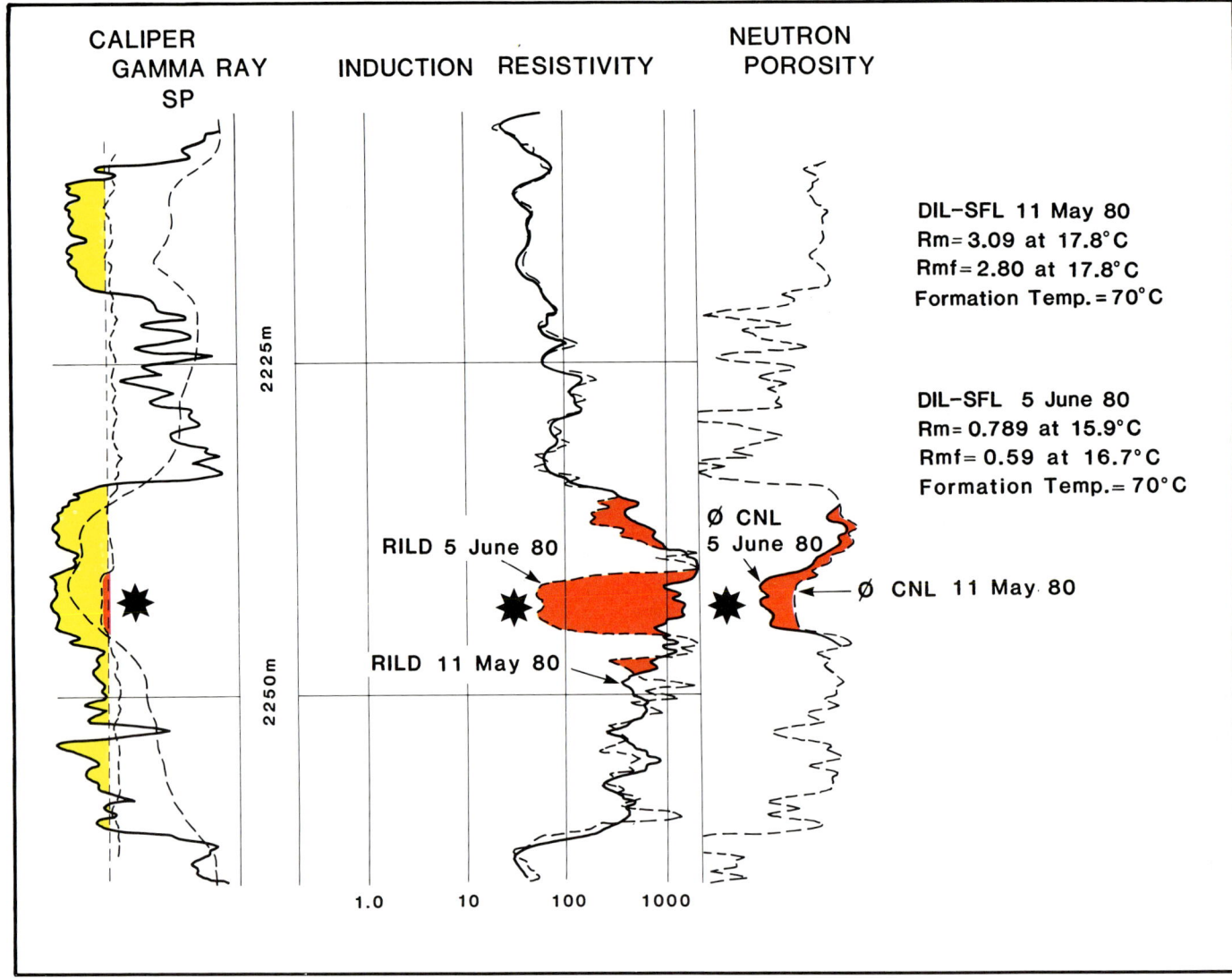

Figure 4. Mud invasion effects on well logs.

Data Type and Quality

Available Well-Log Data—In the Elmworth area many older wells had only electrical logs and micrologs whereas later wells had acoustic-gamma and induction logs. The basic logging program now used to evaluate the Deep basin Cretaceous section consists of: (1) bore-hole-compensated acoustic device; (2) compensated neutron/formation density log; (3) dual induction-laterolog device; and (4) microlog-shallow resistivity device.

Fluid Invasion Problems—Multiple logging runs have eliminated many problems associated with drilling-fluid invasion and bore hole rugosity. The effects of fluid invasion change (with time) can be seen in Figure 4. In this example the same formation has been logged twice over a 25-day period. Note the effect of prolonged invasion over the zone flagged with a star. In the first run on May 11, 1980, the resistivity log read 1,000 ohms. On June 5, 1980, however, because of prolonged contact with drilling mud, the zone is deeply invaded. The apparent formation resistivity has dropped to 55 ohms due mostly to invasion of the more saline mud filtrate. The later neutron log shows a higher porosity due to removal of gas saturation near the well bore. Accurate log analysis is best achieved, therefore, by logging zones of interest as soon as possible after penetration.

Borehole Enlargement—In the early stages of the Deep basin development it was noted that formation density logs were often of poor quality due to bore-hole rugosity. From four-arm caliper measurements of dipmeter tools there were many indications of elliptical bore holes. This phenomenon was noticed by several authors (Bell et al, 1981; Connolly, 1974; and Cox, 1970) and appears to be related to stress directions in the subsurface.

Figure 5 shows a dual caliper, presented in hourglass fashion, with opposing pairs of caliper arms labelled D1-3 and D2-4 (Connolly, 1974). Note the excessive hole enlargement, or "breakout," as shown by caliper pair D2-4 versus the apparent undergauge hole as shown by caliper pair D1-3.

Integrated Rock-Log Calibration, Elmworth Field

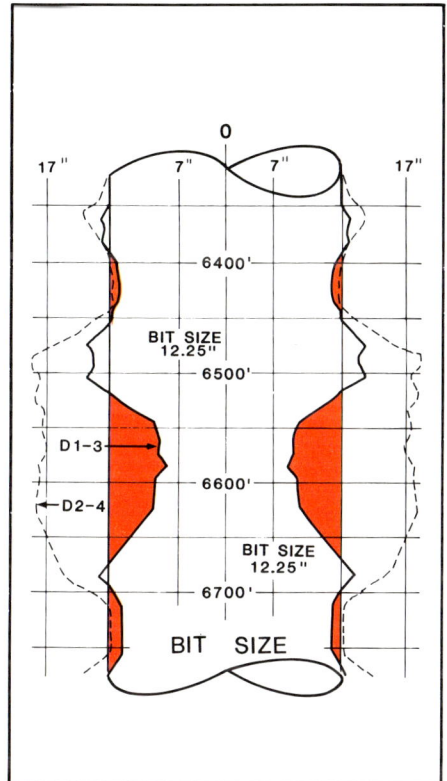

Figure 5. Four-arm caliper-hole orientation study.

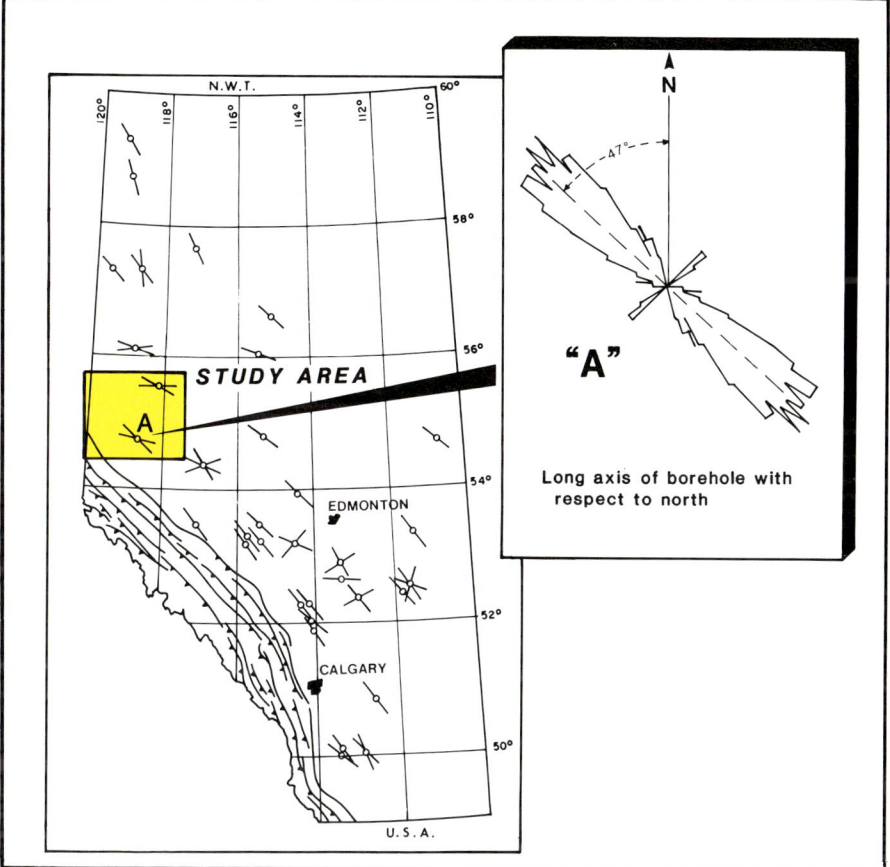

Figure 6. Orientation of long axis of bore hole.

It is this hole elongation that creates difficulty for the density log, and causes serious density-compensation problems that destroy the usefulness of the log readings. An example of hole "breakout" is given in Figure 6 along with a map of direction of "hole breakout" for Alberta, Canada (Bell et al, 1981). Note the green-shaded area which represents the Elmworth field of the Deep basin.

The close co-operation of Canadian Hunter and Schlumberger personnel resulted in the building of a prototype offset density tool to alleviate this problem. The aim of this configuration was to realign the density-measuring device into the short axis of an elliptical bore hole. This change greatly improved the quality of density-log measurements. The current configuration as used in the field is shown in Figure 7 (Schlumberger, 1981).

Figures 8 and 9 illustrate the difference in density-log curves with a change in the orientation within the bore hole using the

Figure 7. Current configuration of dual-axis caliper with 90° offset.

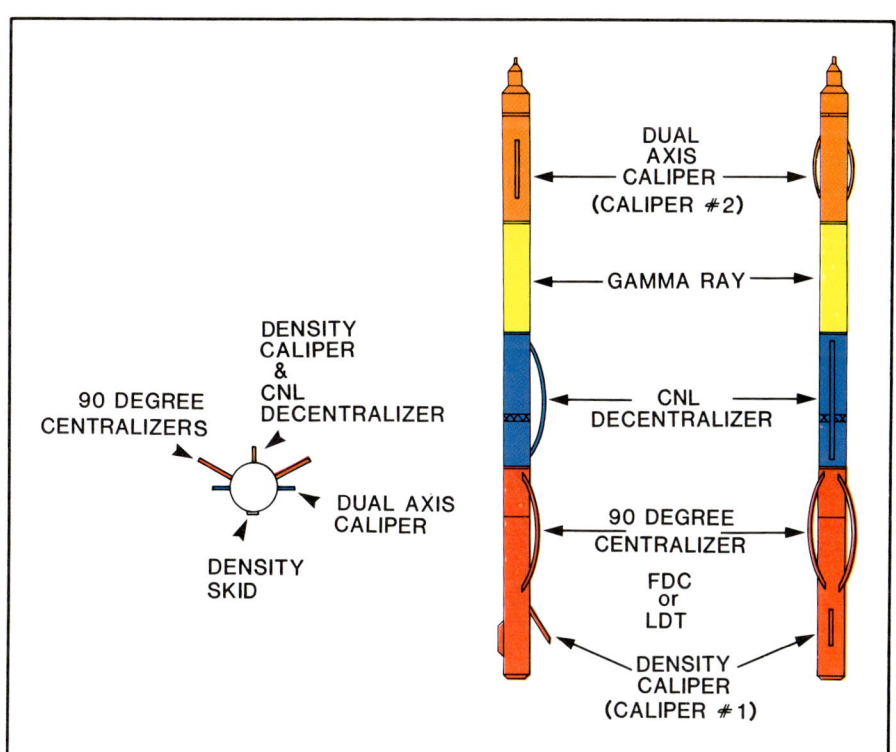

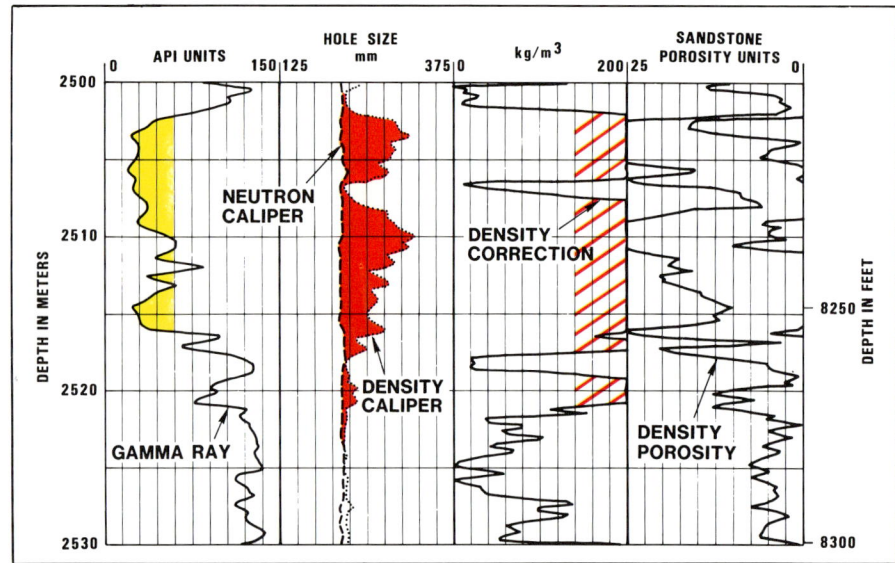

Figure 8. Density logging-pass no. 1.

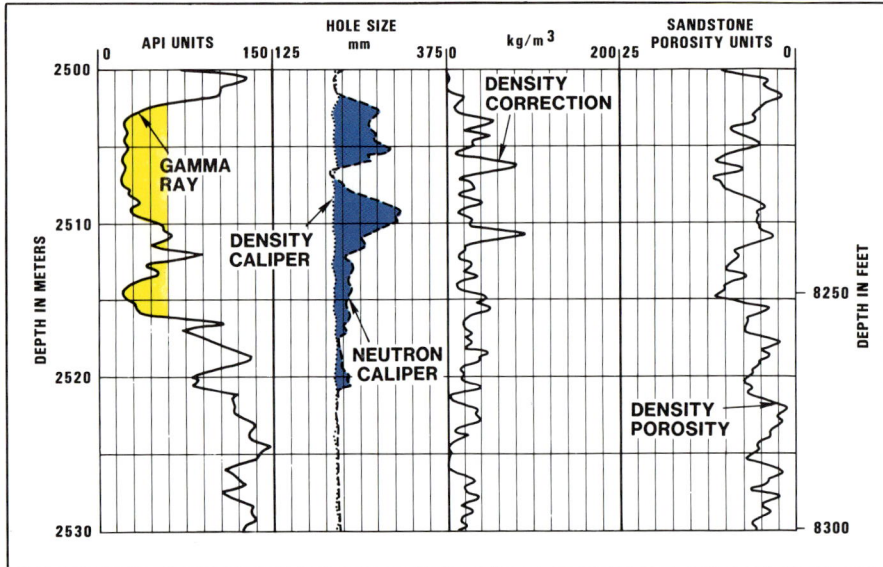

Figure 9. Density logging-pass no. 2.

new device. In Figure 8 the density caliper and measuring device are in the rugose large axis of the bore hole. Note the large density-correction values and the useless density-porosity data. Figure 9 illustrates another logging pass with the density caliper and the measuring device in the short bore-hole axis. The density correction is now within acceptable limits, and the derived porosity values can be utilized for further interpretations.

Application of New Measurements and Methods

Special circumstances and research-related programs have given rise to a supplemental logging program using new or research-stage logging tools. These tools provide useful data to identify lithologies, pore space and contained fluids. The following tools and techniques are being used as an addition to regular evaluation methods: (1) Nuclear Magnetism Logging (NML)*; (2) Natural Gamma-Ray Spectroscopy (NGT)* (NGRSL)*; (3) sidewall cores, and (4) identification of recovered waters.

Nuclear Magnetism Log (NML)*—Log analysis techniques currently in use by the industry arrive at permeability through indirect methods. The nuclear magnetism log (Herrick, 1979) provides some hope of making in situ measurements related to permeability. The nuclear magnetism tool has been used in low-porosity Deep basin sands to indicate porosity and permeability levels. Figure 10 (Best, 1983) illustrates computed permeabilities from a NML* device versus core permeabilities on a Deep basin development well with varying lithology.

Natural Gamma-Ray Spectroscopy (NGT)* (NGRSL)*—Gamma-ray curves have been used in the past to indicate apparent shale volumes. Natural gamma-ray spectroscopy (Westaway et al, 1980; Fertl et al, 1982) has given indications of being a quantitative rather than a qualitative clay measurement technique.

Horizontal shale beds between reservoir units often are barriers to vertical fluid flow. Gamma-ray responses normally interpreted as shale barriers between sands and conglomerates may be misleading. This tool differentiates shales that are true barriers from those that appear as shales due to mineralogical differences. Figure 11 illustrates such a case. The total gamma ray, on the left, shows an increase at the base of a porous and permeable conglomerate. This is not a shale, but the increased gamma-ray response is due to the presence of a radioactive isotope of uranium (Zone A). The gamma-ray spectroscopy data indicate low levels of potassium and thorium and the porosity-permeability plots show that no shale barrier exists. A core photo of the contact shows conglomerate and sand to be present at this interface.

The NGT* tool also has been useful in defining porous carbonates of the Triassic Baldonnel Formation that previously were interpreted as shales interbedded with dolomites. The cased-hole version of this tool can be applied to older wells in mature basins to identify potential by-passed reservoirs.

Sidewall Cores—Sidewall cores can provide information on the presence of conglomerates by carefully examining the condition of the recovered core barrels.

Note: NML and NGT are trademarks of Schlumberger; NGRSL is a trademark of Dresser Atlas. Where these registered trademarks appear in the text, an asterisk* will appear.

Many times a shattered or lost core barrel in a given clastic unit is a good conglomerate indicator. Local experience is needed in evaluating this data.

Identification of Recovered Waters—At times a Drill Stem Test (DST) will recover significant volumes of fluids that appear to indicate that the zones tested have recovered formation water. Figure 12 (Connolly and Reed, 1983) shows data from a well with a DST recovery of 2,430 ft (740 m) of water over a zone of interest. By plotting the chemical analysis of the drilling-mud filtrate water versus the analysis from the recovered water at different drill-pipe intervals, one can determine the presence and amount of mud filtrate. In this example most of the recovery was mud filtrate water. Mud filtrate and formation water is mixed together in the recovered fluid column.

An internally developed system also checks relative density, pH, refractive index, and resistivity (Rw) for each sample. An example of this form of identification is given in Figure 13 where water recoveries from a drill-stem test are compared to a sample of water from the drilling mud. Examination of the data clearly defines this test recovery to be formation water, and the Rw from this zone may be used for log analysis.

DATA MANAGEMENT AND PROCESSING

The digital-log database is the collection point for formation-evaluation data. Each log database contains information on a large geological interval for a particular well. The reason for creating a log database is that large volumes of data are constantly being reassessed, and as new development areas are identified, information is added and existing data are updated.

Sources of Input to the Log Database

Each log database uses a well-location-derived naming convention for easy reference. The well-log database contains digital open-hole well-log curves which are consistently quality-checked and corrected as required. As laboratory-derived core measurements become available for a well the information is entered into the system to be used for comparisons and calculations. Interpreted rock parameters such as rock type, visual porosity, and clay estimates can be added to the digital-log database. Intervals that have definitive production or drill-stem tests are noted with respect to the type, amount of fluid flow, and pressure response associated with the tests. Cased-hole or production-log information, in the form of continuous curve data, can be entered and stored in the file. A schematic of the database and sources of input is given in Figure 14.

All the information to be processed must be carefully checked for quality and depth before it is used in integrated subsurface calibrations.

Core-Log Database Creation

Because of the large amount of core and log data taken in evaluating the Deep basin, a flexible and easily accessible data system was established. The digital core-log system was designed to allow graphical presentation of the core measurements, which would then become usable for log calibrations. Information sources for the log database are illustrated in Figure 15.

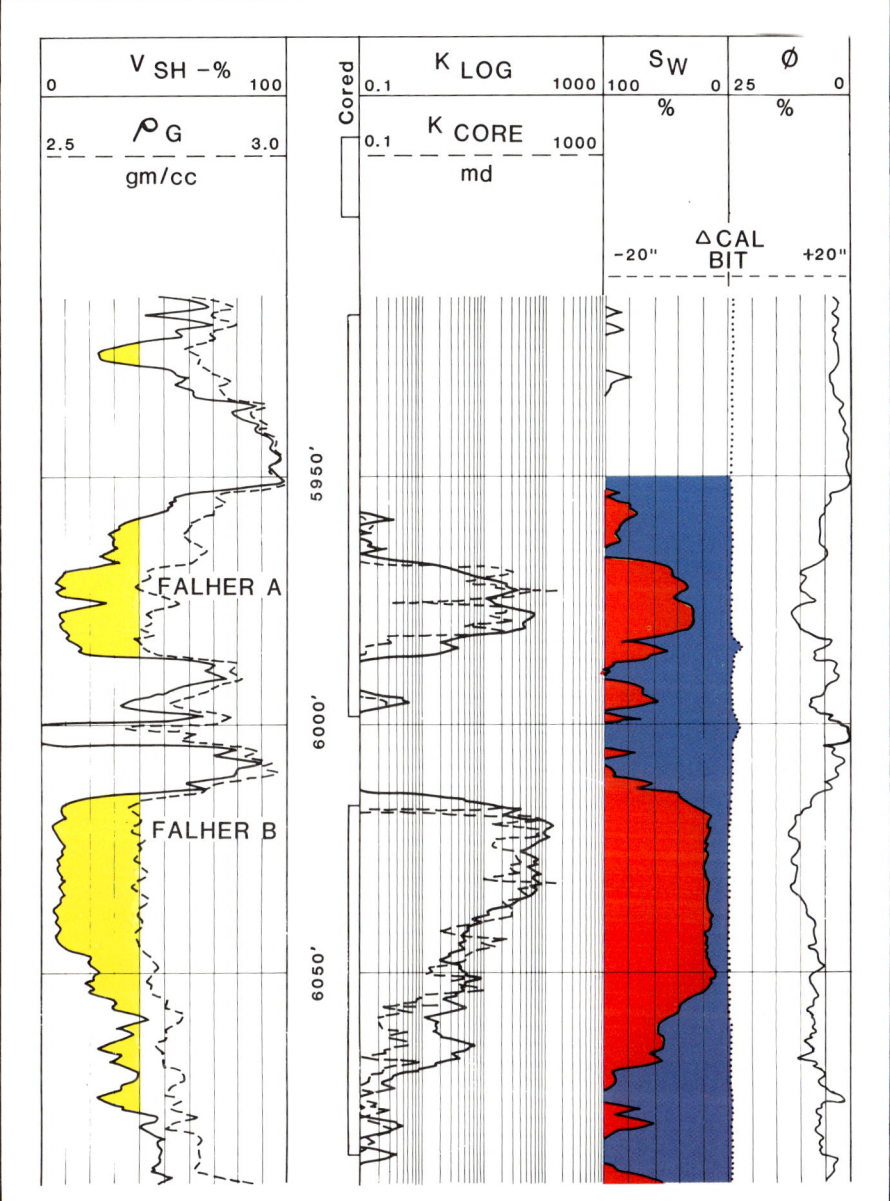

Figure 10. Computed NML permeability versus core permeability.

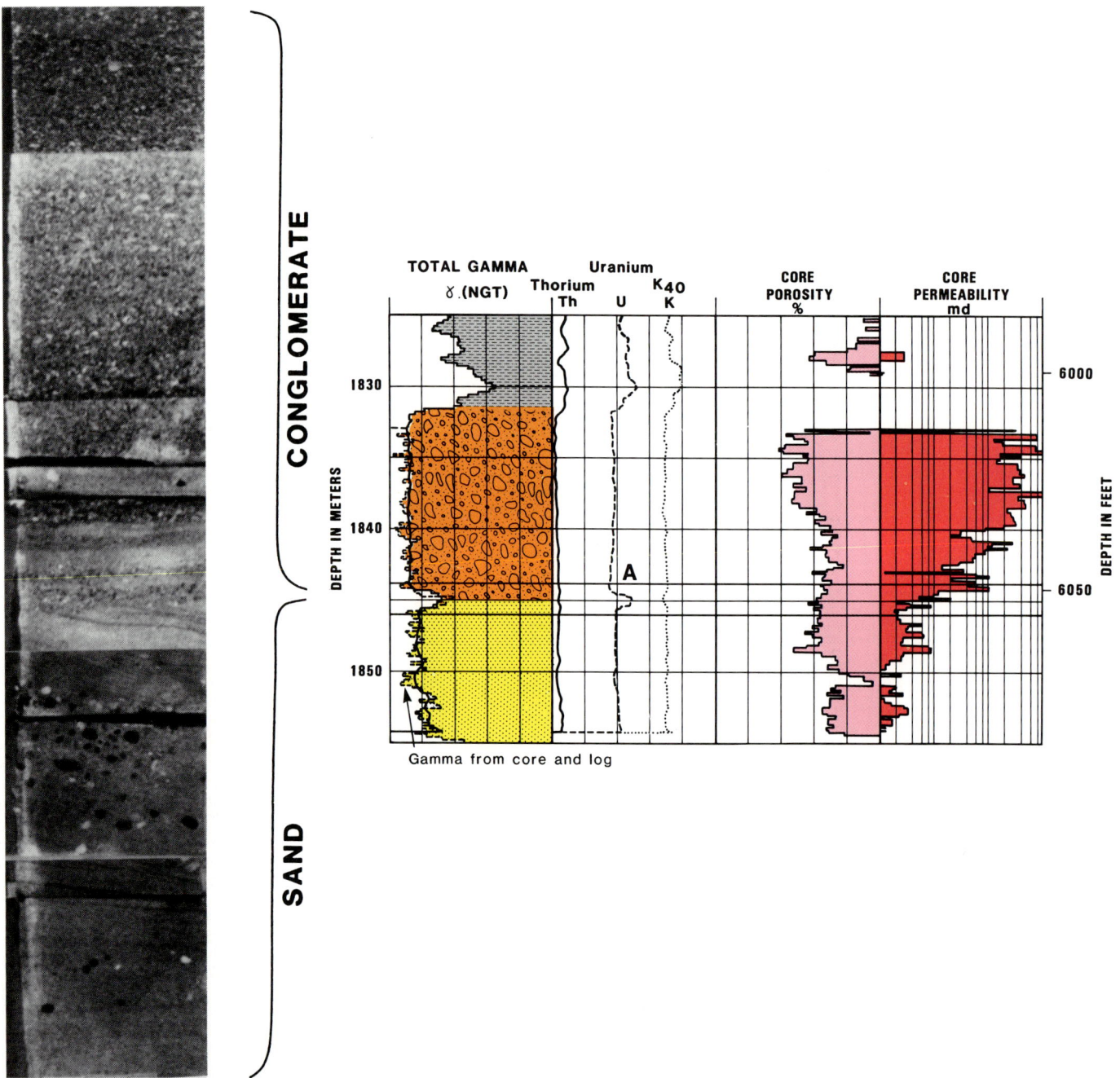

Figure 11. Natural gamma ray spectra over conglomerate/sand interface.

As the service company laboratory report on a core is completed the results are immediately incorporated in the computer system by means of standard lithology-coding techniques. The core data plots can be utilized as subsurface wireline logs are being recorded. Core data plots at the well site have proved invaluable in obtaining better-quality logs.

The log database format is kept very simple with data being referenced by four character names and related to industry-accepted API numbers. The digital core and log databases have an identical computer file format, with only the increment of data storage differing. This allows easy merging and combination of core and log curves. Physical core parameters can be averaged over various vertical intervals to be compatible with the vertical resolution of a given logging tool.

Core-analysis Data Integration—As the core is taken at the well site it is described by the supervising well-site geologist who uses methodology developed by the company. The core is measured for porosity, maximum permeability, permeability 90° to maximum, vertical permeability, bulk density, grain density, oil saturation, and

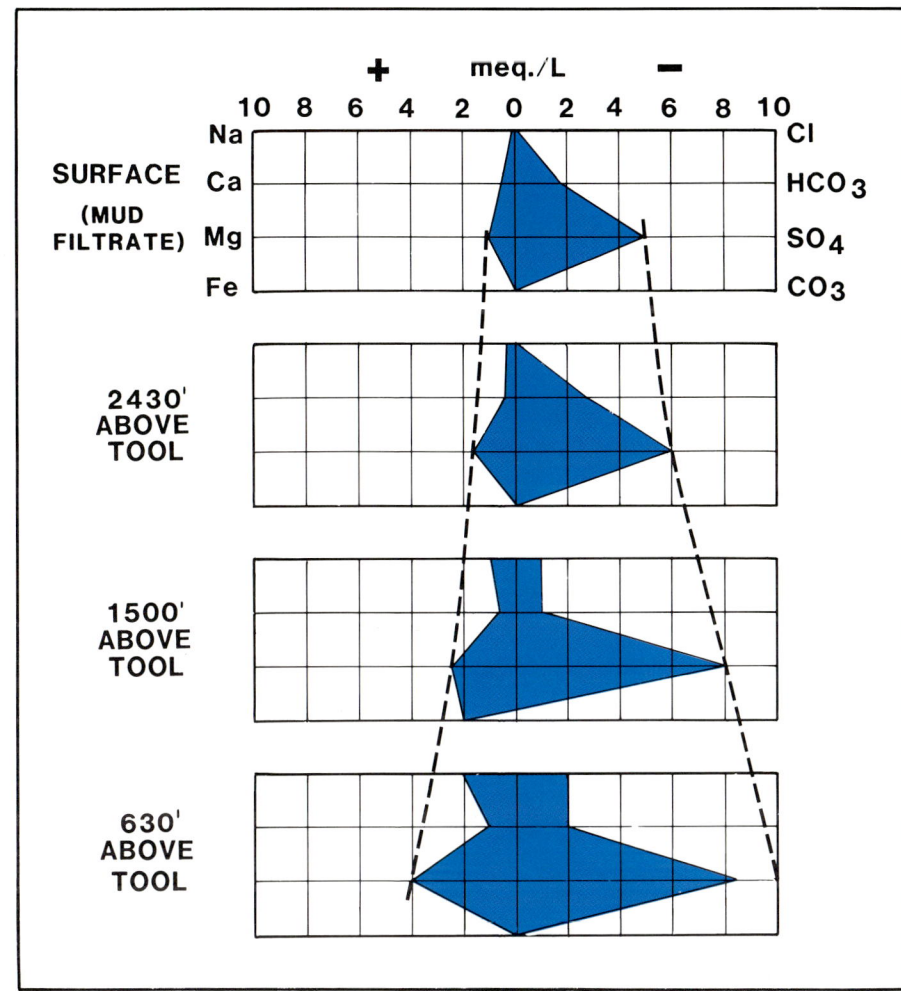

Figure 12. Stiff analysis of D.S.T.

water saturation. Plug and whole-core analysis techniques are used depending on the rock lithology involved.

The first pass of a two-plot standard core analysis presentation is given in Figure 16. This figure indicates data from a commercial core analysis on a five track presentation comprising information on porosity, permeability, lithology, density, and fluid saturations.

Beginning on the left side of the plot the core porosity in track I is shown increasing to the left. Permeability, in track II, increases to the right. A prospective productive zone shows an hourglass effect on this curve layout. An easy-to-use but informative lithology description is given in track III. A detailed explanation of the numeric values for curves A, B, C, and D from Figure 16 is given in Figure 17 (Connolly and Reed, 1983). Curve A illustrates the primary lithology of the interval and curves B and C give the minimum and maximum grain size, while curve D contains remarks.

The fourth track of Figure 16 contains bulk- and grain-density information. Bulk density can be used to check against log measurements, and it should be noted in this illustration that there is a difference between the grain density of the conglomerate and that of the sandstone. Typical conglomerate grain densities are 2,620 kg/cu m to 2,650 kg/cu m (2,062 to 2,072 m) whereas sands in this area (below 2,072 m on diagram) have densities of 2,680 kg/cu m to 2,710 kg/cu m. This matrix density difference, and a calibration to neutron-density log response in uncored wells, can assist in conglomerate identification. Water saturations measured from cores cut in water-base mud systems are used in a qualitative manner only; they are shown in track V.

Permeability-Lithology Plot—The plot shown in Figure 18 supplements the one in Figure 16. This plot is used primarily for visual comparisons of the various permeability measurements. Track I again contains the core porosity. Tracks II, III, and IV show the different permeability measurements: maximum permeability, permeability 90° to maximum, and vertical permeability, respectively. Only the primary lithology curve is shown in track V of this plot. This type of plot is useful in identifying fractures, both natural and artificially induced. Often it has been found that the permeability 90° to maximum correlates best to porosity.

Core-Data Checks—In whole-core analysis measurements periodic conventional measurements should be made where possible to check the homogeneity of large samples. Also, on occasion a whole-core piece should be broken to check internally for clay plugging, as clay particles may have washed out of the exterior pore spaces during coring. This applies mainly to well-sorted pebble conglomerates, but is also applicable to sands.

Well-Log Input—All log information taken on wells where Canadian Hunter is either the operator or a participant is automatically recorded in a digital format at the well site. Logging company field or computer center tapes are reformatted by a digitizing service company to Canadian Hunter specifications. Well-log information that has to be manually digitized is similarly reformatted. This procedure was adopted to allow company personnel to focus on interpretation problems as opposed to data editing.

Log-Calibration Procedures and Techniques

Properly calibrated and consistent well-log data are essential for accurate and comprehensive formation evaluation of a given area. Methods and techniques used to provide such data for the Deep basin include: (1) on-site supervision; (2) casing-signal checks by acoustic devices; (3) checks on log responses for known lithologies; (4) data histograms in consistent marine shales; and (5) crossplots.

It was found that the best way to monitor log quality is to have an experienced log analyst present as field data are acquired. The functions of the field log analyst should include verifying tool calibration procedures, identifying zones of

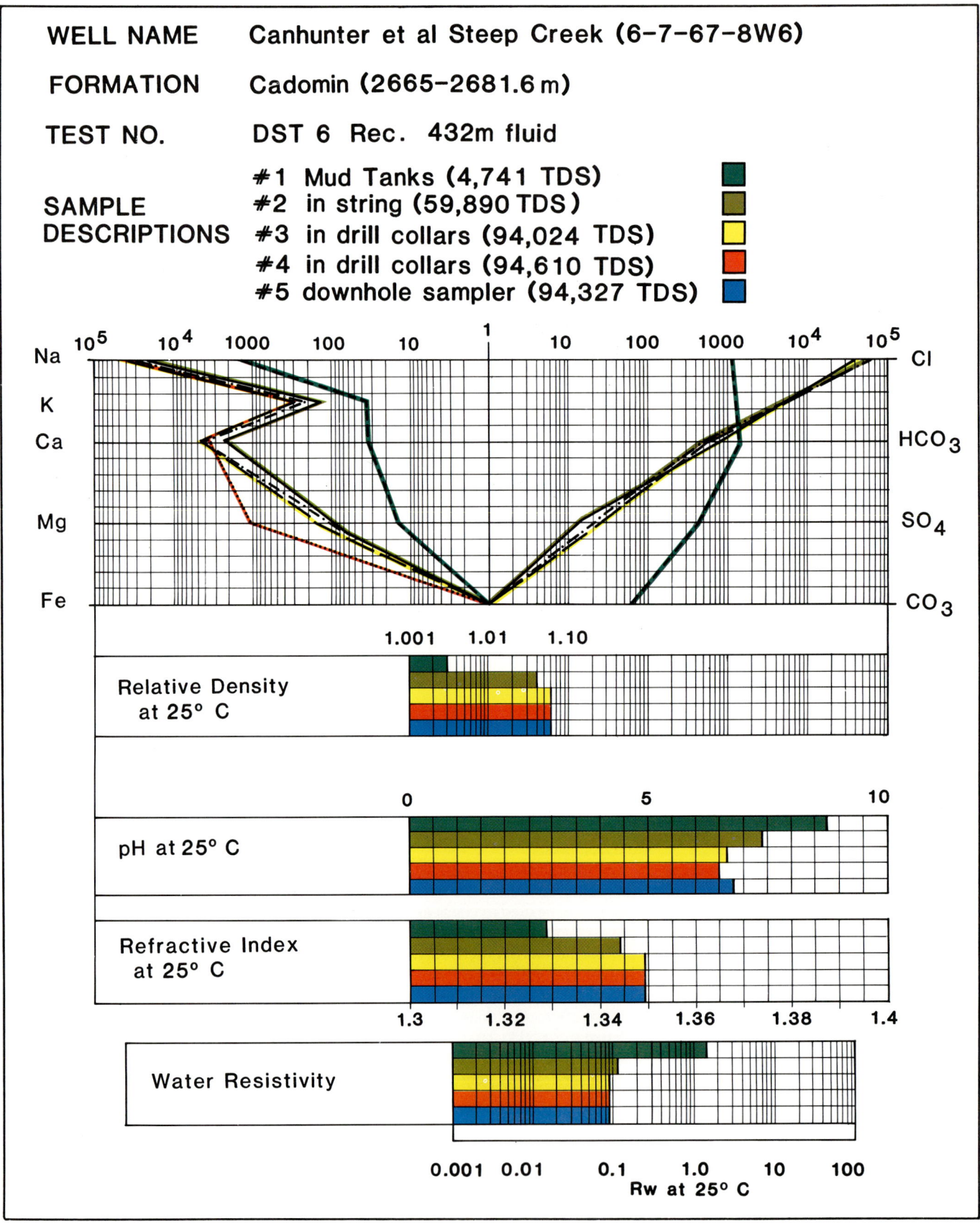

Figure 13. Composite water identification graph.

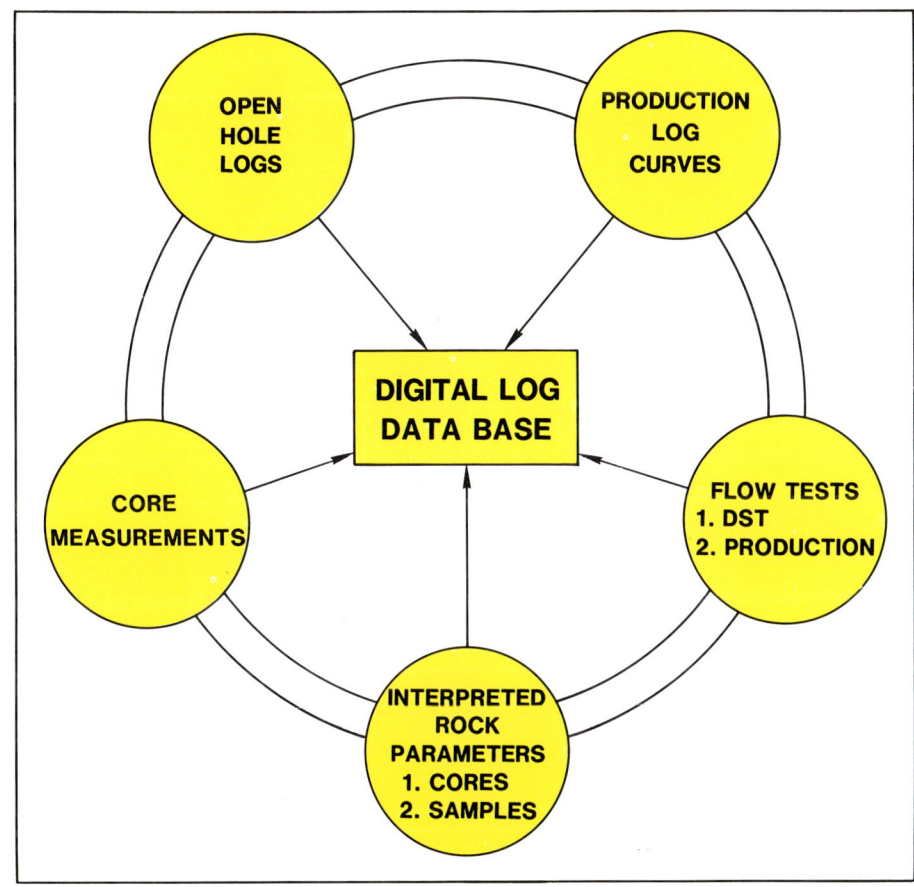

Figure 14. Input to log database.

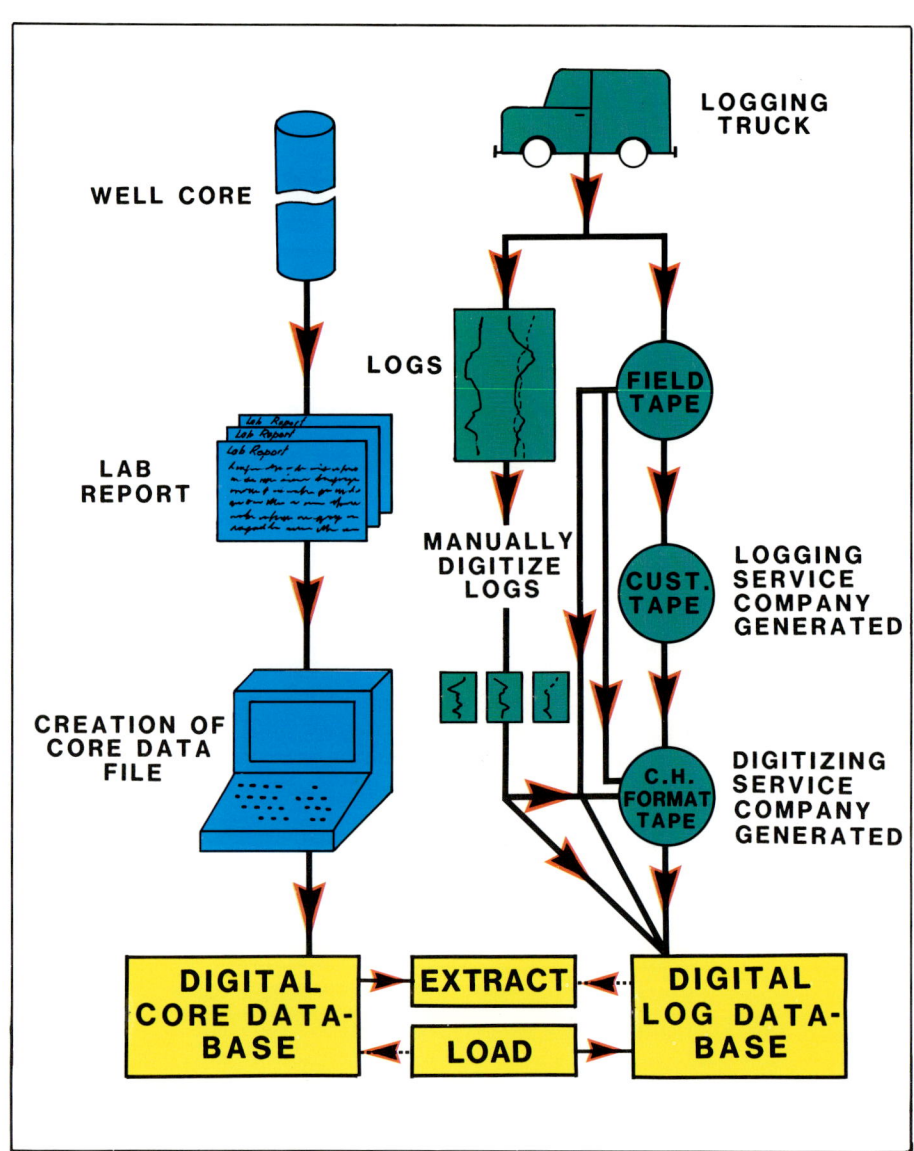

Figure 15. Database information sources.

Integrated Rock-Log Calibration, Elmworth Field 227

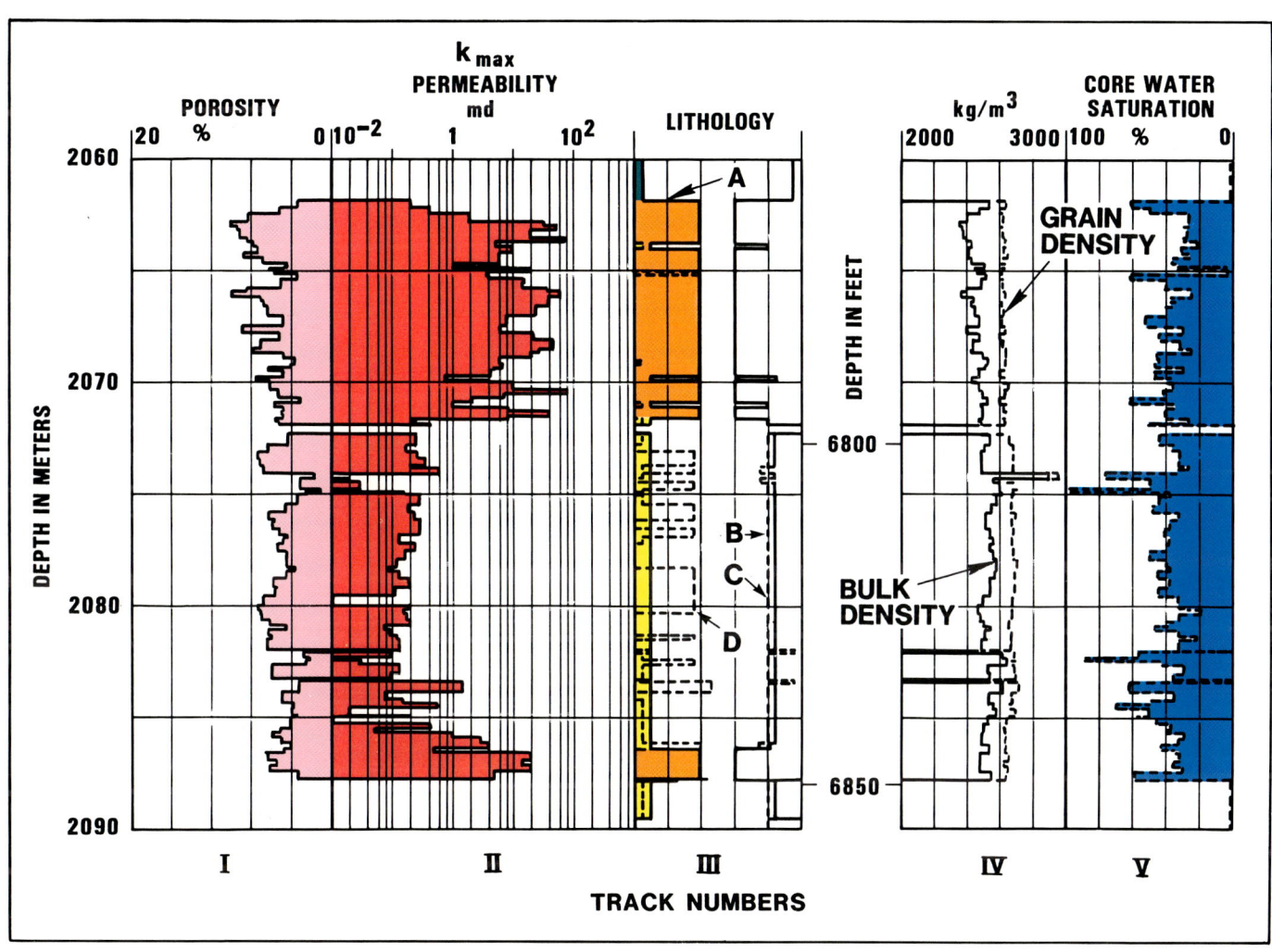

Figure 16. Core analysis plot A.

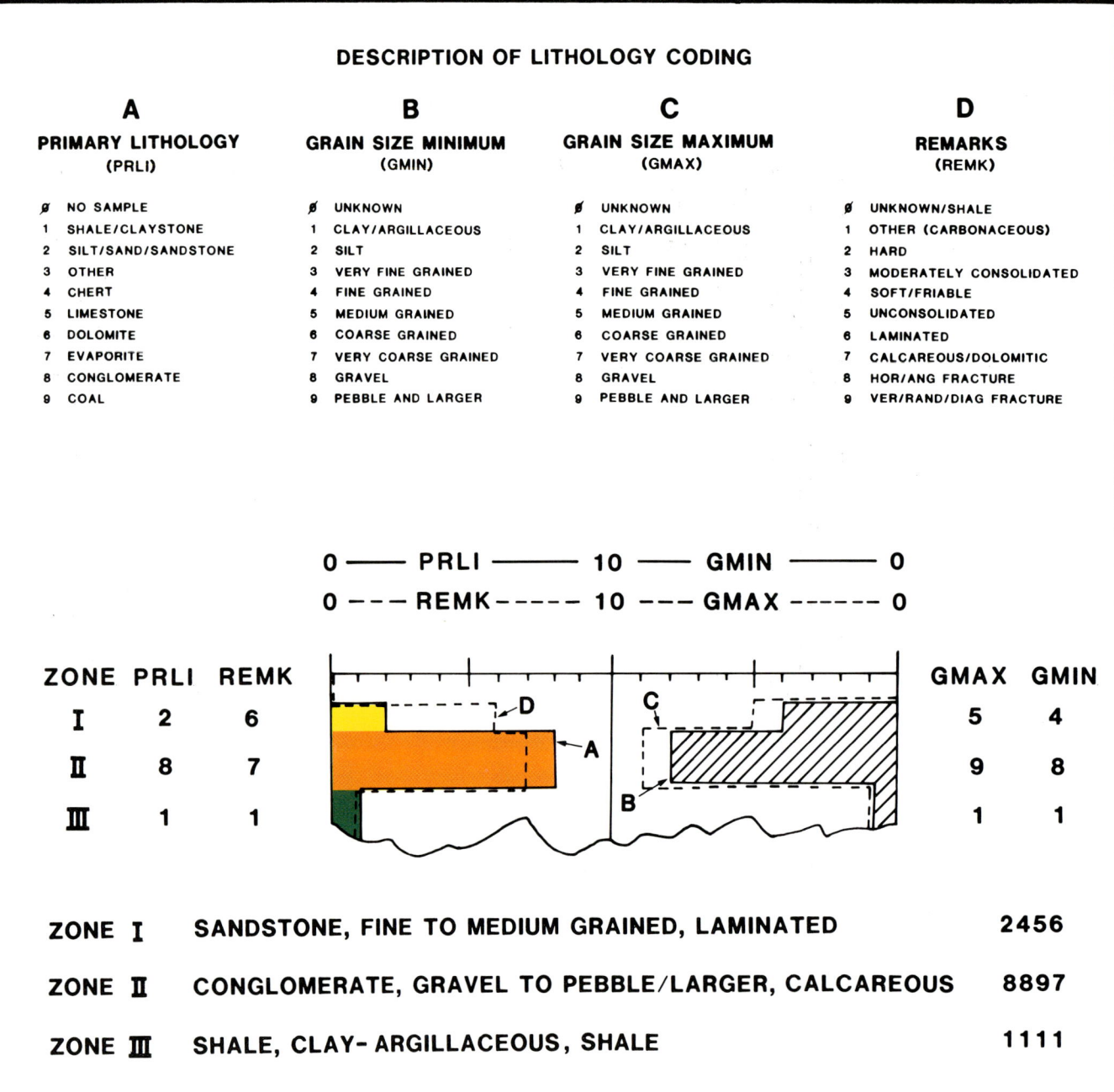

Figure 17. Description of lithology coding.

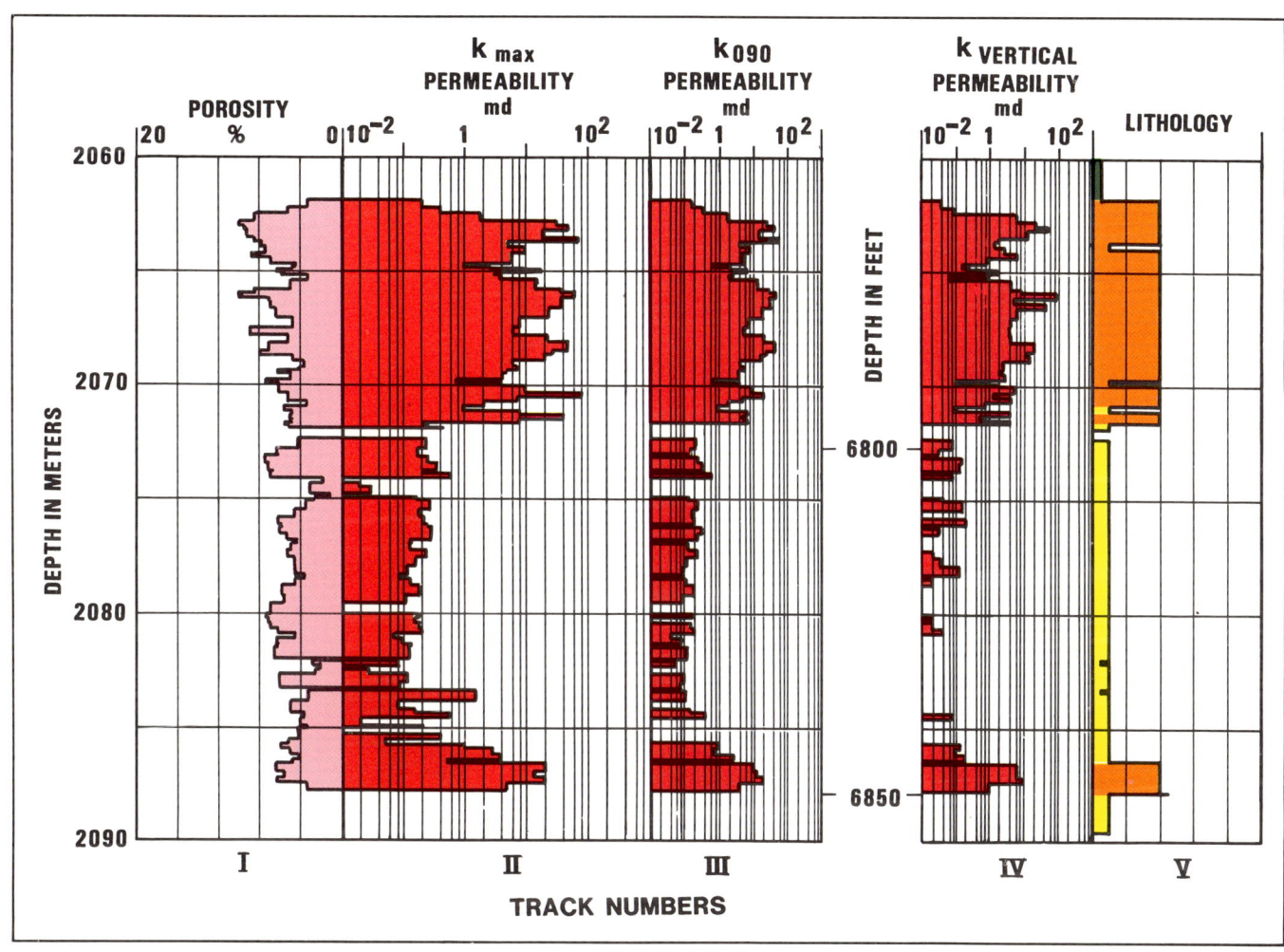

Figure 18. Core analysis plot B.

interest for possible repeat runs, and performing preliminary field well-log analysis.

Acoustic Calibrations—Because the acoustic log serves as one of the three porosity logs in the evaluation procedures casing-signal checks are made for quality-control of this log. A minimum of 200 ft (61 m) of unbonded casing is logged with the acoustic tool and the resulting values are checked against the known response. Figure 19 illustrates the use of the acoustic casing signal for quality control.

The Triassic of the Deep basin contains a number of dense carbonate and anhydrite units which provide known lithologic calibration points. Once the logs have been calibrated to the deeper known lithologies, shallower uniform marine shale responses are noted. These can then serve as semi-quantitative calibration points for surrounding shallower wells.

Neutron-Acoustic Calibration Checks—Neutron and acoustic measurements provide data for fundamental porosity solutions used in the Deep basin because these logging tools can be relied on more often than combinations of neutron and density data. The incorporation of the new "offset" density log adds to the confidence in porosity calculations as the density response is used to check neutron and acoustic porosities.

An example of a neutron-acoustic crossplot versus clay volume from gamma-ray and other logs is given in Figure 20. In this example, plotted from the digital log database, the difference in porosity between neutron and acoustic ($\phi n - \phi s$) is plotted versus log-derived clay volume. The effects of clay (shale) on the two logs provides a matrix-intercept at zero clay that aids in log-porosity calibrations. Note the effect of the apparent clay content as shown by the Z-axis on the basis of gamma-ray response. Both upper and lower limits of the divergent data grouping define the zero value for a porosity calibration check.

Histograms—In the Deep basin the Harmon shale is a thick, relatively consistent marine unit. Histograms of tool responses are checked in this shale section, with digital-log data being used to correct for miscalibrations. A gamma-acoustic log across the Cadotte-Harmon-Notikewin sequence is shown in Figure 21 and a histogram for the acoustic response of the Harmon shale appears on the left of this diagram. Similar histograms are prepared for other logging curves. Sediment compaction due to original burial depth must be taken into account in relating a given well to the general data set.

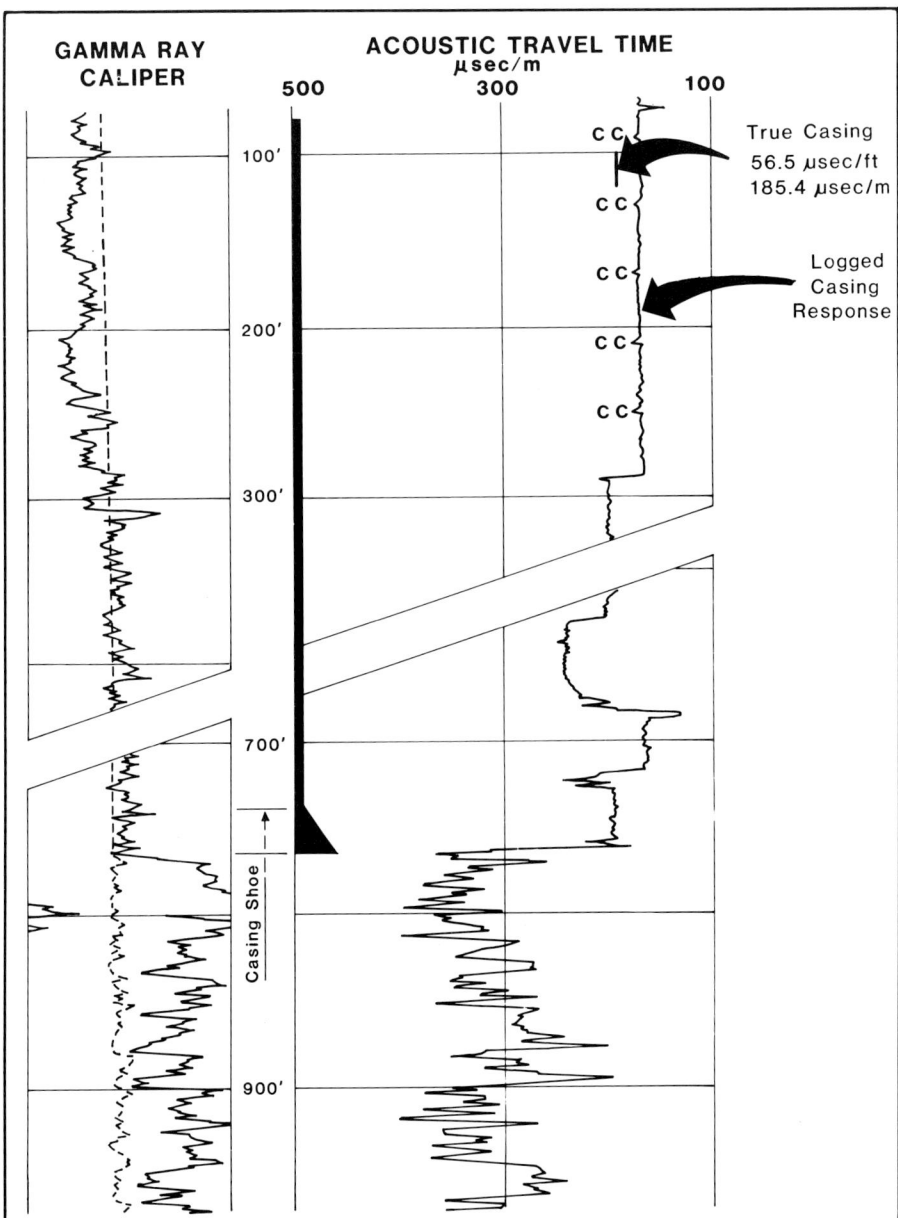

Figure 19. Acoustic log calibration in casing.

ANALYSIS AND DISPLAY

General Interpretation Techniques

"Anomaly-seeking" or porosity-difference techniques have been used to identify conglomerates in the study area. Simple overlays of porosity logs will show lithology changes that are significant (e.g., conglomerates versus sandstones). Resistivity levels may also be used as qualitative quick-look indicators for potential hydrocarbon zones. Figure 22 (Connolly et al, 1983) shows an example of these two techniques using density-acoustic and density-resistivity overlays. Zones A and B both have conglomerates on top of sands. Positive microlog separation would also highlight the better-quality conglomerates. The log responses and drilling criteria enabling the identification of conglomerates may be summarized as follows: (1) slower drilling time; (2) low gamma-ray response (3) high resistivity; (4) low acous-

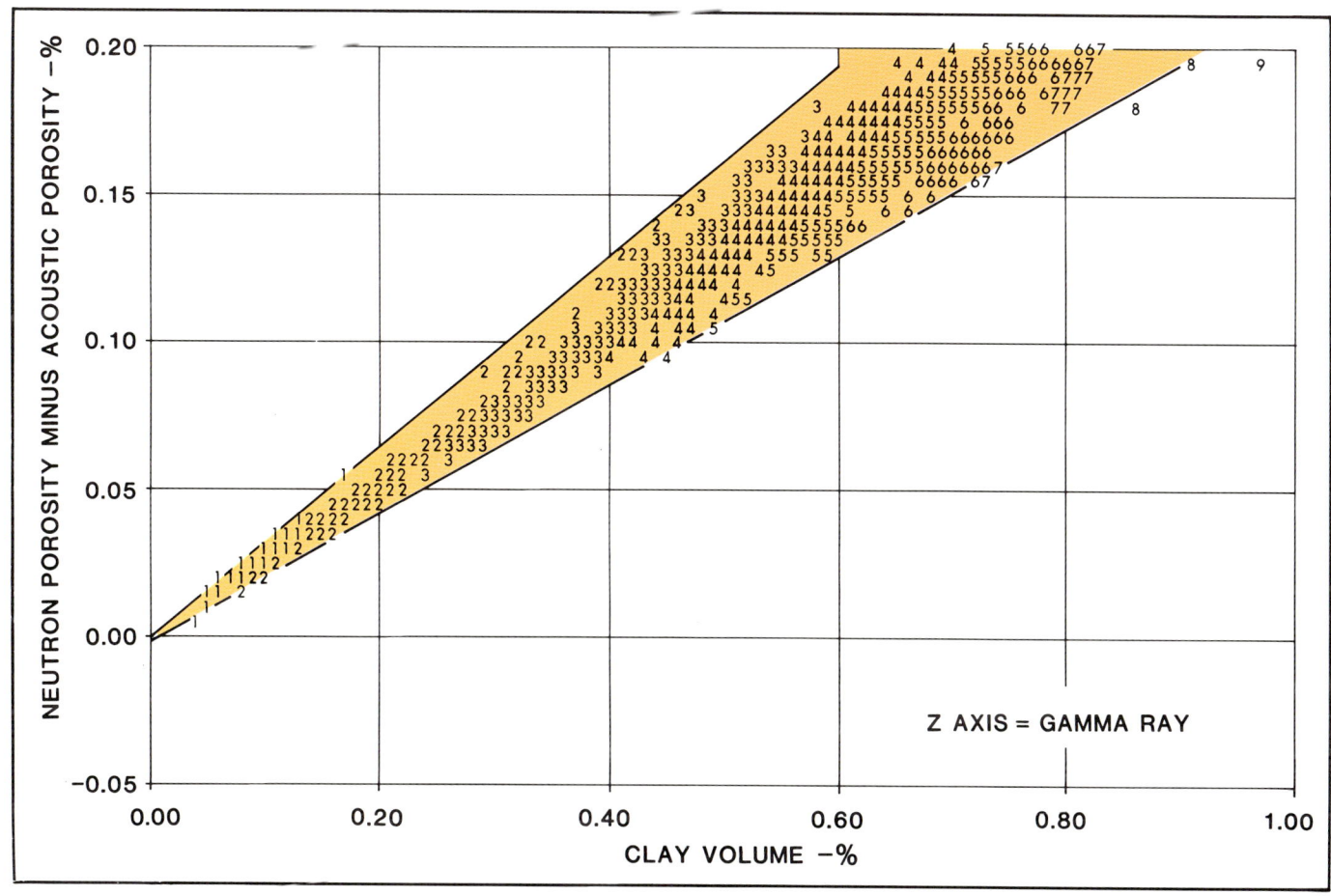

Figure 20. Porosity difference versus clay volume.

tic traveltime; (5) density and neutron porosity greater than acoustic; (6) gauge bore hole with or without mudcake buildup; and (7) good microlog separation.

Figures 23 and 24 illustrate the conglomerate identifiers in the above list. Zones A and B of Figure 23 are good porous and permeable pay zones that flow-tested at 3.7 mmcf/d. Note that both zones in Figure 24 indicate coarsening-upward cycles on the gamma curve. Lithology interpretations based on gamma- and porosity-log response can assist in interpreting depositional environment. Data plotting, as illustrated in Figure 24 (Connolly and Reed, 1983), enables comparisons of gamma response, drilling time, lithology, and physical core measurements that are related to productivity.

Digital Analysis Techniques

Use of Database Crossplots—Crossplots of raw data and calculated values can be examined to help interpret porosity, permeability, and water-saturation cutoffs. In the database crossplot output, the number of data points of any coordinate is given a number or letter at that point. This form of data presentation indicates the data-frequency distribution; a third variable referred to as a Z-axis component may also be incorporated on a crossplot. Low numerical values indicate low Z-axis values, higher numeric values an increasing Z-axis.

When the lithology-coding identifier of a core is incorporated as a Z-axis value lithological discrimination will result. A porosity/permeability plot showing a bimodal data distribution is given in Figure 25. This bimodal character is due to the presence of conglomerates and sandstones and, as can be seen from the plot, for a given porosity the associated conglomerate permeability may be approximately two orders of magnitude larger than that of the sand.

Proper calibration to rock type is very important in establishing flow characteristics for porosity cut-offs.

Matrix and Fluid Parameters—A plot of acoustic-log response versus measured core porosity for a pebble-supported conglomerate is illustrated in Figure 26. Two lines are given as reference. Line #1 is the industry-accepted expression for the variation of acoustic traveltime with porosity, and line #2 is an empirical lithology-adjusted conglomerate response line. It can be seen from this illustration that failure to recognize the presence of conglomerate would lead to a substantial underestimation of acoustic-derived porosity. The company's quick response to the significance of the Elmworth discovery well was due to recognition of the presence of porous conglomerates rather than only fine-grained sands.

A typical crossplot of acoustic traveltime versus core porosity for a Falher sandstone unit is given in Figure 27. Line #1 is the industry-accepted equation of acoustic response to porosity in a sandstone unit, while line #3 is the relationship currently used in the Falher

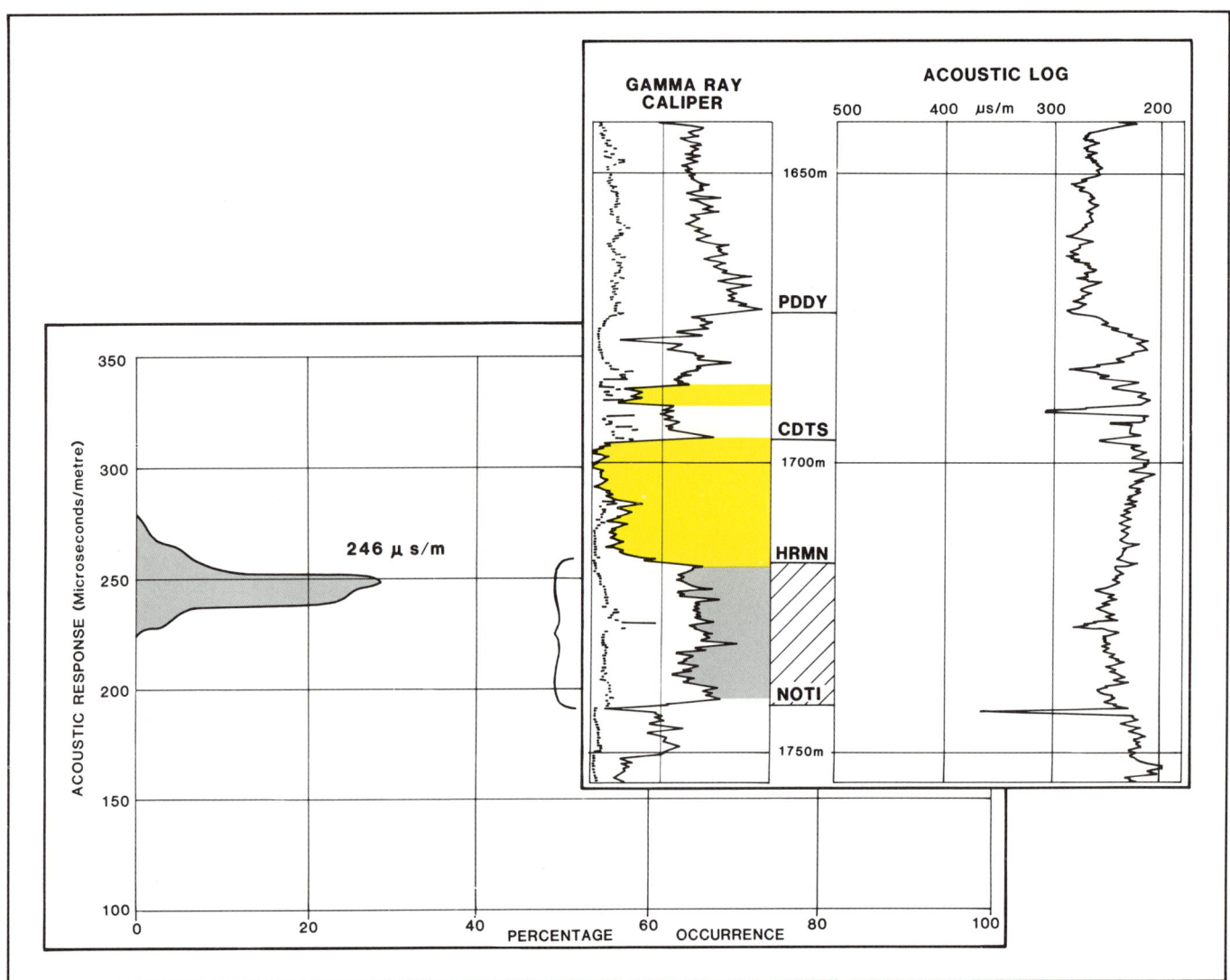

Figure 21. Harmon shale histogram of acoustic velocity.

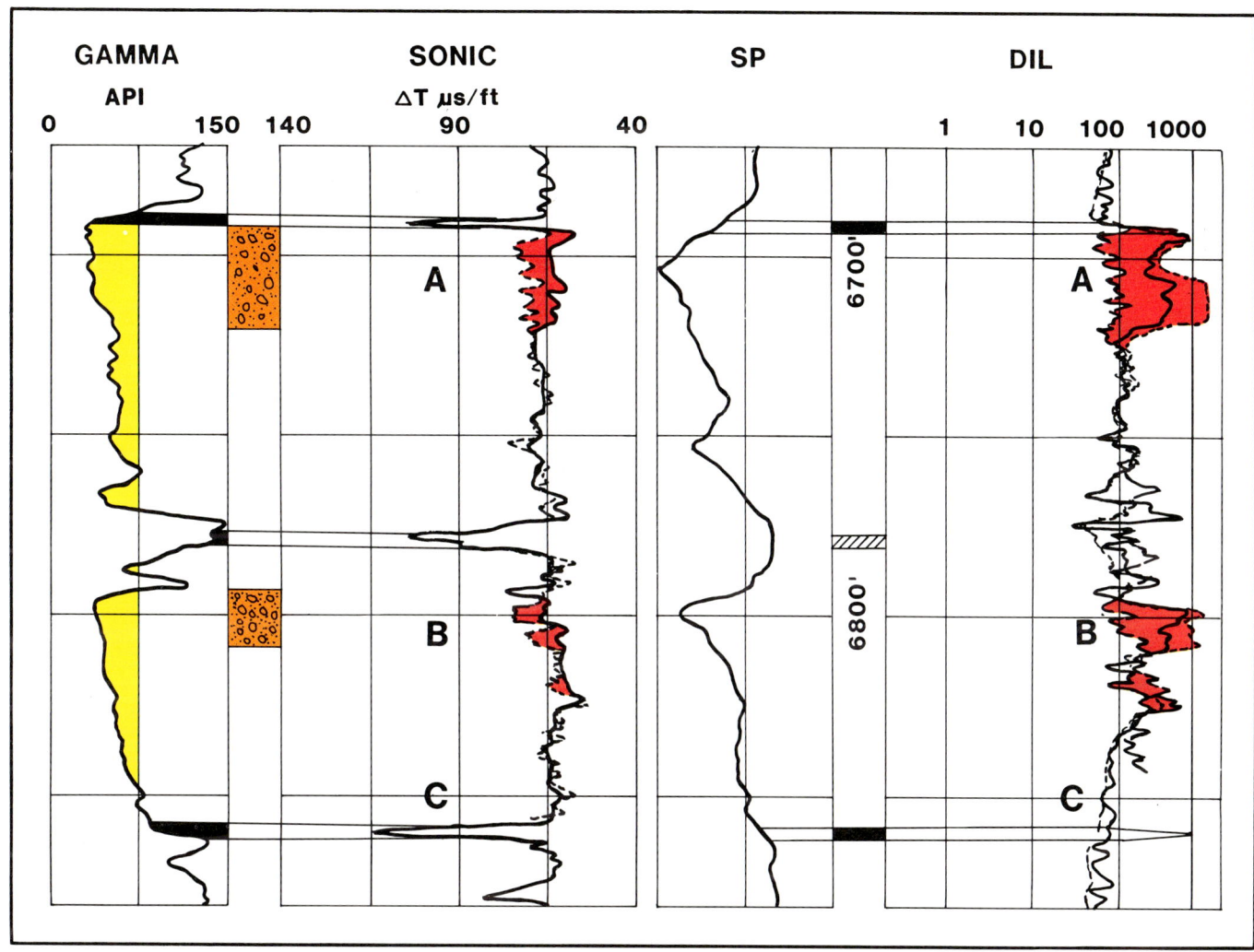

Figure 22. Anomaly seeking density versus sonic and density versus resistivity.

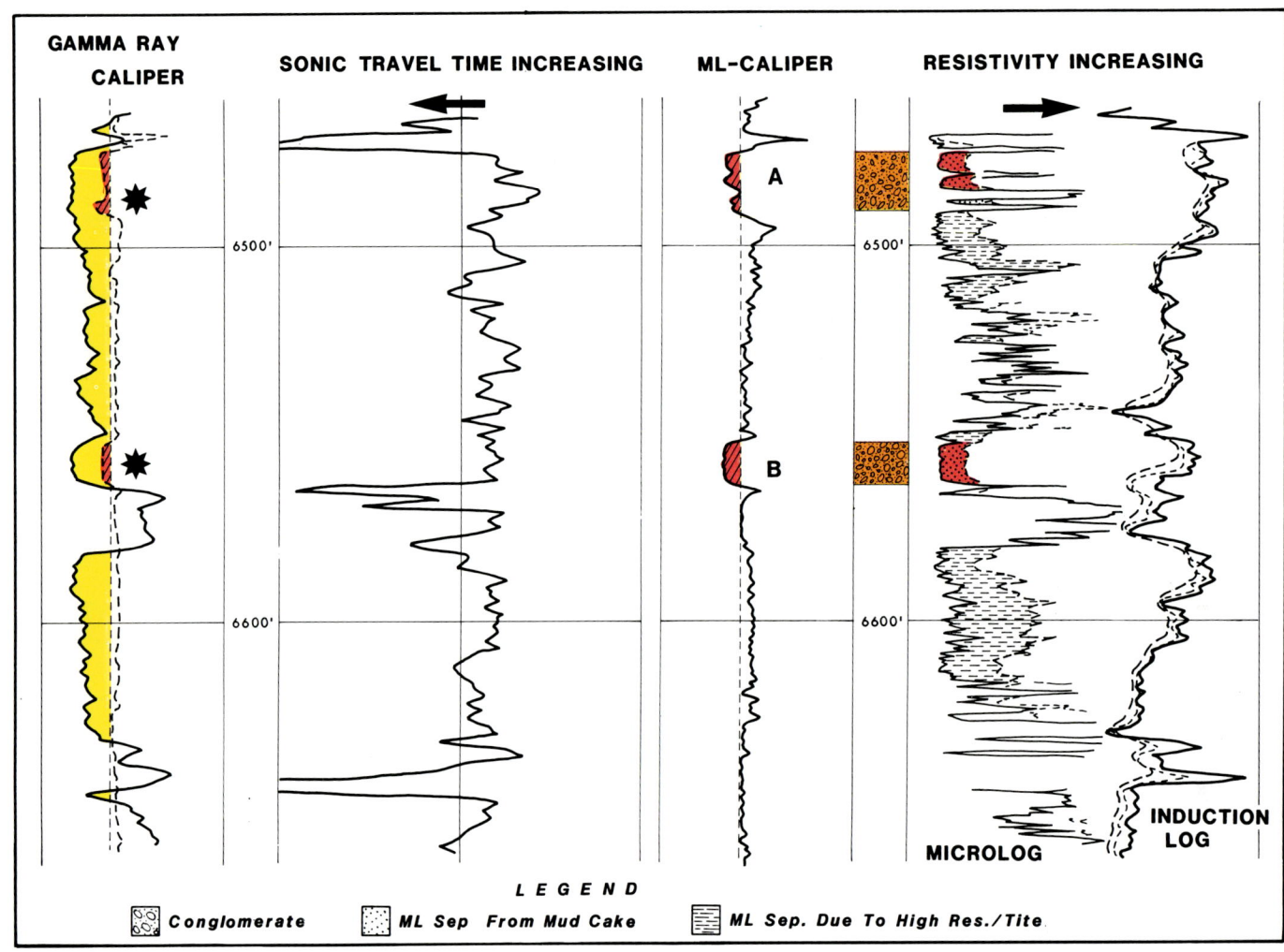

Figure 23. Conglomerate log responses.

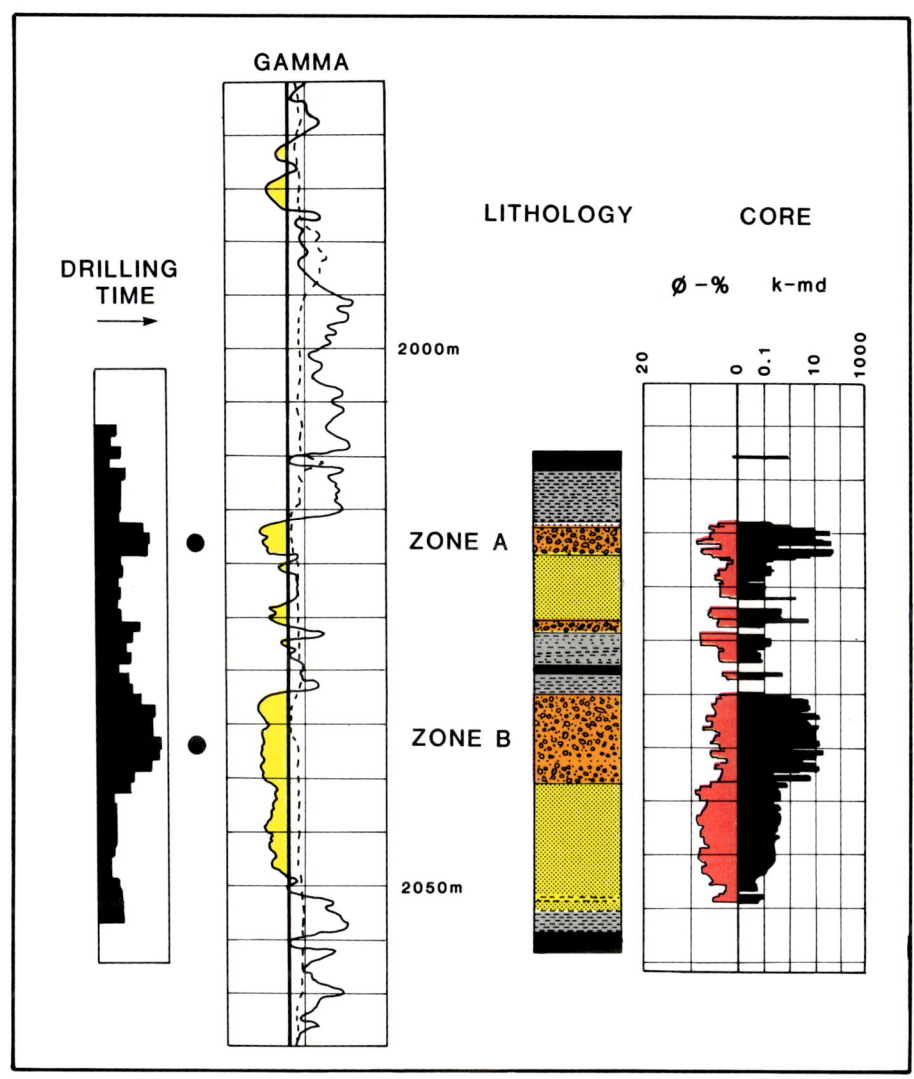

Figure 24. Conglomerate identifiers.

Figure 25. Core porosity versus core permeability with Z-axis code.

Figure 26. Conglomerate acoustic velocity versus core porosity.

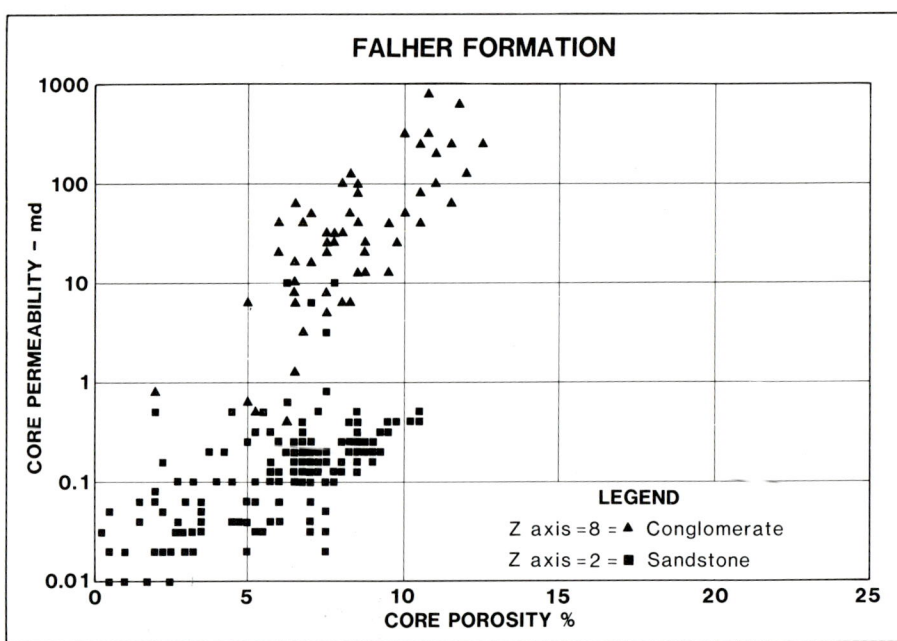

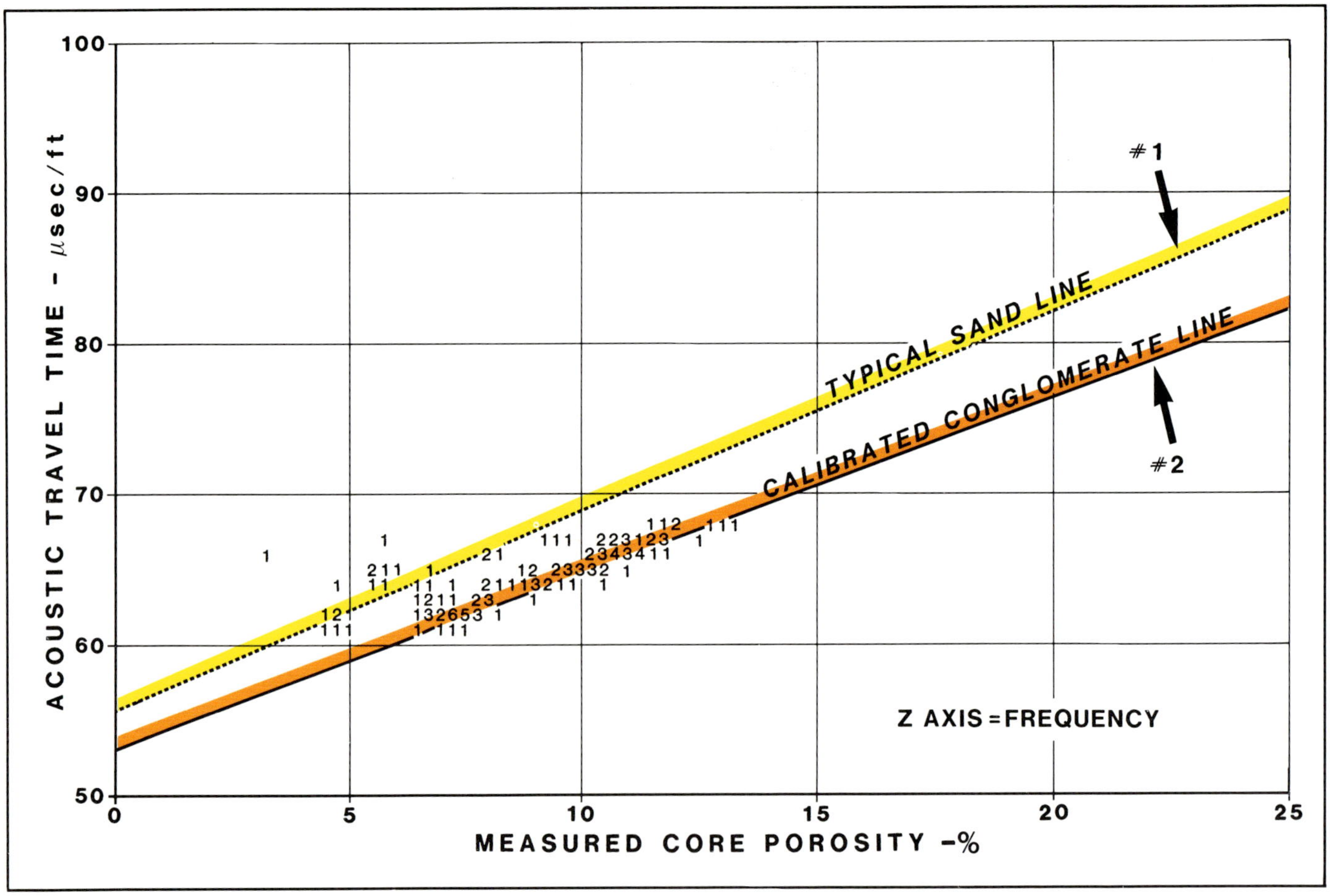

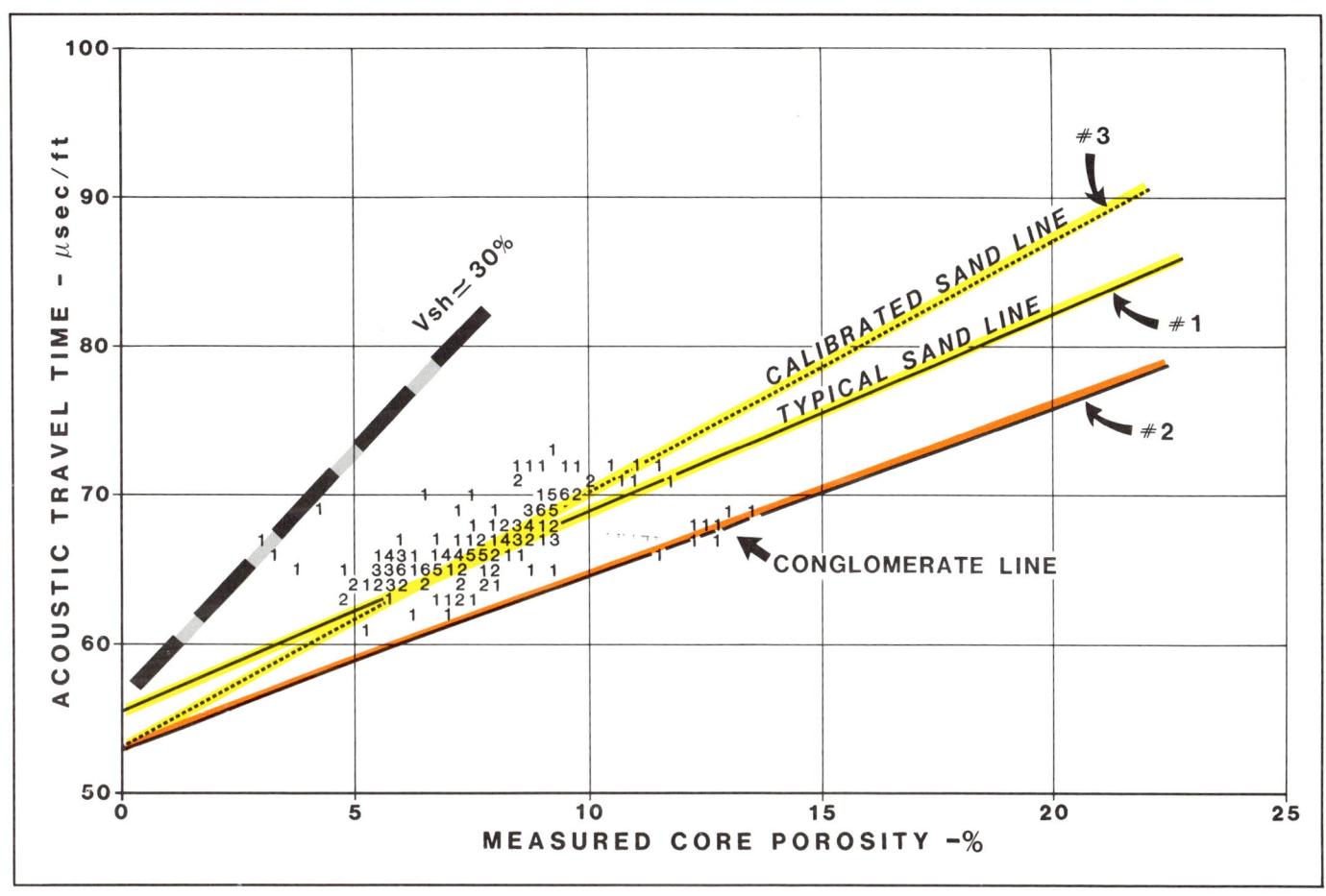

Figure 27. Sandstone acoustic velocity versus core porosity.

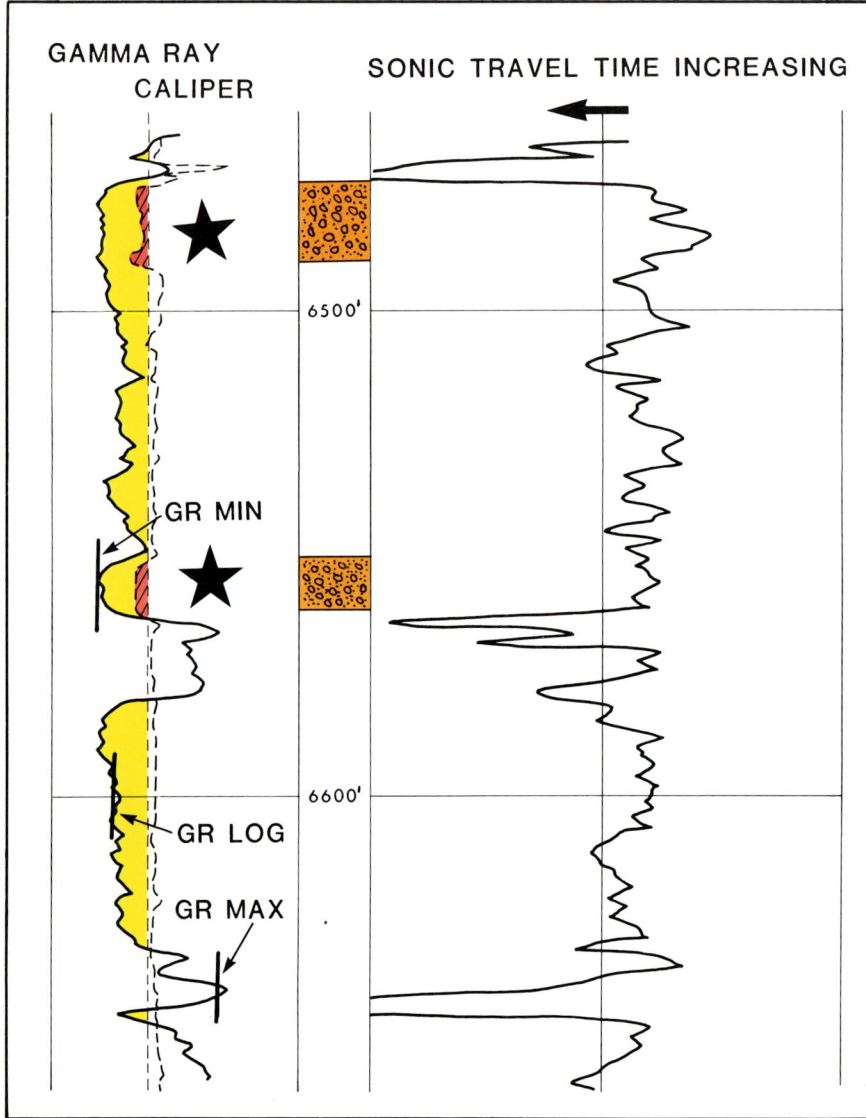

Figure 28. Normalized gamma-ray values.

The final calibrated log porosities as compared with measured core porosities are shown in Figure 29, illustrating a reasonable correlation. The porosity plot is composed of sandstone, conglomerates, and shaly sandstone lithologies.

Shaly-Sand Core-Log Calibration

Clay Types Encountered—The predominant clay types in the sands and conglomerates of the Deep basin are illite and kaolinite. Figure 30 illustrates the presence of both clays, and it can be seen from these SEM photos that both exist together in the pore system. Kaolinite in some cases acts as a pore-filling clay while the illite can occur in the pore-bridging mode. This variation in clay morphology affects permeability levels of the pore system. There is good agreement between the measured and calculated clay volumes as can be seen from the clay volume comparison of Figure 31.

The Shaly-Sand Model—Figure 32 shows the basic components used to describe a shaly-sand model that accounts for the presence of bound water and dry clay minerals. The difference between the total and the effective porosity is a function of both the type and the amount of clay present, and is due to water bound both chemically and electrically to the clays (Hill et al, 1979). Free water is a combination of both the irreducible water held in the pore system and the water capable of physical movement under suitable pressure-drop conditions.

Early work on the calculation of water saturation by G. E. Archie resulted in the now familiar "Archie" equation (Archie, 1942, 1950). Figure 32 diagrammatically expresses the difference between the Waxman-Smits shaly-sand model and the earlier clay-free Archie model. The Archie solution is a single-path conductance model with the formation water serving as the conducting medium. In the Waxman-Smits approach (Waxman et al, 1968) a second path of conductance, the

evaluation procedure. Typical conglomerate responses are shown by Line #2.

Normalized Gamma Ray for Shale (Clay) Volume—An empirical relationship of normalized gamma ray is used to distinguish conglomerates from sandstones. An example of a gamma-ray log recorded across a conglomerate and sand interval was given in Figure 23. This same log, now showing values of maximum gamma, gamma minimum, and gamma log, is given in Figure 28. The normalized gamma-ray index is defined as:

This frequently-used term is incorporated to vary the slope of the response line on an acoustic traveltime versus porosity plot. Note that on the example seen in Figure 27 a clean-sand line and a 30% shale volume line were given. This shale volume of 30% was derived from the normalized gamma ray. The presence of shales in most clastic sequences has the effect of rotating the calibrated sand line counterclockwise; the shale data points (colored in green) illustrate this effect in Figure 27.

$$NGRI = \frac{\text{GAMMA LOG} - \text{GAMMA MINIMUM}}{\text{GAMMA MAXIMUM} - \text{GAMMA MINIMUM}}$$

where: Gamma Maximum is the average shale reading for the log; Gamma Minimum is the cleanest clastic reading; and Gamma Log is the reading for the zone of interest.

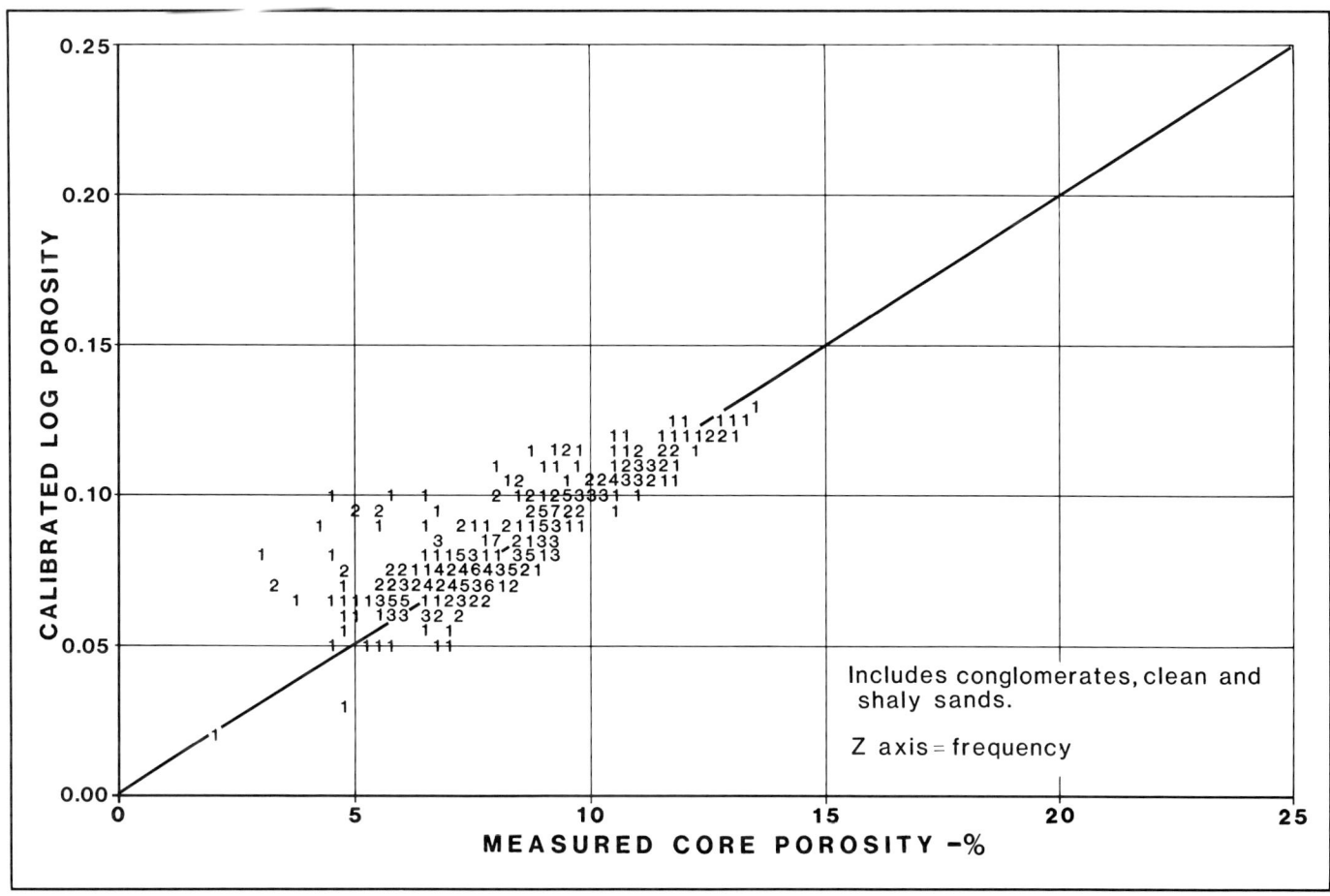

Figure 29. Calibrated porosity versus core porosity.

dispersed clays, is accounted for. The clay component becomes more significant as formation waters become fresher and the clays become more electrically active.

It should be kept in mind that the shaly-sand model reduces to the Archie relationship in clay-free cases.

Dual-Clay Analysis—The solution of a dual-clay system requires the incorporation of gamma ray, neutron porosity, and total porosity. An assumption is made that shales are composed of two-thirds clay and one-third other minerals. It is further assumed that the two clays present have different gamma-ray responses. Neutron-porosity response parameters are needed for the solution of a shaly-sand system with two types of clay having different hydrogen indices. The difference between total and effective porosity is a result of clay-bound water which is immobilized in the formation. The two clay types are calculated and graphically shown in each of the wells in the appendix.

The detailing of these calculations is a subject in itself and is beyond the scope of this paper.

Oil-base Core Verification—Oil-base core data were acquired in various formations in the Deep basin for the purpose of water saturation calibrations. The oil-base core water saturations are used as a standard to compare with calibrated-log analysis models. The oil-base core data is used to observe the net effect of varying the input parameters of clay volume, clay type, water salinity, and formation factor relationships. Figure 33 is a plot of core porosity and Dean-Stark water saturations from oil-base core data in a Falher sandstone unit. Figure 34 is a depth-plot of digital-log analysis water saturation values compared with oil-base core-water saturations. There is good saturation agreement over the interval of varying lithology and clay volume.

Also, numerous wells have been drilled with inverted mud programs utilizing a drilling fluid with less than 10% water. The water saturation data from this broader invert mud program can be used for semi-quantitative comparisons.

Final Rock-Log Data Presentation

The computer-assisted rock-log calibrated formation analysis and accompanying data listings represent the final product in the rock-log integration on a particular well.

These data can be displayed graphically in many ways. The analytical results of six zones and their graphical displays are given in Part III, which follows. In these wells, digitally computed results from logs are compared with core analysis and rock-typing classifications.

CONCLUSIONS

Rock-log calibrated evaluation techniques have provided a sound quantitative

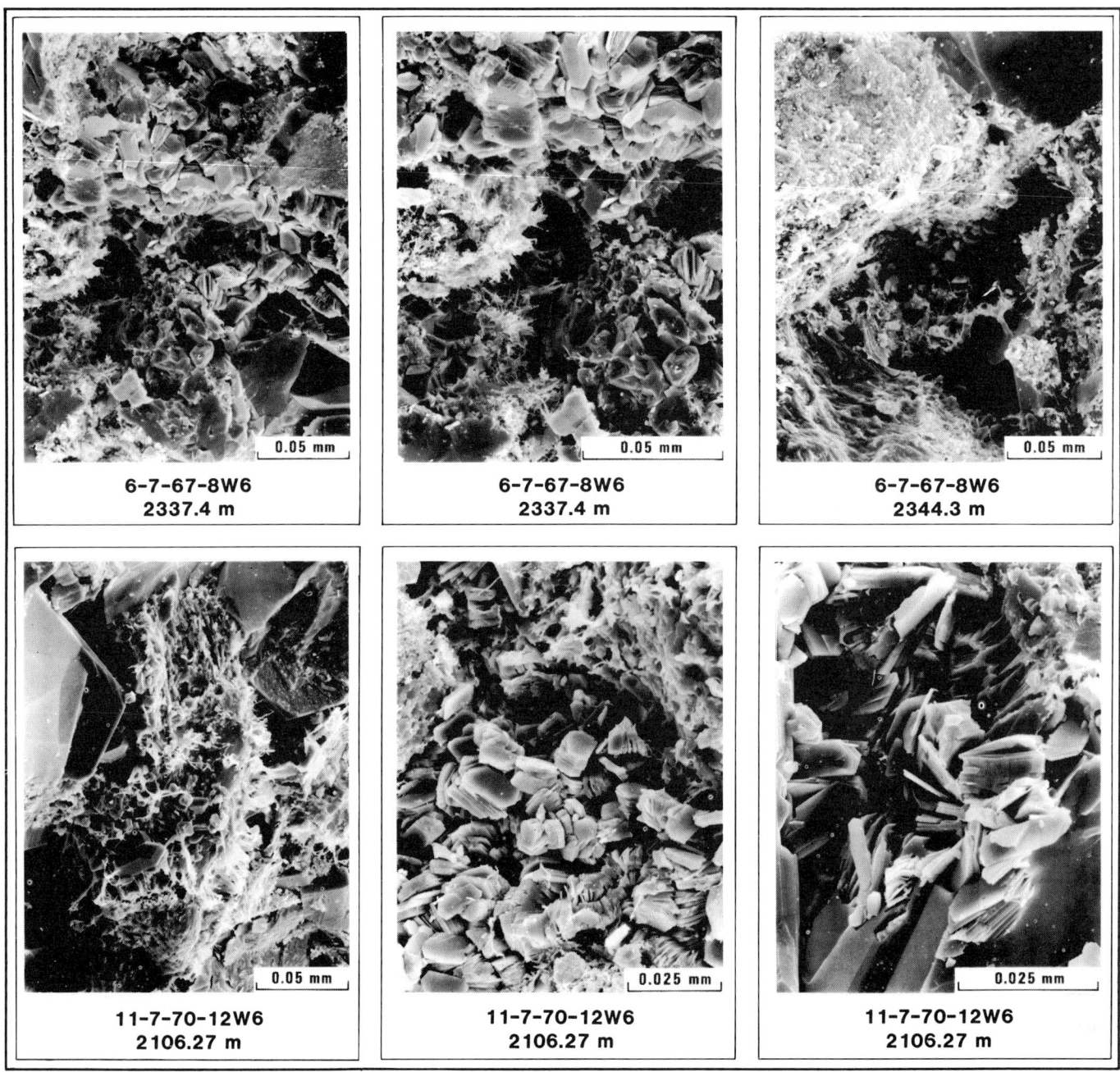

Figure 30. Scanning electron microscope photographs 6-7-67-8W6 and 11-7-70-12W6.

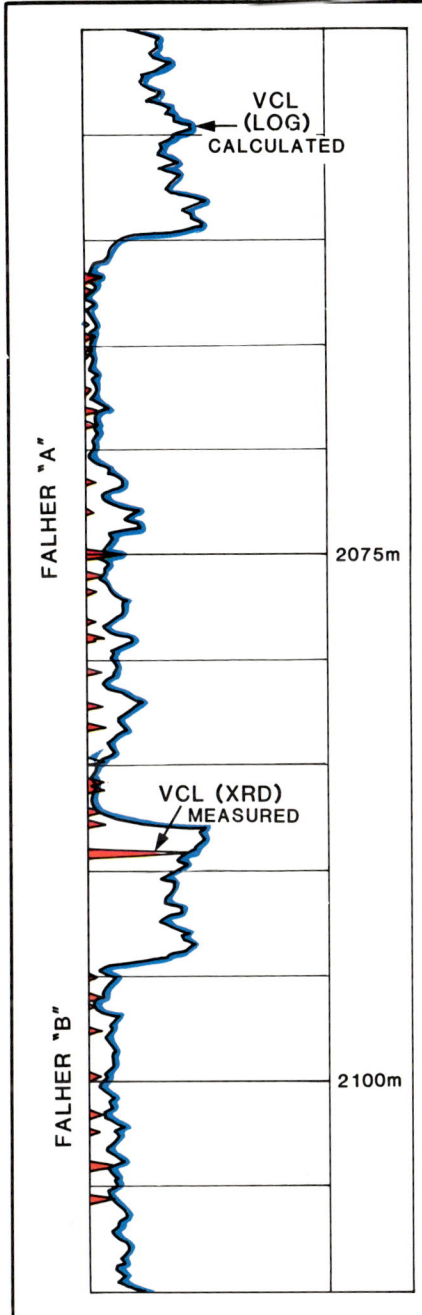

Figure 31. Clay volume (X.R.D.) versus clay volume (log) Falher formation 11-7-70-12W6.

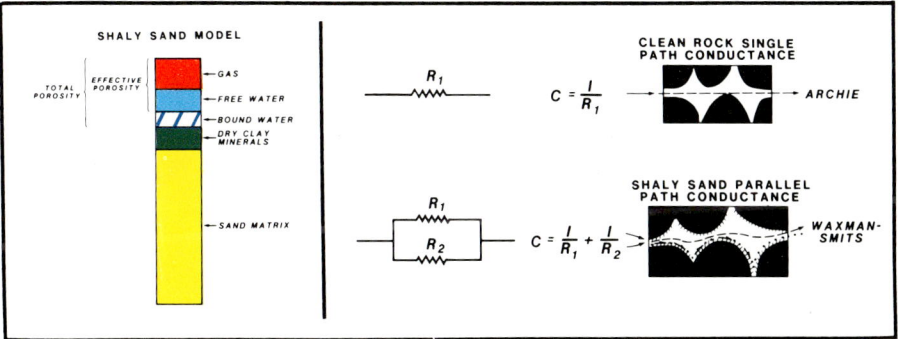

Figure 32. Shaly sand model showing electrical conductance.

assessment of the hydrocarbons in the study area. The following points summarize the main benefits of the methods and techniques reviewed.

(1) "Quick-scan" analysis techniques can be applied successfully to large area exploration projects.

(2) The application of current rock-log calibration techniques to older wells enables one to incorporate the data into present area studies.

(3) Rock-calibrated well-log curves can be utilized for lithologic discrimination in a clastic sequence.

(4) The understanding of rock-derived log relationships enhances the estimate of potential reservoir productivity by the identification of conglomerates from sandstones.

(5) Lithology-adjusted log relationships enable one to accurately define the productive reservoir units in an area.

(6) Computer-aided data manipulation leads to data refinement and improved interpretation quality.

(7) The collection of a broad data bank is important to detailed calibrations and can serve the purpose of an analog for exploration in other areas.

(8) The use of empirical relationships derived from rock and log data has proven to be a valuable tool in formation evaluation.

REFERENCES

Archie, G. E., 1950, Introduction to petrophysics of reservoir rocks: AAPG Bulletin, v. 34, p. 943-961.

——, 1947, The electrical resistivity log as an aid in determining some reservoir characteristics: AAPG Bulletin, v. 31, p. 350.

Bell, J. S., and D. J. Gough, 1981, Intraplate stress orientations from Alberta oil wells: in Evolution of the Earth: in American Geophysical Union-Geological Society of American Geodynamic Series, v. 5, p. 96–109.

Best, D. L., 1983, Using the nuclear magnetism log to predict permeability: Schlumberger internal publication.

Connolly, E. T., 1974, Digital log analysis; recognition and treatment of field recording errors: Society of Professional Well Log Analysts, Annual Logging Symposium.

——, and P. Reed, 1983, Full spectrum formation evaluation: Canadian Well Logging Society Journal, v. 12, no. 1, p. 23–69.

Cox, J. W., 1970, High resolution dipmeter reveals dip-related borehole and formation characteristics: Society of Professional Well Log Analysts, Annual Logging Symposium, Paper D.

Fertl, W. H. and E. Frost, 1982, Experiences with natural gamma ray spectral logging in North America: Society of Petroleum Engineers of American Institute of Mechanical Engineers Annual Conference, SPE Paper 11145.

Herrick, R., S. H. Couturie, and D. Best, 1979, An improved nuclear magnetism logging system and its applications to formation evaluation: Society of Petroleum Engineers of American Institute of Mechanical Engineers Annual Conference, SPE Paper 08361.

Hill, H. J., O. J. Shirley, and G. E. Klein, 1979, Bound water in shaly sands–its relation to Ov and other formation properties: The Log Analyst, v. 20, no. 3, p. 3-19.

Masters, J. A., 1982, Deep basin gas trap, western Canada: AAPG Bulletin, v. 63, p. 152-181.

Schlumberger, 1981, Schlumberger dual axis caliper applications: Internal paper.

Smith, R. D., 1984, Gas reserves and production performance of the Elmworth/Wapiti area of the deep basin, in J. A. Masters, ed., Deep basin gas: AAPG Memoir 38, this volume.

Sneider, R. M., et al, 1981, Methods for

Figure 33. Water saturation versus porosity - Falher A sandstone.

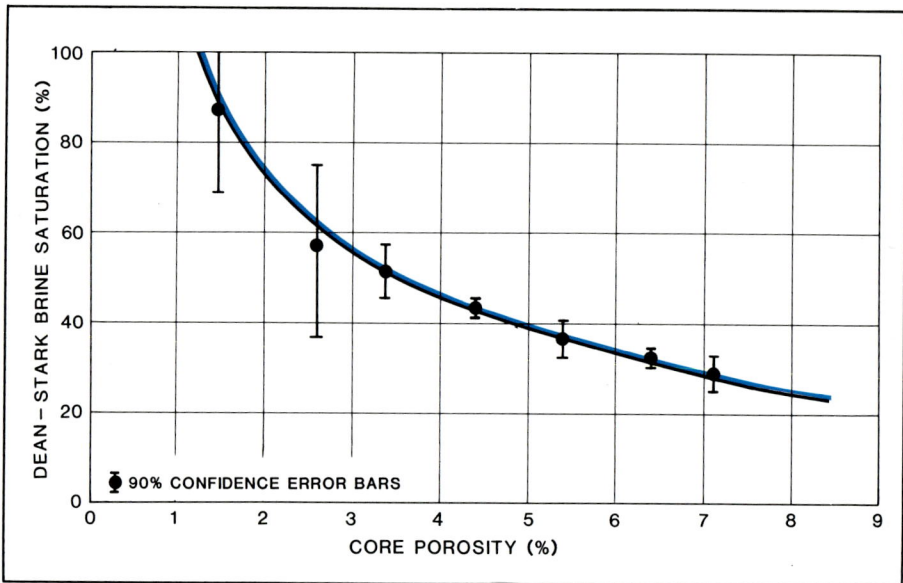

detection and characterization of reservoir rock, Deep basin gas area, western Canada: Society of Petroleum Engineers of American Institute of Mechanical Engineers, SPE Paper 10072.

Waxman, M. H. and L. J. M. Smits, 1968, Electrical conductivities in oil-bearing shaly sands: Society of Petroleum Engineers Journal, v. 243, p. 107-122.

Westway, P., R. Hetzog, and R. E. Plaset, 1980, The gamma ray spectrometer tool in elastic and capture gamma-ray spectroscopy for reservoir analysis: Society of Petroleum Engineers of American Institute of Mechanical Engineers Annual Conference, SPE Paper 9461.

Figure 34. Log Sw versus oil base core Sw.

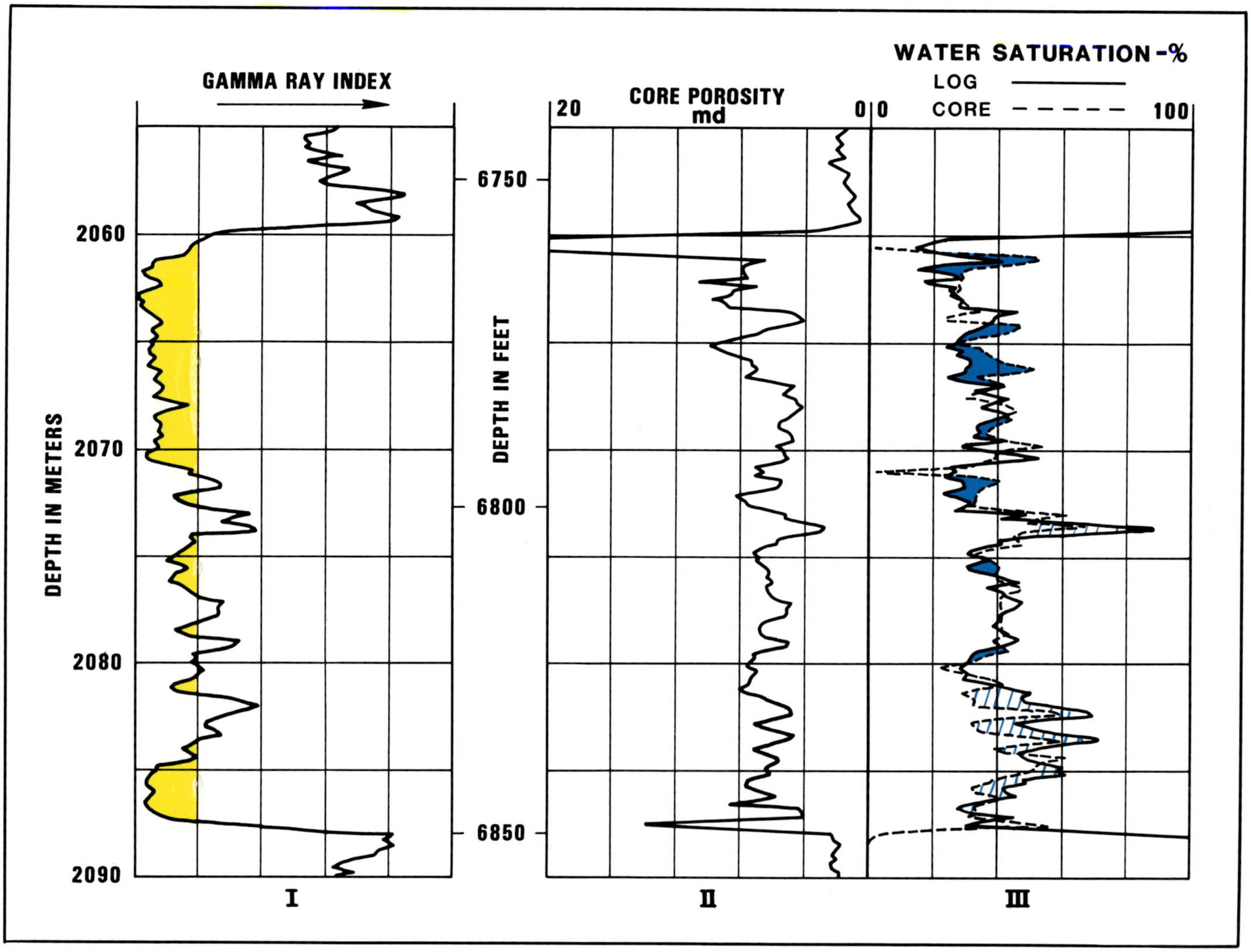

Part III

Selected Examples Illustrating Integration of Rock-Log Data to Determine Reservoir Petrophysical Characteristics

R. W. Hietala
E. T. Connolly
H. R. King
Canadian Hunter Exploration, Ltd.
Calgary, Alberta

R. M. Sneider
Robert M. Sneider Exploration, Inc.
Houston, Texas

Edited by: J. D. Loren
Loren and Associates, Inc.
Houston, Texas

Application of the methodology presented in Parts I and II has led to a comprehensive understanding of reservoirs with diverse characteristics in the Lower Cretaceous of the Elmworth field. The merging of petrophysical, geological, and engineering data has resulted in a two-fold benefit: the accuracy of reserve estimates has been greatly increased and the ability to reliably estimate well productivity from logs and cuttings has been realized. The objective of Part III is to demonstrate by selected examples some of the more important benefits which can be achieved through an integrated rock-log approach to reservoir characterization.

As a means of demonstrating key points, log and rock data and interpreted results are presented for six successfully completed zones which span the stratigraphic interval from the Paddy to the Cadomin. These chosen examples, five of which were cored, have flow rates ranging from 0.1 to 47 mmcf/d, and have rock and log characteristics which are compatible with these diverse flow rates. Some of the specific points which the examples illustrate are:

(1) The Sneider rock classification system provides a basis for estimating well deliverability;
(2) rock, log, and drilling characteristics of conglomerates are identified;
(3) excellent agreement between core porosity and log-derived porosity in rocks with a wide range in clay content and lithology is demonstrated;
(4) variable water saturations are shown to be consistent with visual rock types and measured core properties;
(5) variations in clay volume and clay type as determined from visual examination and from logs relate to water saturation and reservoir flow characteristics; and,
(6) a log-derived reservoir quality index supplements visual rock typing to provide a means of estimating rock permeability.

Accurate values of porosity, water saturation, lithology, and productivity are benefits realized by integration of rock and log data.

Each of the six selected example zones has raw and interpreted data presented in a format consisting of: (1) an open-hole suite of logs, (2) a continuous display of interpreted log and rock information, and (3) scanning electron microscope (SEM) and thin section, photographs, and mercury-injection capillary pressure curves. The depth position of each sample photograph is indicated by a numbered arrow on each of the log and core diagrams. The arrows have been depth adjusted to match log metric depths. For the five wells which were cored a continuous display of measured porosity and permeability is included with the interpreted rock and log display. The well deliverability information for each zone is the maximum sustained test rate after stimulation by acidizing and fracture treating. The open-hole log data, interpreted data, and selected sample data are all presented in a consistent format for each of the six zones.

The first panel for each example is the open-hole log suite, which includes, if available, gamma ray, SP, caliper, acoustic transit time, neutron and density porosity on a sandstone index, deep induction resistivity, drilling rate curve and a microlog. A flag curve is included in each raw data set to denote zones of excessive density-log correction where the density tool is judged unreliable. It is instructive to examine each of the open-hole log data panels with the knowledge that each is a successfully completed zone and to note the following observations:

(1) All but one of the examples have apparent density/neutron crossover commonly taken as an indicator of gas. Yet, examination of the flag curve shows the density log to be unreliable over many intervals, thereby negating this technique as a reliable indicator of hydrocarbons in the Elmworth area;
(2) Many of the zones where the density is reliable, as evidenced by lack of the flag curve, have neutron porosities which are significantly higher than the density porosity, thus denoting the presence of clays, particularly clays with a relatively high hydrogen index;
(3) The acoustic transit-time curve does not exhibit wide variations, giving confidence that accurate formation properties are measured and supporting use of this device as a primary means for determining porosity; and,
(4) Resistivity levels opposite reservoir units range from a low of 30 to a high of 1,000+ ohm-meters and encompass variations in acoustic transit-time (porosity), as well as variations in gamma-ray and neutron response (clay content). This suggests the use of a shaly-sand model for saturation calculations.

General observations such as these are important in formulating the approach for development of a reliable interpretation technique.

The second panel presented for each of the six examples consists of interpreted log and rock data, as well as measured core properties. Four display sections on each panel from left to right include: (1) log-derived reservoir properties and core porosity; (2) log and rock permeability information, lithology, and perforations; (3) rock typing using the Sneider rock classification system; and (4) visual estimates of porosity and clay volume.

Some specific items relating to the various display curves that are important to note are as follows:

(1) Clay volumes range from nil to 30% and are composed of clay A, which is predominantly illite, and clay B, which is predominantly kaolinite;
(2) Porosities range from 3 to 18% with the effective porosity component controlling fluid flow;
(3) Water saturations range from 10 to 50% and consist of a bound-water component and a connate water component;
(4) Two of the examples are completed within a conglomerate zone, the remainder are completed within sandstone intervals; and,
(5) Sneider rock types range from almost 100% Type I to almost 100% Type III with an associated wide range in visual estimate of porosity and clay volume.

The wide range in interpreted reservoir properties summarized in Table I was the basis for selecting six examples included in this paper.

The last series of panels within each of the six selected examples includes scanning electron and thin section microphotographs and a mercury-injection capillary pressure curve for two to five rock samples to illustrate specific points. For each rock type, three SEM photographs (magnification of 25, 100, and 250×, photos A, B, and C) and two thin section photographs (magnifications of 32 and 128×, photos D and E) are shown to illustrate overall rock fabric as well as details of pores, pore interconnection, and clay distribution. The mercury-injection capillary pressure information provides a means of characterizing the pore system and provides a bridge to the water saturation calculated from logs.

The type and distribution of clays in the Cretaceous sandstone and conglomerate reservoirs in the Elmworth field are important in the log calculations and in permeability/deliverability estimates. Based on X-ray diffraction and scanning electron microscopic analyses, the clay minerals present, in order of decreasing abundance are: kaolinite, illite, chlorite, mixed-layered illite/chlorite and illite/smectite, and smectite. In the six zone examples, X-ray analysis shows that the total volume of clay ranges from 3 to about 30% of the total rock volume, with

Table 1

Zone	Formation or Member	Deliverability mmcf/d	Lithology	Clay Volume Log	Clay Volume Visual	Clay Type	Porosity Range	Porosity Comparison Core/Log	Porosity Comparison Core/Visual	Permeability ROI Correlation	
1	Paddy	3.1	Sandstone	Low	Low	Illite	9-18%	Excell	Excell	Fair	10-15%
2	Cadotte	0.75	Sandstone	V. Low to V. High	Mod.	Illite Kaolinite	3-11%	Excell	Excell	Good	30-95%
3	Falher A	20.0	Conglomerate	Low	Low	Kaolinite Illite	4-9%	Excell	Excell	Fair	20-40%
4	Falher B	0.10	Sandstone Sandstone	High Low to Moderate	Low Low	Illite Kaolinite Illite Chlorite	3-7% 2-8%	Good Excell	Good Excell	Fair Good	30-60% 25-65%
5	Falher C	47.0	Conglomerate	Low to Moderate	Low	Kaolinite Illite	5-18%	Fair	Fair	Good	20-50%
			Sandstone	Low to Moderate	Low	Kaolinite Illite	6-11%	Excell	Good	Good	30-40%
6	Cadomin	0.65	Sandstone	Low	Low	Kaolinite	3-6%	—	—	—	20-35%

Table 1. Summary of zone characteristics.

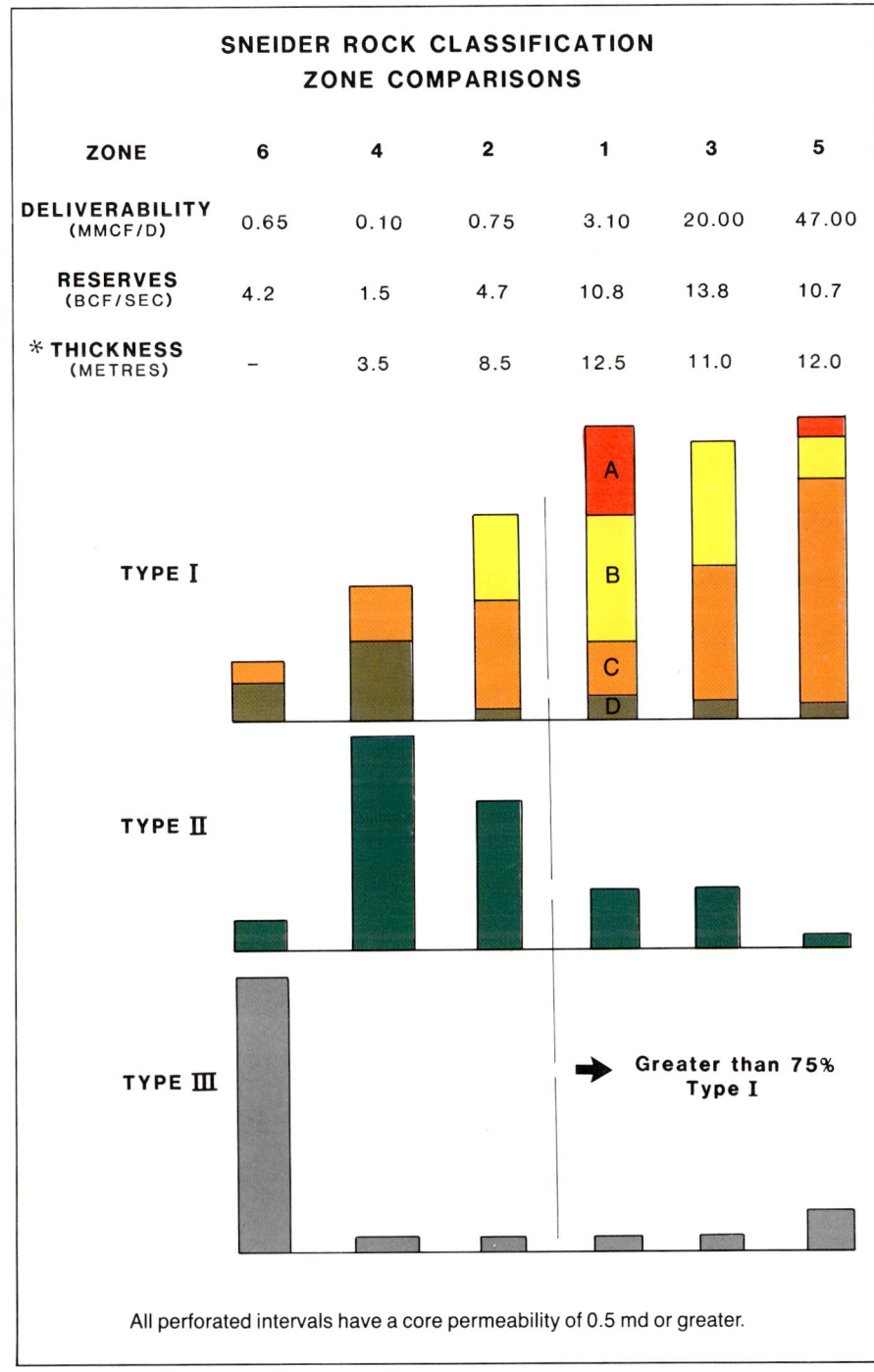

Figure 1. Sneider rock classification zone comparisons.

stone consists of two components, clay-bound water and intergranular connate water. The clay-bound water is chemically tied to the clays and therefore does not contribute to fluid flow. The connate water typically is mobile water and therefore influences fluid-flow characteristics of the rock. However, rocks may contain a micropore system which retains part of the connate water and renders it immobile. Examination of SEM photographs and thin sections of the reservoir rocks in Elmworth frequently reveals the presence of a micropore system within deteriorated cherts as well as a micropore system within and between clay minerals and aggregates. Such rocks exhibit abnormally high values of log-derived water saturation which at first glance may appear to conflict with observed water-free production performance. However, detailed rock examination usually reveals the presence of a micropore system and the associated trapping of connate water, thereby explaining the water-free production at values of calculated water saturation which are higher than normal.

The distribution of rocks according to the Sneider rock classification technique in each of the six example zones is shown in Figure 1. Bar graphs in each of the rock-type categories represent the percentage of that particular rock type in the completion interval based upon visual examination of the core. In zone 6 where no core is available the rock classification work is based upon examination of drill cuttings, as is typically the case in most evaluation work. The arrangement of examples on Figure 1 is from a low percentage on the left to high percentage of Type I rock on the right.

The importance of recognizing Type I rock is vividly illustrated by observing the general correlation of increasing well deliverability with increasing percentage of Type I rock on Figure 1. In these examples, greater than 75% Type I rock correlates with well deliverability which ranges from 3 to 47 mmcf/d. Other factors such as the formation damage ratio and the actual permeability of Type 1A rock explain the variation between the 3 and 47 mmcf/d examples. At the other extreme, zone 6, with only a small percentage of Type I rock, has become an economically successful completion. The rock typing of this zone was done only from drill cuttings and was of primary importance.

There is a good correlation between core

3 to 12% being typical for the Type I rocks. Kaolinite is by far the most abundant clay in the example zones, and typically is 60 to 85% of the total clays. Illite ranges from 10 to 30% of the total clays, in general, but may be up to 40 to 50% in a few samples. The clay minerals usually occur partially filling pores, as partial coatings on detrital grains, and on and within grains of shaly rock. Under SEM examination most of the clay-mineral aggregates contain a microporosity system between the particles. These micropores no doubt contain water tightly held by capillary forces.

The water saturation of a shaly sand-

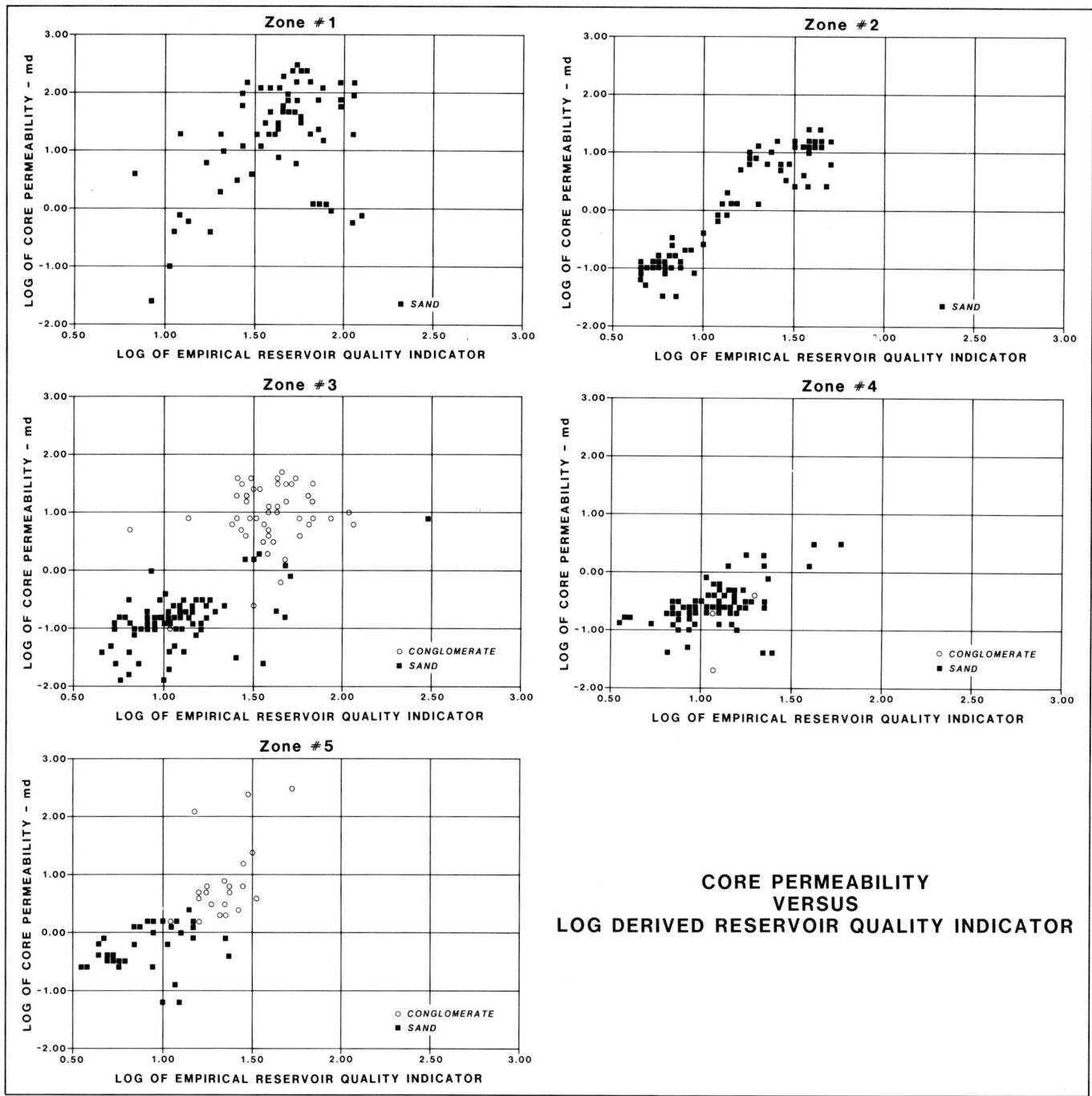

Figure 2. Core permeability versus log-derived reservoir quality indicator, Zones 1 through 5.

permeability and the empirical reservoir-quality indicator derived from log data. In gas-saturated formations the reservoir-quality indicator which is a function of clay volume, and formation resistivity has served a valuable function in comparing both sandstones and conglomerates of varying reservoir characteristics. The relationship between these two parameters can be seen in Figure 2 for the five selected examples which had core control.

Zone 1

This example (Figures 3 to 7) was chosen as representative of a sandstone of excellent reservoir quality. The following observations can be made:

(1) The completion interval is composed primarily of Type I sandstones, with Type IA and Type IB predominating. Type IC and ID, as well as some Type II sands, increase in amount at the top and the lower part of the interval. The 3.1 mmcf/d well deliverability is low due to formation damage in effect after completion.

(2) Clay type and volume are consistent between log calculations and X-ray diffraction work on the three selected samples. The continuous log calculations indicate that clays are predominantly illites. Clay quantity shows a slight upward increase and ranges from nil at the base to just under 10% near the top, with the exception of the very uppermost meter where clay volumes reach 18%. X-ray diffraction work on the three samples confirm the low level of clay and shows that illite volumes do not exceed 7% and that kaolinite volumes remain less than 2%.

(3) Porosity as determined from logs compares very favorably with core porosity and decreases from a high of 18% near the base to a low of 8% near the top. As clay volumes are low, the effective porosity is extremely close to total porosity. The visual estimates of porosity compare favorably to core and log porosity.

(4) The very low water saturation shows a systematic variation with porosity and ranges from 10% near the base to 20% near the top. Due to the low volume of clay, very little of the water saturation is the result of bound water. The rocks are characterized by large pores and the absence of a micropore system. Log-derived water saturations are thus consistent with rock observations.

(5) Core permeability correlates with the reservoir quality index and decreases upward.

(6) The Type I rocks have a well-developed primary intergranular macropore system; some pores are enlarged by partial solution of grains and cement. However, the importance of grain size in controlling pore structure is evident by comparing samples at 1,668.73 and 1,678.34 m (5,474.8 and 5,506.4 ft). Both samples have essentially the same porosity, but differ in permeability by two orders of magnitude (0.8 versus 90 md.). The SEM and thin section photographs and capillary pressure curves show the finer pores and pore interconnection of Type ID compared with the IB rock.

(7) Pore-size distribution, as portrayed by the mercury-capillary pressure curves and the rock photographs, illustrates that the Type IA and IB have a very low entry pressure and a high percentage of large- and medium-size pores compared to the Type ID example which has a high entry pressure and a relatively higher percentage of fine- and medium-size pores.

(8) Rocks and logs both indicate the overall reservoir quality to be excellent and demonstrate that the slight decreases in rock quality from base to top as observed form the SEM, thin section, and mercury-injection capillary pressure curves are all compatible with the variation in porosity, clay volumes, and water saturation deduced from logs.

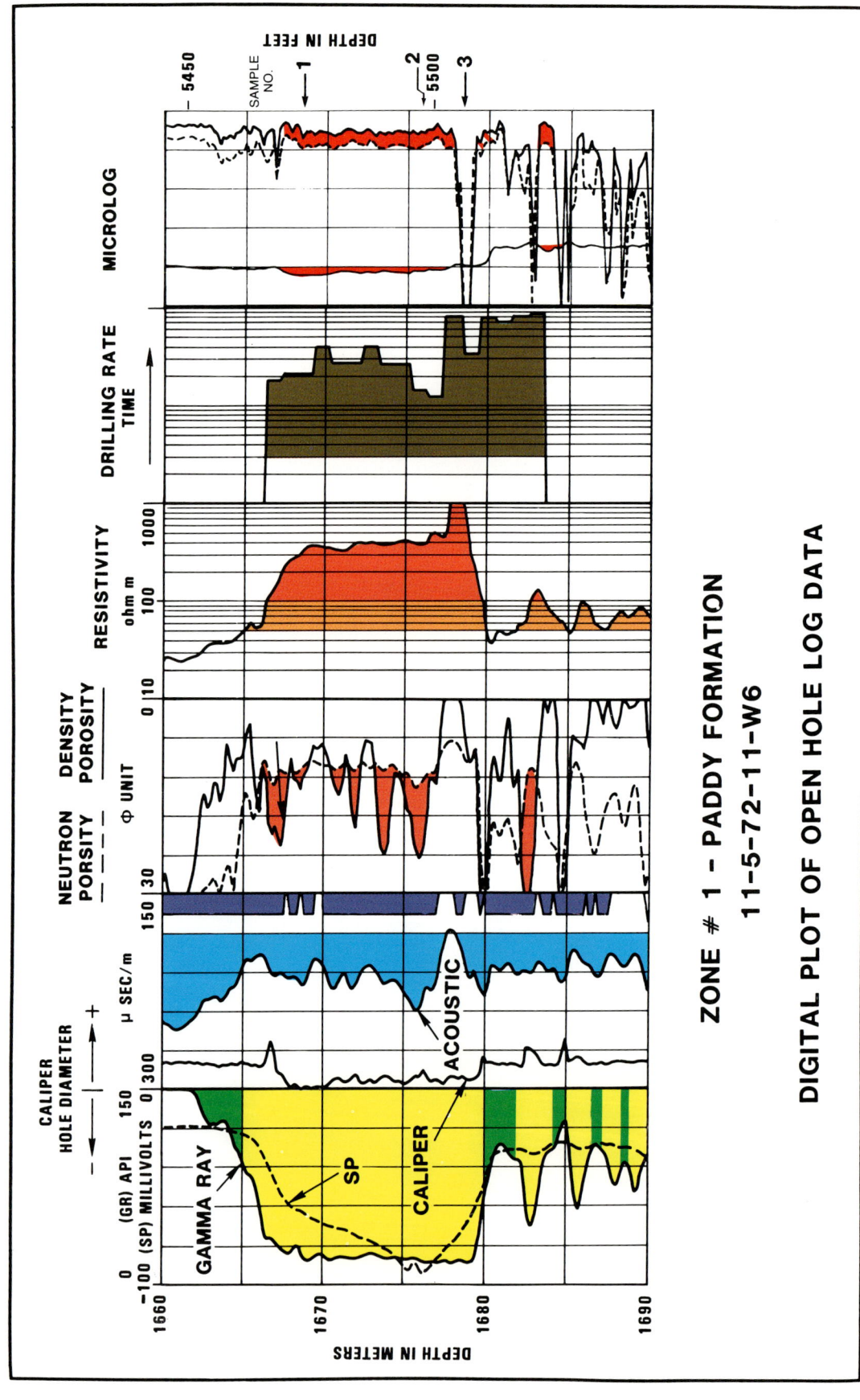

Figure 3. Digital plot of open-hole log data, Zone 1 — Paddy Formation 11-5-72-11-W6.

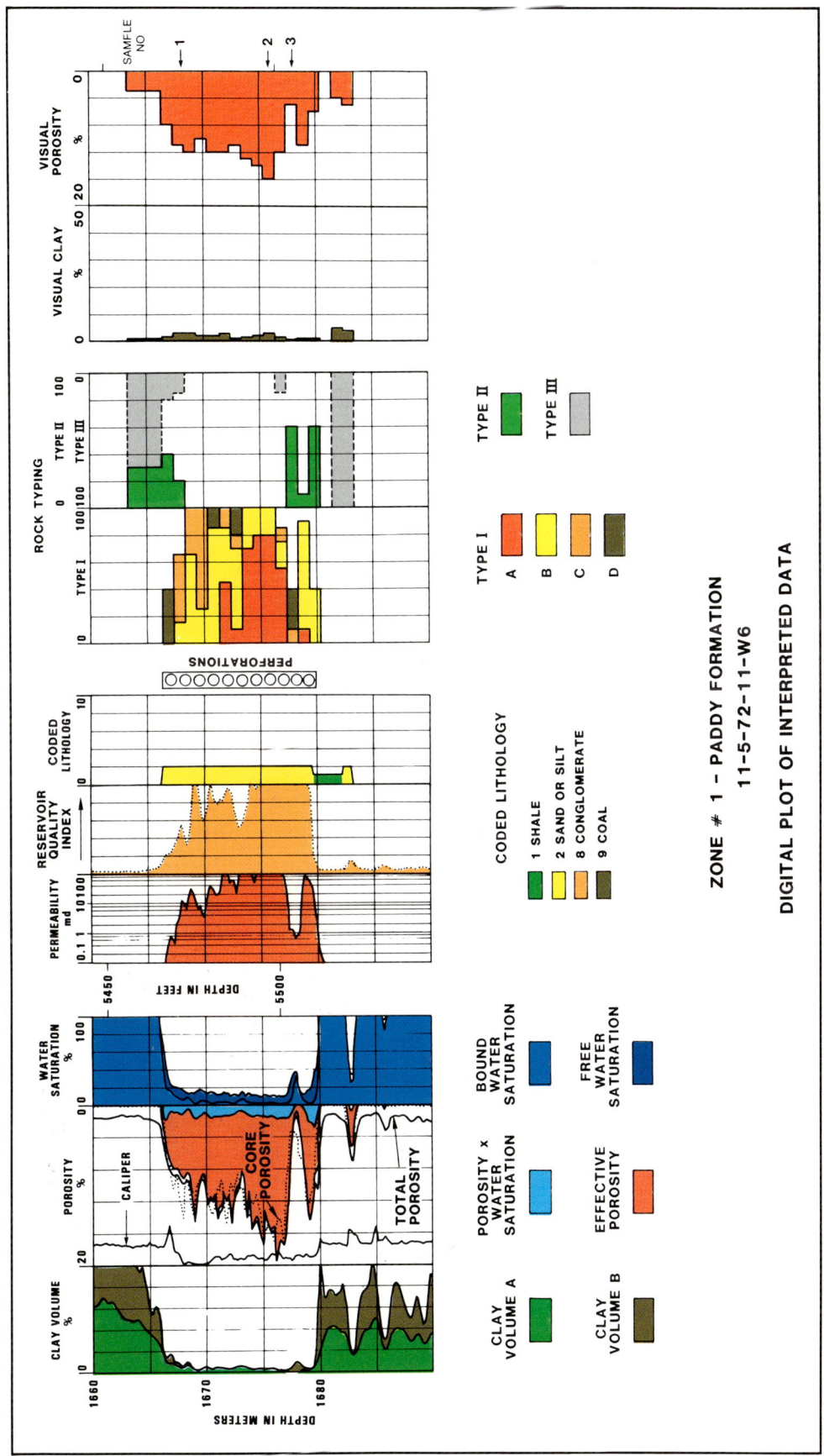

Figure 4. Digital plot of interpreted data, Zone 1 — Paddy Formation 11-5-72-11-W6.

SAMPLE 1

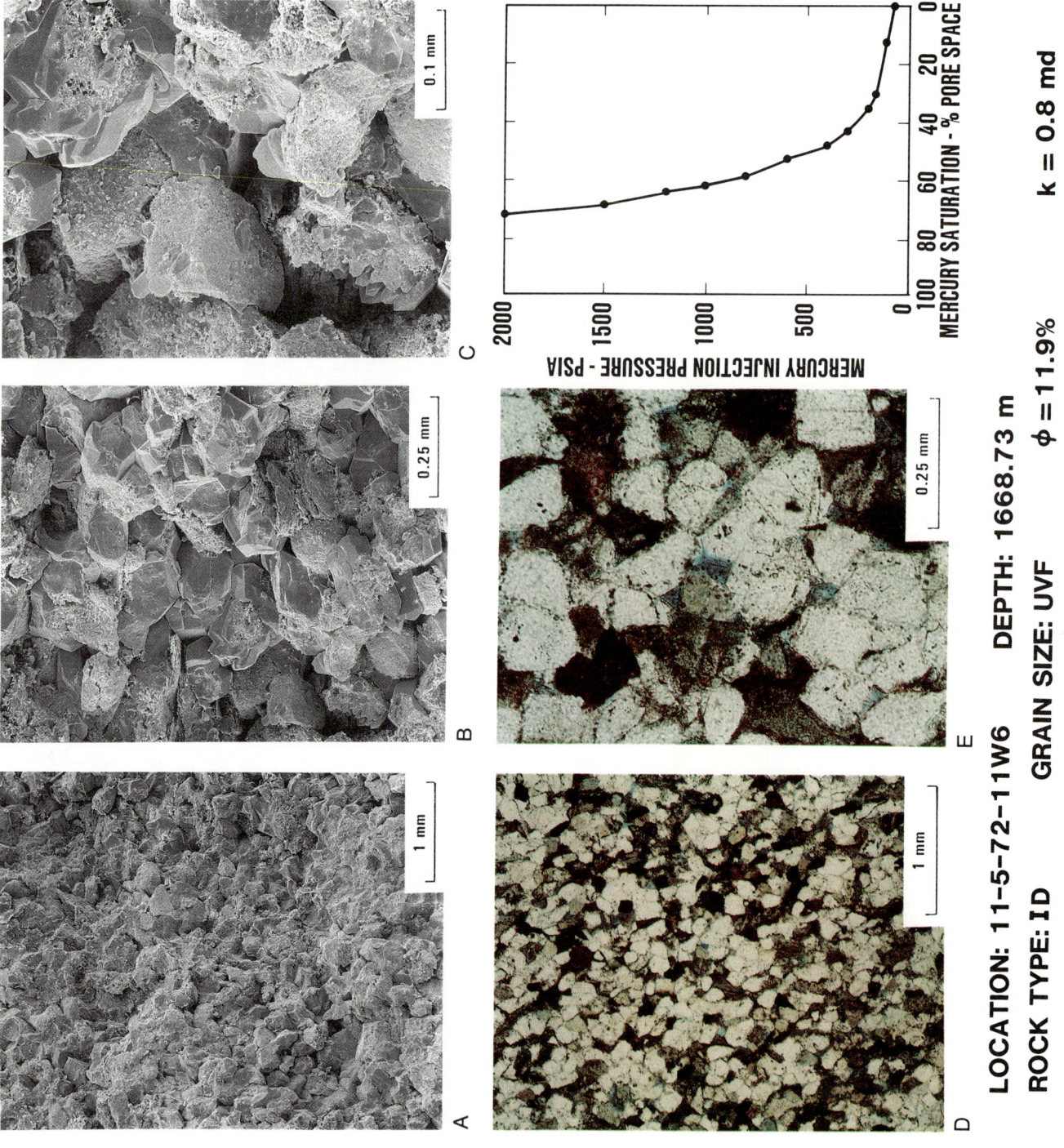

LOCATION: 11-5-72-11W6 DEPTH: 1668.73 m
ROCK TYPE: ID GRAIN SIZE: UVF ϕ = 11.9% k = 0.8 md

Figure 5. Scanning electron and thin section microphotographs and a mercury-injection capillary pressure curve for rock samples of Zone 1 – Paddy Formation 11-5-72-11-W6, illustrating overall rock fabric, at a depth of 1,668.73 m.

SAMPLE 2

LOCATION: 11-5-72-11W6 DEPTH: 1676.04 m
ROCK TYPE: IB GRAIN SIZE: LC $\phi = 11.0\%$ k = 90 md

Figure 6. Continued rock sample data for Zone 1 — Paddy Formation 11-5-72-11-W6, at a depth of 1,676.04 m.

SAMPLE 3

LOCATION: 11-5-72-11W6   DEPTH: 1678.34 m
ROCK TYPE: I A   GRAIN SIZE: UC   φ = 17.9%   k = 150 md

Figure 7. Continued rock sample data for Zone 1 — Paddy Formation 11-5-72-11-W6, at a depth of 1,678.34 m.

Zone 2

This example (Figures 8 to 11) of a sandstone reservoir has properties less favorable than Zone 1 and is more typical or the sandstone reservoirs found in the Elmworth field. The considerable variation in rock quality within the completion interval is useful in illustrating the variation in certain log-derived reservoir parameters. The following key points are important:

(1) The completion interval includes three packages of sandstones: Type II with minor amounts of Type ID, IC and III in the upper portion; Type IB and IC in the middle; and mainly Type II in the lower part. The well deliverability of 0.75 mmcf/d is consistent with these rock observations.

(2) Clay volumes range from very low to 30% and consist primarily of illite and kaolinite as determined from rock-log analyses. The visual estimates of clay give low values throughout the interval and do not show any increase in clay content toward the base of the sand as the logs indicate. Visual estimates of clays, especially in fine-grained rocks, generally are low compared with that inferred from logs because of the inability to see the fine clay particles with the low magnification powers of the binocular microscope. Excellent estimates of clays come from X-ray, thin-section, or SEM analysis. The two samples selected for SEM work are representative of the better reservoir rock and have clay volumes of 4 to 6% of illite and kaolinite. Log-derived clay volume is slightly higher but not inconsistent with the X-ray diffraction data when sampling and measurement accuracy are considered.

(3) Log-derived porosity shows excellent agreement with core porosity over a porosity range of 3 to 10%. It is important to note that effective porosity rather than total porosity is required in order to arrive at agreement with core porosity within rocks containing significant volumes of clay. Visual porosity compares favorably with core and log porosity.

(4) The variation in rock type between the upper, middle, and lower parts of the reservoir causes variations in total water saturation as well as proportionate changes in bound and connate components of water saturation. The best rock has water saturation at 35% with the bound-water component only 5%. As the percentage of Type I rock decreases in the upper portion of the reservoir, the total water saturation increases to 50% primarily due to the increase in the bound-water component (close to 20%). Finally, in the lower part of the reservoir which is predominantly Type II, the total water saturation ranges from 45 to 95% with attendant variations in the bound-water component of 20 to 50%. Consideration of the bound-water component of calculated water saturations is consistent with the observed water-free production.

(5) The reservoir quality index compares favorably with core permeability.

(6) The two rock types illustrated for this zone example are medium-grained, well-sorted IC and IC/B sandstones with very low amounts of clays. Most of the pores are intergranular and many pores show enlargement by dissolution. Some of the detrital shaly rock and chert grains contain both a micro and macro pore system. Comparisons of the capillary-pressure curves and SEM and thin section photographs of these two samples with those of the Type IA and IB rocks in the Zone 1 example show that the IC rocks have a higher percentage of medium and small pores and higher entry pressure. The five rock samples from both the Zone 1 and 2 examples illustrate all four Type I rock types. Comparison of all four rock types show there is good agreement between the expected pore-size distribution from the capillary pressure curves, permeability, and the visual appearance of pores and pore throats illustrated in the photomicrographs.

(7) Drilling time correlates with the rock type.

(8) Neutron porosity is higher than density porosity and this difference increases as the gamma ray response increases. Simultaneous consideration of these two variables forms the basis for determination of relative volumes of two different clay species.

(9) The highest core porosities and permeabilities, log porosities and reservoir-quality index, and intervals of mud cake and the lowest water saturations correspond with the middle-rock package which consists of Type IB and IC rock.

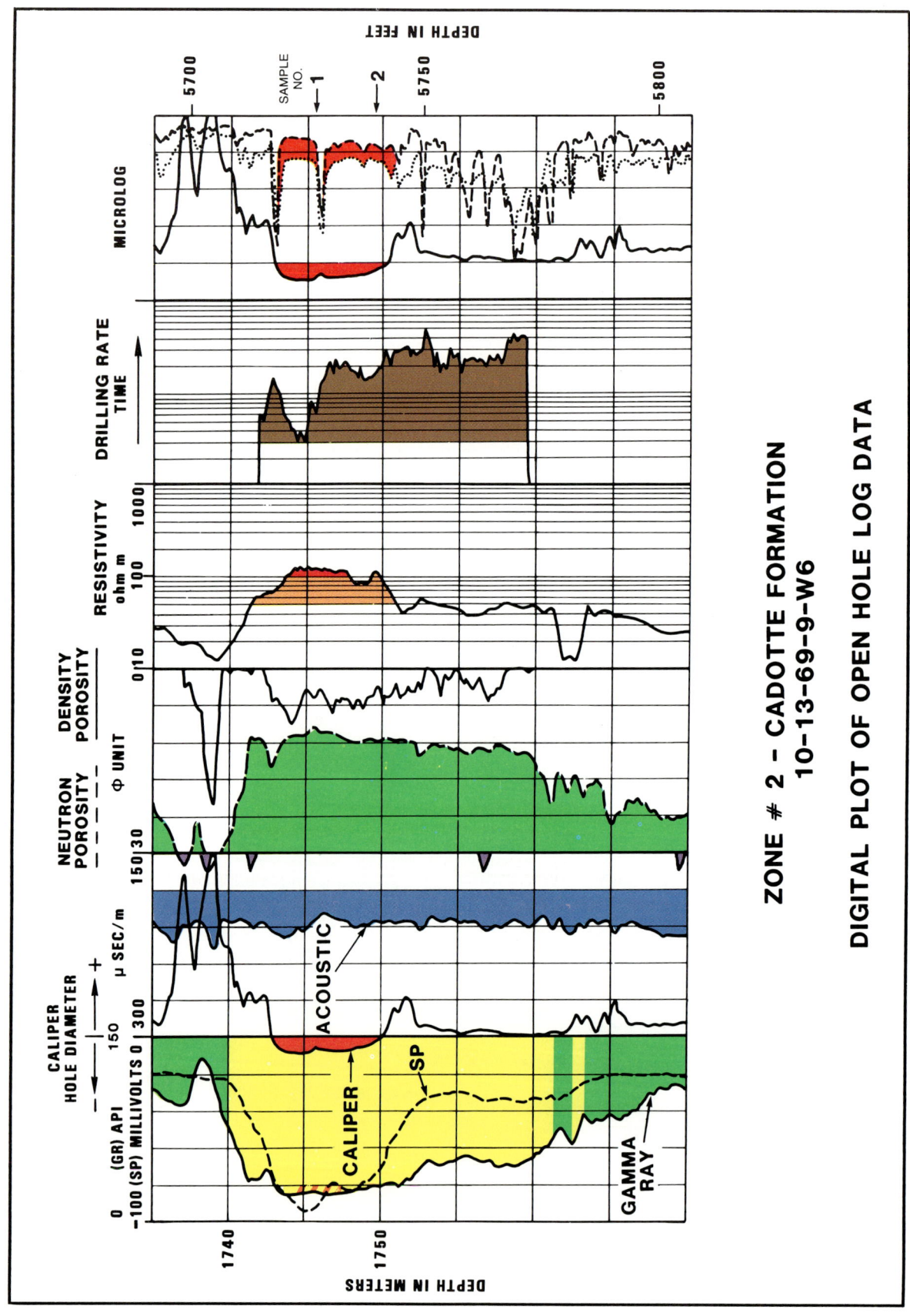

Figure 8. Digital plot of open-hole log data, Zone 2 — Cadotte Formation 10-13-69-9-W6.

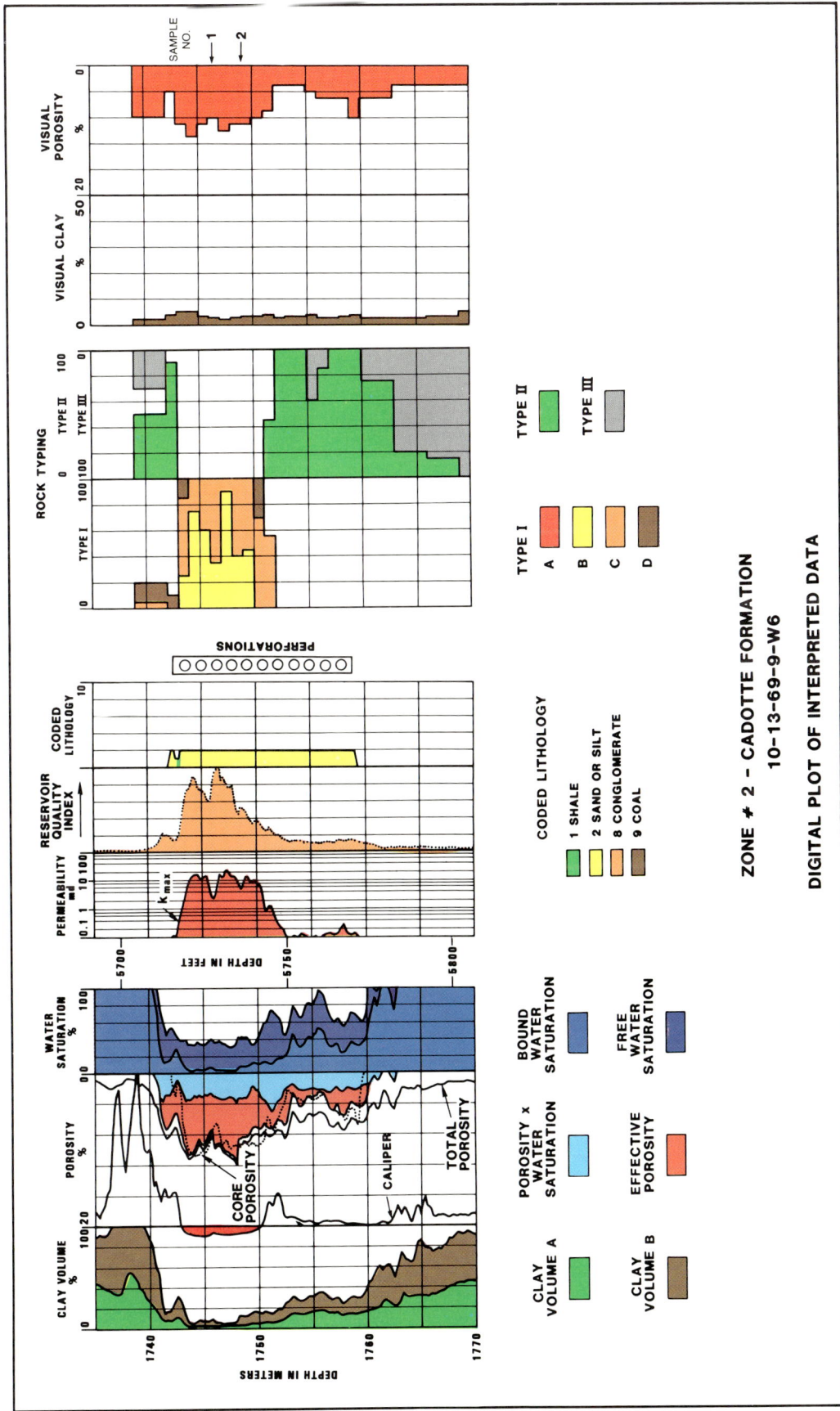

Figure 9. Digital plot of interpreted data, Zone 2 — Cadotte Formation 10-13-69-9-W6.

Figure 10. Scanning electron and thin section microphotographs and a mercury-injection capillary pressure curve for rock samples of Zone 2 — Cadotte Formation 10-13-69-9-W6, at a depth of 1,746.9 m.

Figure 11. Continued rock sample data for Zone 2 — Cadotte Formation 10-13-69-9-W6, at a depth of 1,750.6 m.

Zone 3

This extremely prolific reservoir (Figures 12 to 18) provides a good example of contrasts between conglomerate units and sandstone units which coexist in the same reservoir. Certain characteristics of conglomerates which are useful for lithology identification are discussed. The presence of a micropore system and its effect on log-derived reservoir parameters is important.

(1) The completion interval encompasses a thick conglomerate unit at the top and a thin conglomerate unit at the base, both of which are Type IA and IB. The well deliverability of 20 mmcf/d is characteristic of conglomerates with the observed rock types. A nonperforated interval between the two conglomerate units is comprised of sandstone which is predominantly Type II.

(2) Within the upper conglomerate unit clay volumes from X-ray diffraction data which range from 6 to 9% are predominantly kaolinite and are compatible with clay volumes and type deduced from logs. The volume of clay in the Type II rock sample increases to 15% as also is shown by logs. However, the log results suggest an illite clay rather than the kaolinite which was observed.

(3) The core porosity agrees very favorably with the effective porosity determined from logs over conglomeratic zones as well as sandstone intervals. Visual estimates of porosity compare favorably with core and log porosity within the conglomeratic intervals. The porosity is principally intergranular to enlarged intergranular. There is some microporosity developed within some detrital grains of chert and shaly-rock grains, as in the Type IB rock at 2,066.62 m (6,780.2 ft). Porosity and water saturation are thus higher than those equivalent rock types of similar permeability.

(4) The water saturations within the conglomerate zones are higher than is normally found in the Elmworth field. This is due to the presence of a micropore system associated with the chert as shown in the scanning electron microscope photographs. Because this water is confined to very fine micropore space it is immobile. Within the Type II sandstones unit the water saturation varies from 40 to 60% and consists of a bound-water component which ranges from 15 to 40%. The sample at 2,072.4 m (6,799.2 ft) is a low-porosity, low-permeability fine-grain sandstone or a Type II rock. Most Type II Lower Cretaceous rocks at Elmworth are the result of compaction of shaly, fine- to very fine-grained rocks, as in the above example of extensive pore infilling by diagenetic clays and/or cements. When clays are the principal cause of pore space reduction to Type II rocks, bound-water content is higher than when the reduction is due to a grain size decrease or increased cementation. The capillary pressure curve of the Type II rock examples shows high entry pressure and a high percentage of small and very small pore size and pore throat sizes, and the tortuous interconnection of the pore system.

(5) The reservoir quality index shows a fair correlation with core permeability.

(6) Certain characteristics of conglomerates are evident by examining the basic logs. They include a reduced gamma-ray reading, the presence of mudcake, an increased resistivity reading, and an acoustic transit time which is abnormally low compared to the neutron-porosity curve. Failure to distinguish conglomerates from sandstones would result in an underestimate of porosity from the acoustic transit-time. The resistivity anomaly would be anticipated to be even greater had a micropore system not been present. These particular characteristics form the basis for the technique to identify conglomerates used in the overall interpretation method.

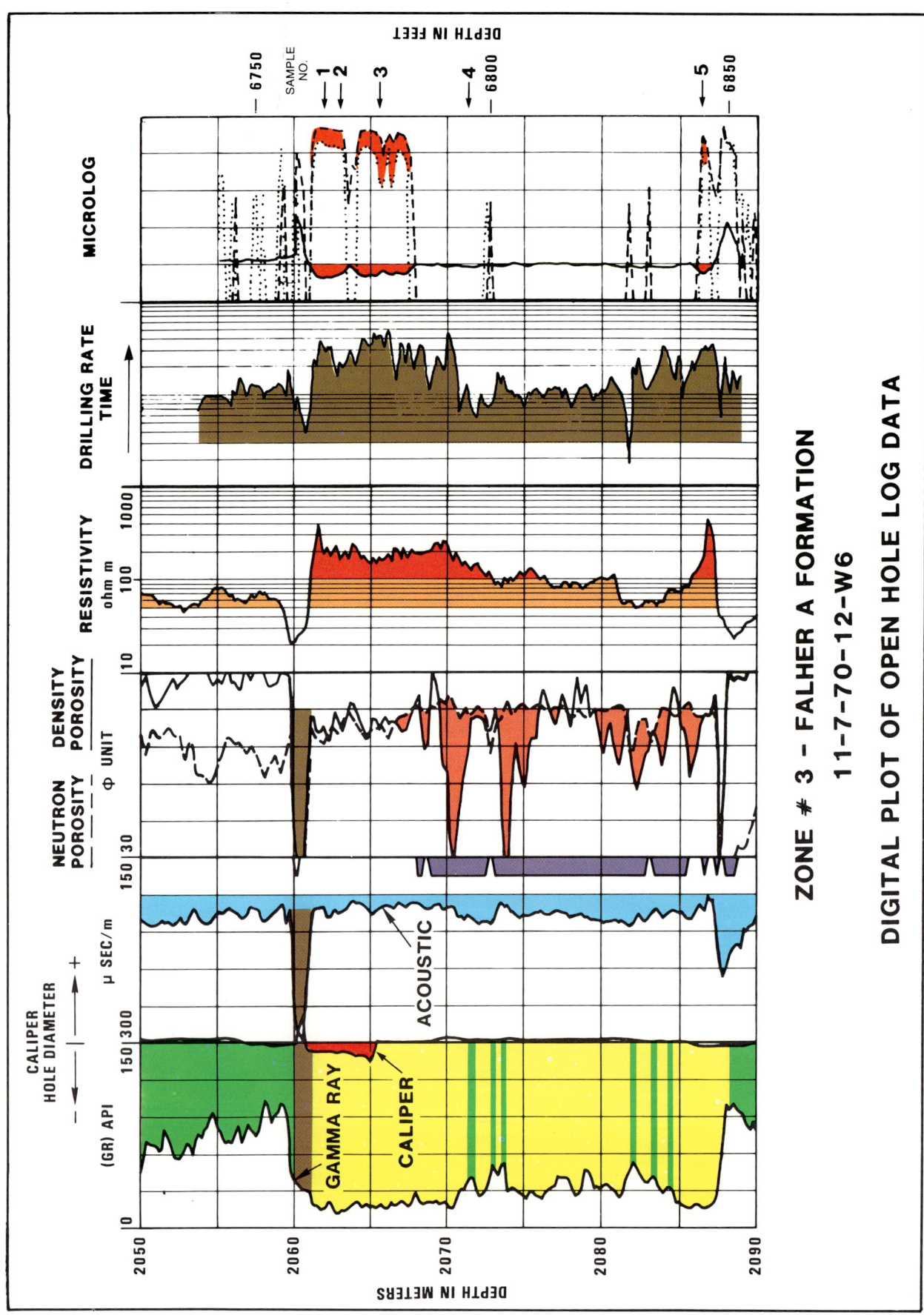

Figure 12. Digital plot of open-hole log data, Zone 3 — Falher A Formation 11-7-70-12-W6.

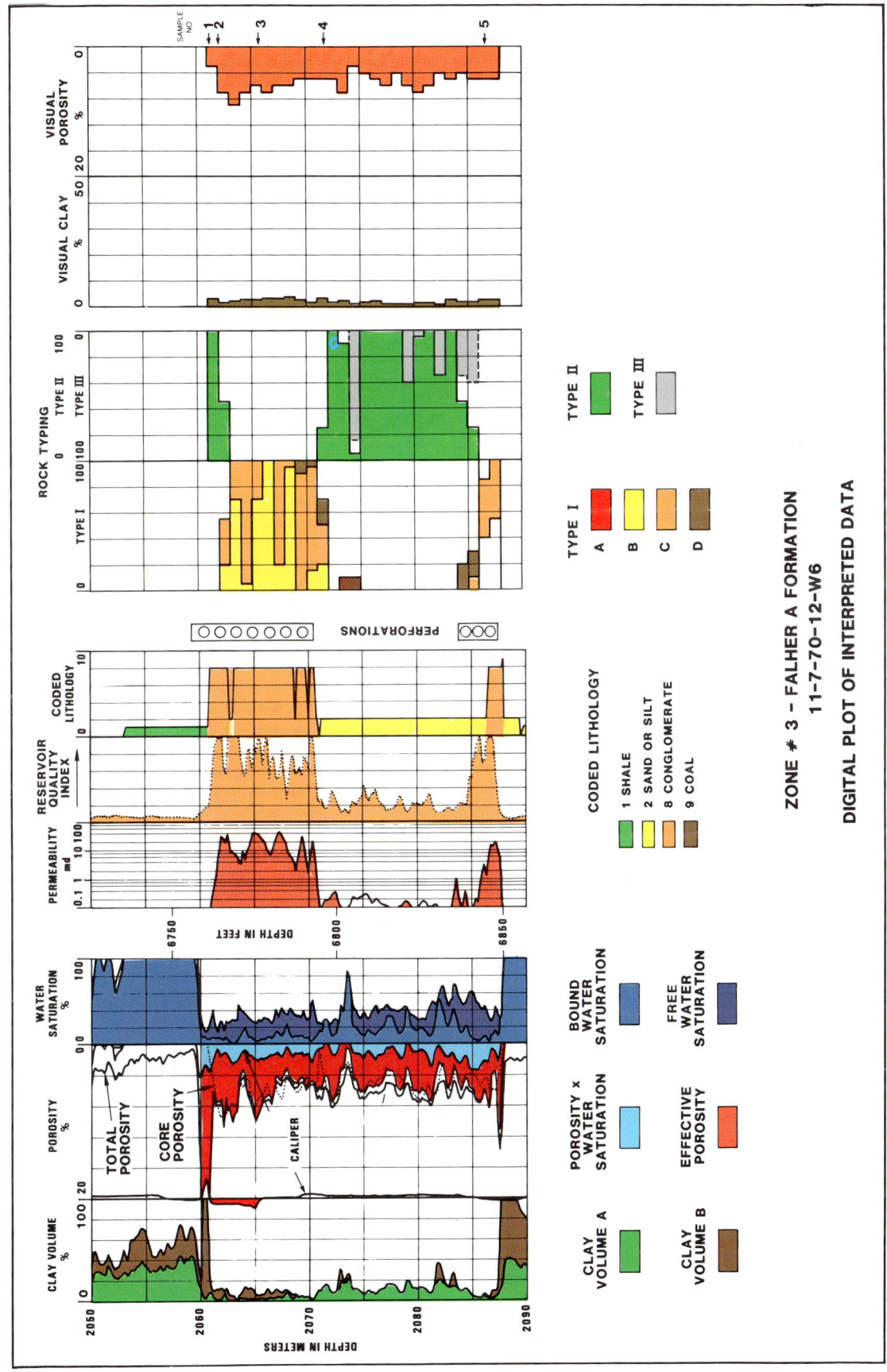

Figure 13. Digital plot of interpreted data, Zone 3 — Falher A Formation 11-7-70-12-W6.

SAMPLE 1

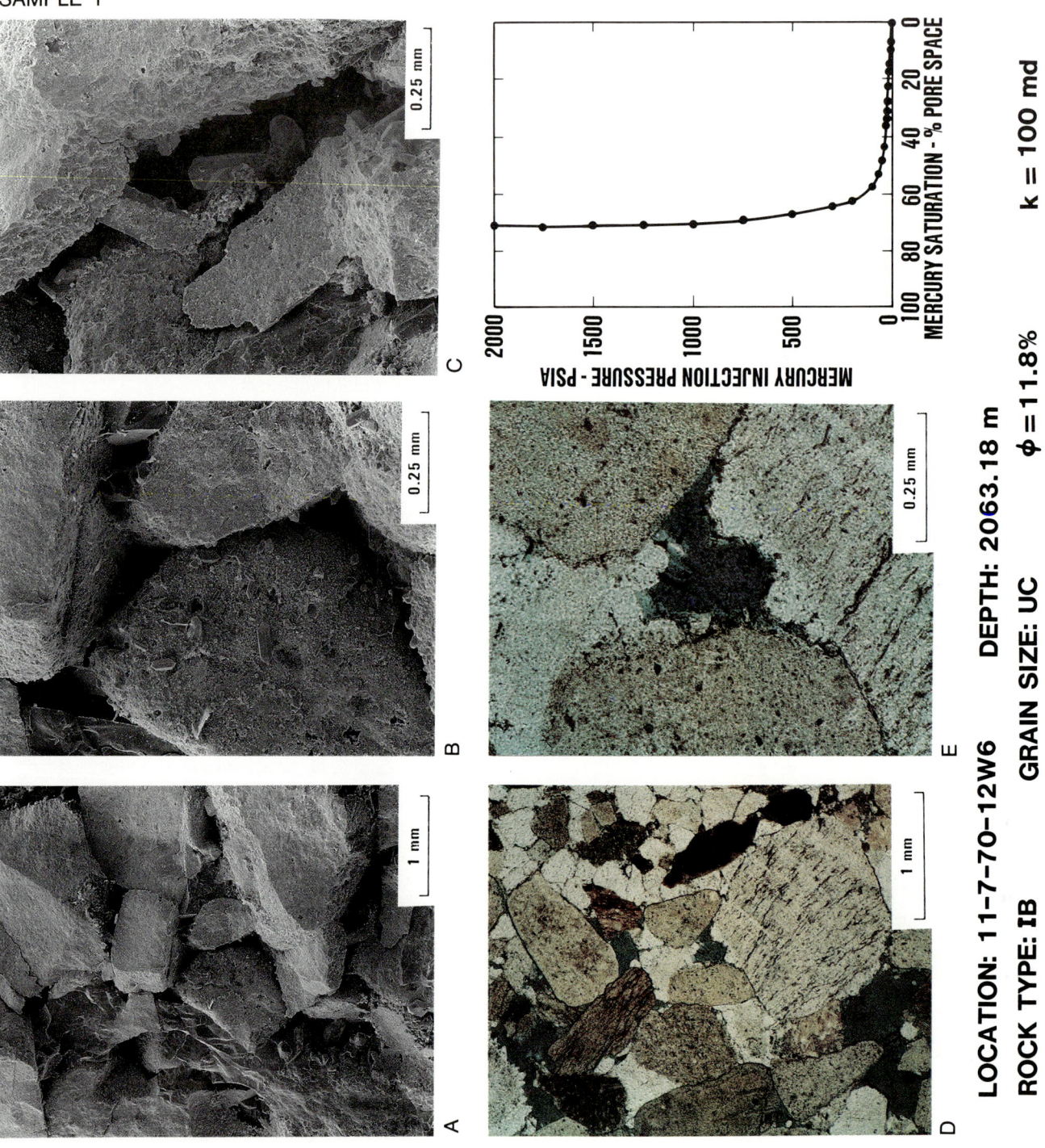

LOCATION: 11-7-70-12W6   DEPTH: 2063.18 m

ROCK TYPE: IB   GRAIN SIZE: UC $\phi = 11.8\%$ k = 100 md

Figure 14. Scanning electron and then section microphotographs and a mercury-injection capillary pressure curve for rock samples of Zone 3 – Falher A Formation 11-7-70-12-W6, at a depth of 2,063.18 m.

SAMPLE 2

LOCATION: 11-7-70-12W6 **DEPTH: 2063.72 m**
ROCK TYPE: IB **GRAIN SIZE: BIMODAL** $\phi = 8.7\%$ $k = 12.2$ md

Figure 15. Continued rock sample data for Zone 3 — Falher A Formation 11-7-70-12-W6, at a depth of 2,063.72 m.

Figure 16. Continued rock sample data for Zone 3 — Falher A Formation 11-7-70-12-W6, at a depth of 2,066.62 m.

Figure 17. Continued rock sample data for Zone 3 — Falher A Formation 11-7-70-12-W6, at a depth of 2,072.4 m.

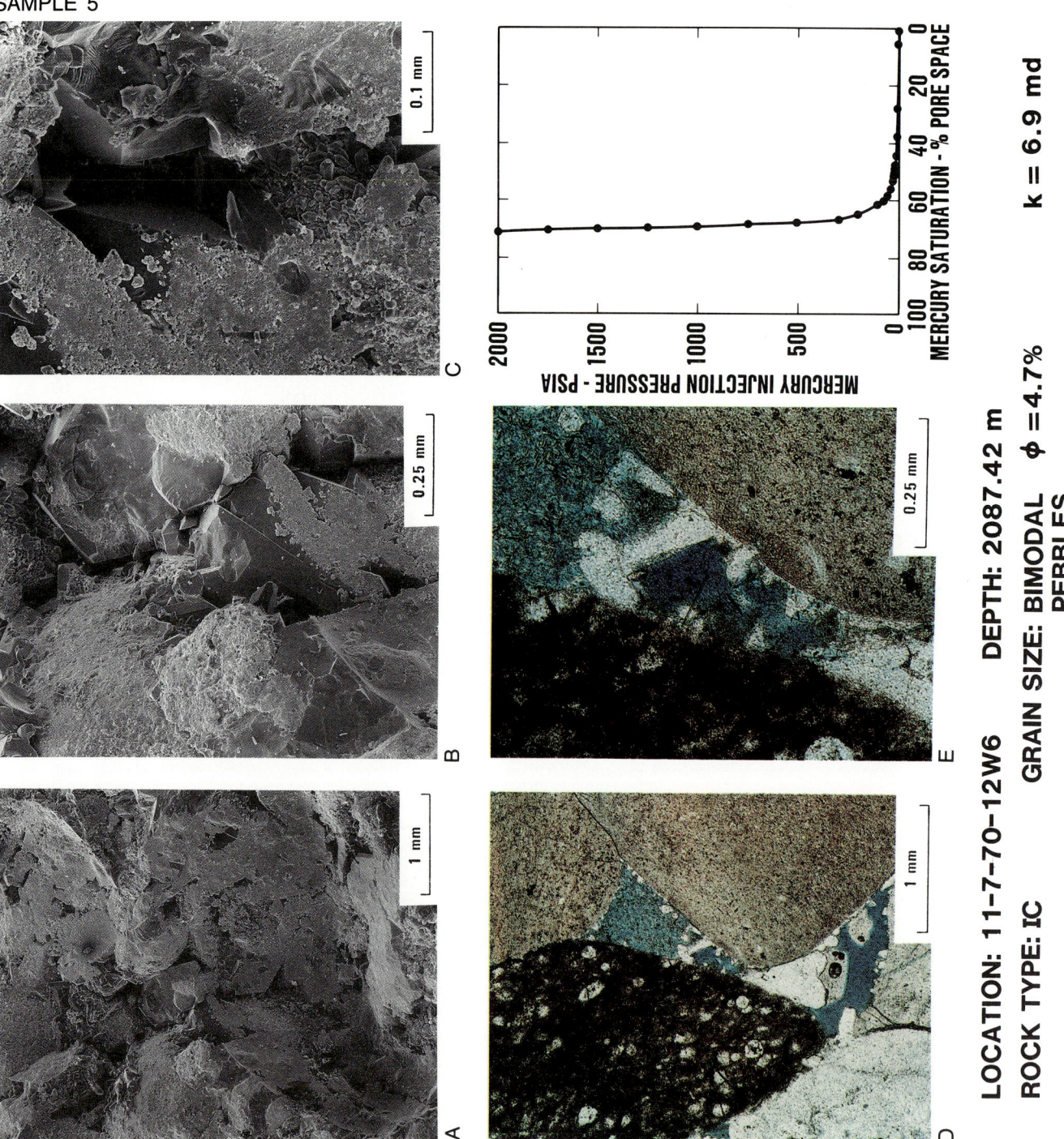

Figure 18. Continued rock sample data for Zone 3 — Falher Formation 11-7-70-12-W6, at a depth of 2,087.42 m.

Zone 4

This zone was selected as representative of some of the lower-quality reservoirs encountered in Elmworth (Figures 19 to 22). It demonstrates the need to consider these poor rock types as well as the better ones as economic reservoirs.

(1) The completion interval is composed of Type IC and ID, predominantly in the upper half and Type II rocks predominantly in the lower half. The well has a deliverability of 100 mcf/d which is consistent with the observed rock types encountered.

(2) Clay volumes from X-ray diffraction range from 9 to 12% and are consistent with log-derived volumes. Kaolinite, illite, and chlorites are encountered in both SEM samples, while log analysis indicates the clays to be primarily illite.

(3) The core porosity and the log-derived effective porosity agree quite well throughout the interval. Porosity is relatively low ranging from 4 to 7%. Visual porosity compares favorably with core and log porosity. SEM photographs of the Type II rock show that porosity is principally primary intergranular with some pores enlarged by grain dissolution. A micropore system is developed within some detrital grains of chert and shaly rock fragments. The low porosity of these Type II rocks is the result of extensive physical compaction of the clay-rich sand grains and interstitial clay, and cementation. Many of the Type II rocks in the Elmworth field are fine- to very fine-grained, clay-rich sandstone, like these two examples.

(4) Water saturation ranges from 20 to 65% and shows a general correlation with porosity. The bound-water component averages 20%. Comparison of the mercury-capillary pressure curves of the two Type II rocks with the Type I rocks illustrated in the three previous examples show that Type II rocks have a very high percentage of very small and small pores and a higher entry pressure. Comparison of the photomicrographs of Type II with Type I rocks shows that Type II rocks have very small pore throats and a very poor interconnection of the pore network. Mercury-injection capillary pressure curves are consistent with log-derived water saturation and range from 30 to 50% thereby further illustrating the presence of an immobile water phase within the very fine pore structure.

(5) The reservoir quality index shows a fair correlation with core permeability.

(6) Rocks characterized by Type IC, ID, and II porosity typically do not show mudcake. This particular zone is an example of this characteristic.

(7) The sand grades from fine grained at the base to coarse grained at the top. Several other variables all contribute to this gradation — namely, clay content decreasing from bottom to top, permeability increasing from bottom to top, and water saturation decreasing from bottom to top.

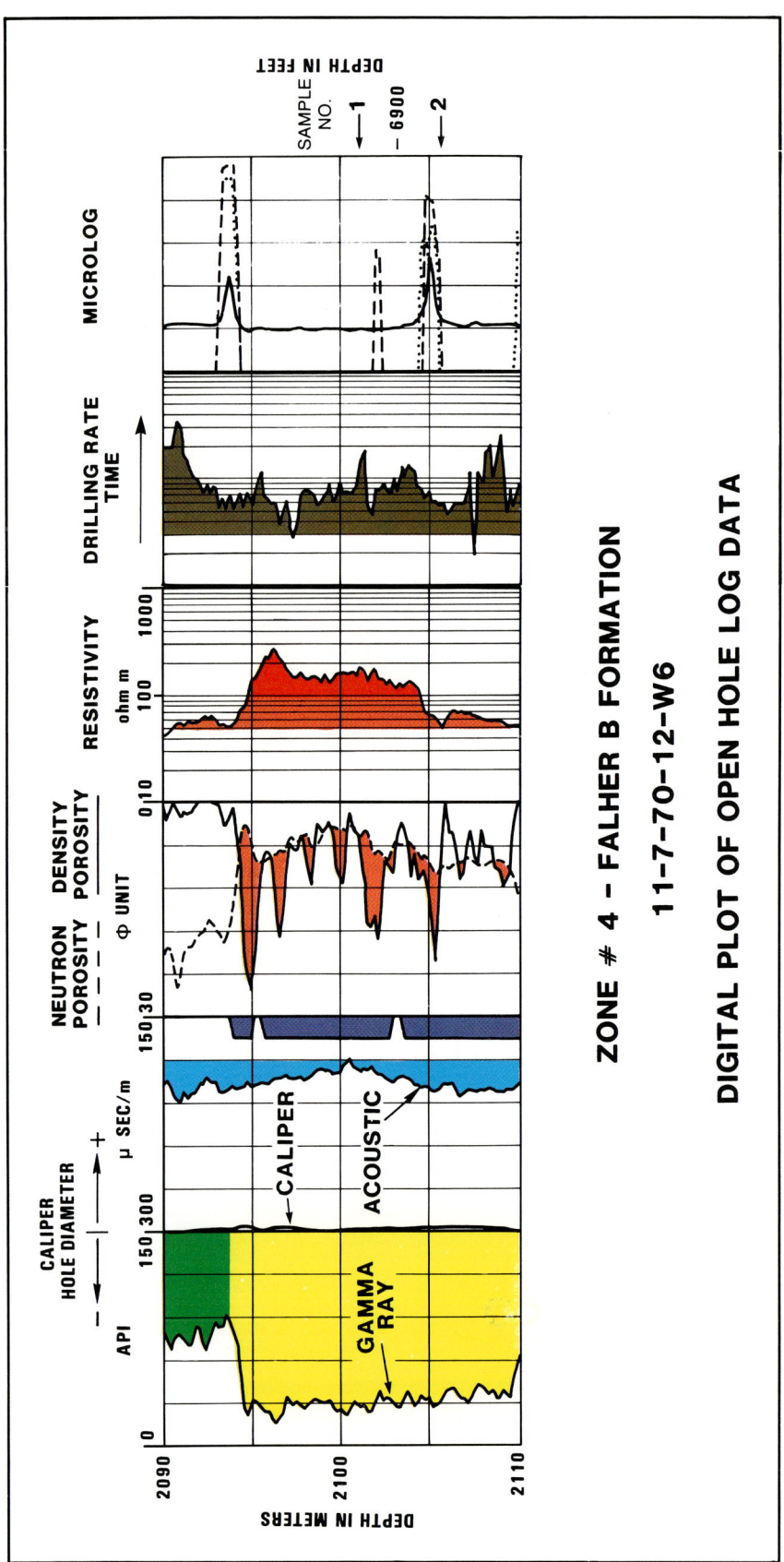

Figure 19. Digital plot of open-hole log data, Zone 4 — Falher B Formation 11-7-70-12-W6.

Integrated Rock-Log Calibration, Elmworth Field 269

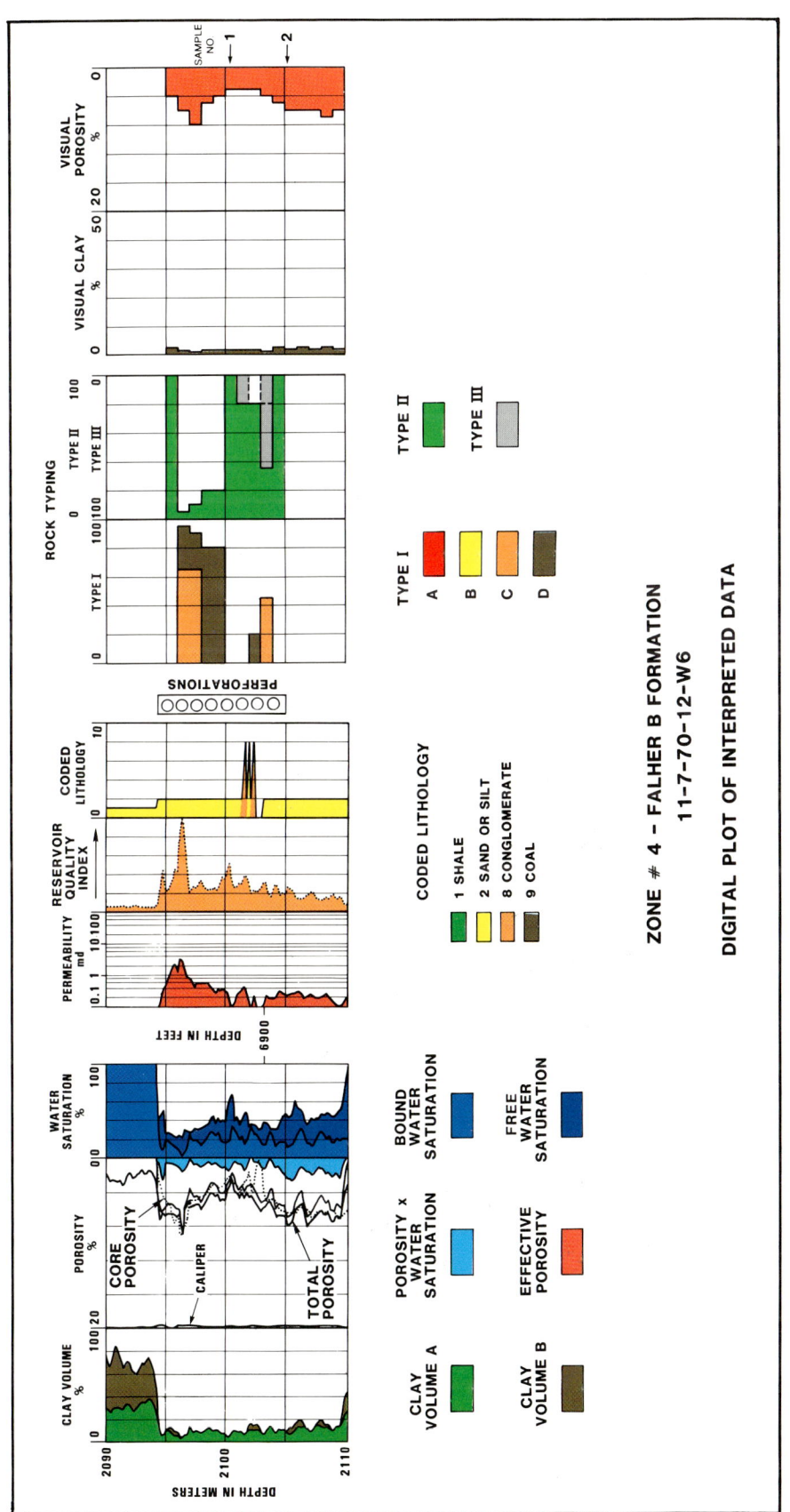

Figure 20. Digital plot of interpreted data, Zone 4 — Falher B Formation 11-7-70-12-W6.

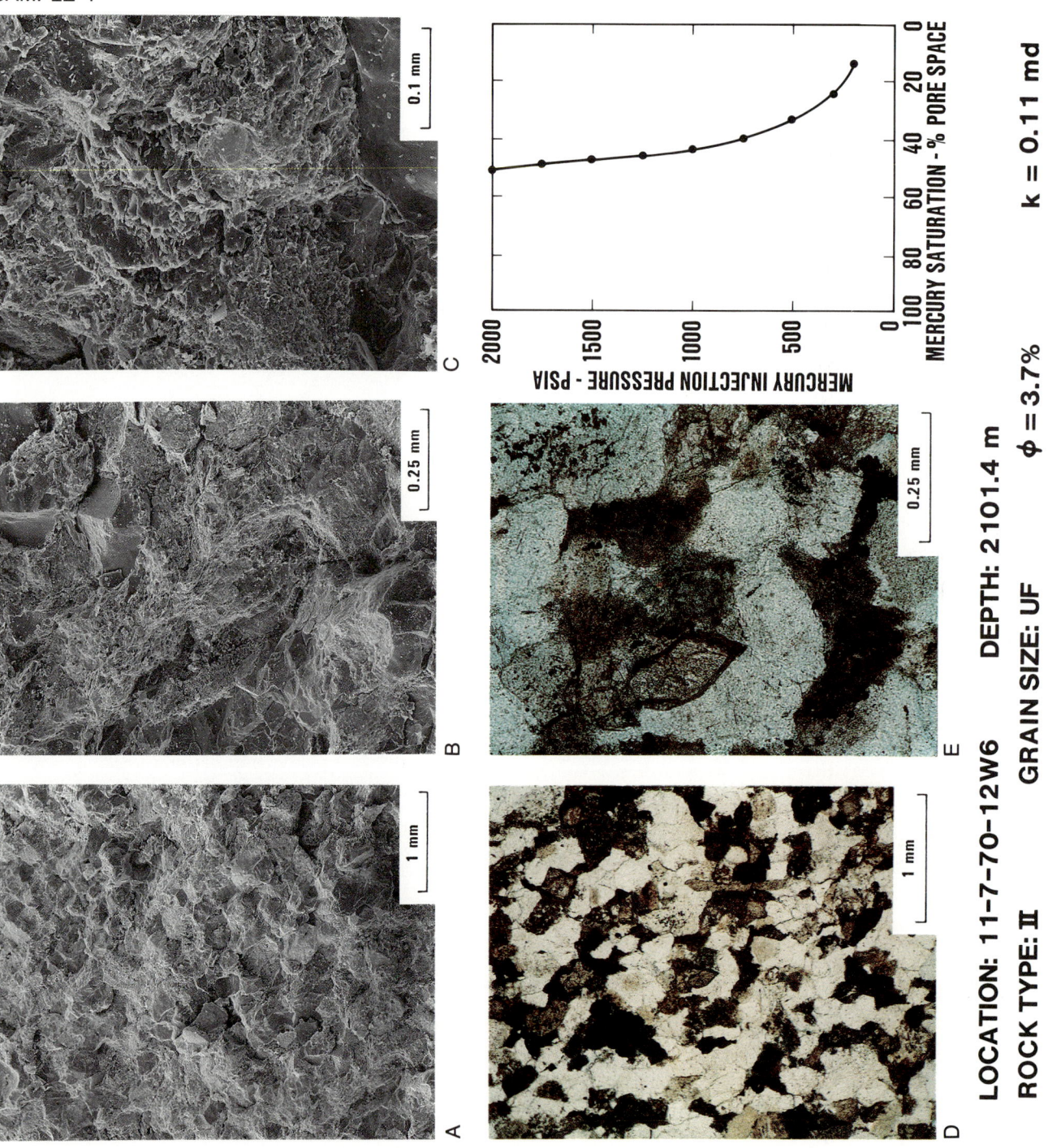

Figure 21. Scanning electron and thin section microphotographs and a mercury-injection capillary pressure curve for rock samples of Zone 4 — Falher B Formation 11-7-70-12-W6, at a depth of 2,101.4 m.

SAMPLE 2

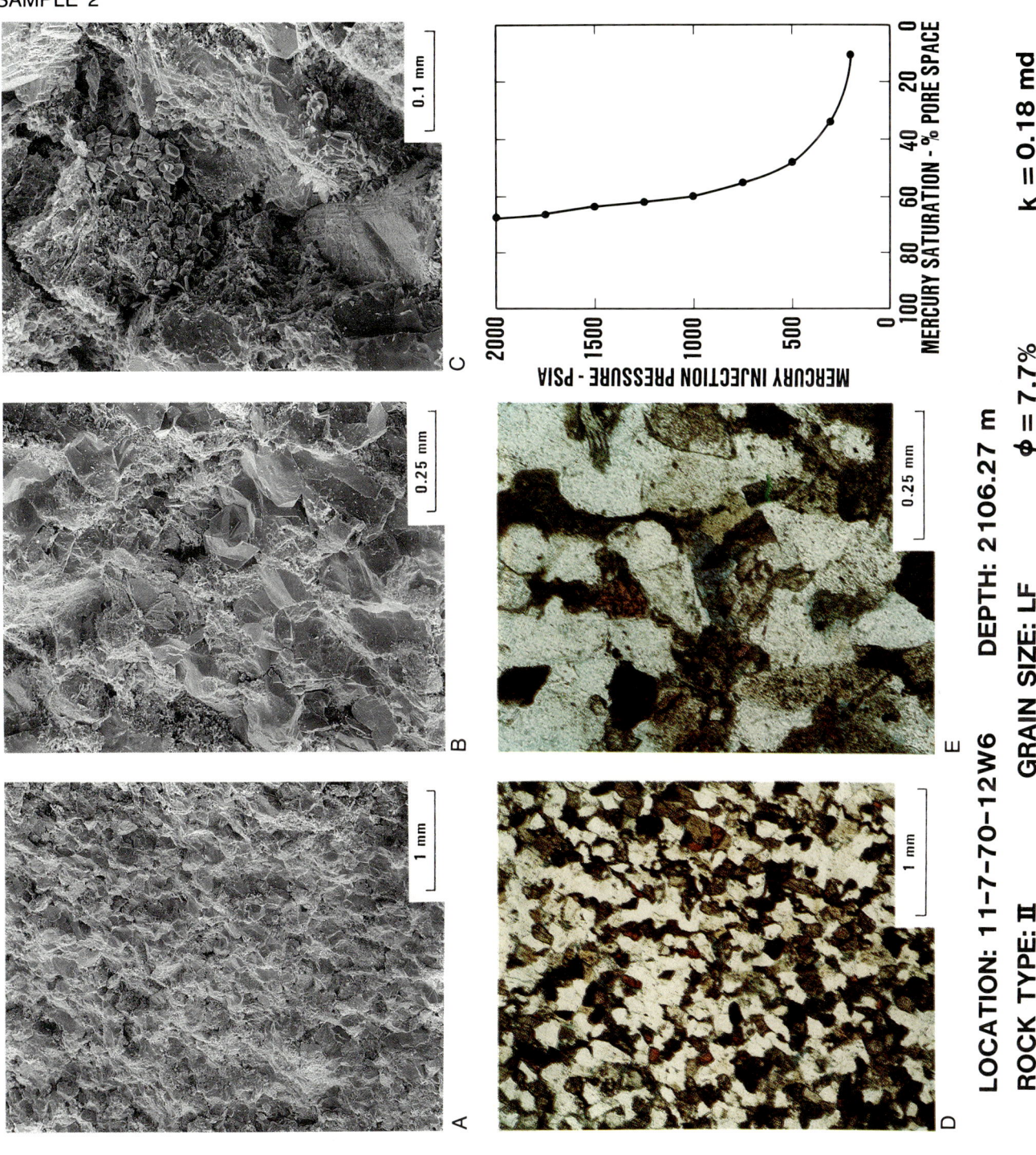

LOCATION: 11-7-70-12W6   DEPTH: 2106.27 m
ROCK TYPE: II   GRAIN SIZE: LF   φ = 7.7%   k = 0.18 md

Figure 22. Continued rock sample data for Zone 4 — Falher B Formation 11-7-70-12-W6, at a depth of 2,106.27 m.

Zone 5

This particular example (Figures 23 to 27) was selected as one of the best producing zones in the field. In addition it contains a wide variety of rock types which makes its examination well worthwhile.

(1) The upper part of the completion interval contains conglomerates of Type I rock. Two rock samples were picked to contrast the pore structure of a pebble-supported conglomerate (IA at 2,332.0 m) and a sand-supported one (IB at 2,332.7 m [7,653.2 ft]). Unfortunately, the microphotographs and capillary-pressure information do not adequately represent the pore structure of these conglomerates because of the sample size used for the SEM. Thin section and capillary-pressure analysis is small compared to the size of the large grains. From a study of the entire core and the smaller samples, one observes that both types of conglomerate have intergranular porosity with some pore enlargement usually from grain dissolution. The pores in the pebble-supported conglomerate are overall much larger and the entry pressure lower than that of the sand-supported conglomerate. The lowermost interval contains sandstones also characterized as Type I. The middle part of the zone which contains no perforations is shalier and contains Types II and III rocks. The excellent Type I rock found throughout the completion interval is consistent with the well deliverability of 47 mmcf/d.

(2) Clay volumes within the conglomerate interval are quite low. Clay volumes within the non-perforated mid-portion of the zone are high ranging from 10 to 40% and predominantly illite. Clay volumes in the lower sandstone interval are low, 10% or less.

(3) Core porosity compares very favorable with log-derived effective porosity except in the extremely porous upper conglomeratic interval. It is highly likely that the core porosity in this upper zone is not representative of in situ porosity and the logs are the more representative. A micropore system is developed within some of the chert grains and between the clay minerals which generally occur as crystalline aggregates attached to grain surfaces. Water trapped in these micropore systems would be immobile.

(4) The high calculated water saturation in the conglomerate zone which reaches 50% is due primarily to the immobile water in the micropore system. This accounts for the water-free production.

(5) The reservoir quality index correlates favorable with core permeability.

(6) Certain characteristics of conglomerates can be observed by examining the raw log data. These are the presence of mudcake, subnormal acoustic transit time as compared to the neutron porosity, lower drilling time, and the relatively low gamma-ray reading.

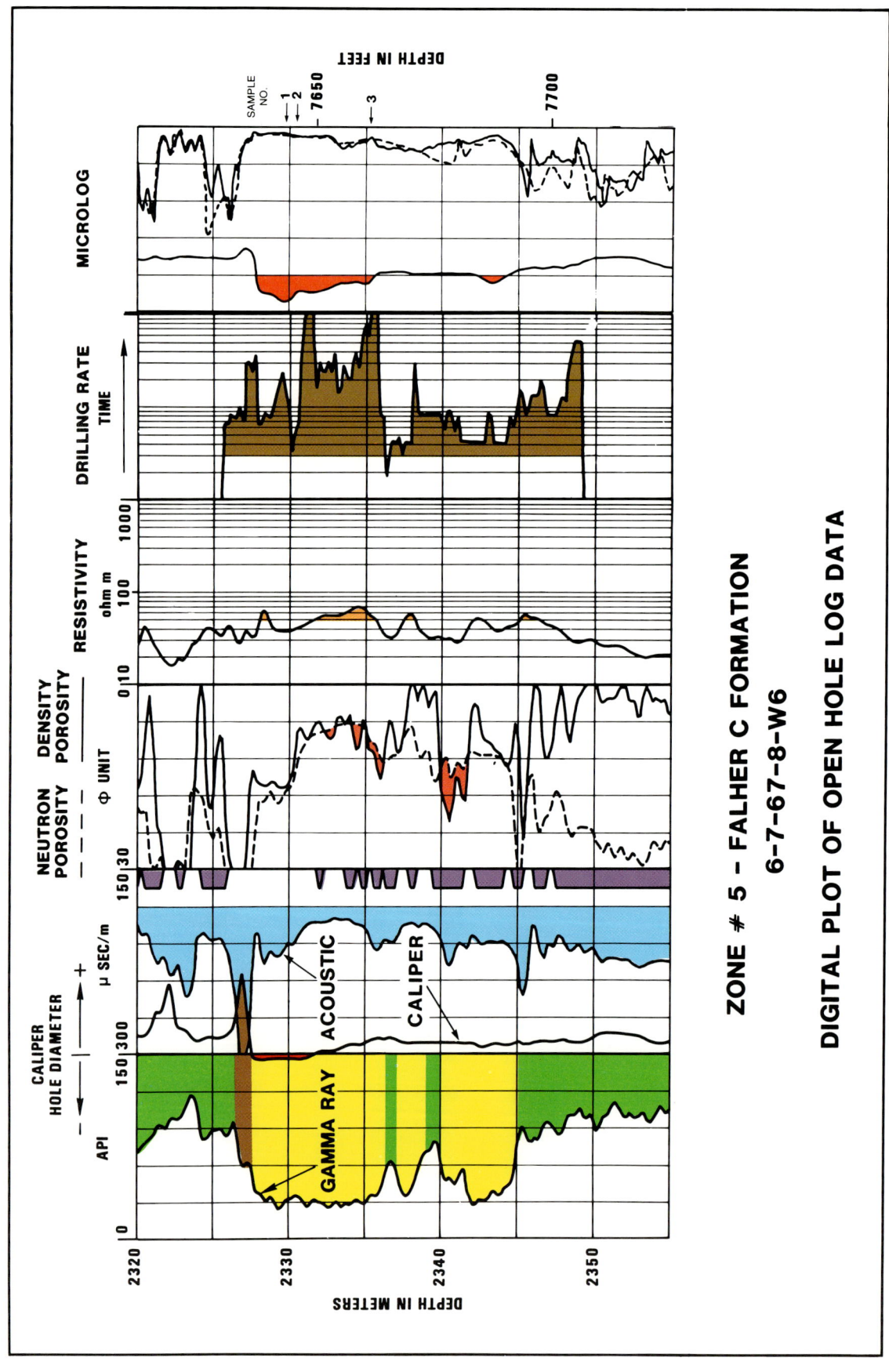

Figure 23. Digital plot of open-hole log data, Zone 5 — Falher C Formation 6-7-67-8-W6.

Integrated Rock-Log Calibration, Elmworth Field 275

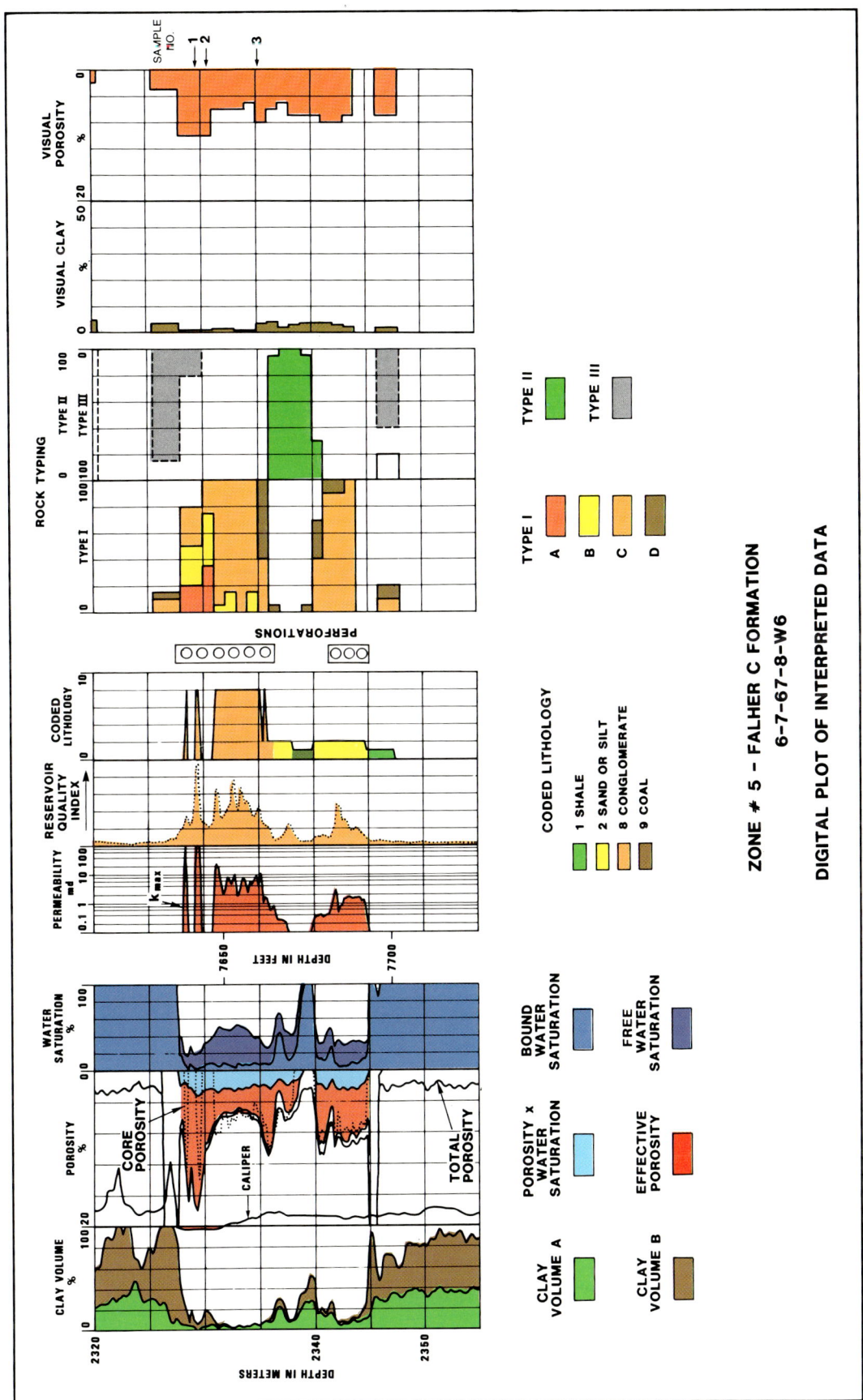

Figure 24. Digital plot of interpreted data, Zone 5 — Falher C Formation 6-7-67-8-W6.

Figure 25. Scanning electron and thin section microphotographs and a mercury-injection capillary pressure curve for rock samples of Zone 5 – Falher C Formation 6-7-67-8-W6, at a depth of 2,332 m.

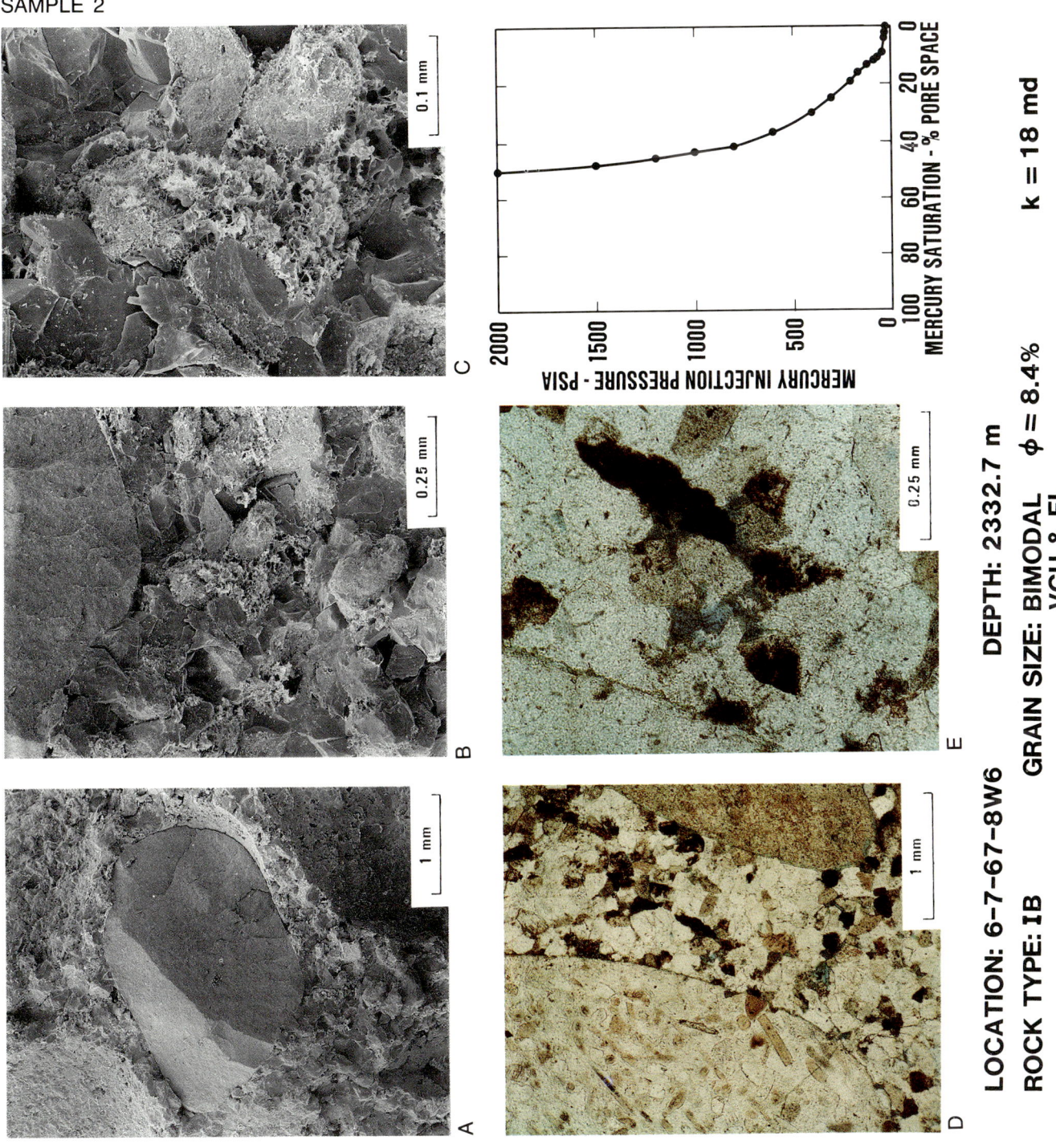

Figure 26. Continued rock sample data for Zone 5 — Falher C Formation 6-7-67-8-W6, at a depth of 2,332.7 m.

SAMPLE 3

LOCATION: 6-7-67-8W6 DEPTH: 2337.4 m
ROCK TYPE: IC GRAIN SIZE: UVF φ = 10.0% k = 1.5 md

Figure 27. Continued rock sample data for Zone 5 — Falher C Formation 6-7-67-8-W6, at a depth of 2,337.4 m.

Zone 6

All of the previous examples had core information over the completion zone. This example (Figures 28 to 30) was chosen to illustrate a successful completion in a low-porosity zone where only rock cuttings and log data were available. Dovetailing the rock and log information, as illustrated by this example, can identify potentially productive low-porosity zones that might be missed by log or rock evaluations done independently.

(1) The rock typing in this example shows overall low percentages of Type I rock. An increase in the amount of Type I rock from the top toward the base of the interval, combined with the presence of 20% Type IC and ID rock, flags the interval as a potential completion candidate even though the overall log porosity is in the 3 to 6% range. The low porosity is the result of coarse grain size with scattered gravel and sand-supported conglomerate (that is, poorer sorting) and pore infilling by cements. The well was completed with a deliverability of 650 mcf/d.

(2) Clay volumes are relatively low, in general 10% or less, thereby compatible with the rock-type observations.

(3) The very low porosity observed from logs and visual methods ranges from 3 to 6% and raises questions concerning the well's productivity. However, detailed examination of the SEM photographs reveals a pore system that is continuous.

(4) Water saturations are relatively low for a rock of this low a porosity, ranging from 20 to 30%, and supporting the Type I observations from cuttings.

(5) The reservoir quality index is at a moderately high level, thereby providing encouragement for a completion attempt.

(6) Mudcake is found throughout the zone, as expected for IB and possibly IC rocks.

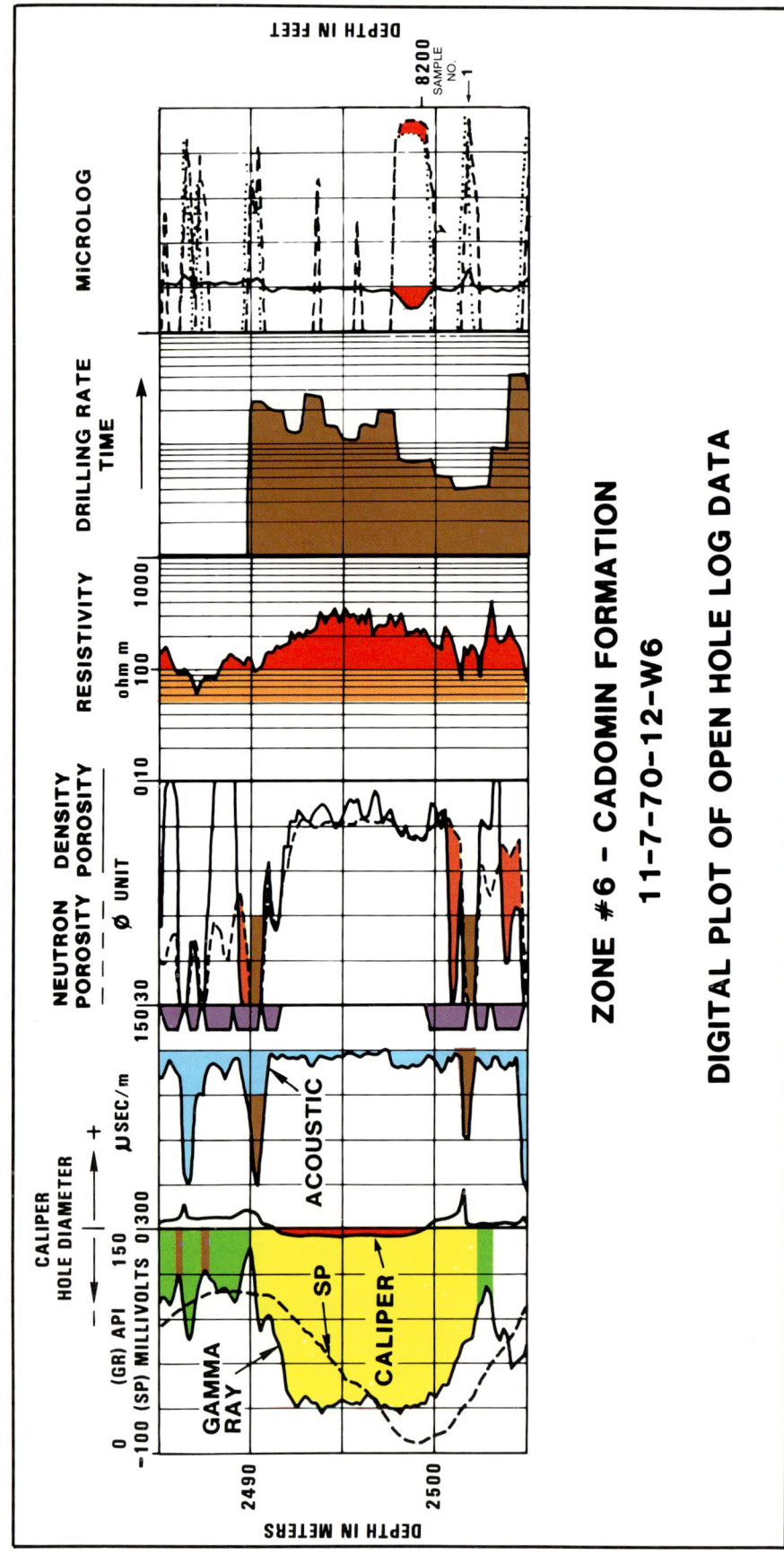

Figure 28. Digital plot of open-hole log data, Zone 6 — Cadomin Formation 11-7-70-12-W6.

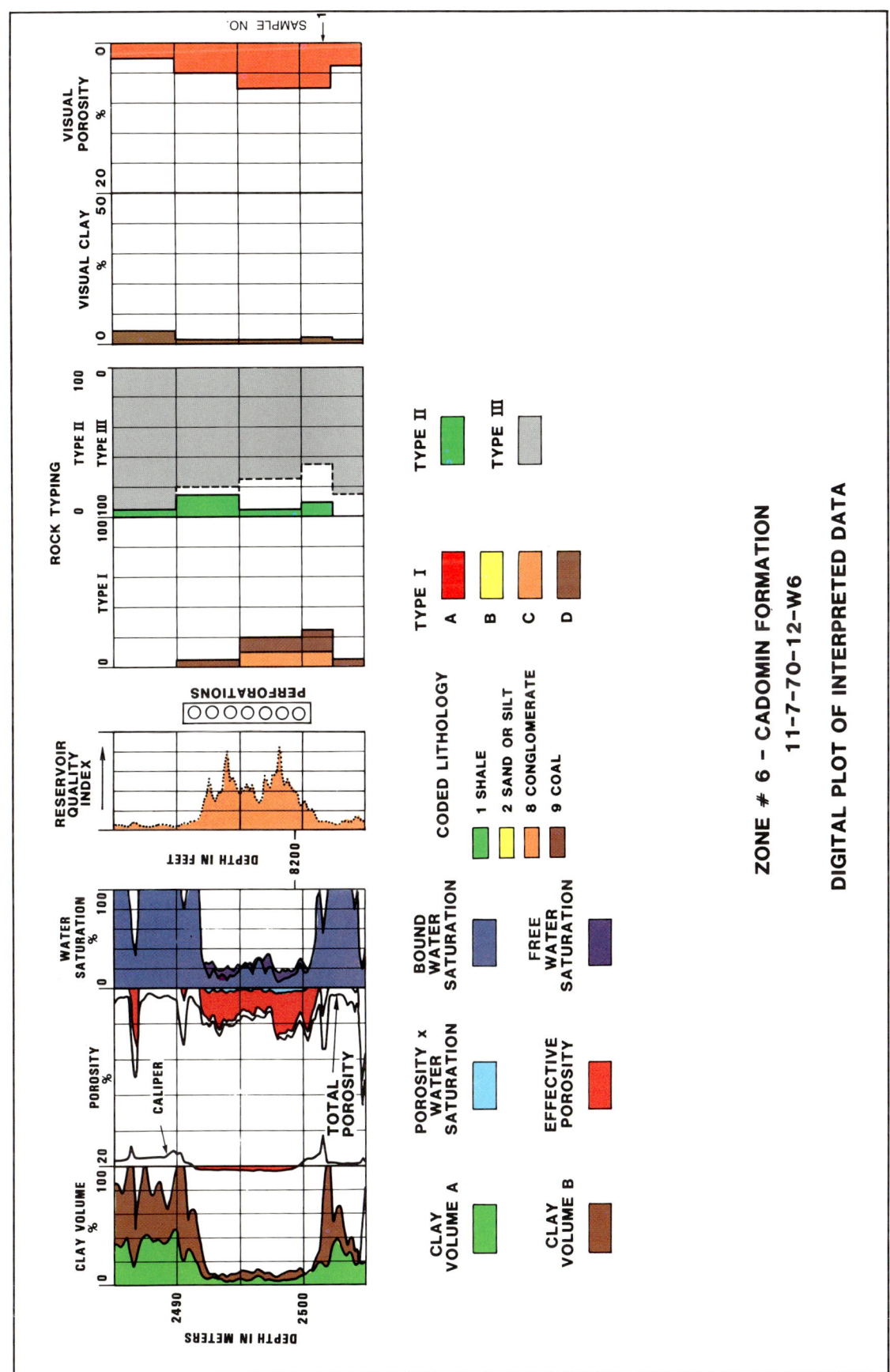

Figure 29. Digital plot of interpreted data, Zone 6 — Cadomin Formation 11-7-70-12-W6.

SAMPLE 1

LOCATION: 11-7-70-12W6   DEPTH: 2502.2 m
ROCK TYPE: IC   GRAIN SIZE: BIMODAL

Figure 30. Scanning electron and thin section microphotographs for rock samples of Zone 6 — Cadomin Formation 11-7-70-12-W6, at a depth of 2,502.2 m.

Drilling in the Deep Basin

Don L. Myers
Canadian Hunter Exploration
Calgary, Alberta

The first objective of Canadian Hunter in drilling is to obtain complete open-hole evaluation including drill cutting samples, cores, drill-stem tests, and logs. The second objective is to install and adequately cement a high-quality production casing that will allow successful completion and stimulation treatments. Predominant field characteristics in the Deep basin which complicate the task are sloughing uphole shales, underpressured formations, and multiple pay zones stacked in each well.

The loss of open-hole data due to severe shale sloughing was a serious problem at one point in our development. We tried different drilling techniques to get around this problem, including the use of intermediate casing to shut off uphole sloughing. Finally, we used oil-based drilling fluid, and this now has become a common practice. This fluid almost totally eliminates shale sloughing, and substantially improves the reliability of all open-hole evaluation operations. An important additional benefit is that we achieve faster penetration rates through these generally underpressured formations due to the lighter mud weights that are used. Greater reliability of primary cementing also results from having in-gauge holes and a light thin mud. However, there are still a number of drawbacks in attempting to evaluate too many zones with one wellbore and a single productive target now appears prudent in some well designs.

INTRODUCTION

The Deep basin gas field lies along the eastern side of the Rocky Mountains, cutting across the British Columbia/Alberta border in an area west of the town of Grande Prairie. The productive trend is at least 150 mi long and 50 mi wide. Since the discovery in 1976, Canadian Hunter has drilled over 250 wells in the field, and other operators have drilled over 450 wells in the general area indicated in Figure 1. Well depths vary from 5,000 ft (1,524 m), for the top of the Lower Cretaceous on the extreme northeast side, to 14,000 ft (4,267 m) to the base of the Triassic in the deepest part of the basin. However, the majority of the production to date has come from zones between 6,000 and 9,000 ft (1,829 and 2,743 m) in the Lower Cretaceous formations.

The objectives which have guided Canadian Hunter in the development of drilling methods in this basin must be clearly understood. The first objective in drilling a well is to obtain adequate open-hole evaluation data to accurately identify the potential gas zones. Next, the well must be cased in a manner that will allow successful completion and stimulation of the various gas zones. The third objective is, of course, to achieve the first two in the most cost-efficient manner and with minimal environmental impact.

As might be expected, an area this large has a variety of drilling problems. Some problems are localized in specific areas of the basin. Included in these are areas of severe lost circulation, deviation control problems, high H_2S concentrations in deep Triassic, isolated pockets of abnormally high reservoir pressure, severe gravel problems near surface, and locations with very difficult surface access. All of these problems have been handled with commonly known drilling technology, and none have had a serious impact on the overall program.

However, there are three predominant factors which have been key concerns in drilling wells in the basin. The first is sloughing uphole shales, which has been by far the most costly drilling problem. The second is the fact that most wells have multiple pay zones stacked over a large interval, making primary cementing difficult. The third point is that the entire basin generally has subnormal formation pressures, which have resulted in large overbalance conditions with normal water-base drilling fluids.

The remainder of this paper will give a more detailed description of the open-hole evaluation program and the three predominant drilling problems mentioned above. It will go on to discuss the development of well design and drilling practices up to the present time, with emphasis on the increasing use of oil-base muds.

OPEN-HOLE EVALUATION

The importance of good open-hole evaluation in this gas field cannot be overemphasized. This information is used to decide what intervals to complete, and in many cases what methods to use. Sometimes the indicators are very subtle.

Canadian Hunter has an excellent Wellsite Geology group which is responsible for the collection of all open-hole data as the wells are drilled. They have developed great skill in cuttings-sample analysis, field core descriptions, and mud/gas log interpretations. They pick the core and DST intervals and supervise most logging operations. Indeed these are the people who first identify the majority of the productive zones, and the close cooperation of this group with all other departments has been essential to the development of the basin.

In addition to cuttings analysis and mud/

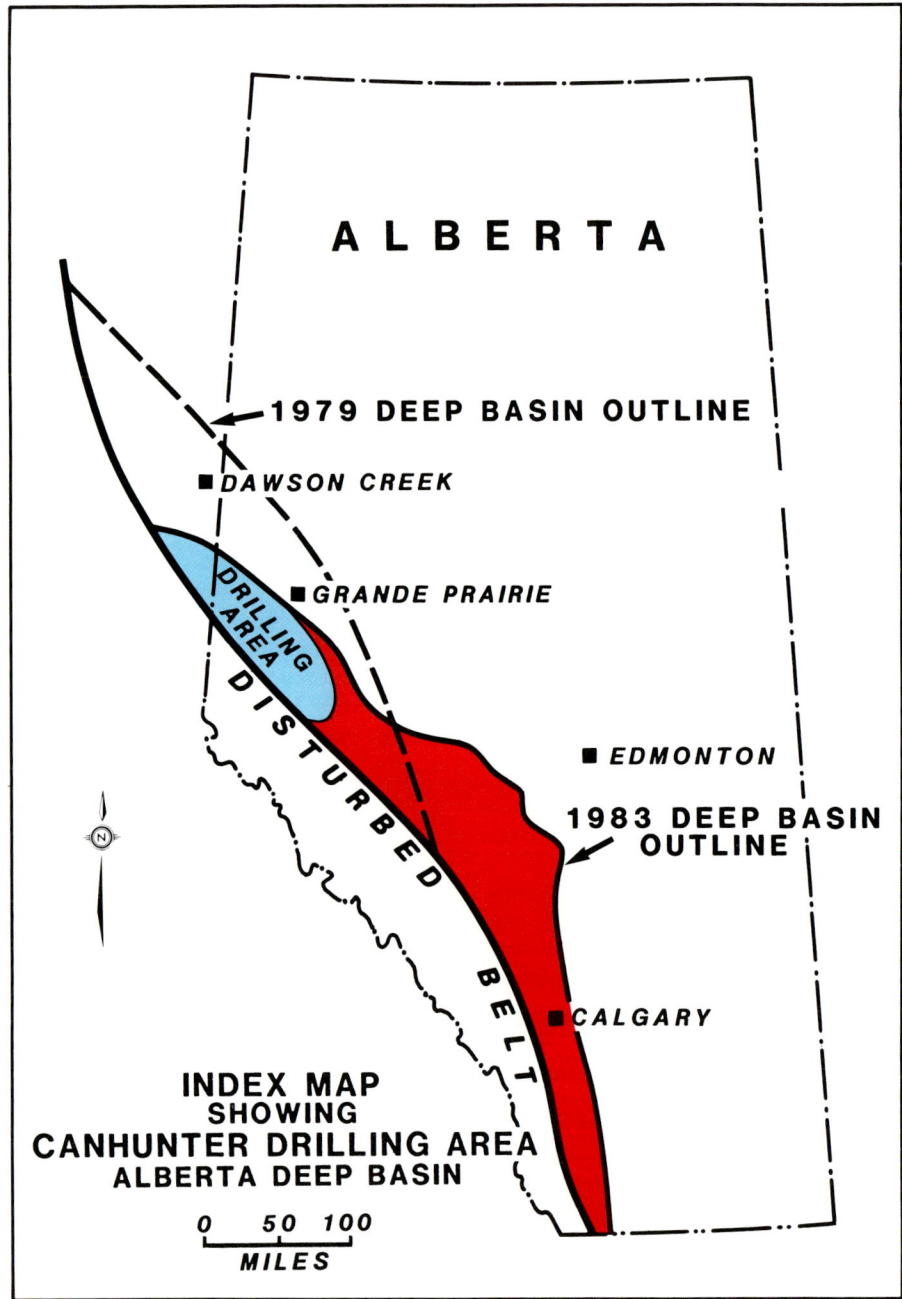

Figure 1. Canadian Hunter drilling area – Alberta Deep basin.

gas logging, a great deal of coring, drill-stem testing, and open-hole logging was required in these wells. Some elaboration on the importance of these operations is worthwhile at this point.

Extensive coring was a key element in the effort to gain a thorough understanding of the Deep basin, as it provided the hard data often necessary to calibrate or confirm other evaluation methods. From 1979 through 1981, Canadian Hunter cut over 20,000 ft (6,096 m) of core in this basin. Many of these cores were cut through some of the toughest coring rock in the world, namely chert-pebble conglomerates. Coring rates of less than 0.5 ft (0.15 m) per hour were not uncommon, nor was the scrubbing of a conventional core bit after making only a few feet. The development of the diamond impregnated core head was a significant breakthrough in this regard. It will drill this type of rock considerably faster, with less severe damage, and much longer bit life. Another common and costly problem during coring operations has been sloughing uphole shales. Reaming through bridges damaged many core heads before they reached bottom, and frequently additional trips with the drilling string were required to condition the hole.

Drill-stem testing provided valuable data for many purposes, ranging from booking gas reserves, to obtaining fluid samples, to helping plan a stimulation job. From 1979 through 1981, over 650 DSTs were attempted in the basin by Canadian Hunter. These tests were almost exclusively conventional bottom-hole tests conducted immediately after penetration of the zone. This is critical because damage generally increases with exposure time. Obtaining accurate bottom-hole pressures during the test and shut-in period is the top priority in most tests. The absence of a good gas blow does not necessarily mean that the zone will not produce. Many tests have been conducted using the closed chamber technique, where the drill pipe is left closed, and the pressure inside is monitored at surface and analyzed on location as an indicator of the downhole response. This information is commonly used to adjust the flow and shut-in times as the test proceeds, which has been useful in obtaining more meaningful data on tighter zones. The main problems encountered when attempting a drill-stem test are the failure to reach bottom due to sloughing shales, and packer-seat failures due to rough or over-gauged hole.

Logging is, of course, a major component of the open-hole evaluation. The minimum suite of logs generally run at TD consists of Borehole-Compensated Sonic, Dual Induction, Neutron-Density, Gamma-Ray, Microlaterolog, and two calipers. Sometimes a full suite of logs are run at an intermediate depth after penetrating the first one or two primary target intervals. This is done so that log quality is not lost due to the filtrate invasion and/or hole enlargement that may occur if the well is to be drilled considerably deeper. Poor hole conditions due to sloughing and bridging have again been the main problem in obtaining logs. Numerous conditioning trips while attempting to log were required in many wells and in some cases all or part of the proposed logs had to be abandoned in order to get casing in the hole. In some cases, log quality also suf-

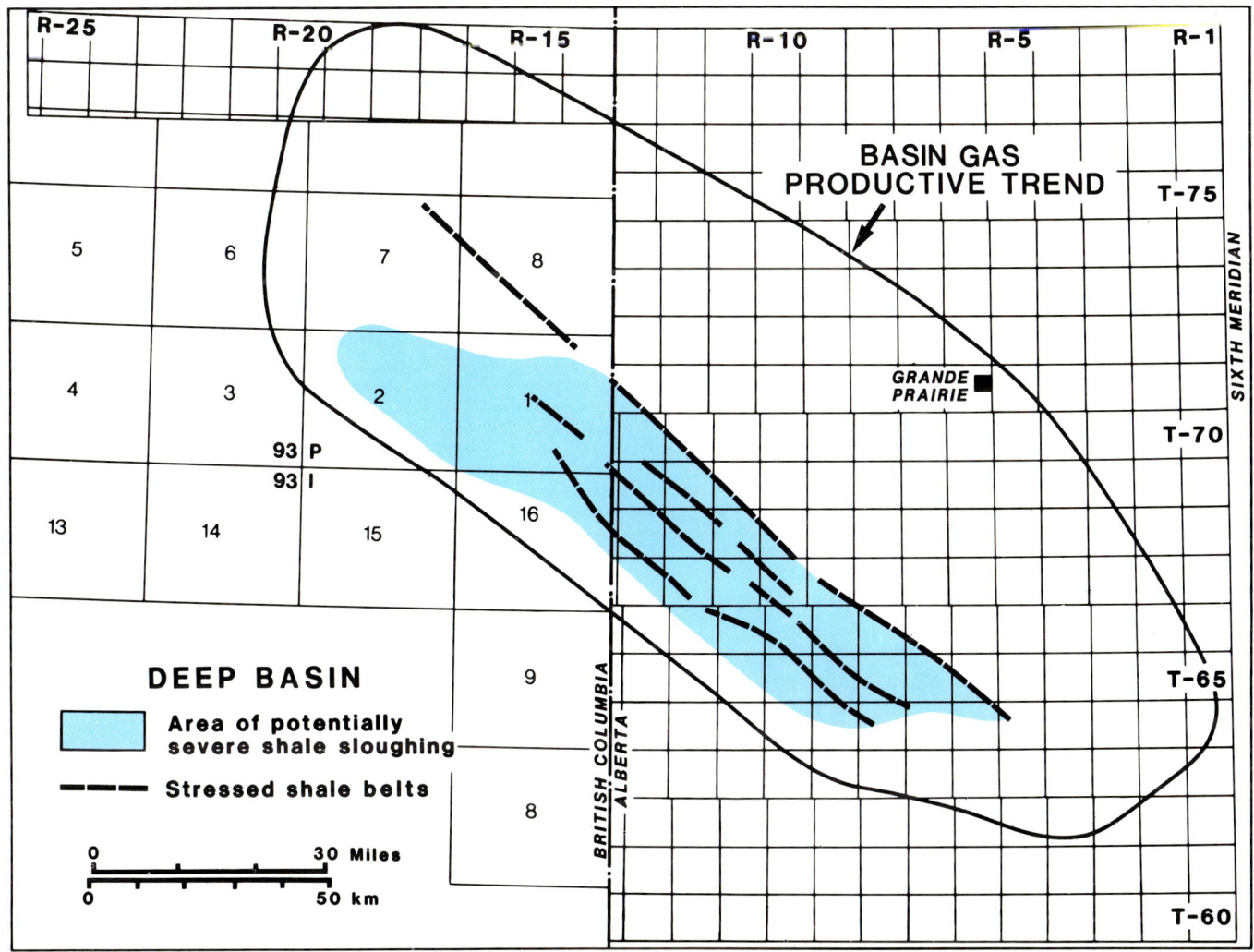

Figure 2. Area of potentially severe shale sloughing – Deep basin.

fered because of enlarged and erratic hole diameter.

SLOUGHING SHALES

Sloughing shale problems have caused considerable grief to most operators, and have been the subject of much debate as to the exact cause and best cure. The following observations were made throughout the drilling experiences of Canadian Hunter.

Firstly, the problem is by no means consistent throughout the basin; some areas are worse than others. In some parts of the field the problem is only minor (see Fig. 2), but there are a number of shaley zones that can slough (see Fig. 3) — the best known is the Shaftsbury, but shales in the Kaskapau, Second Whitespecks, Dunvegan, and Fernie are others that can be very troublesome.

problem. Often, considerable time has elapsed, and considerable shale has been unloaded before the well stabilizes even with the higher mud weight. In one case where the mud weight was increased in advance, sloughing still occurred. The overpressured theory is also somewhat inconsistent with the fact that the permeable intervals both above and below the shales are generally subnormally pressured throughout the basin.

We noted that the worst areas for shale sloughing generally correspond to a series of stressed shale belts identified by geophysical work. Along these stress belts, the shales have been compressed and thickened in an eyebrow effect, which results in multiple mini-fractures, a highly stressed condition, and even a localized change in dip angle. These stress belts are as narrow as half a mile in width, but miles (or tens

When drilling with a freshwater mud, these shales often start sloughing a few days after first penetration. Many operators have simply labelled them water-sensitive. However, shale sample analysis has failed to identify significant amounts of hydrateable clays. Numerous attempts have been made to use KCl polymer muds to inhibit the shales. Results have been mixed, and some wells still experienced serious sloughing.

Pore-pressure plots made from sonic logs have indicated mild overpressuring in some cases, particularly in the Shaftsbury Shale. Increased mud weight has been used with some success. However, mud weight has usually been a remedial step along with high viscosity, used after the fact in an attempt to clear up a sloughing

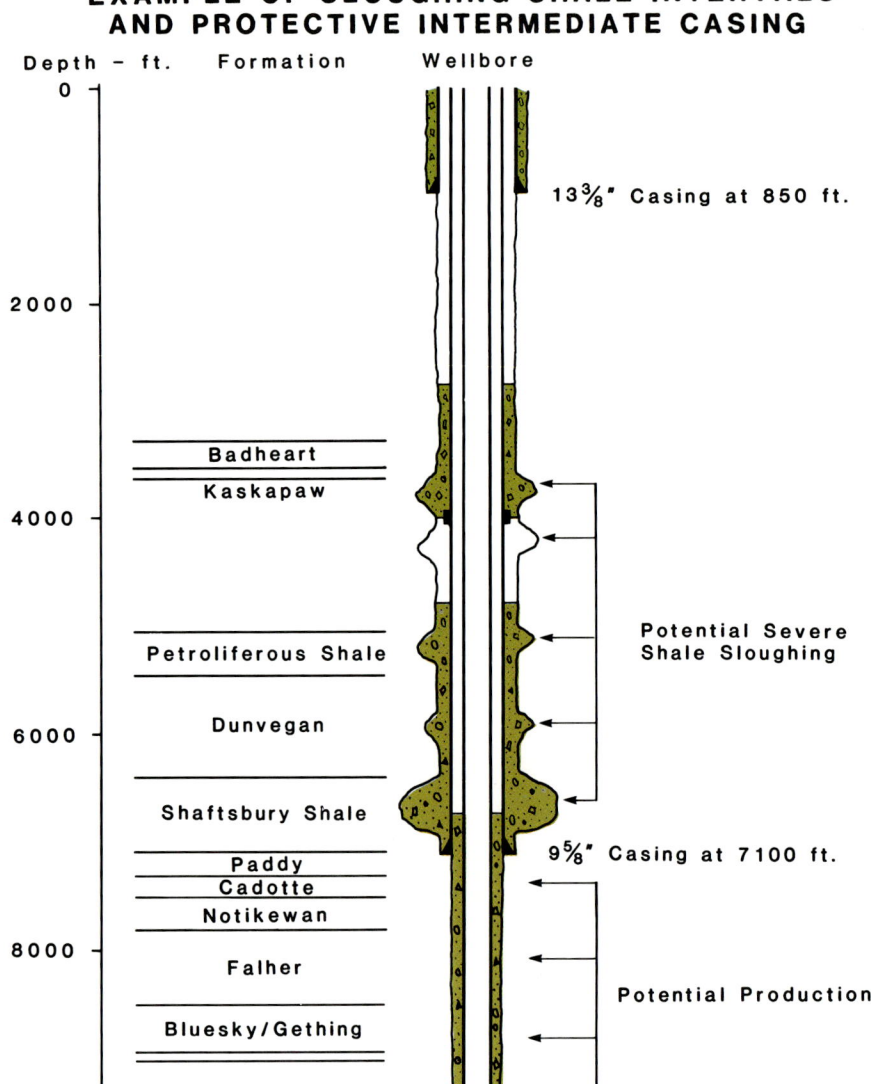

Figure 3. Example of sloughing shale intervals and protective intermediate casing.

of miles) in length. There are many individual belts, all running through the field roughly parallel to the mountains as depicted in Figure 2. These belts can exist in one shale package such as the Shaftsbury, without being present in the shallower shales, and vice versa. This model seems to agree with most of the field experience with these shales, namely:

1. Mechanical factors such as pipe tripping often initiate the sloughing.
2. The bigger the hole size, the worse the problem.
3. A fairly sudden increase in deviation often occurs while drilling through intervals that slough a few days later.
4. The shale is often cleaned out in small angular chunks.
5. Once initiated, the sloughing often accelerates with a "house of cards" effect.

PRIMARY CEMENTING OF MULTIPLE PAYZONES

One of the happy problems that has to be dealt with in the field is that often two or three, and sometimes as many as five or six productive zones may exist in a given wellbore. These zones can span a total section of 3,000 ft (914 m) or more. The production casing must be cemented adequately to isolate each of these zones and provide containment of any stimulation work. In addition, regulations require shallow porous zones must be covered by at least 300 ft (91.4 m) of cement which may bring cement nearly back to the surface casing. An example is shown in Figure 4.

A typical production casing job is 8,000 to 10,000 ft (2,438 to 3,048 m) of 5½ in, 17 #/ft., N80 grade casing inside 8¾-in hole. The higher strength pipe gives the burst rating required to allow sand fracture stimulation jobs to be pumped straight down the casing. This has been an important element in the high operational success ratio of these fracture treatments. Cementing is done in two stages. The first stage consists of perhaps 1200 sacks of Class "G" cement with turbulence inducer and retarder. A displacement rate of 10 BPM is used, and the pipe is reciprocated throughout. A second stage of perhaps 600 sacks of similar slurry is placed after waiting eight hours for an initial set on Stage 1.

It is difficult to achieve complete success in a situation like this. The two-stage job is required since cement cannot be brought high enough in a single stage to cover the shallow porosity without losing returns or causing serious damage to deep producing zones. However, two-stage cementing has a significant mechanical risk, and indeed malfunctions on the stage collars are not uncommon. Even with the two-stage job, Stage 1 can be up to 4,000 ft (1,219 m) of cement, which puts a high differential pressure on the deepest zone, possibly causing some cement invasion and damage. On the other hand, the shallow potential zones are subject to long exposure times during drilling, and could be damaged prior to cementing. In some areas of the basin, it has become apparent that development wells must be designed specifically to produce one or two primary zones at moderate depth rather than jeopardize those zones by deeper drilling and extensive evaluation of other zones.

Early in the development of the field, primary cement job failures during well stimulation work were quite common. This was greatly reduced by a concentrated effort to ensure good basic practices on each job, including centralizing the pipe, reducing the mud viscosity, maximizing cement displacement rates, and reciprocating the pipe during displacement. However, even consistent good practices won't ensure a good cement job if the hole is in poor condition.

The sloughing shale problems in the area can be detrimental to good cementing in several ways. Firstly, it is difficult to fully displace the mud in an over-gauged hole. Secondly, the thick and heavy fresh-water gel muds which usually result from trying to stabilize the sloughing shale are difficult to displace with cement. Also, there is a tendency to minimize the number of centralizers and scratchers run in poor hole conditions for fear of getting the casing stuck. In fact, the casing may get tight and sometimes cannot be reciprocated during cement displacement.

FORMATION PRESSURES

The entire Deep basin is generally underpressured, with virgin reservoir pressures ranging from as low as 0.30 psi/ft to about 0.38 psi/ft in most producing zones. There are exceptions, and a few areas even have isolated pay zones with gradients as high as 0.55 psi/ft The low formation pressures have a significant impact on drilling operations. Even a light gel-water mud with a density of say 9.0 ppg may be in excess of 2.0 ppg overbalanced. At a depth of 8,000 ft (2,438 m), this equates the 832 psi mud column overbalance to the formation. Excessive overbalance is, of course, detrimental to penetration rate. Formation damage is also of great concern, and gas shows are generally suppressed.

DEVELOPMENT OF DRILLING METHODS

Early wells in the field were drilled in the Elmworth area, and generally along the northeastern side of the field. A very basic well design was used, consisting of 9 5/8 in. surface casing set at about 1,000 ft (305 m), and 8 3/4-in main hole to TD. Freshwater-extended gel muds were used. Shale problems are not generally as serious in that part of the field as in other areas. Sloughing occurred in many wells, but the problem was not thought to be critical. Total drilling time for an 8,000 ft (2,438 m) well varied from 30 to 50 days, depending on the shale problems.

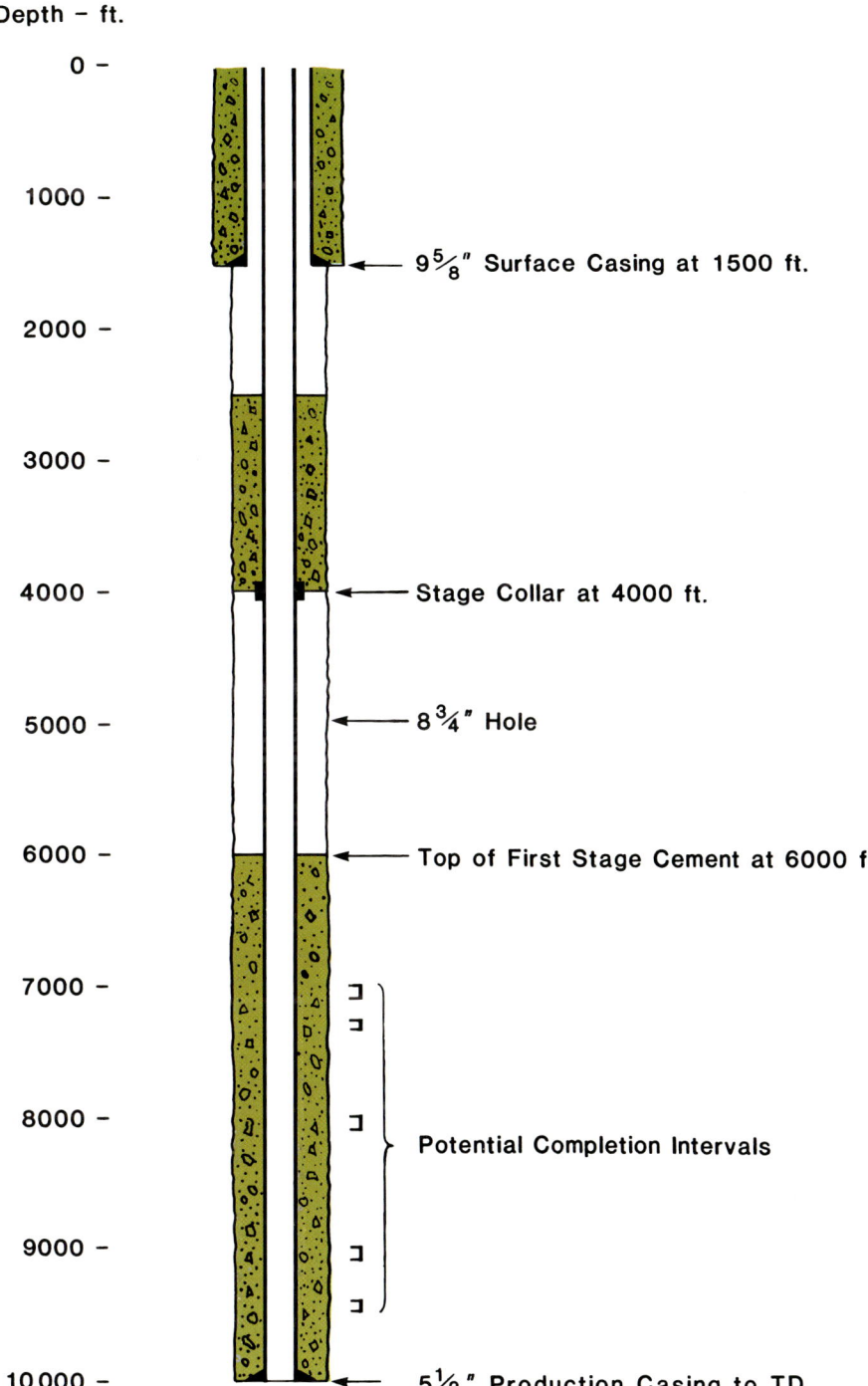

Figure 4. Example of productive casing and cement requirements – Deep basin gas well.

As exploration of the field pushed south and west, the frequency and severity of the shale problem increased, along with the depth of the wells. The cost of the problem in terms of poor open-hole evaluation required a new approach. A larger well design was tried, with 9 5/8-in intermediate casing set throughout the upper shales to

Table 1. Oil-base mud performance.

Location	Total Depth (feet)	Total Costs (millions)	Costs in 1981 Dollars (millions)	Days
A. Wells drilled September 1979 through June 1980 (all with intermediate casing)				
10-19-66-19 W6M	12,464	$ 2.77	$ 3.19	79
11-18-66-10 W6M	11,546	2.77	3.19	104
6- 7-67-8 W6M	11,414	2.57	2.96	95
10-12-67-10 W6M	10,945	2.26	2.60	85
7-15-67-10 W6M	11,240	2.26	2.60	89
10-25-67-11 W6M	10,922	2.22	2.55	109
10-30-67-11 W6M	9,997	2.37	2.73	71
6-16-68-11 W6M	10,430	2.22	2.55	85
7-14-68-13 W6M	9,702	2.10	2.42	80
6-28-68-13 W6M	11,660	2.20	2.53	79
7-23-69-13 W6M	9,420	2.06	2.37	99
6-26-69-12 W6M	9,164	1.56	1.79	66
15-27-66-10 W6M	10,594	3.42	3.93	130
Totals	139,498	$30.78	$35.41	1171
Averages	10,731	$ 2.37	$ 2.72	90
B. Wells drilled June 1980 to August 1981 with oil-base drilling fluid				
11- 7-68-12 W6M	10,430	$ 2.08		68
7- 1-68-12 W6M	10,004	1.96		56
6-21-67-10 W6M	9,289	1.58		42
6-28-69-12 W6M	8,574	1.31		43
6-19-68-12 W6M	9,758	1.41		42
15-16-68-13 W6M	10,165	1.47		40
10-26-66-11 W6M	10,378	1.87		43
9-10-66-10 W6M	10,430	2.07		44
7- 4-68-11 W6M	9,856	1.33		35
Totals	88,884	$15.08		413
Averages	9,876	$ 1.68		46

Note: Total costs include road and lease construction and production casing.

the top of the Paddy Formation as shown in Figure 3. Once this pipe was set, 8½-in hole was drilled through the remaining 3,000 to 4,000 ft (914 to 1,219 m) containing various potential zones without the interference of the sloughing shales. Good mud and hole conditions were easily maintained resulting in good open-hole evaluation. An additional benefit was that the production casing could be cemented with a single stage, eliminating the problems associated with collar failures. However, the cost of this approach was high. The price of the intermediate casing and cementing was in the order of $250,000 per well. Penetration rates in the 12¼-in intermediate hole were much slower than 8¾-in hole, increasing drilling time by a minimum of 6 to 10 days per well. The total drilling operation on a 10,500 ft (3,200 m) well was about 70 days if no problems occurred, but serious shale sloughing in the intermediate hole often caused considerable delays before the 9⅝-in casing could be set.

As was mentioned earlier, KCl polymer muds were also tried with mixed results. Some believe that these systems will do the job if a high KCl concentration is used, and if the system is very carefully controlled. However, these muds appeared to be expensive, difficult to properly maintain, and gave undependable shale control.

In the fall of 1980, an oil-base mud system was tried on a well located in an area of severe shale sloughing. The results were excellent; as no sloughing occurred, no intermediate casing was run, and considerable drilling time was saved. By September 1981, nine wells had been drilled in this problem area, using oil-base muds. The results were compared to the 13 previous wells drilled with intermediate casing and the results are shown in Table I. The average time from spud to rig release was reduced from 90 to 46 days and the average cost was reduced by $1 million per well. The quality of open-hole evaluation was excellent. Since that time, the mud system and operating practices have been refined, resulting in substantial improvements and further cost savings. Two of the more recent wells drilled to 10,000 ft (3,048 m) in this area were completed in 35 days each. The most recent well in the area had a total depth of 9,436 ft (2,876 m), and was completed in only 27 days.

OIL-BASE MUD SYSTEM

The mud system now being used consists of an emulsion of about 10% water in 90% diesel fuel oil. Viscosity is increased when necessary with organophilic clay, and the system is weighted as required with barite. The maximum density at TD is often less than 8.3 ppg, and much of the upper part of the wells are drilling with weights of 7.9 ppg or less. The plastic viscosity is generally 5 to 8, and gel strengths are practically nonexistent.

The primary advantage is the elimination of the sloughing shale. This allows the wells to be drilled without intermediate casing and still obtain complete openhole evaluation.

As the use of oil-base muds expanded, several other advantages became obvious, and were taken advantage of:

1. Many wells can be drilled with substantially lower mud weight than with water-based mud, primarily because the basin is so underpressured. Overbalance pressures are reduced, penetration rates increased, and formation damage may be less severe.
2. Surge and swab pressures are minimal with this light, thin fluid, and tripping speeds can be increased without creating downhole problems.
3. The possibility of any clay swelling within the formation is eliminated.
4. Cores, logs, and DSTs can be run virtually trouble free because the hole is always in-gauge with no sloughing or fill. Packer seats are practically guaranteed on tests, provided high-quality oil-resistant packer rubbers are used.
5. The reliability of production casing jobs is greatly improved because the hole is in-gauge from top to bottom. The pipe can be run with ease and without concern that centralizers may cause a hang up. It is very easy to displace the light thin mud with cement. The pipe can always be reciprocated freely during cementing.

One disadvantage that has been noticed is that drilling cuttings are often smaller and more difficult to analyze than those obtained with gel-water muds. This appears to be due to the annular flow pattern of these very thin muds being turbulent rather than laminar in many cases.

A few cautions are, of course, necessary in the use of oil-base muds. To a large degree, the tremendous success of the system in the Deep basin is due to the low formation pressures. It is very important to know what pressures will be encountered in each well as large gas kicks can be even more dangerous with an oil-mud than they are with water-based systems. It should also be remembered that for any mud weight higher than normal, an oil-mud will contain more solids than a water-based system, because of the lighter base fluid. Lost circulation can be very expensive with oil-mud, and must be considered in planning each well. To date, lost circulation has not occurred with oil-muds in the Deep basin even in areas where this was previously a problem. This is attributed to the much lower mud weight generally used.

It does not appear that the oil-base mud has any detrimental effects on the productivity of the pay zones. Completion results to date have been the same as those obtained from similar zones drilled with water-base gel muds.

CONCLUSIONS

The close working relationship enjoyed among the various disciplines at Canadian Hunter has been an important factor in the evolution of optimum drilling techniques in the Deep basin. Thorough openhole evaluation and high-quality production casing and cementing have been the top priorities when drilling. This has led to the maximum amount of gas being discovered and successfully produced.

The use of oil-based drilling fluid has emerged as the most significant advance in drilling practices. The elimination of sloughing shale problems and increased penetration rates in the underpressured formations are the main reasons why this fluid has been so successful. Substantial savings in total well costs can be realized in many parts of the Deep basin.

Improved reliability on primary cement jobs has resulted from better hole conditions and lighter muds achieved with the oil-based muds. However, obtaining adequate cement bonding across multiple pay zones, especially when most jobs are two stages, still requires close attention. In some parts of the field, it is prudent to design wells to produce specific primary zones rather than attempt to evaluate the entire section.

Canadian Hunter drilling operations have not been marked by technical sophistication so much as by careful attention in the field to established, sensible practice. It is important to choose good men, give them clear-cut directions, monitor their operations carefully, and try to be responsive to surprises. Responsible and competent field men are as essential to a good drilling operation as intelligent engineering.

Completion Practices in the Alberta Deep Basin

John A. Stayura
Canadian Hunter Exploration Ltd.
Calgary, Alberta

Canadian Hunter has established a routine procedure to select completion intervals for every company well drilled in the Alberta Deep basin. This procedure uses a committee of representatives from various disciplines within the company to thoroughly examine all drilling data, relate previous experience in the area, and establish the economic criteria.

The company has also developed a standardized procedure for completing these wells which utilizes pressure transient data extensively to evaluate the various stages of completion operations, predict zone performance, and economically justify additional stimulation. The completion procedure is outlined and completion results from representative wells are tabulated.

INTRODUCTION

The Alberta Deep basin has more than a dozen potential pay sands containing large quantities of gas in different quality reservoirs. A number of submarginal zones were completed during the early years of exploring and developing this area because it was necessary to know the minimum performance levels. A major effort has been made by Canadian Hunter to properly document all pertinent drilling data and test results so that analogies can be used to predict zone performance. The analogy method works well for completions in the same formation but it has limitations when applied to formations of different depositional environments, even though porosity and permeability measurements may be the same. As we evaluate new formations, some empirical testing is still required to establish patterns of response.

Canadian Hunter's major objectives in developing the Alberta Deep basin are to define reserves and to optimize recovery. All wells drilled to present have been completed as single or dual completions, although there is potential for multi-horizon completions. More than adequate gas deliverability levels have been achieved to meet current reserve-based gas purchase contract commitments. Thus, all current development in this area involves completing zones to prove reserves. At this stage of development in the Alberta Deep basin, Canadian Hunter has endeavored to bring onstream the reservoirs with maximum economic returns. Future development will ultimately include some component of comingled completions to permit gas recovery from poorer quality reservoirs. Typically, a Deep basin well may have one or two deeper zones completed, tested, and plugged back and the well left as a dual producer from two uphole zones. Many wells will eventually be recompleted to bring onstream the deeper previously plugged back zones when the area gas deliverability levels fall to gas purchase contract commitments. Finally, Canadian Hunter has completed and suspended a large number of wells which, under current economic conditions, are marginal or noncommercial. Many such currently suspended wells will be brought onstream in the future, if producer economics improve, and the need for additional deliverability arises.

COMPLETION INTERVAL SELECTION

Canadian Hunter routinely uses a committee of specialists from various disciplines within the company to select completion intervals. Typically, the committee is comprised of the geologist who recommended the well, a well log analyst, the well-site geological supervisor, a completions engineer, and a reservoir engineer. Each member of the committee has certain responsibilities with respect to analyzing the drilling data obtained from a well. Because it may not be economically or operationally feasible to have a complete suite of drilling and testing data on every potential zone, previous experience and analogies play an important part in the selection process.

Fundamental in using analogies and relying on previous experience is a complete and well-organized documentation of previously obtained drilling data together with completion and testing results. A major effort was made by Canadian Hunter regarding not only the documentation of this information but the recycling of new information back to all committee members. Several hundred zones have now been completed in the Deep basin area and future development will involve the drilling and completion of several hundred more wells. Prediction of zone performance prior to completion, based on documented analogies, is now done on a regular basis with a reasonable degree of accuracy.

The importance of the committee approach in an area like the Deep basin, with multi-zone potential in every wellbore, cannot be overemphasized. Because of the very high cost of completion operations, there is a tremendous advantage in being able to accurately predict zone deliverability. There is also a real advantage to having field operations people included in the interval selection process. The completions engineer not only derives a much better understanding of why a zone is to be completed, what risks are involved, and what can be expected in productivity, but his regular interfacing with the other disciplines on the completion committee provides the very necessary feedback of field results.

COMPLETION PROCEDURE

All wells in the Alberta Deep basin area are candidates for acid and/or sand fracture stimulation because of unavoidable formation damage occurring during drilling and cementing operations. Type and size of treatment varies with rock character and with results of pre-fracture evaluations. Canadian Hunter's completion technique for wells in the Deep basin has now become a standardized procedure, following extensive lab and petrophysical studies, together with documentation and review of field results. Pressure transient data is used extensively to evaluate the various stages of completion, to predict zone performance, and to justify additional stimulation.

A small-volume 750 gal (2,839 ℓ) acid wash and squeeze is performed after perforating, in order to break down mud-plugged perforations. The acid is usually 15% HCl containing suitable additives. This initial acidizing is required for cleanout of mud in and immediately adjacent to the perforations and to provide communication of the formation to the wellbore. A larger acid treatment of 5,000 gal (18,927 ℓ) follows the small acid job unless the initial acidizing provides positive information that the zone is water bearing. Nitrogen is included in the larger treatment to facilitate cleanup, and alcohol is added where water retention could be a problem in tighter formations. Diverting agents, such as ball sealers, are also used to allow relatively uniform acidizing of the entire zone open to the wellbore.

In the past, temperature logs have been run immediately after the large acid treatment to determine if the stimulation is in zone. Recently, however, recently these large acid treatments have been run with a small amount of radioactive sand to allow tracing with a radioactive log. This technique provides an advantage over using temperature logs in that the well can be immediately flowed back to fully utilize the nitrogen cleanup, especially in tighter formations. A pressure transient test is conducted to obtain a measure of the skin damage and kh in the zone after the well has cleaned up following the large acid treatment. A decision is then made regarding further stimulation of the zone after the results of the pressure transient test have been analyzed.

Low productivity may be a result of severe permeability reduction around the wellbore and/or low formation permeability. In highly permeable formations, drill stem test results can sometimes be used to estimate formation permeability. Many times, however, formation damage can severely affect test results. Additionally, in tight formations, the very shallow radius of investigation of a short-duration drill stem test usually precludes any type of quantitative analysis. Although core analysis provides an estimate of formation permeability; laboratory measurements of permeability may be up to 100 times higher than the in situ permeability of tight gas sands due to saturation and compaction effects. The results of a pressure transient test can provide a much better estimate of the reservoir permeability and can also provide an estimate of the extent of formation damage around the wellbore. Usually, a spectacular increase in productivity can be expected from fracturing in a badly-damaged, highly-permeable reservoir. However, the increase in productivity that can be expected by fracturing in an undamaged, highly permeable reservoir is slight; usually the expected productivity after a fracture is in the order of 1.5 to 2 times the before-fracture productivity. The expected productivity increase is much higher proportionally, in formations of low permeability, than in those of high permeability.

Wellhead pressures are obtained for the pressure transient test and then converted to downhole pressures, in certain situations. However, subsurface pressure recorders are usually used together with a shut-off plug run in the tubing near the tubing bottom. The use of this plug reduces the wellbore volume to minimize afterflow which dominates the early time-buildup pressure response. The shut-off plug is used if the well flows less then 300 mcf/d (8,500 cu m/d) after the large acid treatment. Typically, meaningful pressure transient information is obtained in a matter of hours instead of days.

The results obtained from a pressure transient test are used to predict the incremental benefit of further stimulation of a zone. The economics associated with this benefit are then considered, because an expensive fracture stimulation may not be justified. Other problems must also be considered. For instance, the predicted incremental benefit may assume a deep penetrating fracture requiring control of the fracture height. The log analyst may advise that it would be difficult to contain the fracture because of the lack of low-porosity shales adjacent to the zone of interest, or the drilling engineer may advise that the cement bond is not competent to hold the stimulation.

Two examples of the uses of pressure transient data, obtained during the completion operations, are shown in the Appendix.

Generally, all of the sand fracture jobs performed by Canadian Hunter in the Alberta Deep basin area are gelled KCl water fractures with total sand in the 15,000 to 100,000 lb (7 to 45 tonne) range. The amount and size of proppant used, the staging of the proppant, the pad volume, and fluid additives vary with zone characteristics and the results of pre-fracture evaluation. Radioactive sand is run with the fracture sand to allow tracing of the fracture. In addition to sand fracturing, large acid treatments — 10,000 gal (37,850 ℓ) — have been used as an effective stimulation tool. Experience with horizons such as the Cadomin zone, where sand fracture effectiveness is reduced because of lithology (large particle size in poorly sorted matrix) and also where there is considerable calcareous cementation, has shown that large volume acid treatments are a viable alternative.

Canadian Hunter has performed a massive hydraulic fracture (MHF) treatment on one well in the Alberta Deep basin. The analysis of the treatment (Wyman, Holditch, and Randolph, 1979) indicated that, although the stimulation was a technical success, treatments of this size are not economically viable. A number of reservoirs may be MHF treatment candidates in the future as lower quality reservoirs are required to maintain required deliverability levels and if producer netbacks are significantly improved.

Canadian Hunter has completed several hundred wells and performed approximately 250 sand fractures in the Deep basin area, with only a handful of fractures (less than 5%) not completed due to sanding-off in the casing. Almost all of the sand-offs occurred in the early stages of developing this area when lack of data and inexperience in the area led to erroneous estimations of reservoir parameters. No sand-offs have occurred due to equipment failure.

There are a number of formations where stimulations are either marginally successful or unsuccessful. These formations are shallow underpressured reservoirs that are badly damaged by conventional drilling fluids and/or conventional cementing methods, and occur in wells with a much deeper drilling objective. The major problem appears to be one of containing the stimulation in zone. Future development drilling for these shallower zones will probably require wells that are not drilled below the target zone, in order to minimize formation damage.

After the well is fractured and has cleaned-up, extensive deliverability testing and additional pressure transient testing is conducted. The post-fracture testing is required not only to obtain the deliverability data from the zone, but for comparison with the pre-fracture prediction.

Table 1 is a summary of the completion and test results for a number of representative wells completed by Canadian Hunter across the Deep basin area. This summary reflects not only the range of deliverabilities obtained, but the number of potential zones in each wellbore.

CONCLUSIONS

Canadian Hunter has established and routinely uses a committee with representatives from various disciplines to select completion intervals. This approach is necessary in an area like the Alberta Deep basin, with the multi-zone potential in each well, in order that all data obtained during the drilling and evaluation phase is thoroughly and systematically examined by different specialists.

A major effort is made by Canadian Hunter to properly document all pertinent drilling data and test data in order to calibrate drilling data with test results. Predicting zone performances, based only on logs and drilling cuttings, is now done on a regular basis with reasonable accuracy.

In the future, Canadian Hunter will be required to bring onstream many currently marginal or uneconomic zones because of deliverability requirements. Many of these zones may be MHF candidates. Thus, the standardized completion procedure, as outlined in this paper, will be modified as required. However, the ultimate goal of hydraulic fracture treatment, regardless of size, will be to make an economic success of a stimulated well.

REFERENCE

Wyman, R. E., S. A. Holditch, and P. L. Randolph, 1979, Analysis of an Elmworth hydraulic fracture, Alberta, Canada: SPE 7935, paper presented at the SPE Symposium on low-permeability gas reservoirs (Denver).

Appendix: Examples Using Pressure Build-up Data

Example 1

Well Location: 6-10-64-2 W6M

Formation: Bluesky/Gething

Completion Summary:

— Perforated intervals 7905-7913 ft KB, 7933-7967 ft KB using 4 inch hollow steel carrier gun at 4 shots per foot.
— Acidized with 750 gals 15% HCl + iron stabilizer. After a clean-up period of 12 hours, flowed at 250 mcf/d @ 80 psig.
— Acidized with 5,000 gals 15% HCl + 25% methanol + iron stabilizer + nitrogen + ball sealers. After a clean-up period of 24 hours, flowed at 600 mcf/d @ 190 psig.
— Conducted bottom-hole pressure build-up survey.
— Fractured with 100,000 lb 20-40 sand using gelled KCl water with methanol and CO_2. After a clean-up period of 48 hours, flowed at 12.0 mmcf/d @ 530 psig with condensate mist.

Pressure Build-up Survey Results:

— Average reservoir pressure: 3,125 psia
— Average effective permeability: 5 millidarcys
— Skin factor: > +20

(Further stimulation of the zone was recommended based on the calculated skin factor indicating significant near-wellbore damage).

Flow Test Summary:

— Flowed at 11.0 mmcf/d @ 850 psig for 9 hours on clean-up prior to multi-point test.
— Flowed 4 point (4 hour flow and 4 hour SI) modified isochronal test with 48 hour extended flow. Final rate on extended flow was 8.7 mmcf/d @ 720 psig with condensate cut at 14 bbls/mmcf.

Flow Test Results:

— Average reservoir pressure: 3,142 psia
— Average effective permeability: 6 millidarcys
— Skin factor: −3.5
— Stabilized sandface AOF potential: 9.0 mmcf/d
— Stabilized wellhead deliverability at 1,100 psig: 7.6 mmcf/d

(The negative skin factor indicated that the well had been effectively stimulated with the sand fracture.)

Example 2

Well Location: 6-10-64-2 W6M

Formation: Cadomin

Completion Summary:

— Perforated interval 8262-8275 ft KB using 4 inch hollow steel carrier gun at 4 shots per foot.
— Acidized with 750 gals 15% HCl + iron stabilizer. Gas rate too small to measure after swabbing tubing dry.
— Acidized with 5,000 gals 15% HCl + 25% methanol + iron stabilizer + nitrogen + ball sealers. After a clean-up period of 36 hours, flowed at 30 mcf/d @ 50 psig.
— Conducted bottom-hole pressure build-up survey.

Pressure Build-up Results:

— Average reservoir pressure: 3,040 psig
— Average effective permeability: 0.03 millidarcys
— Skin factor: −3.0

(The negative skin factor indicated that

the well had been effectively stimulated with the large acid treatment. Additional stimulation was not recommended because improvement in productivity was expected to be very marginal.)

Conversion Factors (Approximate)

1 psi = 6.894 kilopascals
1 gallon = 3.785 liters
1 pound = 0.4536 kilogram
1 cubic foot = 0.02832 cubic meter

Table 1. Summary of completion and test results for representative wells completed by Canadian Hunter across the Deep basin area.

Location	Formation	Completion Results		
		After 750 Gals Acid	After 5,000 Gals Acid	After Frac
6-10-64-2 W6M	Cadomin	TSTM	0.03 MMcf/d @ 50 psig	
	Bluesky/Gething	0.25 MMcf/d @ 80 psig	0.60 MMcf/d @ 200 psig	12.00 MMcf/d @ 530 psig (100,000 lb sand)
10-4-65-2 W6M	Bluesky	NGTS	0.40 MMcf/d @ 60 psig	14.40 MMcf/d @ 1,060 psig (100,000 lb sand)
10-23-65-5 W6M	Bluesky	TSTM	TSTM	0.03 MMcf/d @ 10 psig (100,000 lb sand)
	Falher A	TSTM	TSTM	TSTM (100,000 lb sand)
	Cadotte	TSTM	TSTM	TSTM (75,000 lb sand)
10-35-65-5 W6M	Nikanassin	TSTM		
	Bluesky	TSTM	TSTM	0.05 MMcf/d @ 20 psig (75,000 lb sand)
	Falher A	TSTM	TSTM	
	Cadotte	TSTM	TSTM	TSTM (50,000 lb sand)
6-32-65-7 W6M	Nikanassin	TSTM	TSTM	
	Cadomin	0.17 MMcf/d @ 100 psig	1.00 MMcf/d @ 160 psig	3.80 MMcf/d @ 650 psig (30,000 lb sand)
5-32-65-9 W6M	Cadomin	0.20 MMcf/d @ 20 psig	0.38 MMcf/d @ 55 psig	
	L. Gething	0.13 MMcf/d @ 75 psig	0.80 MMcf/d @ 130 psig	4.04 MMcf/d @ 680 psig (50,000 lb sand)
	Falher F	0.44 MMcf/d @ 60 psig	0.89 MMcf/d @ 145 psig	1.03 MMcf/d @ 175 psig (50,000 lb sand)
11-7-66-4 W6M	Bluesky	TSTM	TSTM	0.87 MMcf/d @ 150 psig (80,000 lb sand)
	Falher B	TSTM	4.56 MMcf/d @ 790 psig	11.80 MMcf/d @ 890 psig (50,000 lb sand)
6-3-66-8 W6M	Cadomin	TSTM	0.26 MMcf/d @ 50 psig	
	L. Gething	TSTM	0.33 MMcf/d @ 230 psig	
	Falher F	TSTM	0.09 MMcf/d @ 65 psig	
8-2-66-9 W6M	Cadomin	TSTM	TSTM	
	L. Gething	TSTM	0.59 MMcf/d @ 40 psig	1.10 MMcf/d @ 180 psig (50,000 lb sand)
	Falher F	0.25 MMcf/d @ 160 psig	0.58 MMcf/d @ 89 psig	1.92 MMcf/d @ 140 psig (25,000 lb sand)
12-31-66-9 W6M	Nikanassin	0.06 MMcf/d @ 30 psig	0.11 MMcf/d @ 60 psig	
	Cadomin	0.03 MMcf/d @ 50 psig	0.78 MMcf/d @ 225 psig	0.80 MMcf/d @ 250 psig (75,000 lb sand)
11-18-66-10 W6M	Halfway	1.47 MMcf/d @ 615 psig		
	Nikanassin	TSTM	0.06 MMcf/d @ 20 psig	
	Cadomin	TSTM	0.44 MMcf/d @ 70 psig	
	Falher F	TSTM	TSTM	
	Cadotte	3.18 MMcf/d @ 530 psig	2.94 MMcf/d @ 900 psig	20.90 MMcf/d @ 1,435 psig (20,000 lb sand)
6-7-67-8 W6M	Belloy	Recovered formation water		
	Nikanassin	TSTM		
	Falher D	6.06 MMcf/d @ 960 psig	5.76 MMcf/d @ 900 psig	58.50 MMcf/d @ 2,000 psig (50,000 lb sand)

Table 1. (Continued)

Location	Formation	Completion Results		
		After 750 Gals Acid	After 5,000 Gals Acid	After Frac
7-29-67-8 W6M	L. Nikanassin	TSTM	0.27 MMcf/d @ 65 psig	
	U. Nikanassin	TSTM	0.25 MMcf/d @ 170 psig	2.07 MMcf/d @ 150 psig (100,000 lb sand)
	Falher F	TSTM	TSTM	
	Falher A	7.49 MMcf/d @ 1,300 psig	8.47 MMcf/d @ 1,470 psig	24.60 MMcf/d @ 1,350 psig (23,000 lb sand)
10-14-67-9 W6M	Gething	0.18 MMcf/d @ 30 psig	1.05 MMcf/d @ 180 psig	2.77 MMcf/d @ 470 psig (50,000 lb sand)
	Falher D	TSTM	0.03 MMcf/d @ 85 psig	
10-12-67-10 W6M	Cadotte	TSTM	TSTM	TSTM (50,000 lb sand)
	Cadomin	0.67 MMcf/d @ 100 psig	1.65 MMcf/d @ 250 psig	
	Falher D	1.54 MMcf/d @ 245 psig	2.38 MMcf/d @ 371 psig	13.90 MMcf/d @ 2,200 psig (50,000 lb sand)
10-30-67-11 W6M	Falher B	TSTM	0.06 MMcf/d @ 65 psig	
	Notikewin	TSTM	0.06 MMcf/d @ 150 psig	
	Cadotte	TSTM	0.07 MMcf/d @ 40 psig	
7-4-68-11 W6M	Nikanassin	TSTM	0.58 MMcf/d @ 30 psig	
	Cadomin	0.20 MMcf/d @ 140 psig		
	L. Gething	TSTM	0.65 MMcf/d @ 120 psig	0.75 MMcf/d @ 135 psig (50,000 lb sand)
	Falher D	9.43 MMcf/d @ 414 psig	13.02 MMcf/d @ 570 psig	38.45 MMcf/d @ 1,530 psig (50,000 lb sand)
11-7-68-12 W6M	Cadomin	TSTM	0.65 MMcf/d @ 100 psig	
	L. Gething	TSTM	0.56 MMcf/d @ 85 psig	1.93 MMcf/d @ 325 psig (75,000 lb sand)
	Falher D	TSTM	1.10 MMcf/d @ 170 psig	2.67 MMcf/d @ 450 psig (75,000 lb sand)
7-8-69-9 W6M	Falher A	3.26 MMcf/d @ 550 psig	5.75 MMcf/d @ 440 psig	58.50 MMcf/d @ 1,075 psig (50,000 lb sand)
10-35-69-10 W6M	Halfway	TSTM	0.27 MMcf/d @ 265 psig	1.03 MMcf/d @ 130 psig (50,000 lb sand)
	Nikanassin	TSTM	0.65 MMcf/d @ 100 psig	
	Falher A	1.77 MMcf/d @ 300 psig	3.88 MMcf/d @ 675 psig	28.40 MMcf/d @ 1,265 psig (50,000 lb sand)
8-13-69-11 W6M	Falher A	3.96 MMcf/d @ 700 psig	9.20 MMcf/d @ 710 psig	36.28 MMcf/d @ 1,250 psig (50,000 lb sand)
10-7-69-12 W6M	Nikanassin	TSTM	0.04 MMcf/d @ 30 psig	
	Cadomin	0.30 MMcf/d @ 50 psig	2.20 MMcf/d @ 160 psig	
	Falher A	0.17 MMcf/d @ 100 psig	2.85 MMcf/d @ 500 psig	12.21 MMcf/d @ 970 psig (50,000 lb sand)
6-28-69-13 W6M	Cadomin	TSTM	0.43 MMcf/d @ 60 psig	
	L. Gething	0.31 MMcf/d @ 40 psig	0.80 MMcf/d @ 130 psig	
	U. Gething	TSTM	0.72 MMcf/d @ 200 psig	0.95 MMcf/d @ 150 psig (50,000 lb sand)
	Falher A	TSTM	0.74 MMcf/d @ 220 psig	1.57 MMcf/d @ 265 psig (50,000 lb sand)
10-17-70-9 W6M	Falher B	6.84 MMcf/d @ 300 psig	7.00 MMcf/d @ 530 psig	43.10 MMcf/d @ 1,224 psig (50,000 lb sand)
6-14-70-10 W6M	Falher B	2.74 MMcf/d @ 450 psig	4.74 MMcf/d @ 800 psig	7.68 MMcf/d @ 1,340 psig (50,000 lb sand)
10-1-70-11 W6M	Halfway	0.50 MMcf/d @ 160 psig	0.75 MMcf/d @ 120 psig	Recovered formation water (50,000 lb sand)
	Nikanassin	TSTM	0.14 MMcf/d @ 30 psig	TSTM (75,000 lb sand)
	Bluesky	TSTM	0.04 MMcf/d @ 30 psig	0.14 MMcf/d @ 20 psig (100,000 lb sand)

Table 1. (Continued)

Location	Formation	Completion Results		
		After 750 Gals Acid	After 5,000 Gals Acid	After Frac
	Falher A	TSTM	0.75 MMcf/d @ 110 psig	4.91 MMcf/d @ 770 psig (50,000 lb sand)
	Cadotte	TSTM	0.13 MMcf/d @ 75 psig	0.16 MMcf/d @ 55 psig (50,000 lb sand)
11-1-70-12 W6M	Cadomin	0.50 MMcf/d @ 25 psig	1.67 MMcf/d @ 270 psig	
	Falher B	4.67 MMcf/d @ 790 psig	5.37 MMcf/d @ 900 psig	38.10 MMcf/d @ 1,300 psig (50,000 lb sand)
11-1-70-13 W6M	Cadomin	0.20 MMcf/d @ 120 psig	1.78 MMcf/d @ 300 psig	2.18 MMcf/d @ 365 psig (15,000 gals acid)
	Falher B	0.64 MMcf/d @ 100 psig	4.22 MMcf/d @ 710 psig	10.28 MMcf/d @ 790 psig (50,000 lb sand)
	Falher A	TSTM	0.96 MMcf/d @ 165 psig	2.39 MMcf/d @ 400 psig (50,000 lb sand)
10-15-71-11 W6M	Dunvegan C	TSTM	TSTM	0.10 MMcf/d @ 25 psig (40,000 lb sand)
	Dunvegan A	TSTM	0.15 MMcf/d @ 50 psig	0.63 MMcf/d @ 200 psig (75,000 lb sand)
16-3-71-12 W6M	Nikanassin	TSTM	TSTM	
	Cadomin	TSTM	0.78 MMcf/d @ 230 psig	
	L. Gething	TSTM	0.34 MMcf/d @ 95 psig	
	Falher B	2.16 MMcf/d @ 365 psig	6.53 MMcf/d @ 1,132 psig	26.67 MMcf/d @ 1,160 psig (50,000 lb sand)
12-2-71-13 W6M	Cadomin	TSTM	1.15 MMcf/d @ 220 psig	
	Falher B	0.83 MMcf/d @ 770 psig	4.56 MMcf/d @ 770 psig	15.12 MMcf/d @ 1,190 psig (50,000 lb sand)
10-35-71-13 W6M	Cadomin	TSTM	1.00 MMcf/d @ 320 psig	1.34 MMcf/d @ 400 psig (15,000 gals acid)
	Falher D	6.53 MMcf/d @ 275 psig	5.96 MMcf/d @ 1,035 psig	40.70 MMcf/d @ 1,800 psig (50,000 lb sand)
10-3-72-12 W6M	Falher D	TSTM	3.99 MMcf/d @ 675 psig	8.25 MMcf/d @ 650 psig (50,000 lb sand)
7-11-72-11 W6M	Falher A	0.32 MMcf/d @ 90 psig	0.35 MMcf/d @ 120 psig	1.09 MMcf/d @ 180 psig (50,000 lb sand)
	Paddy	5.93 MMcf/d @ 950 psig	5.61 MMcf/d @ 890 psig	6.50 MMcf/d @ 1,000 psig (50,000 lb sand)
6-29-72-13 W6M	Halfway	TSTM	0.08 MMcf/d @ 100 psig	
	Cadomin	TSTM	1.00 MMcf/d @ 160 psig	1.32 MMcf/d @ 220 psig (30,000 lb sand)
	Notikewan	TSTM	0.42 MMcf/d @ 70 psig	3.51 MMcf/d @ 600 psig (50,000 lb sand)

The Giant Hoadley Gas Field, South-Central Alberta

Kam K. Chiang
Sundance Oil Company
Denver, Colorado

The Hoadley gas field is a giant gas condensate accumulation discovered in November 1977, by Sundance Oil Company. The field covers approximately 1,500 sq mi (3,885 sq km) in south-central Alberta, Canada. The producing zone in the Lower Cretaceous Glauconitic formation comprises 25 to 80 ft (7.6 to 24.4 m) sandstone pay. The sand was deposited as an extensive marine barrier bar complex with an approximate width of 15 mi (24 km) and length of more than 130 mi (209 km), trending southwest-northeast. The middle and southwestern portion of the barrier bar (approximately 100 mi, or 161 km, long) is entirely saturated with gas and natural gas liquids, trapped laterally by impermeable shale and updip by shale-filled tidal channels. Of more than 140 Glauconitic gas wells completed within this section of the barrier bar since discovery, none have tested or produced formation water. The field is estimated to contain an ultimate potential recoverable reserve of 6 to 7 tcf of gas and 350 to 400 million barrels of associated natural gas liquids.

The Hoadley gas field is a buried example of a modern barrier bar. The barrier bar has emerged and submerged several times during seaward progradation. Principal paleogeographic features and depositional facies recognized in the studied area are described herein with well log examples.

INTRODUCTION

The Hoadley gas field is a "giant" gas condensate accumulation discovered in 1977 by Sundance Oil Company. It fits the category of both a giant gas field (defined as one trillion cubic feet [tcf] or more of combustible gas by T.A. Fitzgerald, 1980, AAPG Memoir 30, p. 1) and a giant oil field (defined as 100 million barrels or more of conventional oil by Fitzgerald, 1980). The recoverable potential Hoadley gas reserve of 6 to 7 tcf is the third largest in Canada next to the Elmworth and the Milk River gas fields (Fig. 1). The associated natural gas liquid (NGL) reserve of 350 to 400 million barrels rank as one of the largest NGL fields in North America.

I first announced to the public the discovery and the geology of the field in a paper delivered to the Canadian Society of Petroleum Geologists in April, 1981. This is an updated version of that presentation.

DISCOVERY WELL

The discovery well, Sundance et al Hoadley (LSD 6-Section 2-Township 45-Range 2-West 5th Meridian), was drilled on acreage farmed out by Hudson's Bay Oil and Gas Company Ltd. and Amoco Canada Petroleum Company Ltd. The well is located 120 mi (193 km) north of Calgary and 48 mi (77 km) southwest of Edmonton, Alberta, (Fig. 1). This discovery well encountered 78 ft (24 m) of sandstone pay at a depth of 5,802 ft (1,768 m). The best sand, with up to 26% porosity, occurs near the top and the porosity gradually decreases to as little as 8% in the lower section (Fig. 2). A drill-stem test run at an interval of 5,790 to 5,820 ft (1,765 to 1,774 m) in the upper section tested gas at a rate of 9.8 mmcf/d with a shut-in pressure of 2,378 psi. The middle and lower sections were perforated at intervals of 5,830 to 5,847 ft (1,777 to 1,782 m) and 5,860 to 5,866 ft (1,786 to 1,788 m) with 4 shots per foot. After fracture treatment, the well flowed gas at a stabilized rate of 26 mmcf/d together with recovery of 60 barrels of natural gas liquids per million cubic feet of gas. Subsequent perforation at intervals of 5,800 to 5,816 ft (1,768 to 1,773 m) increased gas flow rate to a calculated absolute open flow (AOF) of 76 mmcf/d. The well has since produced at rates up to 34.5 mmcf/d during production. This was one of the most significant discoveries in Canada in 1977, and indicative of a high-performance gas reservoir with a very high gas liquids content.

STRATIGRAPHY

The Lower Cretaceous Glauconitic sand, producing zone of the Hoadley gas field, is overlain by the continental sediments of the Blairmore Formation consisting of sandstone, siltstone, shale, and coalbeds and is underlain by the calcareous shale and lime beds of the marginal marine Ostracod zone (Fig. 3). Although the name Bluesky formation has been commonly used in British Columbia and in areas west of the Sixth Meridian in Alberta, in central and southern Alberta, the term "Glauconitic" sand is commonly used. Due to the consistent presence of glauconite in the formation and the wide distribution of the sandstone in central and southern Alberta, subsurface geologists have generally adopted "Glauconitic" as the name of the formation. "Glauconitic" has also been used by the Energy Conservation Board of Alberta in the correlation chart of southern Alberta. Since the discovery well is located in south-central Alberta, the writer chooses the term "Glauconitic" as the formation name of the producing zone in this paper.

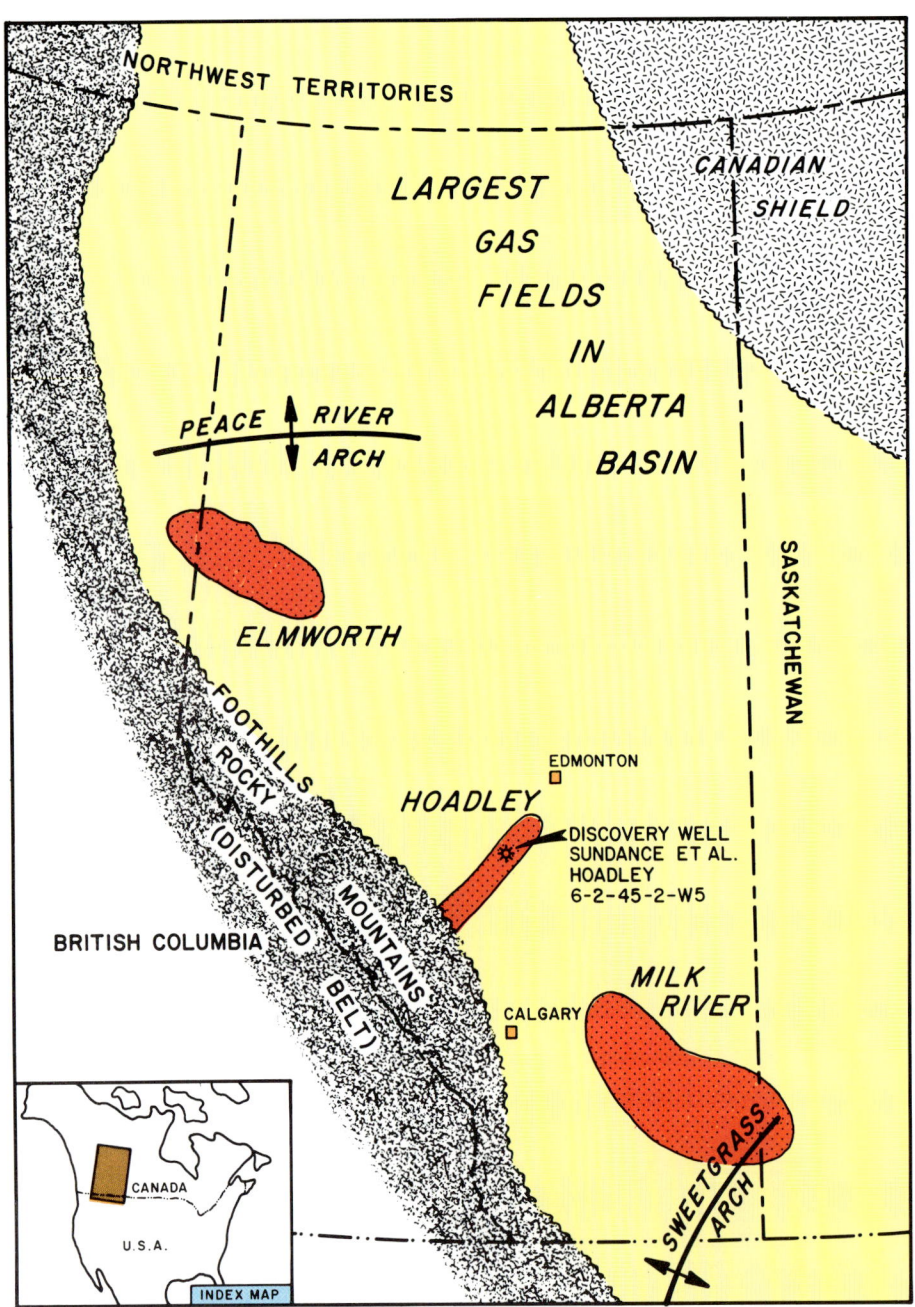

Figure 1. General map showing largest gas fields in Alberta basin, Canada.

belt, foothills area, and Peace River Arch.

The present-day structure south of the Peace River Arch of the Alberta basin shows a simple configuration of northwest-southeast strike with southwest regional dip. This setting is not only supported by the subsurface structural mapping of many different formations, but is also supported by the northwest-southeast trend of both Devonian and Mississippian subcrop edges, and by the well-known Cretaceous Viking sand bars and Cardium conglomerate bars which also trend in a northwest-southeast direction as oil and gas fields.

This northwest-southeast trend concept is deeply buried in the minds of Alberta subsurface geologists, and has affected mapping and interpretation. Many of the Lower Cretaceous Glauconitic sand bars have also been mapped in the same northwest-southeast direction.

The extensive study of the Hoadley gas field indicates that the Hoadley Glauconitic barrier bar trends in a southwest-northeast direction across the entire south-central Alberta basin, essentially normal (i.e. 90°) to the characteristic trend of the Viking and younger Cretaceous sand bars. This study also found that, at that time land was situated to the southeast of the bar and open marine to the northwest during deposition of the Glauconitic sand. This southwest-northeast bar trend contrasts with the predominant northwest-southeast Cretaceous sand trends. The new paleogeographic picture of land and sea during a part of Early Cretaceous time may generate new ideas and prospects in the Alberta basin.

GEOLOGY OF THE HOADLEY BARRIER BAR COMPLEX

The producing sand of the Hoadley gas field was deposited as a marine barrier bar complex. The northeastern end is now buried at 3,500, or 1,200 ft subsea (1,067 m, or 366 m subsea) in the Edmonton area and dips southwestward to a depth of 13,000 ft (3,962 m), or 8,000 ft (2,438 m) subsea, where it plunges under the disturbed belt. The bar is approximately 15 mi (24 km) wide and more than 130 mi (209 km) long, trending southwest to northeast. The original depositional length of the bar is unknown because the southwest end extends under the thrust faults in the disturbed belt to the west

REGIONAL GEOLOGY AND SETTING

The Hoadley barrier bar is located in the south-central area of the Alberta basin which extends from the Northwest Territories southeast to the Canada/United States border. The basin is bounded on the east by the Canadian Shield, on the south by the Sweetgrass Arch, and on the west by the Rocky Mountains (Fig. 1). East of the mountains, Paleozoic and Mesozoic formations dip regionally southwestward toward the disturbed belt of the Rockies at average dips of 50 ft/mi (9.5 m/km) increasing to 150 ft/mi (28.4 m/km) near the foothills area. Locally, structures occur related to the underlying Mississippian erosional topography and to the differential compaction of the deeper Devonian Ireton shale over Leduc reefs. Major faulting appears to be limited to the disturbed

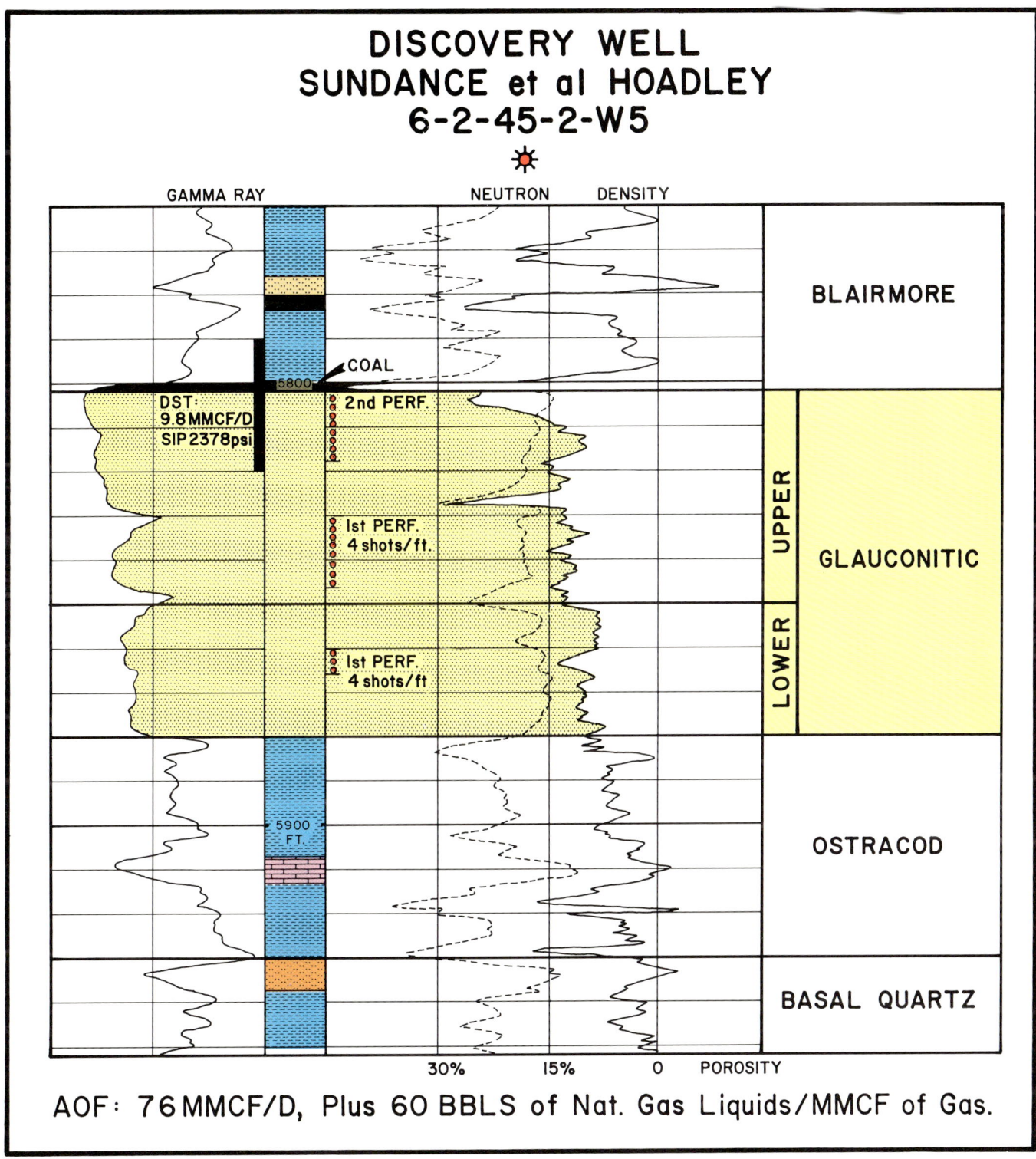

Figure 2. The discovery well, Sundance et al Hoadley 6-2-45-2-W5, showing Compensated Neutron Formation Density Log of the producing Glauconitic sandstone.

whereas the northeastern extremity is truncated by later fluvial deposits of the Blairmore formation. It is reasonable to believe that the original bar deposition extended beyond the mapped area at both ends.

An open-marine shale facies occurs to the northwest of the barrier bar, and bay facies occurs to the southeast (Fig. 4). Principal facies recognized on the barrier bar complex include eolian sand dune, tidal channel, levee, interbar lagoon, and

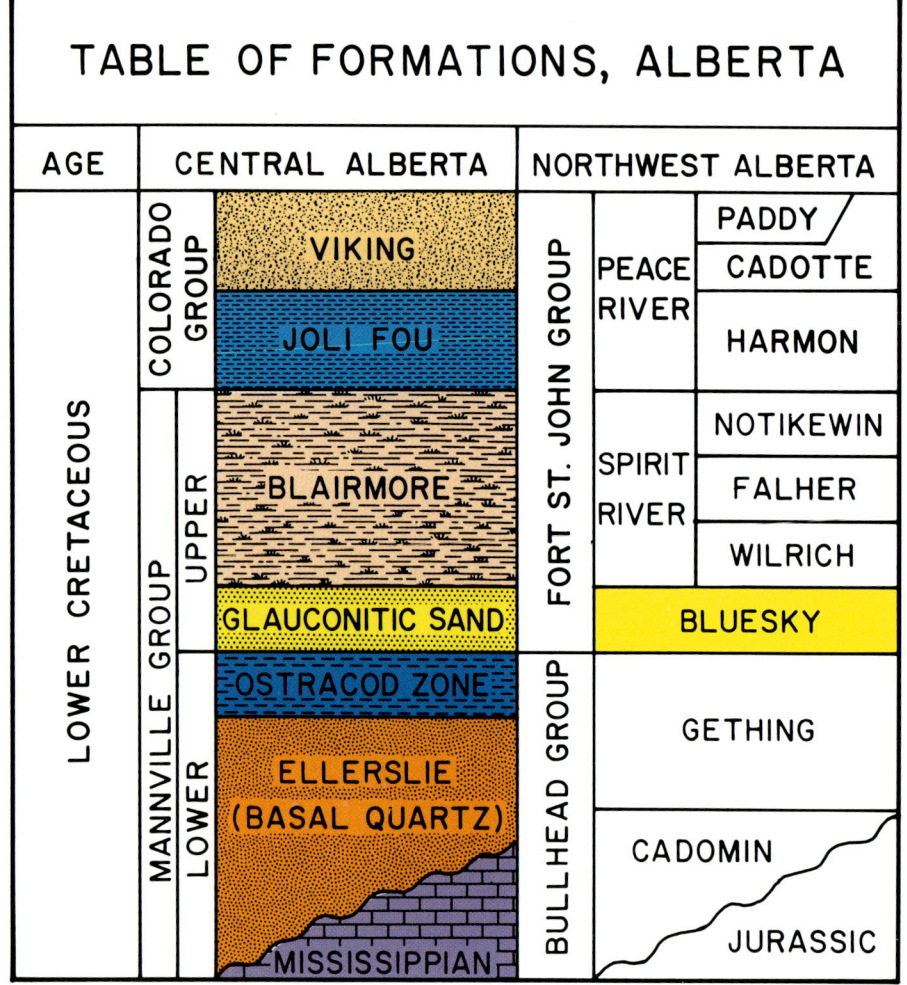

Figure 3. Correlation table of Lower Cretaceous stratigraphy in the Alberta basin (after Energy Resources Conservation Board, March 1976).

back bar washover deposits. The Medicine River delta complex (Chiang, 1981) is found immediately to the southeast of the barrier bar in the Medicine River area (Townships 38-41, Ranges 1-5, West 5th Meridian; Fig. 4) and that delta may have contributed sediments to it. Principal facies recognized within the delta complex include distributary channel, abandoned channel, and deltaic deposits.

The recognition of these facies provides criteria for future exploration on the barrier bar. The characteristics and distribution of these facies are described with well log examples as follows.

Barrier Bar Sandstone Facies

Sandstone and conglomerate make up the barrier in a variable thickness up to 110 ft (33.5 m), but generally less than 80 ft (24.4 m). The bar is broadly stratified into an upper and lower unit. The lower unit consist of 25 to 30 ft (7.6 to 9.1 m) of relatively dirty and low-porosity sandstone averaging 9% porosity and 0.3 of a millidarcy permeability. The upper unit comprises 45 to 60 ft (13.7 to 18.3 m) of fine- to medium-grained sandstone ranging from 8 to 16% porosity. The upper Glauconitic sand is characterized in the well logs by upside-down bell-shaped gamma ray and density curves both of which reflect a gradual increase of porosity and sorting from the base upward (center of Fig. 5), thus displaying the typical log character of a marine barrier bar. The entire unit can also be subdivided into a descending order of A, B, C, and D sand facies based on gamma ray and porosity curve character. These facies may actually reflect contemporaneous deposition in different environments during seaward progradation of the barrier bar (Fig. 6). The average porosities and permeabilities of the four sand facies are as follows: "A" sand is 15% and 10 or more millidarcys; "B" is 13% and 1 to 10 millidarcys; "C" is 11% and 0.5 to 5 millidarcys; "D" is less than 8% and not usually of reservoir quality.

The "A" sand facies is a remarkably clean sand with little clay and absent or rare occurrence of glauconite which suggests that the sand was formed in a subaerial environment, such as a beach or eolian dune sand. The "B" sand has a definite increase of clay matrix as compared with the overlying "A" sand facies. Glauconite is present, but not abundantly. The "B" sand facies probably was deposited in water of 0 to 20 ft (0 to 6.1 m) deep based on the thickness of the sand unit and also by comparison with the modern analogy of the Galveston Barrier Island (Bernard et al, 1959). The "C" sand facies contains notably more clay matrix than the overlying "B" sand and also contains more mica fragments, suggesting the "C" occurs in deeper water (about 20 to 35 ft, or 6.1 to 10.7 m) and a lower energy environment than the "B." The "D" facies represents a transitional zone between the barrier bar sand facies and marine shale facies. The "D" facies and marine shale facies were probably deposited in water depth greater than 30 ft (9.1 m). Such depositional water depths support the interpretation that the Upper Glauconite barrier bar grew seaward by accretion of A, B, C, and D sand facies simultaneously.

Marine Shale Facies

Marine shale was contemporaneously deposited in the area northwest of the barrier bar. Its uniformity as a lateral seal is indicated by a recessive gamma ray curve (Fig. 5). Coalbeds and porous sandstone beds are noticeably absent. The total thickness of the marine shale facies is about one-half the thickness of the equivalent barrier bar sand facies, suggesting finer sediment and slower deposition in the off-bar marine environment as well as greater subsequent compaction of finer-grained sediment upon burial.

Bay Facies

The bay facies consists of a mixed layering of shale, siltstone, sandstone, and coal (Fig. 5) which represents the unsorted mud, seaweed, plant debris, and a minor amount of sand similar to present-day back

Giant Hoadley Gas Field   301

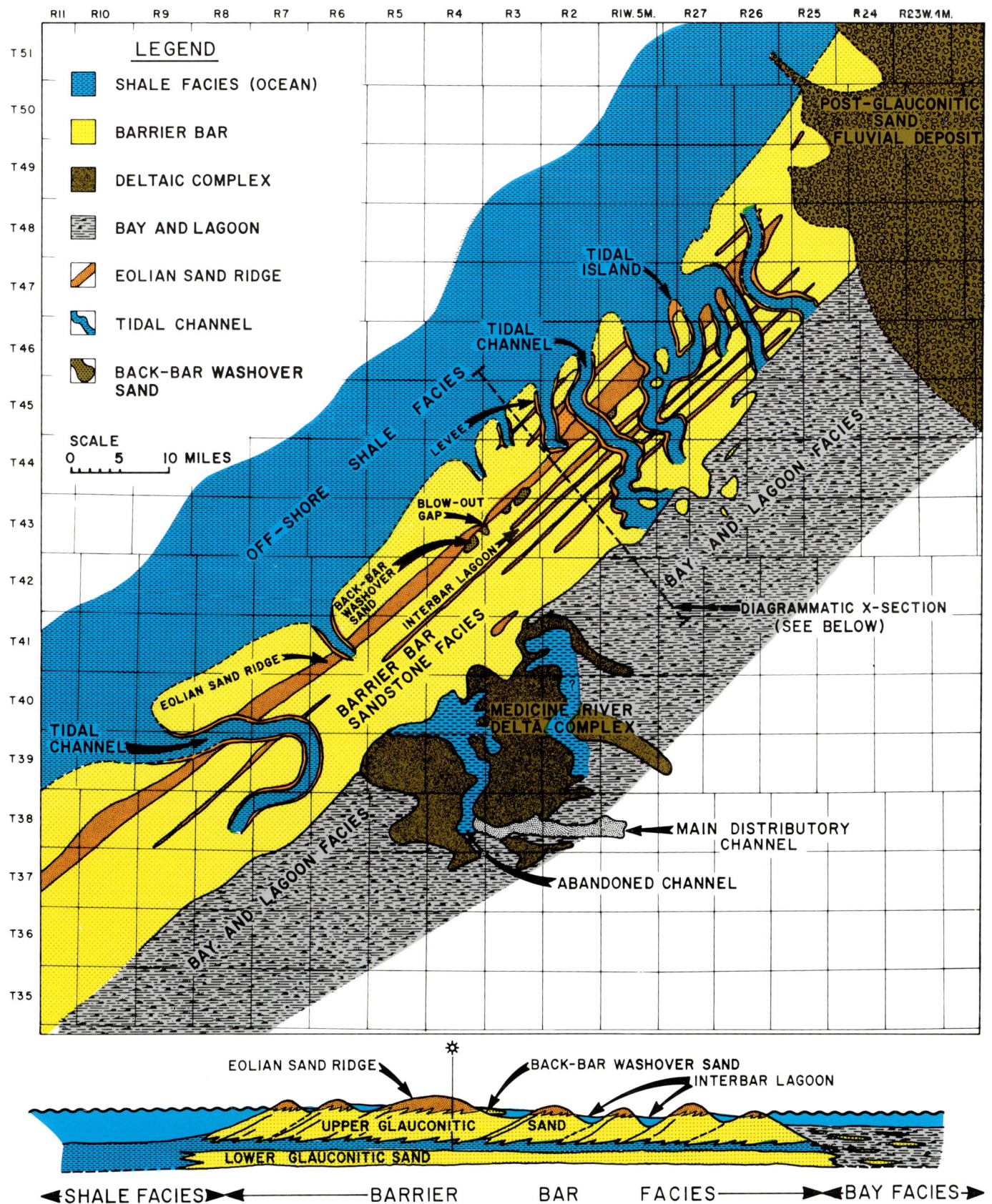

Figure 4. Map showing facies of the Hoadley barrier bar complex and the Medicine River delta complex in south-central Alberta.
Note: The diagrammatic cross section of the barrier bar shown below the map has exaggerated vertical scale.

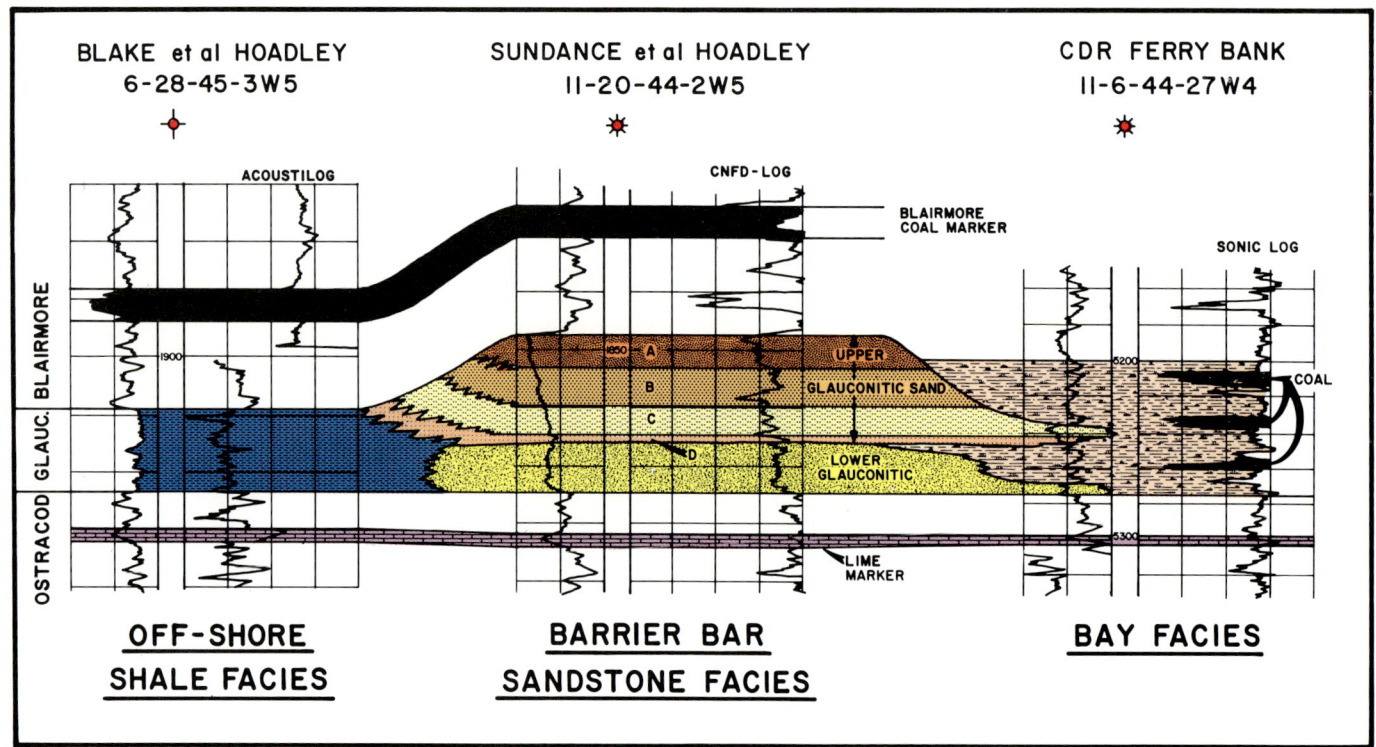

Figure 5. Compensated Neutron Formation Density Logs showing barrier bar sandstone, marine shale and bay facies.

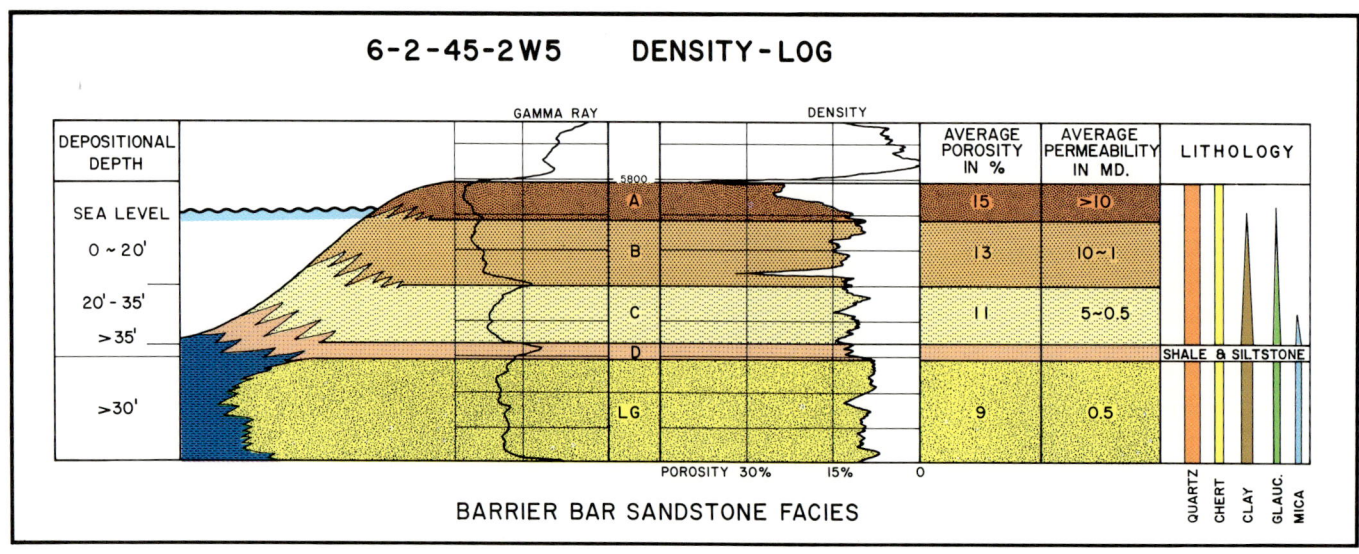

Figure 6. Diagrammatic illustration of the barrier bar sandstone facies showing that A, B, C, D sands are contemporaneous facies of different depositional environments including water depth, reservoir parameters and mineral composition.

bar sediments. Bay sediment on one side and marine shale on the other formed the lateral seal for the hydrocarbons in the barrier bar sandstone.

Eolian Sand Facies

The eolian sand facies consists of fine to medium quartz and chert grains with little clay and absence or rare occurrence of glauconite. It forms the uppermost part of the barrier bar sandstone (the "A" sand unit in Figs. 6 and 7). It occurs in narrow and long ridges on the barrier bar (Fig. 4). Four such parallel eolian sand ridges have been identified along the Hoadley barrier bar, and more are postulated.

Despite the fact that the porosity of the eolian sandstone is higher than 13% and the permeabilities reach several hundred millidarcys, there is evidence that diagenesis has locally altered both porosity

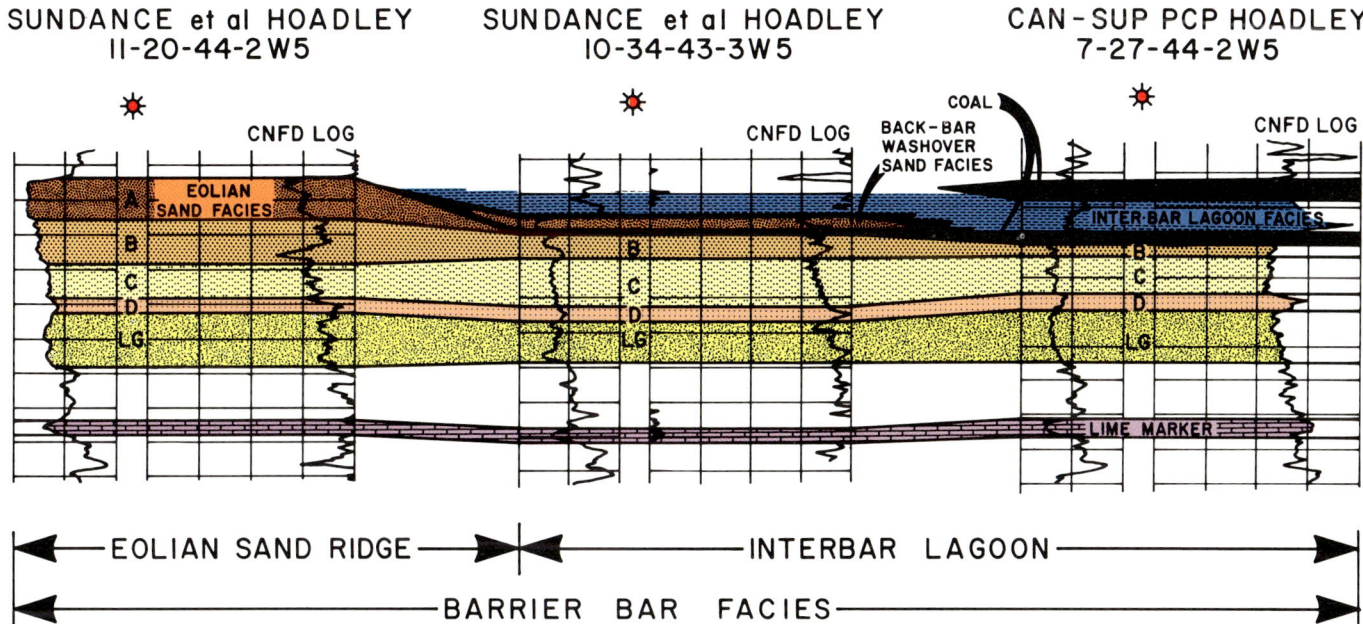

Figure 7. Well logs showing eolian sand, washover sand, and inter-bar lagoon facies of the barrier bar complex.

and permeability. Well samples show euhedral quartz crystals which can only come from post-depositional silica.

Delineation of the eolian sand ridges is important to exploration along the Hoadley barrier bar. The presence and absence of the eolian sand facies can make a difference in well deliverability of several million cubic feet of gas per day and several billion cubic feet of attributable recoverable reserve per spacing unit.

Interbar Lagoon Facies

The interbar lagoons found on the barrier bar between eolian sand ridges appear to have been low areas with vegetation mats now converted to coal. They frequently lie above the lower section of barrier bar sandstone (for example "C" and "D" sand facies; Fig. 7). The interbar lagoon sediment serves to laterally trap the "A" and "B" sand gas.

Back Bar Washover Sand Facies

The back bar washover sand occurs immediately behind the eolian sand ridge in the interbar lagoonal areas (center of Fig. 7). Sand appears to have been washed or blown into the lagoonal area by storms as washover fans. This facies is recognized when a thin porous sand occurs between two coal seams or overlies carbonaceous sediment in the interbar lagoon. Back bar washover sands are commonly found behind modern barrier bar islands (see Fig. 16). At Hoadley, back bar washover sands may produce gas at high initial rates, but deplete very quickly because of the limited reservoir extent. However, if such a sand body is connected to an "A" sand ridge, it can be a thin but excellent producer.

Tidal Channel Facies

Tidal channels, a common feature of modern barrier bar complexes, have been identified along the 130-mi-long (209-km-long) Hoadley barrier bar. Major tidal channels cut transversely through the barrier bar connecting the ocean and bay, whereas the smaller tidal channels cut into the bar for a limited distance. They subdivide the barrier into several separated reservoir compartments and also serve as an updip seal for hydrocarbons.

Tidal channels are usually filled with a uniform shale devoid of coalbeds (Fig. 8). Thinly cross-bedded sandstone may be found in tidal channels of the Hoadley barrier bar. Winkelmolen and Veenstra (1974) reported that sand movement in modern tidal inlets by wave-induced currents is restricted to the shallow area and sediment is relatively immobile in the deeper water.

Levee Facies

Levee occurs along both sides of a tidal channel. It consists of very tight sand with an average porosity of less than 6% (Fig. 8) and is characterized by an expressionless gamma-ray curve with no indication of sorting or upward cleaning. The presence of both levee and tidal channel facies in the same well suggests meandering or lateral movement of the tidal channel (center well log, Fig. 8). Kumar and Sanders (1974) reported that a tidal inlet located south of Long Island, New York, on Fire Island has, due to wave movement, migrated at a mean rate of 64 m/yr (210 ft/yr) during the period 1825 to 1940. Figure 8 illustrates how gamma-ray curves differ between the barrier bar sand and levee sand. The former shows response to a cleaner sand, hence sorting upward whereas the latter shows no change in character.

Deltaic Sand Facies

Back bar deltaic (Fig. 9) in the Medicine River delta complex is generally a thin sheet of wide lateral distribution representing the lower delta plain of a mixed fluvial-wave-tide-dominated delta (Reichenbach, 1982). The sand continuity is occasionally broken by shale-filled, abandoned channels sometimes forming isolated reservoirs. The Medicine River oil field produces oil from the back bar deltaic sand (14 million barrels of recoverable oil) trapped updip by a shale-filled abandoned channel. Additional similar traps containing oil may be found in other parts of the Medicine River delta complex behind the barrier bar.

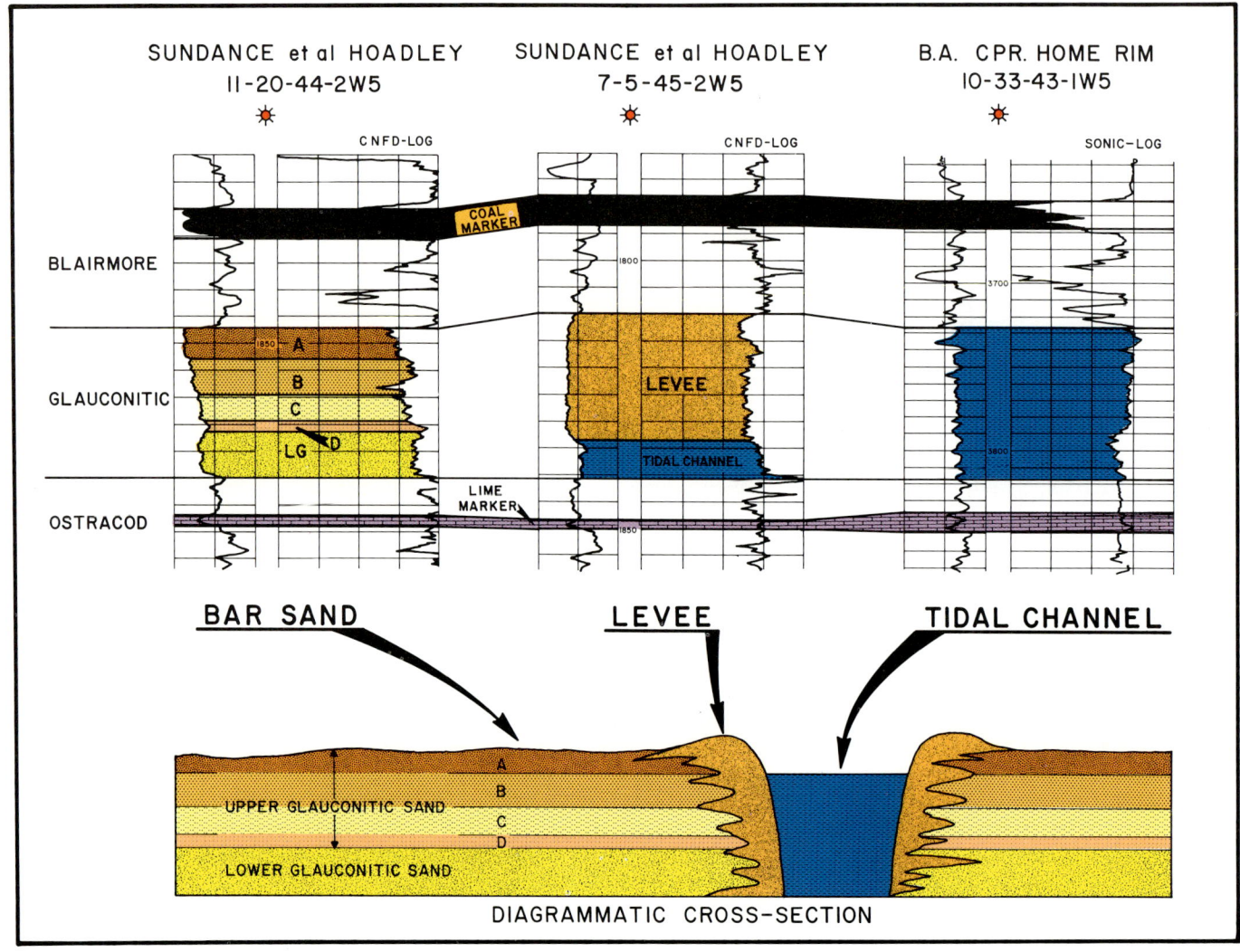

Figure 8. Well logs showing bar sand, levee, and tidal channel facies.

Distributary Channel Facies

A distributary channel about 1 mi (1.6 km) wide and 15 mi (24.1 km) long is found in the Medicine River delta complex, trending east to west in Ranges 1 to 4, Township 38N, West 5th Meridian. Up to 120 ft (36.6 m) of sand was deposited in this main distributary channel forming the reservoir of the Sylvan Lake gas field.

This distributary channel sand facies (Fig. 10) also has a typical bell-shape gamma-ray curve with a sharp cut at the bottom, indicating a gradual upward increase of shale matrix and a resulting decrease in porosity.

Abandoned Channel Facies

An abandoned channel forms when a distributary changes the course of its flow. Abandoned channels filled by fine sediment are found in the Medicine River delta complex as shale bodies (Fig. 11). They also separate the porous deltaic sand into isolated reservoir compartments, and serve to form stratigraphic traps in the delta complex.

DEPOSITIONAL ENVIRONMENT AND HISTORY

The recognition and distribution of different landform features and resulting facies in the studied area help to reconstruct the paleogeographic picture and depositional environment of the Hoadley barrier bar complex. During the deposition of the Upper Glauconitic barrier bar complex, the land mass was situated to the southeast of the barrier bar and the open seaway lay to the north. The occurrence of conglomerate in the southwestern portion of the barrier bar forming pebble beach deposits in the Strachan area (southwest of the Medicine River delta) suggests proximity to source. The conglomerate in the Strachan area probably did not originate from the Medicine River delta, but appears to have been brought in by rivers and strong longshore currents from a southwestern source. A second source is indicated by sand alone being supplied by the Medicine River deltaic system and transported northeastward by longshore currents and deposited in the Wilson Creek-Hoadley-Bonnie Glen area (geographic names are shown in Fig. 17). I interpret the Lower Glauconite as a earlier near-shore coastal blanket of fine, shaley, glauconitic-bearing sand deposited in a

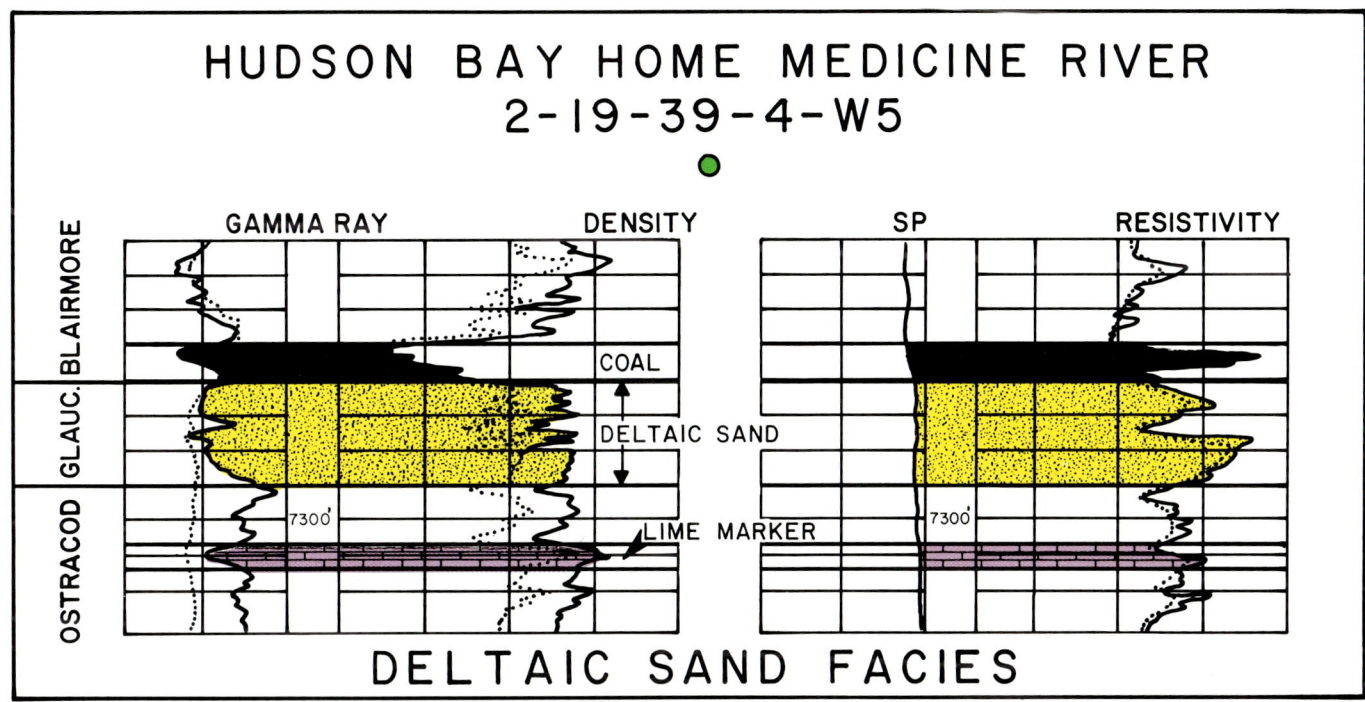

Figure 9. Well logs showing deltaic sand facies.

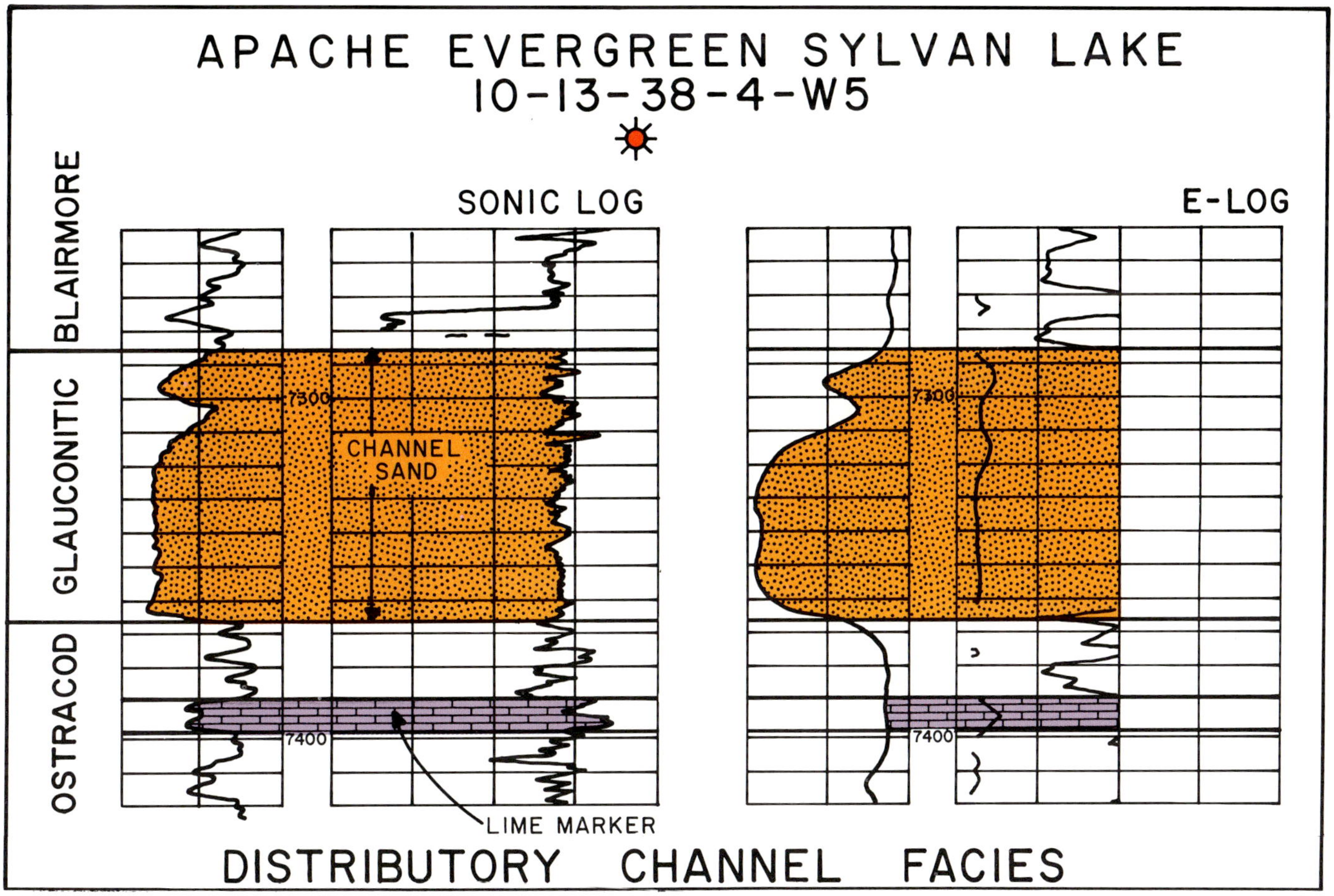

Figure 10. Well logs showing distributary channel facies.

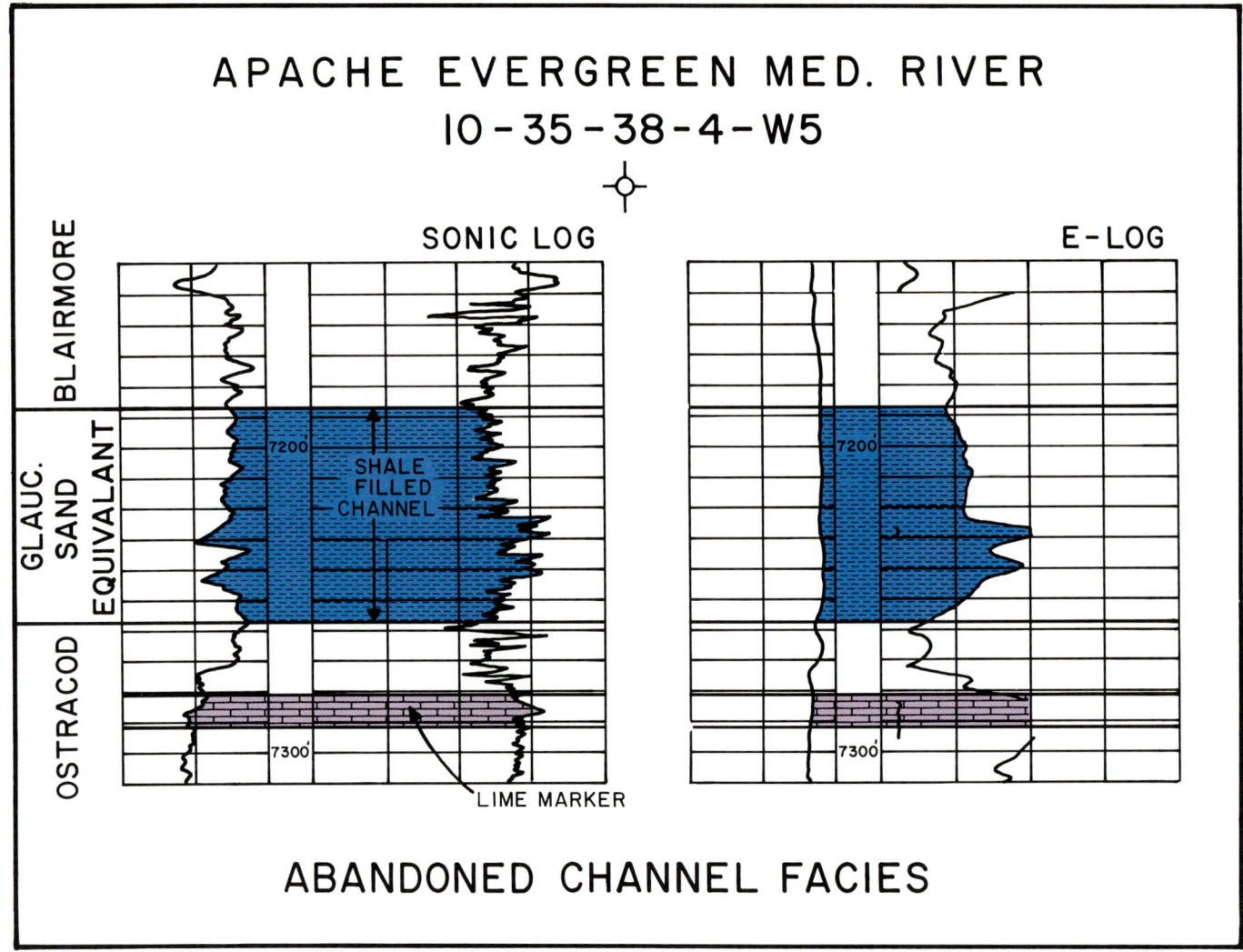

Figure 11. Well logs showing abandoned channel facies.

lower energy marine environment of 30 ft (9.1 m) or more water depth (first stage in Fig. 12). This formed a platform for the subsequent deposition of the Upper Glauconitic barrier bar.

The Upper Glauconitic barrier bar was emerged and submerged several times during seaward progradation, as indicated by the off-lap sequence of parallel eolian sand ridges and interbar lagoons. The emergence and submergence of the sand bar represents a depositional cycle of alternating rapid and slow deposition. During a period of rapid deposition, the barrier bar grew seaward by accretion of A, B, C, and D sand facies simultaneously, forming the emerged barrier bar with eolian sand ridges (second stage in Fig. 12). During a period of slow transportation and deposition, the bar was submerged under water and only B, C, and D sands were deposited (third stage in Fig. 12). When rapid deposition returned, the bar emerged and formed a new regressive "A" sand ridge oceanward. The shallow water area between the first and second sand ridges becomes an interbar lagoon where trapped mud and silts accumulated with vegetation (fourth stage in Fig. 12). As the progradation of the barrier bar continued, successive sand

Figure 12. Diagrammatic cross sections showing depositional history of the Hoadley barrier bar in six stages.
First Stage: Deposition of the Lower Glauconitic sand in water about 30 ft (9.1 m) deep.
Second Stage: Early deposition of the Upper Glauconitic sand formed emerged barrier bar separating the bay from the ocean — seaward progradation by deposition of A, B, C, and D sand facies simultaneously.
Third Stage: Barrier bar submerged under sea level during period of slow deposition — seaward progradation by accretion of B, C, and D sand facies simlutaneously.
Fourth Stage: The barrier bar emerged above sea level when rapid deposition returned, forming a new sand ridge and an interbar lagoon behind.
Fifth Stage: Repeat of rapid and slow depositional cycle formed new sand ridge and new interbar lagoon.
Sixth Stage: Deposition of washover sand in the interbar lagoon after storms during seaward progradation of the barrier bar.

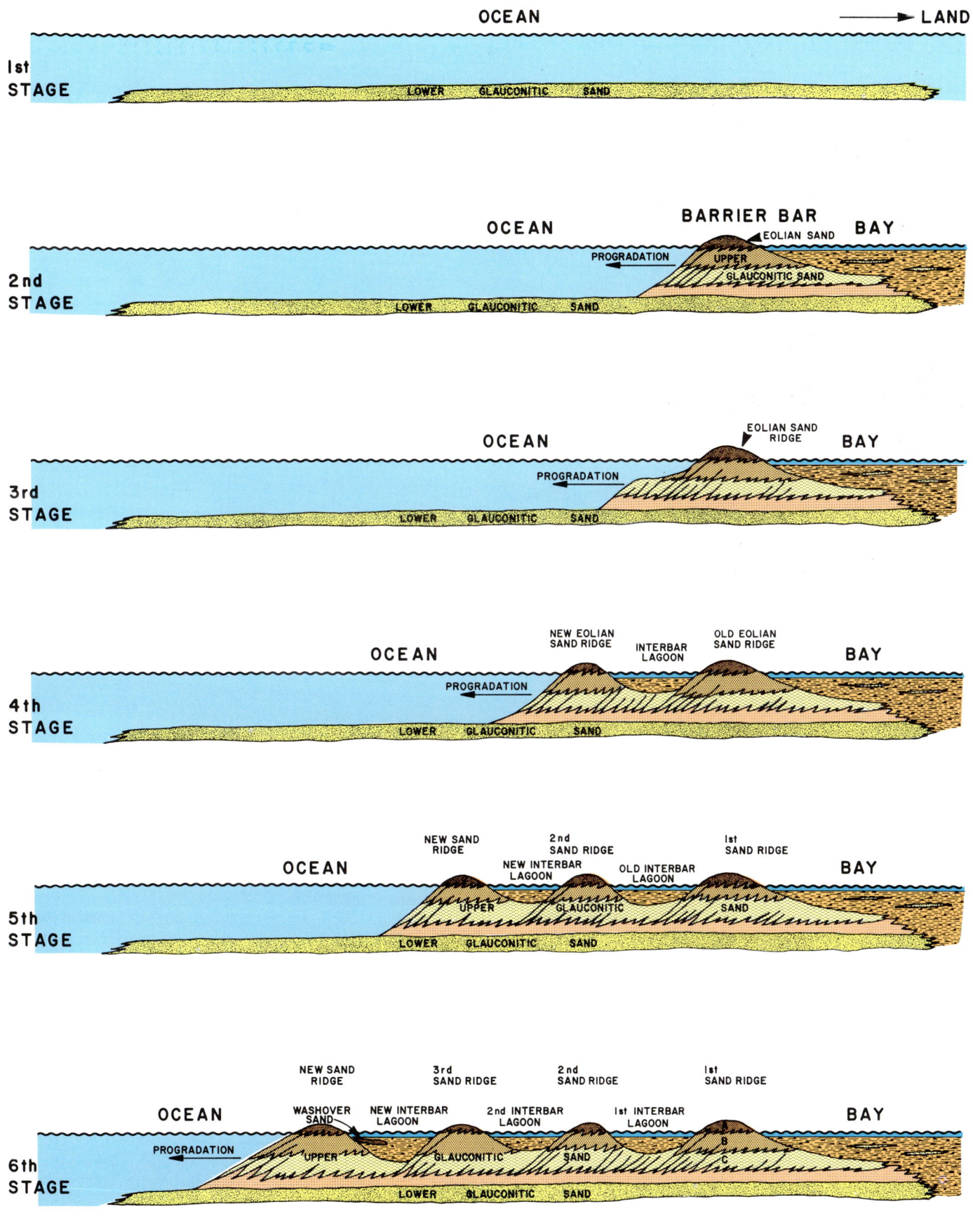

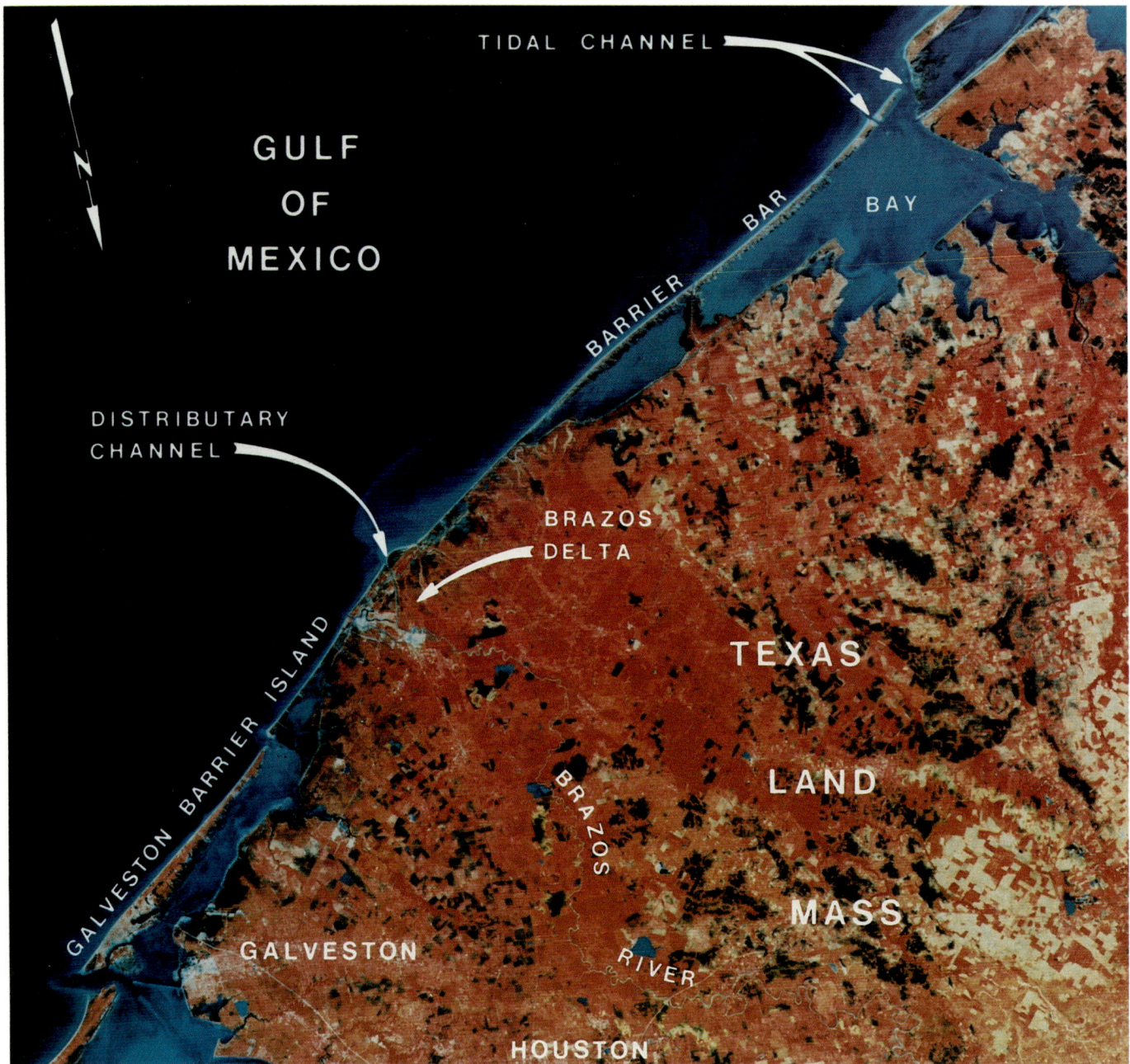

Figure 13. Satellite picture of the Texas Gulf Coast showing barrier bar, ocean, bay, tidal channel, Brazos delta and distributary channel. Note: The picture is rotated 170° for comparison with the Hoadley barrier bar complex. Photo courtesy of the USGS EROS Data Center.

ridges and interbar lagoons were formed (fifth stage in Fig. 12). There are as many as seven sand ridges that can be postulated across the 15-mi-wide (24.1-km-wide) barrier bar. Since these sand ridges are in offlap sequence due to seaward progradation of the bar, they are more or less parallel to each other trending southwest to northeast.

The climate is interpreted to be moderately warm and vegetation lush from the presence of glauconite in the sandstone and the occurrence of coalbeds in the lagoonal areas. Occasional storms caused destruction of the eolian sand ridges as indicated by the occurrence of blowout gaps along the ridges and washover fans found in the interbar lagoons (Fig. 4).

When the Hoadley barrier bar had grown to a width of approximately 15 mi (24.1 km), a major regression took place which terminated the deposition of the marginal marine barrier bar at Hoadley and created new barriers further north-west. The Hoadley area was covered by swamp deposits of the Blairmore Formation. During this period, the northeastern portion of the Hoadley bar suffered fluvial truncation (Fig. 4) and no vestige of its existence remains.

COMPARISON WITH A MODERN BARRIER BAR

A modern analogy can be found on the Texas Gulf coast where a series of barrier

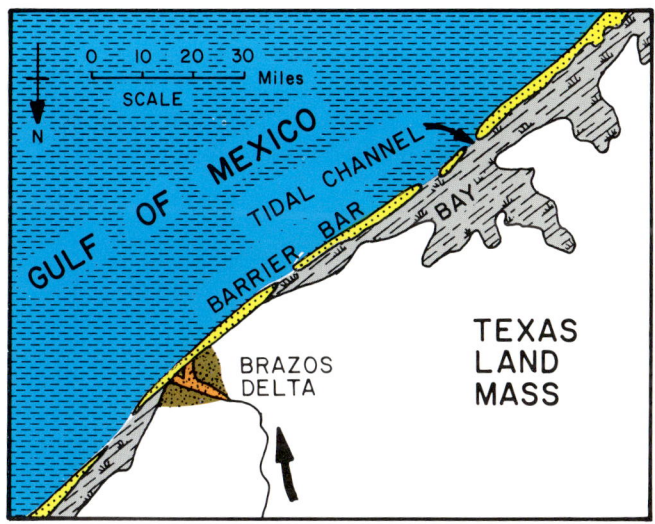 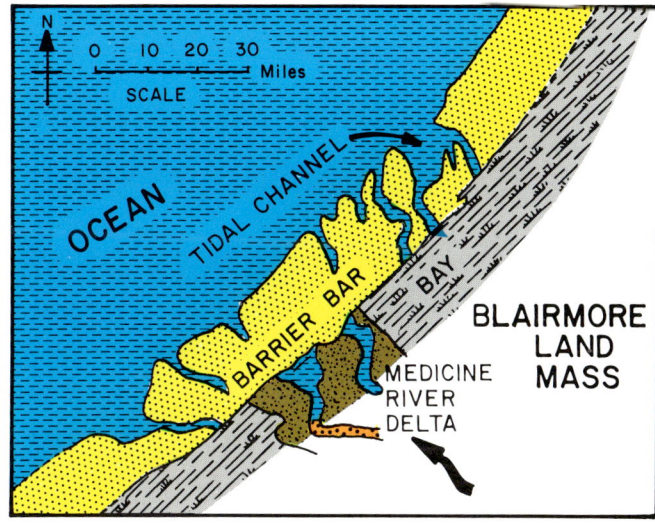

Figure 14. Comparison of the modern Texas Gulf Coast barrier bar and the ancient Hoadley barrier bar in same scale. **A.** Upside down view of Texas Gulf Coast. **B.** Simplified view of Hoadley barrier bar.

bars are forming today. Sand bars 0.5 to 1.5 mi (0.8 to 2.4 km) wide extend several hundred miles along the Gulf coast. These sand bars are separated from the shore by lagoons and bays which are eratically connected to the open ocean by narrow tidal channels. The Brazos River, as an example, forms a delta complex immediately behind the barrier bar (see satellite picture in Fig. 13). If we look at this part of the Texas Gulf coast rotated 180°, it would look like the picture in Figure 14A. Comparison with the ancient Hoadley barrier bar complex in Figure 14B, emphasizes the similarity.

Study of the Galveston barrier bar by Bernard et al, 1959, indicated progradation of the sand bar by accretionary deposition seaward. This was supported by the radiocarbon dating from drilled core samples. Time lines of the Galveston barrier bar are shown in Figure 15A. The inner (landward) edge of the bar is 3,500 years old and becomes younger seaward. Recent accretionary deposition is still taking place along the ocean front edge of the bar. Bernard's study also shows a decrease in grain size downward in the barrier bar sandstone.

If the Galveston barrier bar sands were subdivided into A, B, C, and D facies based on general lithology and water depth, it would be a modern analogy of the Hoadley Upper Glauconitic sand bar facies (Fig. 15B).

Detailed mapping of the Padre Island barrier bar of the Texas Gulf coast by Hunter and Dickinson (1970) shows two sand ridges on the 1.5-mi-wide (2.4-km-wide) barrier island. The older eolian sand ridge occurs along the inner edge of the barrier bar, while the recent eolian sand ridge occurs along the seaward edge of the bar. The area between the two eolian sand edges is a water-covered (0 to 15 ft, or 4.6 m, deep), low area with vegetation which is comparable with the interbar lagoon facies of the ancient Hoadley barrier bar.

Hurricanes and other storms affecting the Gulf coast often cause destruction of the ocean front sand ridge as indicated by the occurrence of the blowout gaps. The sand is carried by storm wave overwash into the interbar lagoon area and forms fans. Such blowout gaps frequently will be repaired by the lateral movement of windblown sand along the ridge and the washover fan may become covered by vegetation.

The aerial photo (Fig. 16) shows an excellent view of a recent barrier bar in the Gulf Coast featuring the beach, an eolian sand ridge, a portion of an interbar lagoon, a blowout gap, and several washover fans. Facies derived from all of these features can be found in the Hoadley barrier bar complex. Thus, it is concluded that the the Hoadley Glauconitic sand bar is an example of a preserved ancient barrier bar complex.

GUIDES TO FUTURE EXPLORATION IN THE HOADLEY BARRIER BAR COMPLEX

Since the middle and southwestern part of the Hoadley barrier bar covering an area of approximately 1,500 sq mi (3,885 sq km) is entirely saturated with natural gas and natural gas liquids, it follows that there is a potential for many more wells. However, it is not true that all wells are currently commercially profitable. Good gas wells are usually found on or near the "A" sand ridges and the conglomerate beaches. Poor gas wells result when drilled in the interbar lagoon areas. It is, therefore, very important to delineate the facies trends, especially eolian sand ridges and conglomerate beaches. If the source of the conglomerate in the Strachan area is further to the southwest, it may provide an area of great potential for future exploration.

Criteria for delineation of "A" sand trends as determined by mapping and extensive drilling are listed as follows:
1. It is important to keep in mind that the regional trend of the Hoadley barrier bar is in a southwest to northeast direction, and each "A" sand ridge is narrow and can be as linear as a ruler edge for a long distance. Exploratory or development wells are best located very carefully on the trend of the "A" sand ridges once established by initial

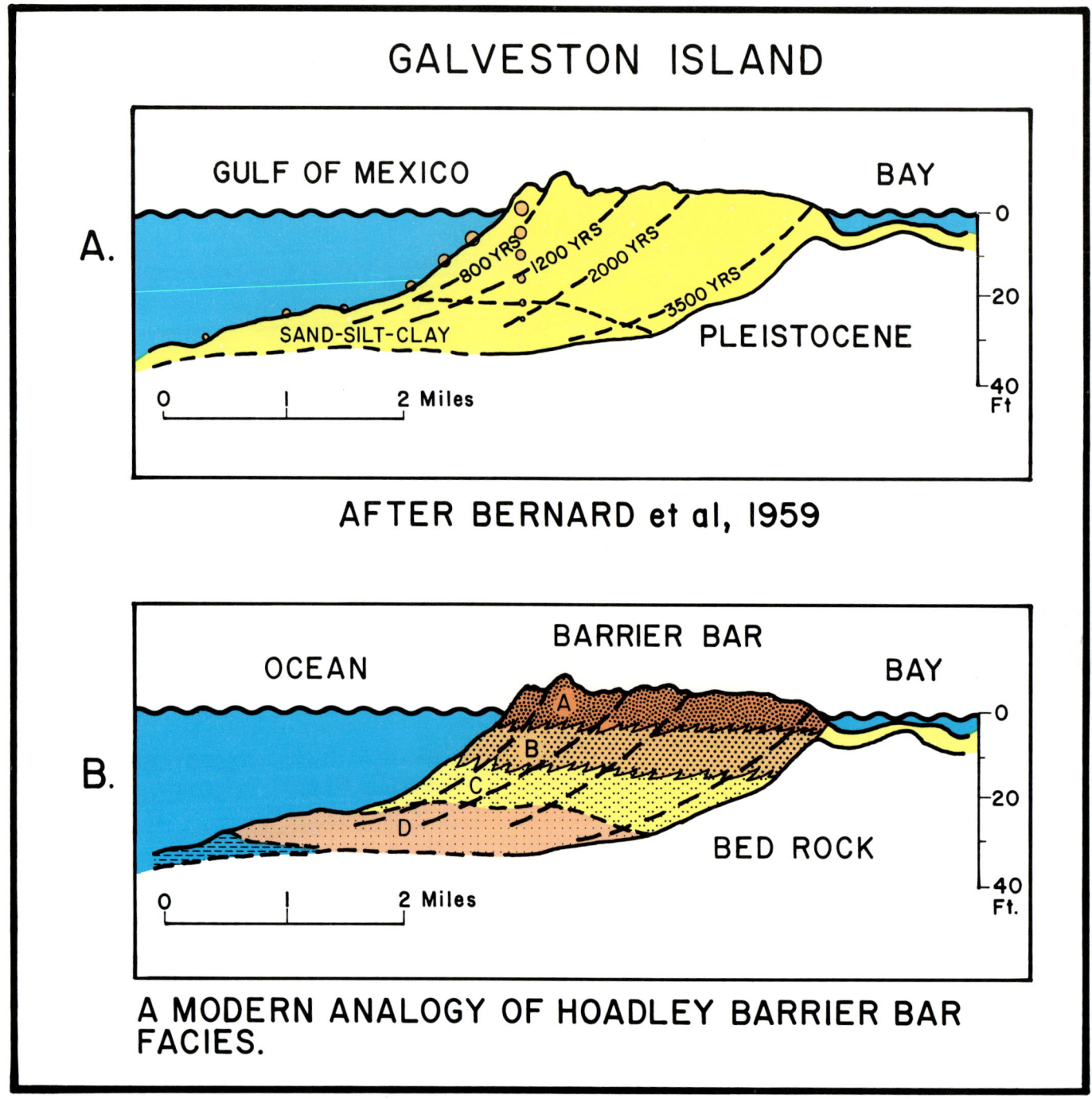

Figure 15. **A.** Galveston barrier island after Bernard et al (1959) showing seaward progradation of the barrier bar as indicated by the dated time lines and grain size distribution. **B.** Subdivision of the modern Galveston barrier bar into A, B, C, and D sand facies for comparison with the ancient Hoadley A, B, C, and D sand facies, suggesting similar depositional water depth.

exploratory wells.
2. Since interbar lagoons are formed between two "A" sand ridges, an interbar lagoon well suggests the occurrence of "A" sand bars to the northwest and southeast of that well, but never on trend to the southwest or northeast.
3. If a back bar washover sand facies is present in an interbar lagoon well, the potential presence of an "A" sand bar is indicated within a half a mile to the northwest (seaward direction) of that well.
4. The occurrence of tidal channels and levees is very difficult to predict. If an

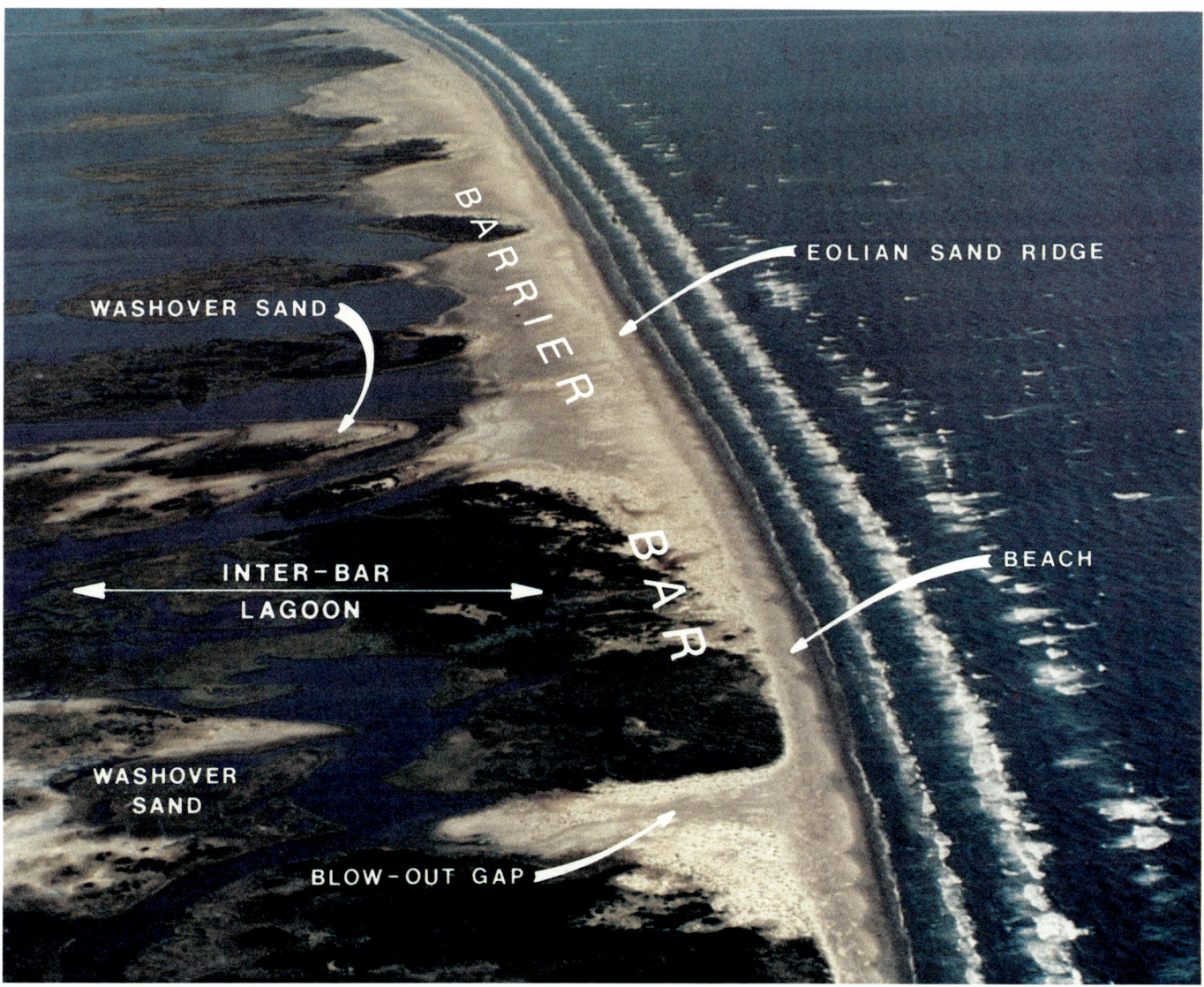

Figure 16. Photographic view of a modern barrier bar in the Gulf coast showing beach, eolian sand ridge, back-bar lagoon, blowout gap, and washover fans. Note the vegetation (in green) in the inter-bar lagoon area. Photo courtesy of David Brooks from Sundance Oil Company, Calgary, Canada.

exploratory well is, unfortunately, drilled in the tidal channel or in the levee zone, the next exploratory well must be located longitudinally along the bar perhaps as much as 2 mi (3.2 km) to the northeast or southwest of the unsuccessful well.

RESERVES OF THE HOADLEY GAS FIELD

Since discovery, more than 140 gas wells have been completed in the Hoadley barrier bar (Fig. 17); none have tested or produced formation water. These wells, together with many old wells drilled in the bar with identified bypassed gas pay, indicate that the middle and southwestern part of the barrier bar (approximately 100 mi, or 161 km, long) is entirely saturated with gas and natural gas liquids. The discovery well, Sundance et al Hoadley (6-2-45-2W5), tested at a calculated AOF of 76 mmcf/d with 60 barrels of natural gas liquid per mmcf of gas. The farthest downdip completed well, Scarboro et al Crimson 10-31-37-8W5 (approximately 58 mi, or 93.3 km, southwest of the discovery well), tested at an AOF of 9.5 mmcf/d with 25 barrels of natural gas liquid per mmcf of gas; and the most updip completed well, Spur et al Bonnie Glen 7-11-46-27W5 (approximately 17 mi, or 27.4 km, northeast of the discovery well), tested an AOF of 21 mmcf/d with 40 barrels of natural gas liquid per million cubic feet (Fig. 17).

The good wells on the Hoadley barrier bar, such as the discovery well, calculated approximately 15 bcf of recoverable gas and 900,000 barrels of recoverable natural gas liquids per 640 acre spacing unit. The poor wells, such as the interbar lagoon wells, indicate about 3 bcf of recoverable gas and 180,000 barrels of natural gas liquids per spacing unit. Assuming an average of 6 to 7 bcf of recoverable gas per section for the gas-saturated portion of the

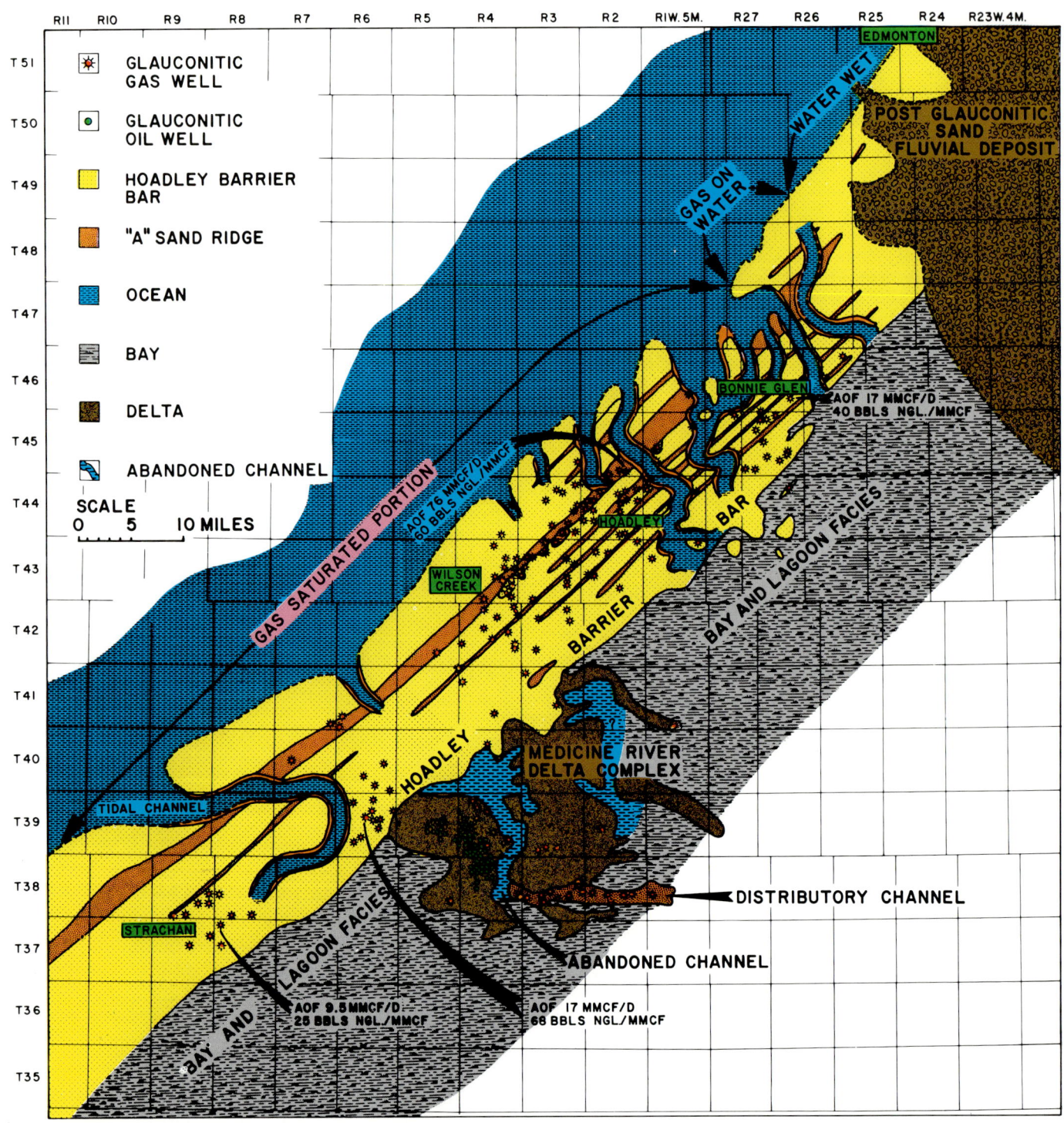

Figure 17. Map showing locations of completed Glauconitic gas wells along the Hoadley barrier bar trend.

bar with more than 25 ft (7.6 m) sand thickness (approximately 1,000 sq mi, or 2,590 sq km, of area), the Hoadley bar complex is estimated to have a potential recoverable reserve of 6 to 7 trillion cubic feet (tcf) of gas and 350 to 400 million barrels of natural gas liquids. In the year of discovery, 1977, the total remaining gas reserve in Canada reported by the National Energy Board was 64.5 tcf, thus the discovery of the Hoadley gas field may have added approximately 10% to the gas reserves of Canada.

Although the Hoadley gas field contains a giant recoverable gas reserve, the ultimate amount of gas that will be recovered from the reservoir is dependent on two main factors:

1. The number of wells encountering "A" sand bar which will probably serve as an underground gathering system for

most of the gas trapped in the lower porosity and less permeable reservoir rock of the "B" and "C" sand.
2. The economic factor: A great reserve of the gas at Hoadley is trapped in tighter sand reservoirs which require expensive fracturing techniques, and about one-third of the reservoir rock is buried at a depth of greater than 9,000 ft (2,743 m). Thus, a large prospective area is economically unattractive on the basis of today's gas price and lack of incentives for tight sand gas in Canada.

Early exploration and development of the Hoadley gas field was mainly concentrated in the shallower part of the bar because of its proximity to the discovery well and less expensive drilling cost. Exploration activity decreases toward the deeper southwestern portion due to higher drilling costs. Much of the potential producing area is unavailable because it is held by production from other formations, and because of the lack of knowledge of the Glauconitic gas potential. Exploration activity was sparked in the middle and downdip part of the barrier bar after the announcement of the Hoadley gas field and its geological area extension, by a paper I delivered to the Canadian Society of Petroleum Geologists in April, 1981. The discovery of conglomerate beaches in the Strachan area providing high gas flow rates has somewhat compensated the expensive drilling cost in the area. More than 40 Glauconitic gas wells have been completed along the Hoadley trend within the two-year period since the announcement. It is logical that more sweet spots may be found along the trend.

The discovery of the Hoadley gas field not only adds a great amount of gas reserve in Canada, but strongly indicates the overlook potential of both tighter sand reservoir rock and sweet spots in it. Prior to discovery, the Hoadley Glauconitic sand bar had been penetrated by hundreds of wells, none recognizing that this sand bar complex contains a giant reserve of recoverable gas. Missing such a giant sand bar, which extends almost across the entire south-central Alberta, seems impossible. I wonder how many more sleeping giants are waiting to be found.

ACKNOWLEDGMENTS

I wish to thank Sundance Oil Company for giving permission to publish this study; Mr. Caswell Silver (President of Sundance) for helpful reading and supportive encouragement; and John Masters and Paul Jackson of Canadian Hunter Exploration Ltd., for their suggestions and review of the manuscript.

REFERENCES CITED

Bernard, H. A., C. F. Major, and B. S. Parrot, 1959, The Galveston barrier island and environs — a model for predicting reservoir occurrence and trend: Transactions Gulf Coast Association of Geological Societies, v. 9, p. 221–224.

Chiang, K. K., 1981, Hoadley — a potential super-giant gas field in south-central Alberta: Canadian Society of Petroleum Geologists Reservoir, v. 8, p. 1–2, (abstract).

Fitzgerald, T. A., 1980, Giant field discoveries 1968-1978; an overview, in M.T. Halbouty, ed., Giant oil and gas fields of the decade, 1968-1978: AAPG Memoir 30, p. 1–5.

Hunter, R. E., and K. A. Dickinson, 1970, Map showing landforms and sedimentary deposits of the Padre Island portion of the South Bird Island, 7.5 minutes quadrangle, Texas: U.S. Geological Survey Miscellaneous Geologic Investigation, Map 1-65-9.

Kumar, N., and J. E. Sanders, 1974, Inlet sequence; a vertical succession of sedimentary structures and textures created by the lateral migration of tidal inlets: Sedimentology, v. 21, p. 491–532.

Reichenback, M. E., 1982, Reservoir lithologies in the Medicine River delta and Hoadley barrier complex, in J.C. Hopkins, ed., Depositional environments and reservoir facies in some Western Canadian oil and gas fields: University of Calgary Core Conference, p. 52–58.

Winkelmolen, A. M., and H. J. Veenstra, 1974, Size and shape sorting in a Dutch tidal inlet: Sedimentology, v. 21, p. 107–126.

Index

A reference is indexed according to its important, or "key," words.

Three columns are to the left of a keyword entry. The first column, a letter entry, represents the AAPG book series from which the reference originated. In this case, ME stands for Memoir Series. Every five years, AAPG will merge all its indexes together, and the letters ME will differentiate this reference from those of the AAPG Studies in Geology Series (S) or the AAPG Bulletin (B).

The following number is the series number. In this case 38 represents a reference from AAPG Memoir 38. The third column lists the page number of this volume on which the reference can be found.

† = titles
* = authors

ME 38   6   ALBERTA CORRELATION CHART, CRETACEOUS-JURASSIC FMS
ME 38 60   ALBERTA CORRELATION/FACIES CHART, LOWER CRETACEOUS
ME 38 292   ALBERTA DEEP BASIN, COMPLETION PROCEDURE
ME 38   5   ALBERTA IN-PLACE OIL/GAS, STRATIGRAPHIC DISTRIBUTION
ME 38 23   ALBERTA OIL FIELDS, LOWER CRETACEOUS
ME 38   9   ALBERTA SOURCE ROCKS, LOWER CRETACEOUS
ME 38 205   ALBERTA, ELMWORTH FIELD, ROCK-LOG CALIBRATION
ME 38 79   ALBERTA, ELMWORTH, CRETACEOUS, PALEOGEOGRAPHY
ME 38 24   ALBERTA, GAS FIELDS, LOWER CRETACEOUS
ME 38 22   ALBERTA, GAS MIGRATION/TRAPPING
ME 38 172   ALBERTA, GAS PRODUCTION CHART, DEEP BASIN
ME 38   4   ALBERTA, OIL/GAS RESERVES, STRATIGRAPHIC DISTRIBUTION
ME 38 206   ALBERTA, PORE-TYPE CLASSIFICATION, ELMWORTH AREA
ME 38 112   ALBERTA, RESERVOIR PERMEABILITY, LOWER CRETACEOUS
ME 38 112   ALBERTA, RESERVOIR POROSITY, LOWER CRETACEOUS
ME 38 243   ALBERTA, RESERVOIR PROPERTIES, ELMWORTH AREA
ME 38   3   ALBERTA, STRUCTURE MAP, PRECAMBRIAN
ME 38   3   ATHABASCA REGION, ALBERTA
ME 38 91   BLUESKY FORMATION PALEOGEOGRAPHY, ALBERTA
ME 38 68   BLUESKY FORMATION, PALEOGEOGRAPHIC MAP, ALBERTA
ME 38 79   BRITISH COLUMBIA, ELMWORTH, CRET., PALEOGEOGRAPHY
ME 38 135   CADOMIN FM. GAS MIGRATION, DEEP BASIN, ALBERTA
ME 38 116   CADOMIN FM. PALEOGEOGRAPHY, ALBERTA
ME 38 136   CADOMIN FM. TRANSMISSIBILITY MAP, DEEP BASIN, ALBERTA
ME 38 84   CADOMIN FORMATION PALEOGEOGRAPHY, ALBERTA
ME 38 115   CADOMIN FORMATION, ALBERTA, GAS TRAP
ME 38 115† CADOMIN FORMATION, GAS TRAP CASE HISTORY
ME 38 107   CADOTTE/PEACE RIVER FM. PALEOGEOGRAPHY, ALBERTA
ME 38 115   CASE HISTORY, DEEP BASIN GAS TRAP: CADOMIN FORMATION
ME 38 286   CEMENTATION, MULTIPLE PAYZONES, DEEP BASIN, ALBERTA
ME 38 297* CHIANG, K. K.-HOADLEY GAS FIELD SOUTH-CENTRAL ALBERTA
ME 38 12   CLEARWATER SHALE ISOPACH MAP, LOWER CRETACEOUS, ALBERTA
ME 38 196   CLINTON FM., EASTERN OHIO, GAS RESERVOIRS
ME 38 174   COAL ANALYSIS, ELMWORTH AREA, ALBERTA
ME 38 175   COAL CHEMISTRY, ELMWORTH AREA, ALBERTA
ME 38 180   COAL PETROGRAPHY, ELMWORTH AREA, ALBERTA
ME 38 173   COAL, GAS RESOURCES, ELMWORTH AREA, ALBERTA
ME 38 185   COAL, LOG IDENTIFICATION, ELMWORTH AREA, ALBERTA
ME 38 291   COMPLETION PRACTICES IN THE ALBERTA DEEP BASIN
ME 38 291† COMPLETION PRACTICES, ALBERTA DEEP BASIN
ME 38 292   COMPLETION PROCEDURE, DEEP BASIN, ALBERTA
ME 38 243* CONNOLLY, E. T.-RESERVOIR PROPERTIES, ELMWORTH FIELD
ME 38 215* CONNOLLY, E. T.-WELL LOG ANALYSIS METHODS, ELMWORTH
ME 38 182   CROSS SECTION, LOWER CRETACEOUS COAL, ELMWORTH, ALBERTA
ME 38 143   CROSS SECTION, LOWER CRETACEOUS, DEEP BASIN, ALBERTA
ME 38 62   CROSS SECTION, MANNVILLE GR. FACIES, DEEP BASIN
ME 38 189* DAVIS, T. B.-PRESSURE PROFILES, GAS-SATURATED BASINS
ME 38   6   DIAGRAMMATIC CROSS SECTION ACROSS CENTRAL ALBERTA
ME 38 37   DIAGRAMMATIC CROSS SECTION, DEEP BASIN, ALBERTA
ME 38 283   DRILLING IN THE DEEP BASIN
ME 38 283† DRILLING IN THE DEEP BASIN, ALBERTA
ME 38 287   DRILLING METHODS, DEEP BASIN, ALBERTA
ME 38 81   ELMWORTH AREA CORRELATION CHART, LOWER CRETACEOUS

ME 38 173   ELMWORTH AREA, COAL, GAS RESOURCES
ME 38 141   ELMWORTH AREA, FALHER A CYCLE, GAS TRAP, FACIES CONTROL
ME 38 79   ELMWORTH AREA, LOWER CRETACEOUS, PALEOGEOGRAPHY
ME 38 36   ELMWORTH FIELD, ALBERTA, GAS GENERATION/MIGRATION
ME 38 155   ELMWORTH/WAPITI AREA RESERVE SUMMARY, LOWER CRET.
ME 38 157   ELMWORTH/WAPITI AREA RESERVOIR DATA, ALBERTA
ME 38 153   ELMWORTH/WAPITI AREA, GAS RESERVES
ME 38 153   ELMWORTH/WAPITI AREA, PRODUCTION PERFORMANCE
ME 38 189   ELMWORTH, DEEP BASIN AREA, ALBERTA, GAS RESERVOIRS
ME 38 298   FACIES ANALYSIS, HOADLEY GAS FIELD, CENTRAL ALBERTA
ME 38 141   FACIES CONTROLLED GAS TRAP, CRETACEOUS FALHER A CYCLE
ME 38 301   FACIES MAP, HOADLEY BARRIER BAR COMPLEX
ME 38 301   FACIES MAP, MEDICINE RIVER DELTA COMPLEX, SOUTH ALBERTA
ME 38 141   FALHER A CYCLE, FACIES CONTROL, GAS TRAP, ELMWORTH AREA
ME 38 141† FALHER A CYCLE, GAS TRAP, FACIES CONTROL
ME 38 76   FALHER A/GRAND RAPIDS B PALEOGEOGRAPHIC MAP, ALBERTA
ME 38 102   FALHER A/SPIRIT RIVER FM. PALEOGEOGRAPHY, ALBERTA
ME 38 144   FALHER A/SPIRIT RIVER FM., MAJOR FACIES, ALBERTA
ME 38 145   FALHER A/SPIRIT RIVER FM., PRESSURE-DEPTH PLOT, ALBERTA
ME 38 75   FALHER B/GRAND RAPIDS C PALEOGEOGRAPHIC MAP, ALBERTA
ME 38 100   FALHER B/SPIRIT RIVER FM. PALEOGEOGRAPHY, ALBERTA
ME 38 98   FALHER C/SPIRIT RIVER FM. PALEOGEOGRAPHY, ALBERTA
ME 38 96   FALHER D/SPIRIT RIVER FM. PALEOGEOGRAPHY, ALBERTA
ME 38 94   FALHER E/SPIRIT RIVER FM. PALEOGEOGRAPHY, ALBERTA
ME 38 73   FALHER G/REX PALEOGEOGRAPHIC MAP, ALBERTA
ME 38 72   FALHER H/LLOYDMINSTER PALEOGEOGRAPHIC MAP, ALBERTA
ME 38 287   FORMATION PRESSURE, DEEP BASIN, ALBERTA
ME 38 44   GAS DIFFUSION MODEL, ELMWORTH FIELD, ALBERTA
ME 38 35   GAS GENERATION AND MIGRATION IN THE DEEP BASIN
ME 38 36   GAS GENERATION/MIGRATION, ELMWORTH FIELD, ALBERTA
ME 38 35† GAS GENERATION, MIGRATION, DEEP BASIN
ME 38 129   GAS PRESSURE, CADOMIN FM., ELMWORTH AREA, ALBERTA
ME 38 153† GAS RESERVES, PRODUCTION, ELMWORTH/WAPITI
ME 38 153   GAS RESERVES, PRODUCTION, ELMWORTH/WAPITI AREA
ME 38 173† GAS RESOURCES IN ELMWORTH COAL SEAMS
ME 38 115   GAS TRAP, DEEP BASIN, CADOMIN FORMATION
ME 38 87   GETHING FORMATION PALEOGEOGRAPHY, ALBERTA
ME 38 57   GETHING FORMATION, PALEOGRAPHY, ALBERTA
ME 38 297   GIANT HOADLEY GAS FIELD SOUTH-CENTRAL ALBERTA
ME 38 115* GIES, R. M.-CADOMIN FORMATION, GAS TRAP CASE HISTORY
ME 38 297   GLAUCONITIC FM., CRETACEOUS, SOUTH-CENTRAL ALBERTA
ME 38 200   GREEN RIVER BASIN, WYOMING, GAS RESERVOIRS
ME 38 243* HIETALA, R. W.-RESERVOIR PROPERTIES, ELMWORTH FIELD
ME 38 215* HIETALA, R. W.-WELL LOG ANALYSIS METHODS, ELMWORTH AREA
ME 38 298   HOADLEY BARRIER BAR COMPLEX, SOUTH-CENTRAL ALBERTA
ME 38 299   HOADLEY GAS FIELD DISCOVERY WELL, SOUTH-CENTRAL ALBERTA
ME 38 297† HOADLEY GAS FIELD SOUTH-CENTRAL ALBERTA
ME 38 297   HOADLEY GAS FIELD, FACIES STUDY, SOUTH-CENTRAL ALBERTA
ME 38 297   HOADLEY GAS FIELD, LOG ANALYSIS, SOUTH-CENTRAL ALBERTA
ME 38 10   HYDROCARBON GENERATION, LOWER CRETACEOUS, ALBERTA
ME 38 10   HYDROCARBON SEALS, LOWER CRETACEOUS, ALBERTA
ME 38 14   ISOPACH MAP JOLI FOU-HARMON SHALE, LOWER CRETACEOUS
ME 38 186   ISOPACH MAP LOWER CRETACEOUS COAL, ELMWORTH, ALBERTA
ME 38 163   ISOPACH MAP, BASAL FALHER A NET PAY, ELMWORTH/WAPITI
ME 38 168   ISOPACH MAP, BASAL FALHER D NET PAY, ELMWORTH/WAPITI
ME 38 84   ISOPACH MAP, CADOMIN FORMATION, CONGLOMERATE MEMBER
ME 38 169   ISOPACH MAP, CADOMIN NET PAY, ELMWORTH/WAPITI
ME 38 160   ISOPACH MAP, CADOTTE NET PAY, ELMWORTH/WAPITI, ALBERTA
ME 38 70   ISOPACH MAP, CLEARWATER FORMATION, ALBERTA
ME 38 12   ISOPACH MAP, CLEARWATER SHALE, LOWER CRETACEOUS
ME 38 17   ISOPACH MAP, CRETACEOUS AND PALEOCENE, ALBERTA
ME 38 146   ISOPACH MAP, FALHER A CAPPING COAL, DEEP BASIN, ALBERTA
ME 38 161   ISOPACH MAP, FALHER A CONGL. NET PAY ELMWORTH/WAPITI
ME 38 162   ISOPACH MAP, FALHER A SDST. NET PAY, ELMWORTH/WAPITI
ME 38 164   ISOPACH MAP, FALHER B CONGL. NET PAY, ELMWORTH/WAPITI
ME 38 165   ISOPACH MAP, FALHER B SDST. NET PAY, ELMWORTH/WAPITI
ME 38 166   ISOPACH MAP, FALHER D CONGL. NET PAY, ELMWORTH/WAPITI
ME 38 167   ISOPACH MAP, FALHER D SDST. NET PAY, ELMWORTH/WAPITI
ME 38 11   ISOPACH MAP, JURASSIC, ALBERTA-BRITISH COLUMBIA
ME 38 59   ISOPACH MAP, LOWER COLORADO GROUP, ALBERTA
ME 38 57   ISOPACH MAP, LOWER MANNVILLE, ALBERTA
ME 38 13   ISOPACH MAP, MANNVILLE AND JURASSIC COAL, ALBERTA
ME 38 28   ISOPACH MAP, MANNVILLE TIGHT SANDS, GAS-SATURATED
ME 38 91   ISOPACH MAP, NET SAND, BLUESKY FORMATION, ALBERTA
ME 38 87   ISOPACH MAP, NET SAND, GETHING FORMATION, ALBERTA
ME 38 122   ISOPACH MAP, TOTAL CADOMIN, LOWER CRETACEOUS, ALBERTA
ME 38 58   ISOPACH MAP, TOTAL MANNVILLE GROUP, ALBERTA
ME 38 10   ISOPACH MAP, TRIASSIC, ALBERTA-BRITISH COLUMBIA
ME 38 49* JACKSON, P. C.-MANNVILLE GROUP, LOWER CRETACEOUS
ME 38 243* KING, H. R.-RESERVOIR PROPERTIES, ELMWORTH FIELD
ME 38 205* KING, H. R.-RESERVOIR ROCK DETECTION, ELMWORTH FIELD
ME 38 221   LOG DATA MANAGEMENT, ELMWORTH FIELD, ALBERTA
ME 38 243* LOREN, J. D.-RESERVOIR PROPERTIES, ELMWORTH FIELD
ME 38   1   LOWER CRETACEOUS OIL AND GAS IN WESTERN CANADA
ME 38   7   LOWER CRETACEOUS RESERVOIR ROCKS, ALBERTA
ME 38 66   LOWER GETHING FORMATION, PALEOGEOGRAPHIC MAP, ALBERTA
ME 38   8   LOWER MANNVILLE FLUVIAL TRENDS, SUMMARY MAP
ME 38 65   LOWER MANNVILLE PALEOGEOGRAPHIC MAP, ALBERTA
ME 38 55   LOWER MANNVILLE, PALEOGEOGRAPHY
ME 38 13   MANNVILLE AND JURASSIC COAL ISOPACH MAP, ALBERTA
ME 38 51   MANNVILLE GROUP OIL/GAS FIELDS, DEEP BASIN, ALBERTA
ME 38 53   MANNVILLE GRP. STRATIGRAPHY, LOWER CRET., ALBERTA
ME 38 49† MANNVILLE GROUP, LOWER CRETACEOUS
ME 38   1* MASTERS, J. A.-OIL, GAS, CRETACEOUS, WESTERN CANADA

ME 38	190	MEDICINE HAT AREA, ALBERTA, GAS RESERVOIRS
ME 38	74	MIDDLE FALHER PALEOGEOGRAPHIC MAP, ALBERTA
ME 38	58	MIDDLE MANNVILLE PALEOGEOGRAPHY, ALBERTA
ME 38	288	MUD PROGRAM, DEEP BASIN, ALBERTA
ME 38	283*	MYERS, D. L.-DRILLING IN THE DEEP BASIN, ALBERTA
ME 38	77	NOTIKEWIN/GRAND RAPIDS A PALEGEOGRAPHIC MAP, ALBERTA
ME 38	105	NOTIKEWIN/SPIRIT RIVER FM. PALEOGEOGRAPHY, ALBERTA
ME 38	32	OIL AND GAS FIELDS, NORTH AMERICA
ME 38	20	OIL MIGRATION/TRAPPING, ALBERTA
ME 38	4	OIL/GAS RESERVES, ALBERTA, STRATIGRAPHIC DISTRIBUTION
ME 38	30	OIL/GAS/TAR FIELDS, WESTERN CANADA
ME 38	1†	OIL, GAS, CRETACEOUS, WESTERN CANADA
ME 38	15	ORGANIC CARBON CONTENT OF SAMPLES FROM CENTRAL ALBERTA
ME 38	109	PADDY/PEACE RIVER FM. PALEOGEOGRAPHY, ALBERTA
ME 38	49	PALEOGEOGRAPHY OF LOWER CRETACEOUS MANNVILLE GRP.
ME 38	79	PALEOGEOGRAPHY, CRETACEOUS, ALBERTA, DEEP BASIN
ME 38	79†	PALEOGEOGRAPHY, CRETACEOUS, ELMWORTH
ME 38	58	PALEOGEOGRAPHY, MIDDLE MANNVILLE, ALBERTA
ME 38	3	PALEOZOIC UNCONFORMITY, ALBERTA
ME 38	189†	PRESSURE PROFILES, GAS-SATURATED BASINS
ME 38	130	PRESSURE-DEPTH PLOT, CADOMIN FM. ELMWORTH AREA, ALBERTA
ME 38	192	PRESSURE-DEPTH PLOT, CADOTTE FM., ELMWORTH, ALBERTA
ME 38	194	PRESSURE-DEPTH PLOT, FALHER A, ELMWORTH, ALBERTA
ME 38	202	PRESSURE-DEPTH PLOT, GREEN RIVER BASIN, WYOMING, USA
ME 38	196	PRESSURE-DEPTH PLOT, MILK RIVER FM., SOUTHERN ALBERTA
ME 38	200	PRESSURE-DEPTH PLOT, RED DESERT BASIN, WYOMING, USA
ME 38	198	PRESSURE-DEPTH PLOT, SILURIAN-CLINTON SAND, OHIO, USA
ME 38	35*	RADKE, M-GAS GENERATION, MIGRATION, DEEP BASIN
ME 38	141*	RAHMANI, R. A.-FALHER A CYCLE, GAS TRAP, FACIES CONTROL
ME 38	197	RED DESERT BASIN, WYOMING, GAS RESERVOIRS
ME 38	311	RESERVES, HOADLEY GAS FIELD, SOUTH-CENTRAL ALBERTA
ME 38	243†	RESERVOIR PROPERTIES, ELMWORTH FIELD
ME 38	243	RESERVOIR PROPERTY DETERMINATION FROM ROCK-LOG DATA
ME 38	205	RESERVOIR ROCK DETECTION AND CHARACTERIZATION
ME 38	205†	RESERVOIR ROCK DETECTION, ELMWORTH FIELD
ME 38	156	RESERVOIR TYPES, ELMWORTH/WAPITI AREA, ALBERTA
ME 38	230	ROCK-LOG ANALYSIS TECHNIQUES, ELMWORTH AREA, ALBERTA
ME 38	205	ROCK-LOG CALIBRATION, ELMWORTH FIELD-ALBERTA, CANADA
ME 38	35*	SCHAEFER, R. G.-GAS GENERATION, MIGRATION, DEEP BASIN
ME 38	285	SHALE SLOUGHING MAP, DEEP BASIN, ALBERTA
ME 38	285	SLOUGHING SHALES, DEEP BASIN, ALBERTA
ME 38	79*	SMITH, D. G.-PALEOGEOGRAPHY, CRETACEOUS, ELMWORTH
ME 38	153*	SMITH, R. D.-GAS RESERVES, PRODUCTION, ELMWORTH/WAPITI
ME 38	79*	SNEIDER, R. M.-PALEOGEOGRAPHY, CRETACEOUS, ELMWORTH
ME 38	243*	SNEIDER, R. M.-RESERVOIR PROPERTIES, ELMWORTH FIELD
ME 38	205*	SNEIDER, R. M.-RESERVOIR ROCK DETECTION, ELMWORTH FIELD
ME 38	9	SOURCE ROCKS, LOWER CRETACEOUS, ALBERTA
ME 38	291*	STAYURA, J. A.-COMPLETION PRACTICES, ALBERTA DEEP BASIN
ME 38	35*	STOESSINGER, W.-GAS GENERATION, MIGRATION, DEEP BASIN
ME 38	156	STRATIGRAPHIC COLUMN, ELMWORTH/WAPITI AREA, ALBERTA
ME 38	5	STRATIGRAPHIC DISTRIBUTION, IN-PLACE OIL/GAS, ALBERTA
ME 38	4	STRATIGRAPHIC DISTRIBUTION, OIL/GAS RESERVES, ALBERTA
ME 38	118	STRATIGRAPHIC SECTION, ELMWORTH DEEP BASIN - MESOZOIC
ME 38	54	STRUCTURE MAP, PRE-MANNVILLE SURFACE, ALBERTA
ME 38	3	STRUCTURE MAP, PRECAMBRIAN, ALBERTA
ME 38	126	STRUCTURE MAP, TOP CADOMIN, GAS ACCUMULATION, ALBERTA
ME 38	18	STRUCTURE MAP, TOP LOWER CRETACEOUS, ALBERTA
ME 38	83	STRUCTURE MAP, TOP NOTIKEWIN, LOWER CRETACEOUS, ALBERTA
ME 38	189	SUBSURFACE PRESSURE PROFILES IN GAS-SATURATED BASINS
ME 38	20	TYPE ELECTRIC LOG, WEST CENTRAL ALBERTA
ME 38	92	TYPE LOG, BLUESKY FORMATION, LOWER CRETACEOUS, ALBERTA
ME 38	85	TYPE LOG, CADOMIN FORMATION, LOWER CRETACEOUS, ALBERTA
ME 38	108	TYPE LOG, CADOTTE, LOWER CRETACEOUS, ALBERTA
ME 38	81	TYPE LOG, ELMWORTH FIELD, LOWER CRETACEOUS, ALBERTA
ME 38	151	TYPE LOG, FALHER A, LOWER CRETACEOUS, ALBERTA
ME 38	103	TYPE LOG, FALHER A, LOWER CRETACEOUS, ALBERTA
ME 38	101	TYPE LOG, FALHER B, LOWER CRETACEOUS, ALBERTA
ME 38	99	TYPE LOG, FALHER C, LOWER CRETACEOUS, ALBERTA
ME 38	97	TYPE LOG, FALHER D, LOWER CRETACEOUS, ALBERTA
ME 38	95	TYPE LOG, FALHER E, LOWER CRETACEOUS, ALBERTA
ME 38	88	TYPE LOG, GETHING FORMATION, LOWER CRETACEOUS, ALBERTA
ME 38	106	TYPE LOG, NOTIKEWIN, LOWER CRETACEOUS, ALBERTA
ME 38	110	TYPE LOG, PADDY, LOWER CRETACEOUS, ALBERTA
ME 38	286	TYPE SECTION, SLOUGHING SHALE INTERVALS, DEEP BASIN
ME 38	67	UPPER GETHING, PALEOGEOGRAPHIC MAP, ALBERTA
ME 38	9	UPPER MANNVILLE LITHOFACIES MAP
ME 38	65	UPPER MANNVILLE PALEOGEOGRAPHY, ALBERTA
ME 38	128	WATER PRESSURE, CADOMIN FM., ELMWORTH AREA, ALBERTA
ME 38	215	WELL LOG ANALYSIS METHODS AND TECHNIQUES
ME 38	215†	WELL LOG ANALYSIS METHODS, ELMWORTH AREA, ALBERTA
ME 38	35*	WELTE, D. H.-GAS GENERATION, MIGRATION, DEEP BASIN
ME 38	173*	WYMAN, R. E.-GAS RESOURCES IN ELMWORTH COAL SEAMS
ME 38	79*	ZORN, C. E.-PALEOGEOGRAPHY, CRETACEOUS, ELMWORTH